BIOREMEDIATION OF CHLORINATED AND POLYCYCLIC AROMATIC HYDROCARBON COMPOUNDS

Edited by

Robert E. Hinchee and Andrea Leeson
Battelle, Columbus, Ohio

Lewis Semprini
Oregon State University, Corvallis, Oregon

Say Kee Ong
Polytechnic University, Brooklyn, New York

Library of Congress Cataloging-in-Publication Data

Catalog record is available from the Library of Congress

This book represents information obtained from authentic and highly regarded sources. Reprinted material is quoted with permission, and sources are indicated. A wide variety of references are listed. Every reasonable effort has been made to give reliable data and information, but the author and the publisher cannot assume responsibility for the validity of all materials or for the consequences of their use.

Direct all inquiries to CRC Press, Inc., 2000 Corporate Blvd., N.W., Boca Raton, Florida 33431.

Lewis Publishers is an imprint of CRC Press

International Standard Book Number 0-87371-983-2
Printed in the United States of America 1 2 3 4 5 6 7 8 9 0
Printed on acid-free paper

CONTENTS

Technical Notes

FOREWORD

Bioremediation as a whole remains an emerging and rapidly changing field. Since the first symposium in 1991, significant advances have been made. In 1991, only one paper was devoted to biofilters, but in 1993 biofiltration was of significant interest to generate an entire session. Natural attenuation, also little discussed in 1991, was of great interest in 1993. Increased interest in surfactant-enhanced biodegradation, bioventing, air sparging, and metals bioremediation also was apparent. Interest in chlorinated solvent bioremediation remained steady but strong, and a trend toward field application has developed. The ratio of laboratory to field studies has clearly moved toward the field. This ratio is an indication of a technology that is beginning to mature, but by no means indicates that bioremediation is a mature technology. Clearly, many knowledge gaps and needs exist. Well-developed relationships between laboratory bench-scale testing and field practice frequently are lacking. Although much encouraging laboratory research work has been done with chlorinated and many other recalcitrant organics, field practice of bioremediation is still largely directed at petroleum hydrocarbon-contaminated sites. The number of well-documented field demonstrations of bioremediation is increasing; however, before bioremediation can mature into a readily accepted and widely understood technology area, many more field demonstrations will be necessary.

This book and its companion volumes, *Hydrocarbon Bioremediation* and *Applied Biotechnology for Site Remediation*, represent the bulk of the papers arising from the Second International Symposium on In Situ and On-Site Bioreclamation held in San Diego, California, in April 1993. Two other books, *Air Sparging* and *Emerging Technology for Bioremediation of Metals*, also contain selected papers on these topics.

The symposium was attended by more than 1,100 people. More than 300 presentations were made, and all presenting authors were asked to submit manuscripts. Following a peer review process, 190 papers are being published. The editors believe that these volumes represent the most complete, up-to-date works describing both the state of the art and the practice of bioremediation.

The symposium was sponsored by Battelle Memorial Institute with support from a wide variety of other organizations. The cosponsors and supporters were:

Bruce Bauman, *American Petroleum Institute*
Christian Bocard, *Institut Français du Pétrole*
Rob Booth, *Environment Canada, Wastewater Technology Centre*
D. B. Chan, *U.S. Naval Civil Engineering Laboratory*
Soon H. Cho, *Ajou University, Korea*
Kate Devine, *Biotreatment News*
Volker Franzius, *Umweltbundesamt, Germany*
Giancarlo Gabetto, *Castalia, Italy*
O. Kanzaki, *Mitsubishi Corporation, Japan*

Dottie LaFerney, *Stevens Publishing Corporation*
Massimo Martinelli, *ENEA, Italy*
Mr. Minoru Nishimura, *The Japan Research Institute, Ltd.*
Chongrak Polprasert, *Asian Institute of Technology, Thailand*
Lewis Semprini, *Oregon State University*
John Skinner, *U.S. Environmental Protection Agency*
Esther Soczo, *National Institute of Public Health and Environmental Protection, The Netherlands*

In addition, numerous individuals assisted as session chairs, presented invited papers, and helped to ensure diverse representation and quality. Those individuals were:

Bruce Alleman, *Battelle Columbus*
Christian Bocard, *Institut Français du Pétrole*
Rob Booth, *Environment Canada, Wastewater Technology Center*
Fred Brockman, *Battelle Pacific Northwest Laboratories*
Tom Brouns, *Battelle Pacific Northwest Laboratories*
Soon Cho, *Ajou University, Korea*
M. Yavuz Corapcioglu, *Texas A&M University*
Jim Fredrickson, *Battelle Pacific Northwest Laboratories*
Giancarlo Gabetto, *Area Commerciale Castalia, Italy*
Terry Hazen, *Westinghouse Savannah River Laboratory*
Ron Hoeppel, *U.S. Naval Civil Engineering Laboratory*
Yacov Kanfi, *Israel Ministry of Agriculture*
Richard Lamar, *U.S. Department of Agriculture*
Andrea Leeson, *Battelle Columbus*
Carol Litchfield, *Keystone Environmental Resources, Inc.*
Perry McCarty, *Stanford University*
Jeff Means, *Battelle Columbus*
Blaine Metting, *Battelle Pacific Northwest Laboratories*
Ross Miller, *U.S. Air Force*
Minoru Nishimura, *Japan Research Institute*
Robert F. Olfenbuttel, *Battelle Columbus*
Say Kee Ong, *Polytechnic University, New York*
Augusto Porta, *Battelle Europe*
Roger Prince, *Exxon Research and Engineering Co.*
Parmely "Hap" Pritchard, *U.S. Environmental Protection Agency*
Jim Reisinger, *Integrated Science & Technology*
Greg Sayles, *U.S. Environmental Protection Agency*
Lewis Semprini, *Oregon State University*
Ron Sims, *Utah State University*
Marina Skumanich, *Battelle Seattle*
Jim Spain, *U.S. Air Force*
Herb Ward, *Rice University*
Peter Werner, *University of Karlsruhe, Germany*
John Wilson, *U.S. Environmental Protection Agency*
Jim Wolfram, *Montana State University*

The papers in this book have been through a peer review process, and the assistance of the peer reviewers is recognized. This typically thankless job is essential to technical publication. The following people peer-reviewed papers for the publication resulting from the symposium:

Jens Aamand, *Water Quality Institute*
Nelly M. Abboud, *University of Connecticut*
Daniel A. Abramowicz, *GE Corporate R&D Center*
Dan W. Acton, *Beak Consultants Ltd., Canada*
William Adams, *Monsanto Company U4E*
Peter Adriaens, *University of Michigan*
C. Marjorie Aelion, *University of South Carolina*
Robert C. Ahlert, *Rutgers University*
David Ahlfeld, *University of Connecticut*
Hans-Jorgen Albrechtsen, *Technical University of Denmark*
Bruce Alleman, *Battelle Columbus*
Richelle M. Allen-King, *University of Waterloo*
Sabine E. Apitz, *NCCOSC RDTE DIV 521*
John M. Armstrong, *The Traverse Group*
Boris N. Aronstein, *Institute of Gas Technology*
Mick Arthur, *Battelle Columbus*
Erik Arvin, *Technical University of Denmark*
Steven D. Aust, *Utah State University*
Serge Baghdikian
M. Talaat Balba, *TreaTek - CRA Company*
D. Ballerini, *Institut Français du Pétrole*
N. Bannister, *University of Kent, England*
Jeffrey R. Barbaro, *University of Waterloo*
James F. Barker, *University of Waterloo*
Morton A. Barlaz, *North Carolina State University*
Denise M. Barnes, *Ecosystems Engineering*
Edward R. Bates, *U.S. Environmental Protection Agency*
Tad Beard, *Battelle Columbus*
Cathe Bech, *SINTEF Applied Chemistry, Norway*
Pamela E. Bell, *Hydrosystems, Inc.*
Judith Bender, *Clark Atlanta University*
James D. Berg, *Aquateam Norwegian Water Technology Centre A/S*
Christopher J. Berry, *Westinghouse Savannah River Company*
Sanjoy K. Bhattacharya, *Tulane University*
Jeffery F. Billings, *Billings & Associates*
James N. P. Black, *Stanford University*
Joan Blake, *U.S. Environmental Protection Agency*
Robert Blanchette, *University of Minnesota*
Bert E. Bledsoe, *U.S. Environmental Protection Agency*
Christian Bocard, *Institut Français du Pétrole*
Gary Boettcher, *Geraghty & Miller, Inc.*
David R. Boone, *Oregon Graduate Center*
James Borthen, *ECOVA Corporation*
Edward J. Bouwer, *Johns Hopkins University*
John P. Bowman, *University of Tennessee*
Joan F. Braddock, *University of Alaska*

A. Braun-Lullemann, *Forstbotanisches Institut der Universität Göttingen, Germany*
Susan E. Brauning, *Battelle Columbus*
Alec W. Breen, *U.S. Environmental Protection Agency*
James A. Brierley, *Newmont Metallurgical Services*
Fred Brockman, *Battelle Pacific Northwest Laboratories*
Kim Broholm, *Technical University of Denmark*
Thomas M. Brouns, *Battelle Pacific Northwest Laboratories*
Edward Brown, *University of Northern Iowa*
Guner Brox, *EIMCO Process Equipment*
Gaylen R. Brubaker, *Remediation Technologies, Inc.*
Wil P. de Bruin, *Wageningen Agricultural University, The Netherlands*
Robert S. Burlage, *Oak Ridge National Laboratory*
David Burris, *Tyndall Air Force Base*
Timothy E. Buscheck, *Chevron Research and Technology Company*
Larry W. Canter, *University of Oklahoma*
Jason A. Caplan, *ESE Biosciences*
Peter J. Chapman, *U.S. Environmental Protection Agency*
Abe Chen, *Battelle Columbus*
G. O. Chieruzzi, *Keystone Environmental Resources*
Soon Hung Cho, *Ajou University, Korea*
Patricia J. S. Colberg, *University of Wyoming*
Edward Coleman, *MK Environmental*
Ronald L. Crawford, *University of Idaho*
Steven L. Crawford, *DPRA Inc.*
Craig Criddle, *Michigan State University*
Jon Croonenberghs, *Coors Brewing Company*
Scott Cunningham, *DuPont Central Research and Development*
Mohamed F. Dahab, *University of Nebraska*
Lois Davis, *Sybron Chemicals, Inc.*
Wendy J. Davis-Hoover, *U.S. Environmental Protection Agency*
Peter Day, *Rutgers University*
Sue Markland Day, *University of Tennessee*
Mary F. DeFlaun, *Envirogen*
Richard A. DeMaio, *Mycotech Corporation*
Dave DePaoli, *Oak Ridge National Laboratory*
Allen Deur, *Polytechnic University*
Kate Devine, *Biotreatment News*
L. Diels, *Vlaamse, Instelling voor Technologisch Onderzoek, Belgium*
Greg Douglas, *Battelle Ocean Sciences*
Douglas C. Downey, *Engineering-Science, Inc.*
David Drahos, *SPB Technologies*
Murali M. Dronamraju, *Tulane University*
Jean Ducreux, *Institut Français du Pétrole*
James Duffy, *Occidental Chemical Corporation*
Ryan Dupont, *Utah State University*
Geraint Edmunds
Elizabeth A. Edwards, *Beak Consultants Ltd., Canada*
Richard Egg, *Texas A&M University*
David L. Elmendorf, *University of Central Oklahoma*
Mark Emptage, *DuPont Company*
Burt D. Ensley, *Envirogen*
Michael V. Enzien, *Westinghouse Savannah River Site*

David C. Erickson, *Harding Lawson Associates*
Richard A. Esposito, *Southern Company Services*
J. van Eyk, *Delft Geotechnics, The Netherlands*
Brandon J. Fagan, *Continental Recovery Systems, Inc.*
Liv-Guri Faksness, *SINTEF Applied Chemistry, Norway*
John Ferguson, *University of Washington*
J. A. Field, *Wageningen Agricultural University, The Netherlands*
Pedro Fierro, *Geraghty & Miller, Inc.*
Stephanie Fiorenza, *Amoco Corporation*
Paul E. Flathman, *OHM Remediation Services Corporation*
John Flyvbjerg, *Water Quality Institute, Denmark*
Cresson D. Fraley, *Stanford University*
W. T. Frankenberger, *University of California, Riverside*
James Fredrickson, *Battelle Pacific Northwest Laboratories*
David L. Freedman, *University of Illinois*
Ian V. Fry, *Lawrence Berkeley Laboratory*
Clyde W. Fulton, *CH2M HILL*
Kathryn Garrison, *Geraghty & Miller, Inc.*
Edwin Gelderich, *U.S. Environmental Protection Agency*
Richard M. Gersberg, *San Diego State University*
John Glaser, *U.S. Environmental Protection Agency*
Fred Goetz, *Mankato State University*
C. D. Goldsmith, *EnvironTech Mid-Atlantic*
James M. Gossett, *Cornell University*
Peter Grathwohl, *University of Teubingen, Germany*
Charles W. Greer, *Biotechnology Research Institute, Canada*
Christian Grøn, *Technical University of Denmark*
D.R.J. Grootjen, *DSM Research BV, The Netherlands*
Matthew J. Grossman, *Exxon Research & Engineering*
Ipin Guo, *Alberta Environmental Centre, Canada*
Haim Gvirtzman, *The Hebrew University of Jerusalem*
Paul Hadley, *California Environmental Protection Agency*
John R. Haines, *U.S. Environmental Protection Agency*
Kenneth Hammel, *U.S. Department of Agriculture*
Mark R. Harkness, *GE Corporate R&D Center*
Joop Harmsen, *The Winand Staring Center for Integrated Land, Soil and Water Research, The Netherlands*
Zachary Haston, *Stanford University*
Gary R. Hater, *Chemical Waste Management, Inc.*
Tony Hawke, *Groundwater Technology Canada Ltd.*
Caryl Heintz, *Texas Tech University*
Barbara B. Hemmingsen, *San Diego State University*
Stephen E. Herbes, *Oak Ridge National Laboratory*
Gorm Heron, *Technical University of Denmark*
Ronald J. Hicks, *Groundwater Technology, Inc.*
Franz K. Hiebert, *Alpha Environmental, Inc.*
E. L. Hockman, *Amoco Corporation*
Robert E. Hoffmann, *SiteRisk Inc.*
Desma Hogg, *Woodward-Clyde Consultants*
Brian S. Hooker, *Tri-State University*

Kevin Hosler, *Wastewater Technology Centre, Canada*
M. Akhter Hossain, *Atlantic Environmental*
Perry Hubbard, *Integrated Science & Technology, Inc.*
Michael H. Huesemann, *Shell Development Company*
Scott G. Huling, *U.S. Environmental Protection Agency*
Jasna Hundal, *CH2M HILL*
Peter J. Hutchinson, *The Hutchinson Group, Ltd.*
Aloys Huttermann, *Forstbotanisches Institut der Universität Göttingen, Germany*
Mary Pat Huxley, *U.S. Naval Civil Engineering Laboratory*
Charles E. Imel, *Ecosystems Engineering*
Danny R. Jackson, *Radian Corporation*
Peter Jaffe, *Princeton University*
Trevor James, *Woodward Clyde Ltd., New Zealand*
D. B. Janssen, *University of Groningen, The Netherlands*
Minoo Javanmardian, *Amoco Oil Company*
Ursula Jenal-Wanner, *Western Regional Hazardous Substance Research Center, Stanford University*
Bjorn K. Jensen, *Water Quality Institute, Denmark*
Douglas E. Jerger, *OHM Remediation Service Corporation*
Randall M. Jeter, *Texas Tech University*
Richard L. Johnson, *Alberta Environmental Centre, Canada*
George Johnson, *Stillwater, Inc.*
C. D. Johnston, *CSIRO, Australia*
Donald L. Johnstone, *Washington State University*
E. Fraser Johnstone, *Exxon Company*
K. C. Jones, *Lancaster University*
Warren L. Jones, *Montana State University*
E. de Jong, *Wageningen Agricultural University, The Netherlands*
Linda de Jong, *University of Washington*
Miryan Kadkhodayan, *University of Cincinnati*
Don Kampbell, *U.S. Environmental Protection Agency*
Yacov Kanfi, *Israel Ministry of Agriculture*
Chih-Ming Kao, *North Carolina State University*
Leslie Karr, *U.S. Naval Civil Engineering Laboratory*
Keith Kaufman, *RESNA Industries, Inc.*
S. Keuning, *Bioclear Environmental Biotechnology, The Netherlands*
T. Kent Kirk, *U.S. Department of Agriculture*
Michael D. Klein, *EG&G Rocky Flats, Inc.*
Calvin A. Kodres, *U.S. Naval Civil Engineering Laboratory*
Simeon J. Komisar, *University of Washington*
Raj Krishnamoorthy, *Keystone Environmental Resources, Inc.*
M. Kuge, *Industrial and Fine Chemicals Division, Environmental Technology*
Debi Kuo, *University of Tennessee*
Bruce E. LaBelle, *California Environmental Protection Agency*
William F. Lane, *Remediation Technologies, Inc.*
Margaret Lang, *Stanford University*
Robert LaPoe, *U.S. Air Force*
Barnard Lawes, *DuPont Company*
Maureen E. Leavitt, *IT Corporation*
Clifford Lee, *DuPont Environmental Remediation Services*
Kun Mo Lee, *Ajou University, Korea*

Michael D. Lee, *DuPont Environmental Remediation Services*
Richard F. Lee, *Skidaway Institute of Oceanography*
Paul LeFevre, *Coors Brewing Company*
Robert Legrand, *Radian Corporation*
Terrance Leighton, *University of California*
Sarah K. Leihr, *North Carolina State University*
M. Tony Lieberman, *ESE Biosciences*
Carol D. Litchfield, *Chester Environmental*
Kenneth H. Lombard, *Bechtel Savannah River, Inc.*
Sharon C. Long, *University of North Carolina*
Charles R. Lovell, *University of South Carolina*
Ja-Kael Luey, *Battelle Pacific Northwest Laboratories*
J. S. Luo, *Center for Environmental Biotechnology*
Stuart Luttrell, *Battelle Pacific Northwest Laboratories*
John Lyngkilde, *Technical University of Denmark*
Ian D. MacFarland, *EA Engineering, Science, and Technology, Inc.*
Joan Macy, *University of California*
Andzej Majcherczyk, *Forstbotanisches Institut der Universität Göttingen, Germany*
David Major, *Beak Consultants Ltd., Canada*
Pryodarshi Majumdar, *Tulane University*
Leo Manzer, *E. I. DuPont, De Nemours & Co., Inc.*
Nigel V. Mark-Brown, *Woodward-Clyde International, New Zealand*
Donn Marrin, *InterPhase Environmental, Inc.*
Dean A. Martens, *University of California, Riverside*
Michael M. Martinson, *Delta Environmental Consultants, Inc.*
Perry L. McCarty, *Stanford University*
Gloria McCleary, *EA Engineering*
Linda McConnell, *Logistics Management Institute*
Mike McFarland, *Utah Water Research Laboratory, Utah State University*
Ilona McGhee, *University of Kent, England*
David H. McNabb, *Alberta Environmental Centre, Canada*
Sally A. Meyer, *Georgia State University*
Kathy Meyer-Schulte, *Computer Science Corporation*
Robert Miller, *Oklahoma State University*
Ali Mohagheghi, *Solar Energy Research Institute*
Peter Molton, *Battelle Pacific Northwest Laboratories*
Ralph E. Moon, *Geraghty & Miller, Inc.*
Jim Morgan, *The MITRE Corporation*
Frederic A. Morris, *Battelle Seattle Research Centers*
Pamela J. Morris, *University of Florida*
Klaus Müller, *Battelle Europe*
Julie Muolyta, *Stanford University*
H. S. Muralidhara, *Cargill, Inc.*
Reynold Murray, *Clark Atlanta University*
Karl W. Nehring, *Battelle Columbus*
Christopher H. Nelson, *Groundwater Technology, Inc.*
Per H. Nielsen, *Technical University of Denmark*
Dev Niyogi, *Battelle Marine Research Laboratory*

Robert Norris, *Eckenfelder, Inc.*
John T. Novak, *Virginia Tech*
Evan Nyer, *Geraghty & Miller*
Joseph E. Odencrantz, *Lavine-Fricke Consulting Engineers*
Laurra P. Olmsted, *Brown & Root Civil, England*
Brian O'Neill, *Dearborn Chemical Company, Ltd., Canada*
Richard Ornstein, *Battelle Pacific Northwest Laboratories*
David Ostendorf, *University of Massachusetts*
Donna Palmer, *Battelle Columbus*
Anthony V. Palumbo, *Oak Ridge National Laboratory*
Sorab Panday, *HydroGeologic Inc.*
Joel W. Parker, *The Traverse Group*
John H. Patterson, *Continental Recovery Systems*
Richard E. Perkins, *DuPont Environmental Biotechnology Program*
James N. Peterson, *Washington State University*
Erik Petrovskis, *University of Michigan*
Brent Peyton, *Battelle Pacific Northwest Laboratories*
Frederic K. Pfaender, *University of North Carolina at Chapel Hill*
S. M. Pfiffner, *University of Tennessee*
George Philippidis, *National Renewable Energy Laboratory*
Peter Phillips, *Clark Atlanta University*
C.G.J.M. Pijls, *TAUW Infra Consult B.V., The Netherlands*
Keith R. Piontek, *CH2M HILL*
Michael Piotrowski, *Biotransformations, Inc.*
Augusto Porta, *Battelle Europe*
Roger C. Prince, *Exxon Research and Engineering Co.*
Parmely "Hap" Pritchard, *U.S. Environmental Protection Agency*
Jaakko A. Puhakka, *University of Washington*
Santo Ragusa, *CSIRO Division of Water Resources, Australia*
Ken Rainwater, *Texas Tech University*
Svein Ramstad, *SINTEF Applied Chemistry, Norway*
Grete Rasmussen, *University of Washington*
Mark E. Reeves, *Oak Ridge National Laboratory*
Roger D. Reeves, *Massey University, New Zealand*
H. James Reisinger, *Integrated Science & Technology, Inc.*
Charles M. Reynolds, *U.S. Army Cold Regions Research and Engineering Laboratory*
Hanadi S. Rifai, *Rice University*
Derek Ross, *ERM Inc.*
J. J. Salvo, *GE Corporate, R&D Center*
Réjean Samson, *Biotechnology Research Institute, Canada*
Erwan Saouter, *Center for Environmental Diagnostics and Bioremediation*
Bruce Sass, *Battelle Columbus*
Eric K. Schmitt, *ESE Biosciences, Inc.*
Gosse Schraa, *Wageningen Agricultural University, The Netherlands*
Alan G. Seech, *Dearborn Chemical Co., Ltd., Canada*
Robert L. Segar, *University of Texas at Austin*
Douglas Selby, *Las Vegas Valley Water District*
Patrick Sferra, *U.S. Environmental Protection Agency*
Daniel R. Shelton, *U.S. Department of Agriculture*
Tatsuo Shimomura, *Ebara Research Co. Ltd., Japan*
Mark Silva, *American Proteins, Inc.*

Thomas J. Simpkin, *CH2M HILL*
Judith L. Sims, *Utah State University*
Rodney S. Skeen, *Battelle Pacific Northwest Laboratories*
George J. Skladany, *Envirogen*
Marina Skumanich, *Battelle Seattle Research Centers*
Lawrence Smith, *Battelle Columbus*
Gregory Smith, *ENSR Consulting and Engineering*
Darwin Sorenson, *Utah State University*
Jim Spain, *Tyndall Air Force Base*
Gerald E. Speitel, *University of Texas at Austin*
D. Springael, *Vlaamse, Instelling voor Technologisch Onderzoek, Belgium*
Thomas B. Stauffer, *Tyndall Air Force Base*
Robert J. Steffan, *Envirogen*
H. David Stensel, *University of Washington*
Jan Stepek, *EA Engineering Science and Technology*
David Stevens, *Utah Water Research Laboratory, Utah State University*
Gerald W. Strandberg, *Oak Ridge National Laboratory*
Janet Strong-Gunderson, *Oak Ridge National Laboratory*
John B. Sutherland, *U.S. Food & Drug Administration*
C. Michael Swindoll, *DuPont Environmental Remediation Services*
Robert D. Taylor, *The MITRE Corporation*
Alison Thomas, *U.S. Air Force*
Francis T. Tran, *Diocese Loire-Atlantique, Seminaire Des Carmes*
Mike D. Travis, *RZA-AGRA Engineering and Environmental Services*
Sarah C. Tremaine, *Hydrosystems, Inc.*
Jack T. Trevors, *University of Guelph, Canada*
Marleen A. Troy, *OHM Remediation Services Corporation*
Mark Trudell, *Alberta Research Council, Canada*
Michael J. Truex, *Battelle Pacific Northwest Laboratories*
Samuel L. Unger, *Groundwater Technology, Inc.*
J. P. Vandecasteele, *Institut Français du Pétrole*
Ranga Velagaleti, *Battelle Columbus*
Albert D. Venosa, *U.S. Environmental Protection Agency*
Stephen J. Vesper, *University of Cincinnati*
Bruce Vigon, *Battelle Columbus*
John S. Waid, *La Trobe University, Australia*
Terry Walden, *BP Research*
Mary E. Watwood, *Idaho State University*
Lenly Joseph Weathers, *University of Iowa*
Marty Werner, *Washington State University*
Mark Westray, *Remediation Technologies Inc.*
David C. White, *University of Tennessee*
Patricia J. White, *Battelle Marine Research Laboratory*
Jeffrey Wiegand, *Alton Geoscience*
J. W. Wigger, *Amoco Corporation*
Peter Wilderer, *Technische Universität München, Germany*
Barbara H. Wilson, *Dynamac Corporation*
John T. Wilson, *U.S. Environmental Protection Agency*
Roger M. Woeller, *Water & Earth Science Associates, Ltd., Canada*
Arthur Wong, *Coastal Remediation*

Jack Q. Word, *Battelle Marine Research Laboratory*
Darla Workman, *Battelle Pacific Northwest Laboratories*
Brian A. Wrenn, *University of Cincinnati*
Lin Wu, *University of California*
Robert Wyza, *Battelle Columbus*
Andreas Zeddel, *Forstbotanisches Institut der Universität Göttingen, Germany*
Gerben Zylstra, *Rutgers University*

The editors wish to recognize some of the key contributors who have put forth significant effort in assembling this book. Lynn Copley-Graves served as the text editor, reviewing every paper for readability and consistency. She also directed the layout of the book and production of the camera-ready copy. Loretta Bahn worked many long hours converting and processing files, and laying out the pages. Karl Nehring oversaw coordination of the book publication with the symposium, and worked with the publisher to make everything happen. Gina Melaragno coordinated manuscript receipts and communications with the authors and peer reviewers.

None of the sponsoring or cosponsoring organizations or peer reviewers conducted a final review of the book or any part of it, or in any way endorsed this book.

Rob Hinchee
June 1993

DEGRADATION KINETICS OF CHLORINATED SOLVENTS BY A PROPANE-OXIDIZING ENRICHMENT CULTURE

J. E. Keenan, S. E. Strand, and H. D. Stensel

ABSTRACT

A suspended enrichment culture of propane-oxidizing bacteria degraded 1,1,1-trichloroethane (TCA) and trichloroethylene (TCE). TCA degradation followed first-order kinetics for TCA concentrations from 0 to 20 mg/L with a first-order rate constant of K = 0.0031 L (mg Total Suspended Solids [TSS] hr)$^{-1}$ without propane. Propane strongly inhibited TCA degradation. However, the data did not fit any of the accepted inhibition models including competitive, noncompetitive, and uncompetitive inhibition. TCE degradation followed Michaelis-Menten kinetics for TCE concentrations from 0 to 9 mg/L, with V_{max} = 0.0016 mg TCE (mg TSS hr)$^{-1}$ and K_s = 0.6 mg TCE/L. Propane inhibited TCE degradation, fitting a noncompetitive inhibition model:

$$V = \frac{0.0016\ S}{(S + 0.6)\left(1 + \frac{P}{10.6}\right)}$$

where V is the TCE degradation rate in milligrams of TCE (mg TSS hr)$^{-1}$, S is the TCE concentration in milligrams of TCE/L, and P is the percentage of propane in the headspace.

INTRODUCTION

The historical use and improper storage, handling, and disposal of chlorinated organic compounds has resulted in the widespread contamination of groundwater. TCE and 1,1,1-TCA are among the most commonly reported solvents contaminating groundwater systems.

Chlorinated hydrocarbons are not used as a primary substrate by most heterotrophic organisms, but they are susceptible to aerobic biological degradation by cometabolism. Cometabolism results from the expression of nonspecific enzymes that degrade the primary substrate and involves the transformation of a compound that does not supply carbon, energy, or reducing power to the organism. The organism requires another compound to supply carbon and energy.

Much work has been done to study and characterize the ability of methane-oxidizing bacteria to cometabolize chlorinated hydrocarbons such as TCA and TCE (Alvarez-Cohen & McCarty 1991, Henry & Grbić-Galić 1989, Oldenhuis et al. 1989, Strand et al. 1990). Bacteria that aerobically degrade toluene (Wackett & Gibson 1988) and ammonia (Vannelli et al. 1990) also have been found to cometabolically degrade TCE.

Researchers have identified a monooxygenase enzyme in organisms grown on propane, called propane monooxygenase (PMO) (Wackett et al. 1989). As with methane monooxygenase (MMO), PMO is thought to catalyze the aerobic transformation of CF, TCA, and TCE (Wackett et al. 1989). To fully exploit cometabolic transformations by propane-oxidizing cultures, the kinetics of the oxidation of the primary substrate (propane) and the degradation of chlorinated hydrocarbons must be better understood.

Several researchers have identified microorganisms that are able to use propane aerobically. Propane oxidizers have been isolated from soil and water samples, including the following genera: *Rhodococcus* (Woods & Murrell 1989); *Mycobacterium* (Perry 1980, Hou et al. 1983); *Arthrobacter* (Stephens & Dalton 1986); and *Pseudomonas, Acinetobacter, Brevibacter, Nocardia,* and *Actinomyces* (Hou et al. 1983). Most isolates can oxidize a broad spectrum of aliphatic hydrocarbons (Perry 1980, Stephens & Dalton 1986, Woods & Murrell 1989), which would imply that the enzyme systems used to degrade propane are nonspecific enough to degrade other aliphatics (Perry 1980). That the PMO is inducible (Perry 1980) contrasts with MMO, which is a constitutive enzyme system.

Wackett et al. (1989), found several strains of propane-oxidizing bacteria able to oxidize TCE when grown on propane, including *Mycobacterium vaccea* JOB-5. Work to date has not covered important aspects of the application of these organisms to a remediation scheme. Nor has the inhibition of TCE degradation by propane been studied. Additionally, the degradation of TCA by propane oxidizers has not been investigated.

This research focused on defining the biological kinetics of a mixed culture of propane-oxidizing bacteria. The specific objectives were to (1) determine propane and oxygen utilization kinetics, (2) determine TCA and TCE biodegradation rates, and (3) determine and describe any inhibition of TCA and TCE biodegradation rates caused by propane.

MATERIALS AND METHODS

A suspended growth reactor was operated at constant conditions to provide a source culture for the batch TCA and TCE degradation tests. The reactor, described by Strand et al. (1990), was maintained at a solids and hydraulic retention time of 10 days with approximately 10% propane in the headspace above the culture with the balance being air. The bacterial seed for the mixed culture came from a combination of sources including soil samples from near a propane storage tank and from the University of Washington in Seattle, Washington, and liquid samples from Lake Washington and a small tributary leading to it, also in Seattle; and two pure cultures, *Rhodococcus rhodochrous* PNKb1 (Woods & Murrell 1989) and *Mycobacterium vaccea* (Perry 1980).

The nutrient medium used for this culture contained 500 mg KNO_3, 200 mg $MgSO_4$, 425 mg Na_2HPO_4, 200 mg KH_2PO_4, 50 mg NH_4Cl, 15 mg $CaSO_4$, 1 mg $FeSO_4$ $7H_2O$, 0.5 mg NaEDTA, 0.1 mg $CuSO_4$, 0.1 mg $ZnSO_4$ $7H_2O$, 0.03 mg $NaMoO_4$ $2H_2O$, 0.02 mg $MnCl_2$ $4H_2O$, 0.02 mg H_3BO_3, 0.02 mg $NiCl_2$ $6H_2O$, and 0.01 mg $CoCl_2$ $6H_2O$ per liter of deionized water, and was buffered at pH = 7.2.

Sampling Methods

Batch tests were prepared by inserting 50 mL of propane-oxidizing bacteria in suspension into 160-mL serum vials fitted with mini-nert screw caps. An abiotic control consisting of organic-free water also was prepared. A measured amount of propane was injected into each vial except the abiotic control and vials that were to be run without propane. Then a measured quantity of TCA- or TCE-saturated water was added and the vials were shaken for 1 minute. A 1-mL sample was taken immediately after shaking and at regular intervals using a 5-mL Luer-lock syringe containing 1-mL of pentane. The vials were shaken and maintained at 20°C throughout the tests. Killed controls were previously tested and showed that there was not a significant level of TCE adsorption to the inactivated biomass.

Analytical Techniques

A pentane extraction procedure described by Bjelland (1989) was used to quantify liquid concentrations of chlorinated organics during batch biodegradation experiments. The pentane was analyzed for TCA and TCE by a gas chromatograph with electron-capture detection (GC-ECD) (Perkin-Elmer Autosystem) using an RTX-1 column (Restex Corp.). The GC was operated isothermally at 80°C with helium as the carrier gas (2 mL/min). Both external and internal (ethylene dibromide [EDB]) standards were used to quantify the concentrations of TCA and TCE. Trichloroethanol has been measured by GC in some research (Alvarez-Cohen & McCarty 1991), however, there were no products of TCA or TCE transformation measured during this research.

Total suspended solids (TSS) measurements were performed by filtering a measured quantity of the reactor contents through 0.45-micron pore size filter paper, drying the filter paper at 105°C for 1 hour, and measuring the difference in weight of the filter paper.

EXPERIMENTAL RESULTS

TCA Degradation Kinetics

Figure 1 shows the time course of TCA degradation for some of the batch tests with varying initial levels of propane. A range of TCA concentrations were

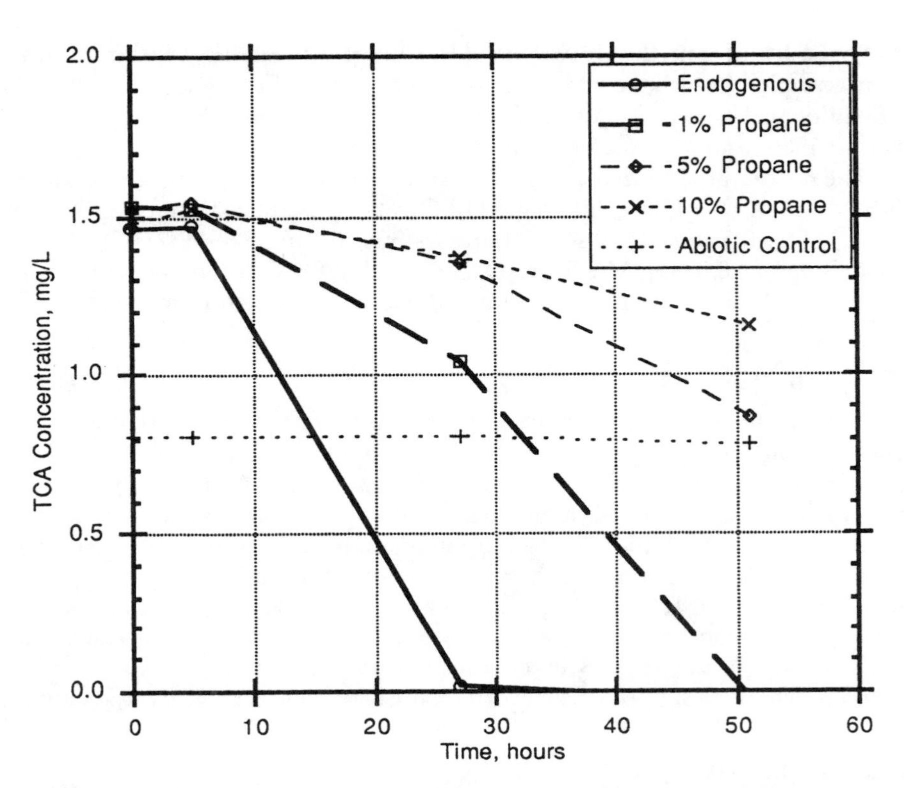

FIGURE 1. Typical time course of TCA degradation with various propane concentrations.

tested in this manner and the initial rates of TCA degradation were measured by taking a representative slope of the concentration versus time early in the course of the batch test and dividing by the biomass concentration to yield an initial rate of degradation with the units of mg TCA (mg TSS hr)$^{-1}$. The endogenous condition is defined as the absence of propane in the test vial.

Figure 2 summarizes all of the TCA degradation rates with a least squares regression plot for each level of propane initially present. The statistical package, SYSTAT 5 (SYSTAT Inc.), was used to fit the TCA data to a simple, first-order rate equation:

$$V = K\ S \tag{1}$$

where: V = velocity (rate) of the reaction, mg TCA (mg TSS hr)$^{-1}$
K = first-order rate constant, L (mg TSS hr)$^{-1}$
S = substrate (TCA) concentration, mg TCA/L

The first-order rate constants are plotted versus the headspace propane level in Figure 3.

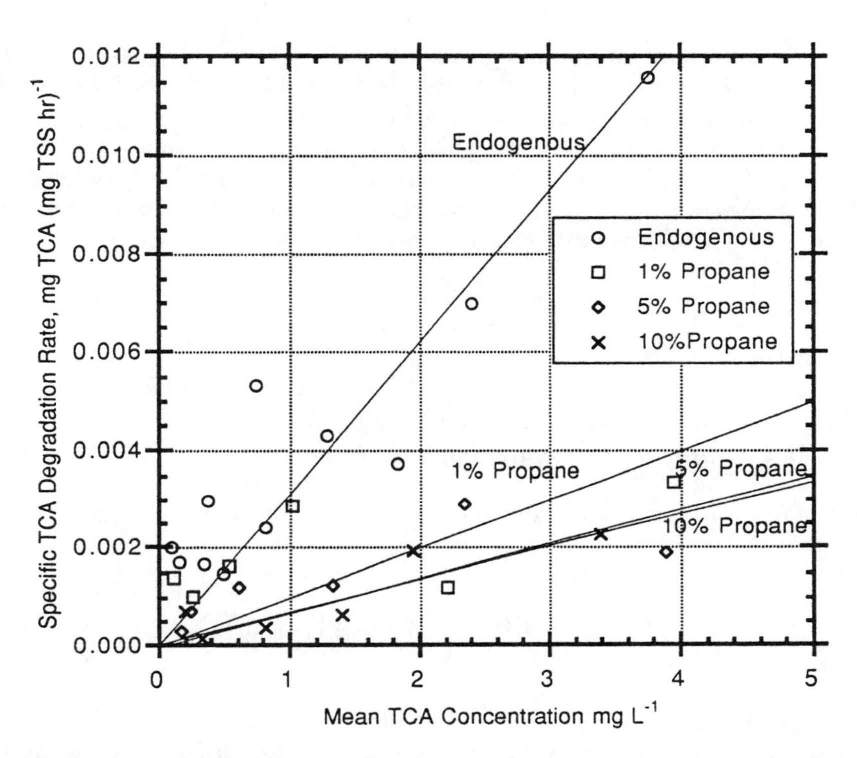

FIGURE 2. TCA degradation rate as a first-order function of the TCA concentration and inversely related to the propane concentration. (The error bars represent the standard error of the mean.)

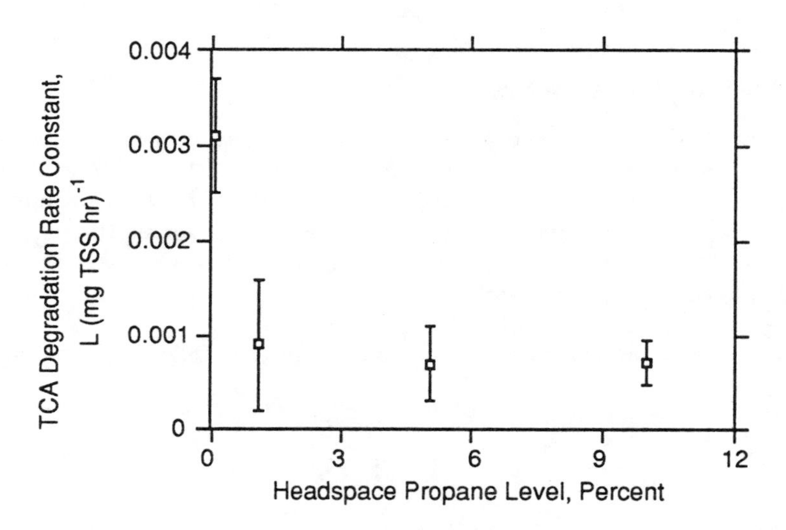

FIGURE 3. First-order rate constant of TCA degradation as an inverse function of propane level. (Correlation coefficients ranged from 0.69 to 0.92.)

The rate constant for the zero propane level (0.0031 ± 0.0006 L [mg TSS hr]$^{-1}$) was significantly different from the rate constants at the 1, 5, and 10% propane levels (0.1% level using Student's t-test), but further increases of propane level greater than 1% did not result in significant changes in the rate constant.

The inhibition of TCA degradation by propane is very apparent from Figures 2 and 3, but does not appear to fit any conventional model. If it were competitive inhibition, the first-order rate could be described by

$$V = KS\left(\frac{I}{I + P}\right) \tag{2}$$

where: I = inhibition constant, % propane
P = propane concentration, % propane

The first-order rate constant can be written as a function of P, the propane level:

$$K(P) = K\left(\frac{I}{I + P}\right) \tag{3}$$

When this analysis was applied to the rate constant values, the inhibition constant, I, was calculated as 0.44 for 1% propane, 1.6 for 5% propane, and 3.1 for 10% propane. Therefore, simple competitive inhibition of the first-order reaction does not adequately explain the mechanism of inhibition of TCA degradation by propane. Similarly, the data do not fit other accepted inhibition models, such as noncompetitive and uncompetitive inhibition.

TCE Degradation Kinetics

Figure 4 shows the time course of TCE degradation for some of the batch tests with varying initial levels of propane. A range of TCE concentrations were tested in this manner, and the initial rates of TCE degradation were measured by taking a representative slope of the concentration versus time early in the course of the batch test and dividing by the biomass concentration to yield an initial rate of degradation with the units, mg TCE (mg TSS hr)$^{-1}$. Again, the endogenous vial did not contain propane.

As can be seen from Figure 5, the data appear to follow Michaelis-Menten kinetics as described by

$$V = \frac{V_{max}S}{K_S + S} \tag{4}$$

where: V = velocity (rate) of the reaction, mg TCE (mg TSS hr)$^{-1}$
V_{max} = maximum velocity of reaction, mg TCE (mg TSS hr)$^{-1}$

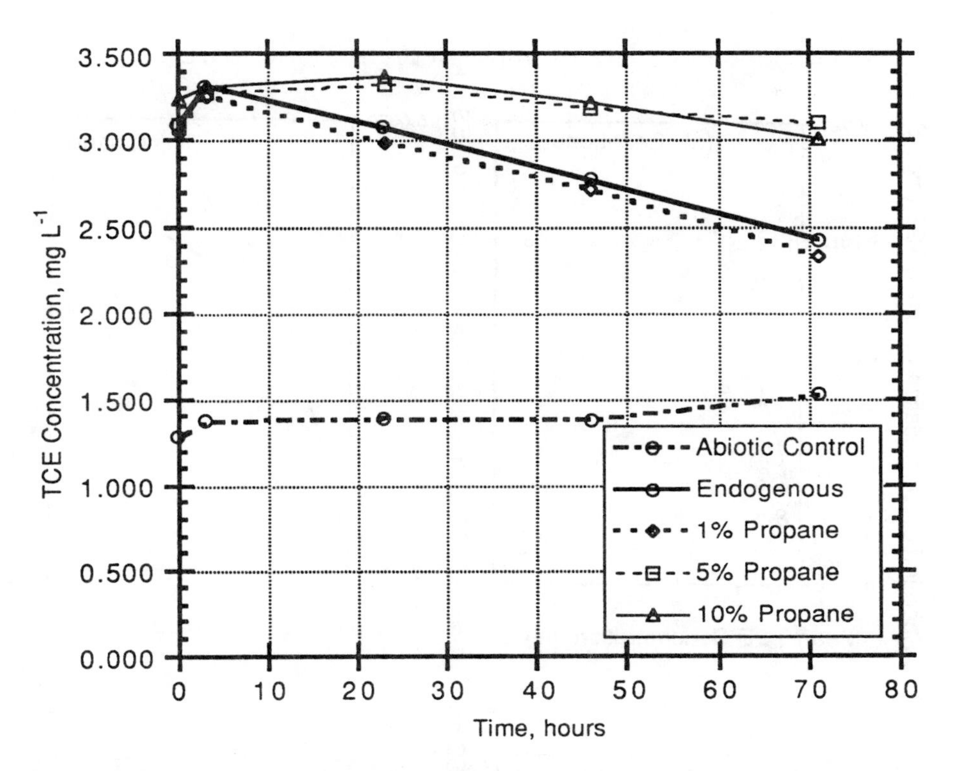

FIGURE 4. Typical time course of TCE degradation with various propane concentrations.

S = substrate (TCE) concentration, mg TCE/L
K_s = half-saturation constant, mg TCE/L

Although TCE is the cometabolite of the propane-oxidizing culture and does not supply the organisms with carbon or energy, TCE was represented by S (substrate) in this model. Conversely, propane, the normal substrate and energy source, was modeled as an inhibitor of TCE kinetics because it inhibited the cometabolism of TCE.

The TCE degradation rate and concentration data were fit to the Michaelis-Menten equation by a nonlinear regression algorithm using SYSTAT by minimizing the sum of squares of the residuals to determine the values of the parameters V_{max} and K_s that best fit those data. Table 1 shows the results of these regressions.

The data fit the Michaelis-Menten equation well, with correlation coefficients $r^2 \geq 0.91$. These data are shown as a function of headspace propane in Figure 6. The parameter estimates were compared using Student's t-test to determine if the values of V_{max} and K_s were statistically different for the four concentrations of propane. The results of this analysis for V_{max} are listed in Table 1.

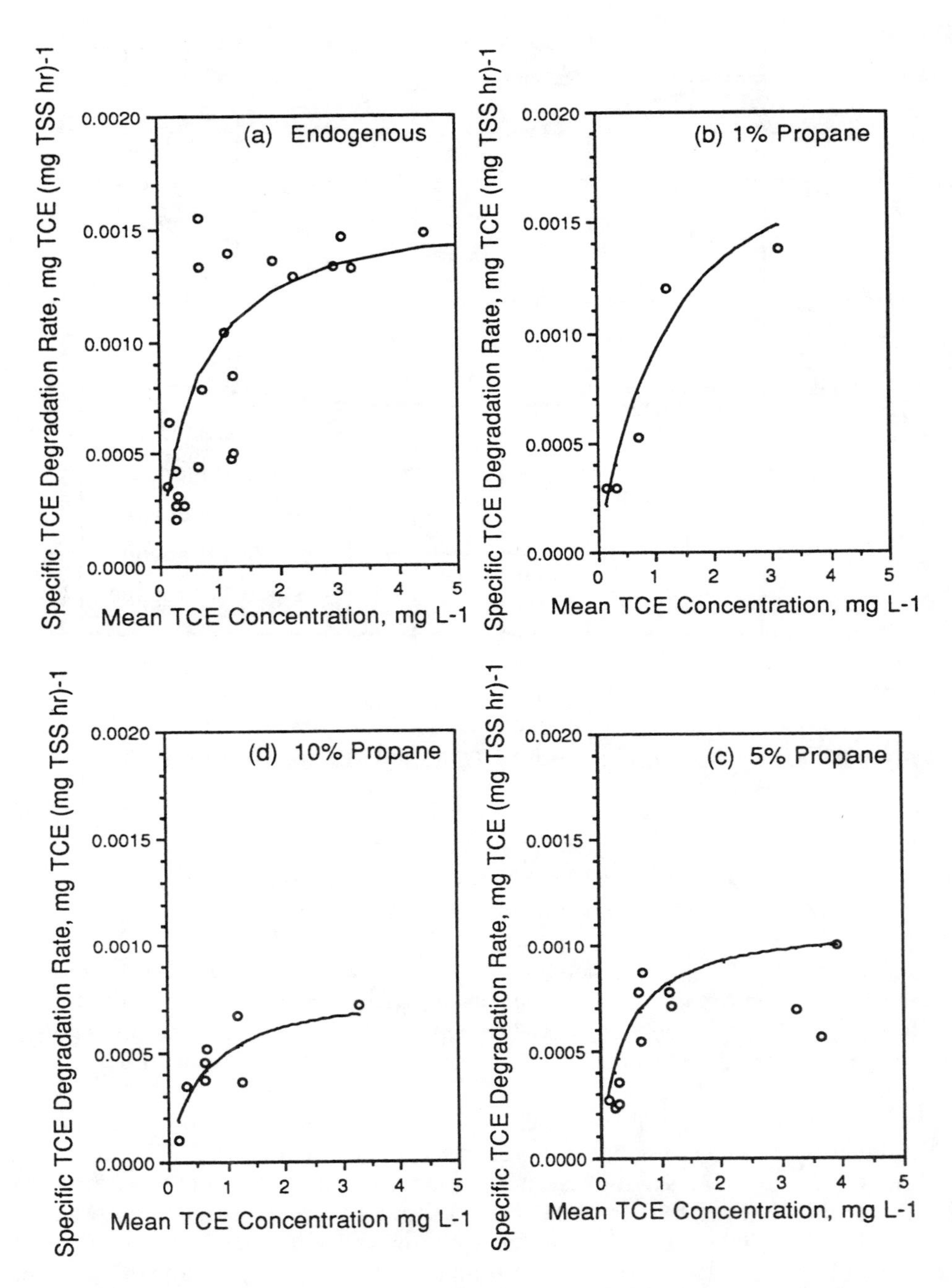

FIGURE 5. TCE degradation rate as a nonlinear function of the TCE concentration and inversely related to the propane concentration.

TABLE 1. Michaelis-Menten parameters for TCE degradation.

Propane Level	V_{max} mg TCE (mg TSS hr)$^{-1}$	K_S mg TCE/L
0% (endogenous)	0.0016±0.0004$^{(a,b)}$	0.6±0.5
1%	0.0021±0.0003	1.3±2.3
5%	0.0011±0.0003$^{(a)}$	0.4±0.5
10%	0.0008±0.0004$^{(b)}$	0.6±0.8

Errors listed represent a 95% confidence interval. (a) Values that are significantly different at the 10% significance level; (b) at the 5% level.

The values of V_{max} for the endogenous condition were significantly different from the 5% and 10% propane values of V_{max} at the 90% or higher confidence level. The K_s values were found by Student's t-test not to be significantly different at the different propane levels. Therefore, it is concluded that the data sets can be combined to get an average K_s value.

Based on the V_{max} being different and the values of K_s being the same at different propane levels, the inhibition is modeled as noncompetitive inhibition in which the inhibitory compound (propane) can combine with either the free enzyme or the enzyme-substrate (TCE) complex. This mechanism can be written as

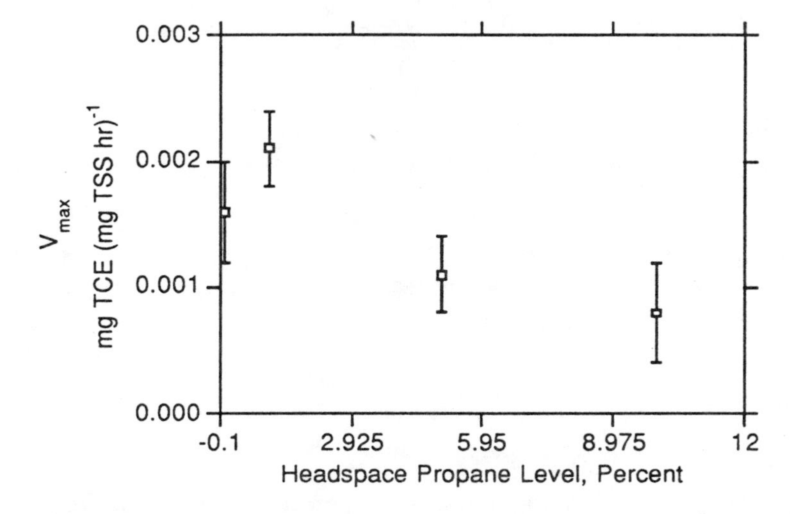

FIGURE 6. The maximum velocity of TCE degradation as a nonlinear function of propane level.

$$
\begin{aligned}
&E + S \leftrightarrow ES \rightarrow Æ\ E + \text{by-product} \\
&E + P \leftrightarrow EP \rightarrow ÆE + \text{propanol} \\
&ES + P \leftrightarrow ESP
\end{aligned}
\tag{5}
$$

where: E = enzyme
S = substrate (TCE)
P = inhibitor (propane)

From this mechanism, the rate equation can be derived as (Bailey & Ollis 1977)

$$V = \frac{V_{max\ Endo} S}{(K_S + S)\left(1 + \frac{P}{K_P}\right)} \tag{6}$$

where: V = velocity (rate) of the reaction, mg TCE (mg TSS hr)$^{-1}$
$V_{max\ Endo}$ = maximum velocity of reaction under endogenous conditions, mg TCE (mg TSS hr)$^{-1}$
S = substrate (TCE) concentration, mg TCE/L
K_s = half-saturation constant for the substrate, mg TCE/L
P = propane concentration, % propane
K_p = half-saturation constant for the inhibitor, % propane

Although K_s remains unaffected, V_{max} is affected by the presence of the inhibitor with this inhibition expressed as

$$V_{max} = \frac{V_{max\ Endo}}{\left(1 + \frac{P}{K_P}\right)} \tag{7}$$

where: V_{max} = maximum velocity of reaction at any propane level, mg TCE (mg TSS hr)$^{-1}$

The V_{max} values for endogenous, 1%, 5%, and 10% propane data sets can be used to determine the value of K_p by rearranging equation (7) to yield

$$K_P = \frac{P}{\left(\frac{V_{max\ Endo}}{V_{max}} - 1\right)} \tag{8}$$

Because the rate at 1% propane was not significantly different from the rate under the endogenous condition, the 5% and 10% propane data sets were used to calculate K_p and are presented in Table 2 with the 95% confidence interval.

TABLE 2. Propane inhibition constant.

% Propane	V_{max}	K_p
0	0.0016	—
5	0.0011	10.3
10	0.0008	10.9
		Average = 10.6±0.8

An explicit expression for the degradation of TCE in the presence of propane can be written as

$$V = \frac{0.0016\ S}{(S + 0.6)\left(1 + \frac{P}{10.6}\right)} \tag{9}$$

where: $V =$ mg TCE (mg TSS hr)$^{-1}$
$S =$ mg TCE/L
$P =$ % propane

Figure 7 shows the curves predicted for 0%, 1%, 5%, and 10% propane level.

CONCLUSIONS

Based on these results, the following conclusions can be drawn:

1. A propane-oxidizing enrichment culture was able to degrade TCA and TCE.
2. Within the range of concentrations tested (0 to 20 mg/L), TCA degradation followed first-order kinetics with the rate constant for endogenous conditions, $K = 0.0031$ (mg TSS hr)$^{-1}$.
3. Propane strongly inhibited TCA degradation. However, the data did not fit accepted inhibition models including competitive, noncompetitive, and uncompetitive inhibition.
4. Within the range tested (0 to 5 mg/L), TCE degradation followed Michaelis-Menten kinetics, with $V_{max} = 0.0016$ mg TCE (mg TSS hr)$^{-1}$ and $K_s = 0.6$ mg TCE/L for endogenous conditions.
5. Propane inhibited TCE degradation and was best described by a noncompetitive inhibition model

$$V = \frac{0.0016\ S}{(S + 0.6)\left(1 + \frac{P}{10.6}\right)}$$

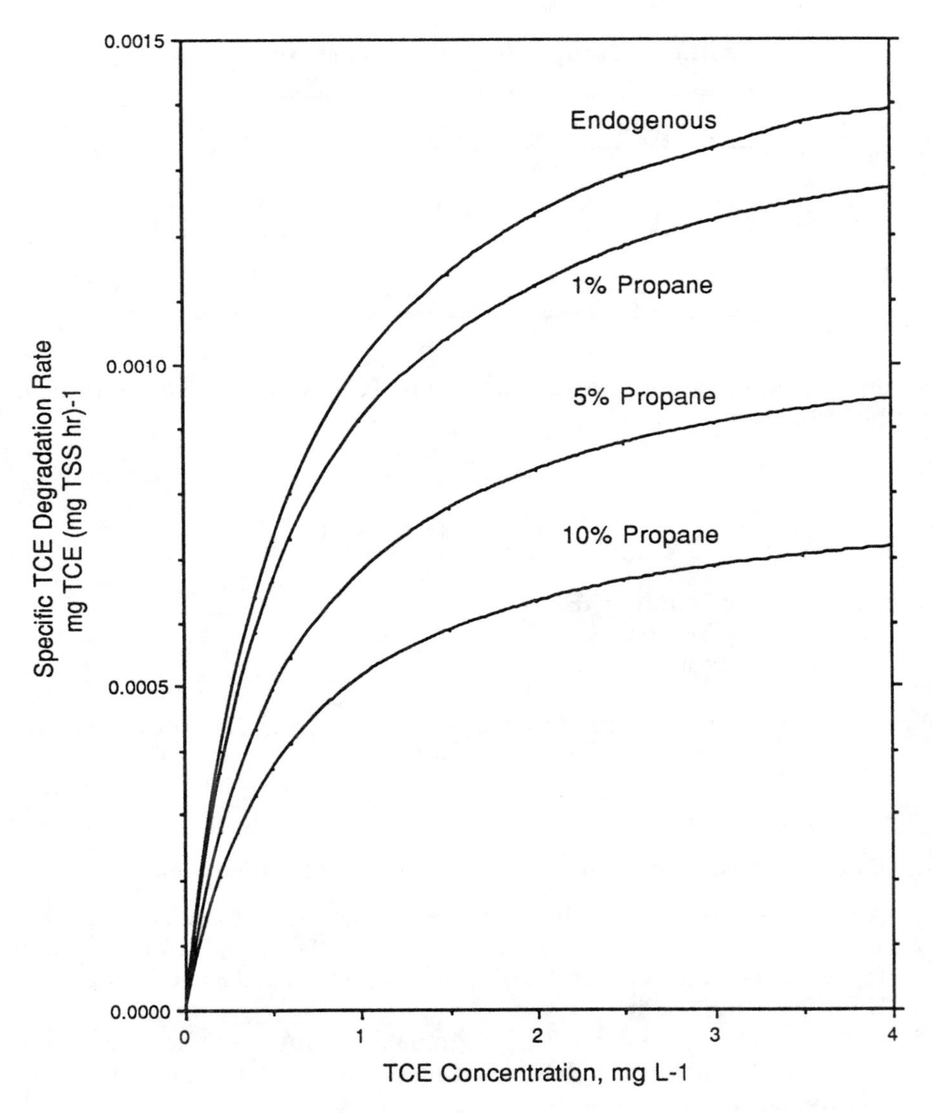

FIGURE 7. Predicted TCE degradation rate as a function of TCE and propane concentrations.

REFERENCES

Alvarez-Cohen, L., and P. L. McCarty. 1991. "Effects of toxicity, aeration, and reductant supply on TCE transformation by a mixed methanotrophic culture." *Appl. Environ. Microbiol. 51*: 228-235.

Bailey, J. E., and D. F. Ollis. 1977. *Biochemical Engineering Fundamentals*. McGraw-Hill Book Company, New York, NY.

Bjelland, M. 1989. "Kinetics of Methanotrophic Biodegradation of Chlorinated Aliphatic Hydrocarbons." M.S. Thesis. University of Washington, Seattle, WA.

Henry, S. M., and K. Grbić-Galić. 1991. "Influence of endogenous and exogenous electron donors and toxicity on TCE oxidation by methanotrophic cultures from a groundwater aquifer." *Appl. Environ. Microbiol. 570*: 236-244.

Hou, C. T., R. Patel, A. I. Laskin, N. Barnabe, and I. Barist. 1983. "Epoxidation of short-chain alkenes by resting-cell suspensions of propane-grown bacteria." *Appl. Environ. Microbiol. 46*: 171-177.

Oldenhuis, R., R. L. J. M. Vink, D. B. Janssen, and B. Witholt. 1989. "Degradation of chlorinated aliphatic hydrocarbons by *Methylosinus trichosporium* OB3b expressing soluble methane monooxygenase." *Appl. Environ. Microbiol. 55*: 2819-2826.

Perry, J. J. 1980. "Propane utilization by microorganisms." *Adv. Appl. Microbiol. 26*: 89-115.

Stephens, G. M., and H. Dalton. 1986. "The role of the terminal and subterminal oxidation pathways in propane metabolism by bacteria." *J. Gen. Microbiol. 132*: 2453-2462.

Strand, S. E., M. D. Bjelland, and H. D. Stensel. 1990. "Kinetics of chlorinated hydrocarbon degradation by suspended cultures of methane-oxidizing bacteria." *Res. J. Water Poll. Control Fed. 62*: 124-129.

Vannelli, T., M. Logan, D. M. Arciero, and A. B. Hooper. 1990. "Degradation of halogenated aliphatic compounds by the ammonia-oxidizing bacterium *Nitrosomonas europaea*." *Appl. Environ. Microbiol. 56*: 1169-1171.

Wackett, L. P., and D. T. Gibson. 1988. "Degradation of trichloroethylene by toluene dioxygenase in whole-cell studies with *Pseudomonas putida* F1." *Appl. Environ. Microbiol. 54*: 1703-1708.

Wackett, L. P., G. A. Brusseau, S. R. H. Householder, and R. S. Hanson. 1989. "Survey of microbial oxygenases: Trichloroethylene degradation by propane-oxidizing bacteria." *Appl. Environ. Microbiol. 55*: 2960-2964.

Woods, N. R., and J. C. Murrell. 1989. "The metabolism of propane in *Rhodococcus rhodochrous* PNKb1." *J. Gen. Microbiol. 135*: 2335-2344.

A FIELD EVALUATION OF IN SITU MICROBIAL REDUCTIVE DEHALOGENATION BY THE BIOTRANSFORMATION OF CHLORINATED ETHENES

R. E. Beeman, J. E. Howell, S. H. Shoemaker, E. A. Salazar, and J. R. Buttram

ABSTRACT

Results have demonstrated the in situ biotransformation of tetrachloroethene (PCE), trichloroethene (TCE), 1,2-dichloroethene (DCE), chloroethane (CA), and vinyl chloride (VC) to ethane and ethene using microbial reductive dehalogenation. These investigations were conducted in a 12.2 × 36.6 m test zone in PCE-contaminated aquifer underlying a plant near Victoria, Texas. Initial concentrations of PCE, TCE, and DCE in the aquifer approximated 10, 4, and 4 µM (1,700, 535, and 385 ppb), respectively. After 2 years of anaerobic treatment, chlorinated hydrocarbons were below detectable levels (BDL) in some monitoring wells using U.S. Environmental Protection Agency (EPA) Method 8240 (PCE, TCE, DCE <5 ppb; 0.03, 0.04, and 0.05 µM, respectively; VC, CA < 10 ppb, <0.16 µM). Microbial reductive dechlorination was accomplished by pumping either a benzoate or sulfate solution into the circulating groundwater. Mass balance estimates using bromide in the test zone approximated 74%. Mass balance estimate of PCE and daughter product formation approximated 65% in the last monitoring well. As a control, two additional wells near the test site were circulated in a similar manner but did not receive benzoate or sulfate addition. No loss of the PCE was observed in the control site. We conclude that in some aquifers, reductive dehalogenation can be used to remove halogenated hydrocarbons from groundwater.

INTRODUCTION

The presence and problems of halogenated organic compounds in aquifers have been well documented. Conventional pump-and-treat technologies may contain or control halogenated organic plume movement. However, pump-and-treat technologies have limited applications for aquifer and groundwater restoration. Innovative technologies that have the potential to remediate aquifers, such as biological reductive dehalogenation, have been widely sought.

This paper summarizes a 2-year investigation of in situ microbial reductive dehalogenation of PCE to ethene and ethane in an aquifer underlying the E. I. du Pont de Nemours and Company (Du Pont) plant near Victoria, Texas. The study was completed in two phases. First, it was demonstrated that PCE could be dechlorinated in situ. Second, in a controlled field experiment, PCE and its daughter products were degraded to ethene and ethane in situ under sulfate-reducing conditions using benzoate as the electron donor. Initial concentrations of PCE, TCE, and DCE in the pilot site approximated 10, 4, and 4 μM, respectively, whereas VC was not detected. After 2 years, aquifer concentrations of PCE, TCE, DCE, VC, CA, and dichloroethane (DCA; <5 ppb; 0.04 μM) were below detection limits (BDLs).

MATERIALS AND METHODS

Study Area

The test aquifer underlies the former Du Pont Plant West Landfill near Victoria Texas. Landfill construction depth approximated 5 m (15 ft; Figure 1) below the surface. Beginning in the early 1950s, the landfill received a variety of solid and liquid wastes from industrial activities for about 20 years. It is underlain by a semiconfined sand aquifer (Zone B), the top of which is located about 18.3 m (60 ft) below the surface. The overlying stratigraphy consists of interbedded sands and clays of late Pleistocene age, leading down to the Zone B sand, which is a relatively continuous water-bearing sand averaging 12.2 to 18.3 m (40 to 60 ft) thick.

The Zone B sand is composed of quartz, plagioclase, potassium feldspars, carbonate rock fragments, and rare mica grains. Analyses of 22 samples from the Zone B sand indicate an average total organic content (TOC) of approximately 0.1% with a range between 0.07% to 0.71%. The hydraulic conductivity of the sand was determined from pump test data to be approximately 30 m/day. The natural groundwater movement is southwesterly at approximately 0.3 m/day with discharge toward a manmade canal (Figure 1) along the southwestern site boundary. PCE and benzene waste placed in the landfill have migrated through the overlying strata and into the Zone B aquifer. Currently, a state-of-the-art, pump-and-treat facility is used to control and contain this groundwater contamination. This facility meets or exceeds all regulatory concerns of the Texas Water Commission.

Pilot Site Description. In Phase I, a 12.2 m x 18.3 m (40 x 60 ft) test site was established in the Zone B sand consisting of recovery, monitoring, and recharge wells (Figures 1 and 2). All wells were screened in the top 4.6 m (15 ft) of the aquifer. The wells do not fully penetrate the Zone B sand.

Wells 4N, 4, and 4S were extraction wells, whereas 1N, 1, and 1S were recharge wells (Figure 2). Wells 2 and 3 were monitoring wells. Pumping rates from

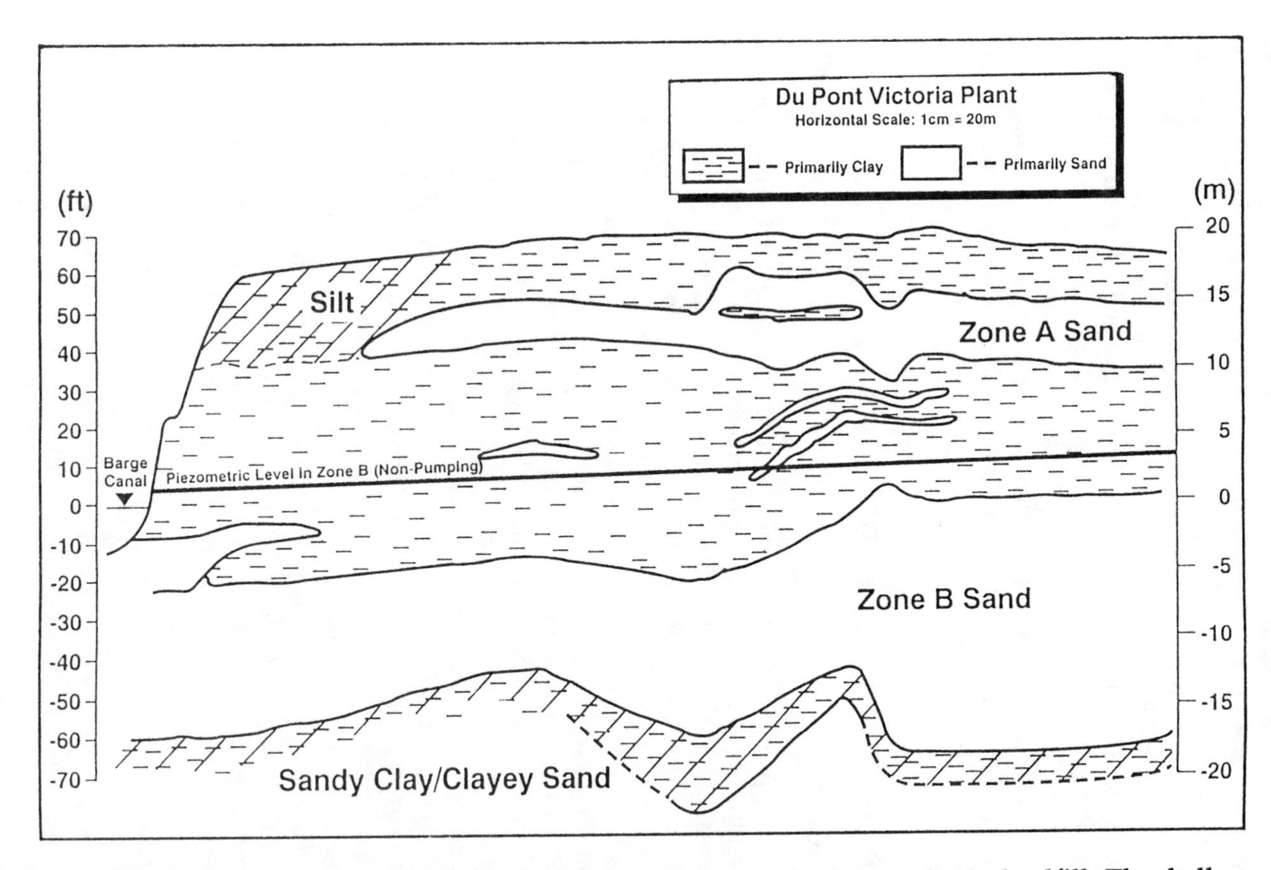

FIGURE 1. Geologic cross section of the West Victoria Landfill. Two sands underlie the landfill. The shallow zone A sand is thin and discontinuous and lies above the water table. The deeper zone B sand is the aquifer where the pilot work was performed.

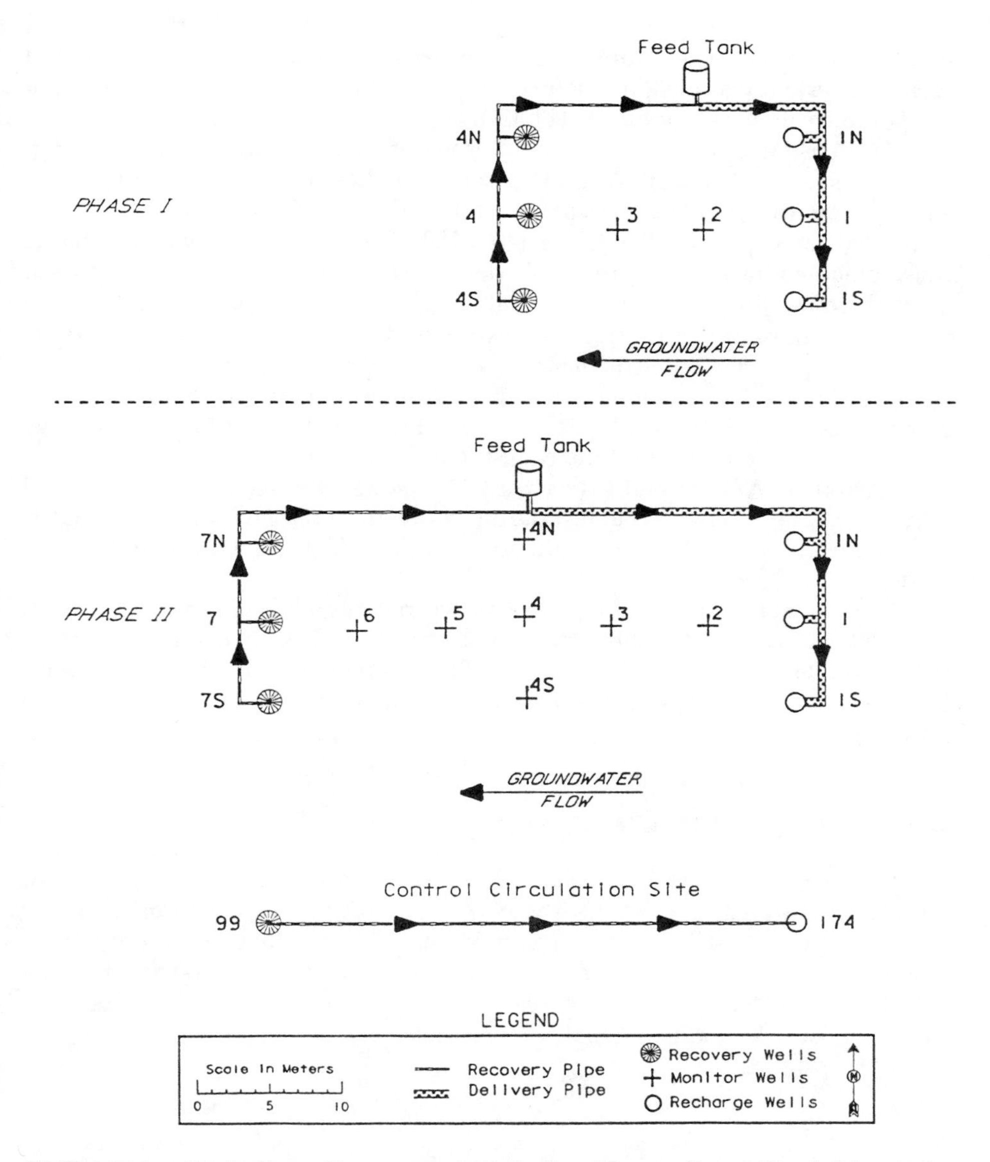

FIGURE 2. Well flow diagrams used during Phases I and II of this study. Note the additional wells used during Phase II (5, 6, 7N, 7, and 7S).

Wells 4N, 4, and 4S were 11.4, 26.6, and 11.4 L/min (3, 7, and 3 gpm), respectively. Wells 1N, 1, and 1S delivered water back to the Zone B sand at 11.4, 26.6, and 11.4 L/min, respectively. As water was brought to the surface from the recovery wells, it passed through a sealed mixing manifold where nutrient additions were made. Delivery from the nutrient feed tank was at 3.8 L/hr. A final dilution concentration of either 0.3 mM (38 ppm) sodium benzoate or 0.4 mM (56 ppm) magnesium sulfate was pumped into the recharge water. The recovery and recharge system was designed to prevent volatile losses due to air stripping of the volatile organic compounds (VOCs).

For Phase II, a new section of aquifer was included in the experiment (Figure 2). This was adjacent to Wells 4N, 4, and 4S. These wells were established at the same depth with the same screened intervals as in Phase I. The site dimensions were enlarged to 12.2 × 36.6 m (40 × 120 ft), and the recovery and recharge rates remained the same as previously described. Wells 7N, 7, and 7S were used as extraction wells, whereas Wells 2, 3, 4N, 4, 4S, 5, and 6 were used as monitoring wells. As before, Wells 1N, 1, and 1S were used to delivered water back to the aquifer. Groundwater movement in the pilot site approximated 1.1 m/day and the site had a hydraulic retention time of about 30 days (well series 1 to 7). For background gas analysis, Well 15 (not shown), which is located in the VOC plume, but outside the hydraulic influence of the pilot site, was sampled.

Separately, Wells 99 and 174 were established near this second site as a control (Figure 2). They were used to test the effects of simple circulation on the VOCs. This site was circulated at approximately 19 L/min (5 gpm) without nutrient or feedstock additions.

Modeling of the circulation pattern between Wells 99 and 174 and the flow of the pilot site revealed significant interaction on both recovery and recharge patterns (data not shown). Therefore, after approximately 4 months of operations, flow between Wells 99 and 174 was discontinued and Well 99 was used as a monitoring well.

ANALYTICAL PROCEDURES

Samples from wells were collected following proper EPA protocol. Briefly, monitor wells were purged for three well volumes and then sampled using dedicated purge and gas bladder pumps. Analysis for halogenated compounds was accomplished using gas chromatography (GC) and mass spectrometry by EPA Method 8240. Analysis for gases was by GC with flame ionization detection following EPA Method 8015.

RESULTS

Phase I: Demonstration of PCE and TCE Degradation In Situ

Aerobic Treatment. Figure 3 displays the variation in PCE, TCE, and DCE groundwater concentration during 1990 in Wells 2, 3, 4N, 4, and 4S (Figure 2). This period encompassed the time before and after the start of anaerobic treatment. Anaerobic treatment began with the addition of benzoate to the pilot site, on day 203.

Efforts to degrade PCE in the pilot site using aerobic techniques were unsuccessful through day 203 (Figure 3). However, the aerobic remediation efforts

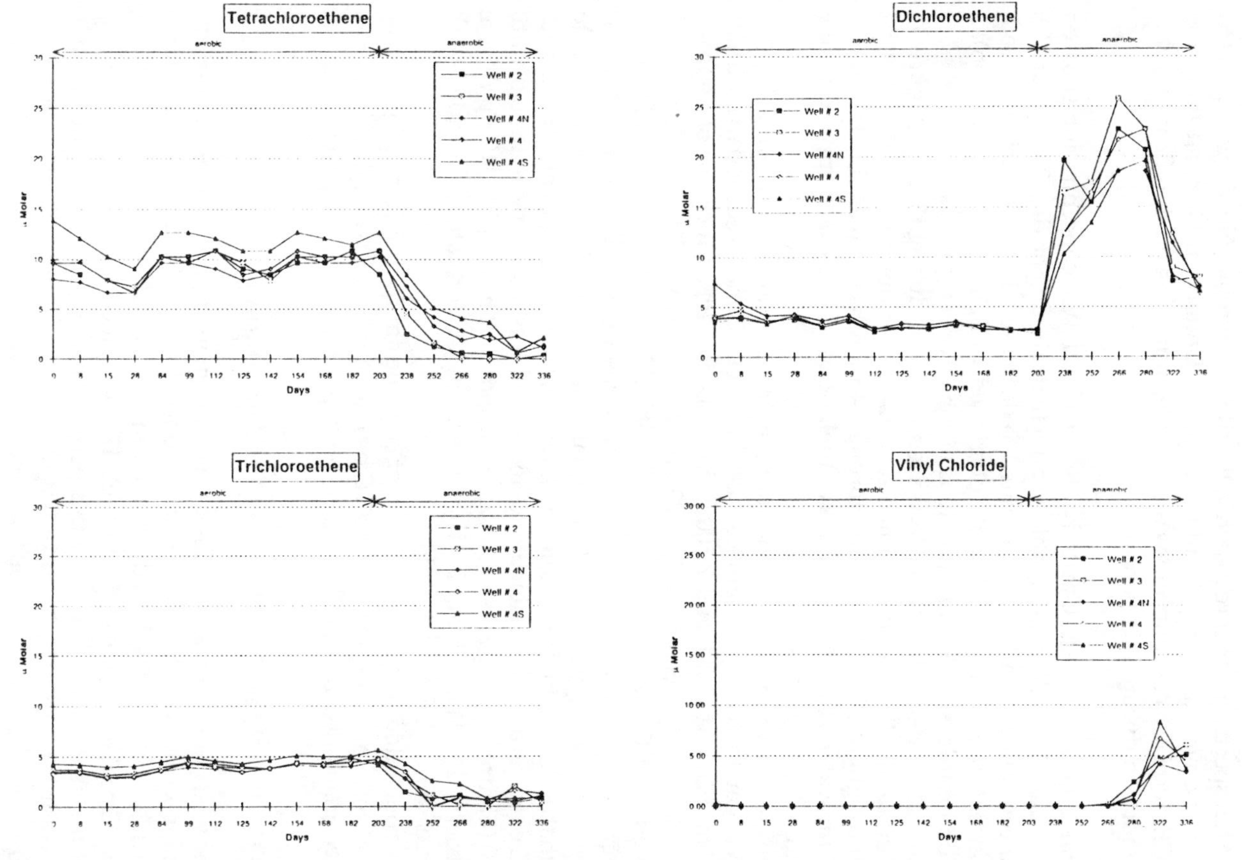

FIGURE 3. Tetrachloroethene, trichloroethene, dichloroethene, and vinyl chloride concentrations during Phase I. Note the pilot site was aerobic from day 0 to 203, then anaerobic from day 203 to 336.

reduced the groundwater concentration of benzene in the pilot site to BDLs (<5 ppb; data not shown).

Groundwater PCE, TCE, and DCE median concentrations approximated 10, 4, and 4 μM, respectively (Figure 3), while the site was aerobic. No VC was detected while the site was aerobic (<.16 μM; <10 ppb). Under aerobic conditions, the PCE, TCE, and DCE concentrations were stable, indicating no significant aerobic biodegradation during this period.

Anaerobic Treatment. Groundwater PCE and TCE concentration showed decreases after 1 month of anaerobic treatment beginning on day 203 (Figure 3). PCE concentrations in all wells continued to decrease with time during Phase I. PCE concentrations in Wells 2 and 3 decreased to BDL (<0.1 μM; <15 ppb), a reduction of at least 98% from previous aerobic concentrations.

The TCE concentrations also decreased with time during Phase I (Figure 3). However, unlike PCE, TCE remained detectable in groundwater from Wells 2 and 3 throughout Phase I. Overall, through 4 months of anaerobic treatment, the groundwater TCE concentrations in Wells 2 and 3 decreased approximately 85% and 89%, respectively.

The groundwater DCE concentrations increased in all the wells until day 280 when it began to decrease (Figure 3). DCE in Wells 2 and 3 reached maximums near 23 and 26 μM (2,200 and 2,500 ppb), respectively, an increase of approximately 7-to 8-fold. Between days 203 and 252 increasing DCE concentrations correlated with the falling concentrations of PCE and TCE. Clearly, PCE and TCE were being biotransformed into DCE. Groundwater DCE concentrations decreased in all wells after day 280 to a final concentration ranging between 2- to 3-fold that of the original DCE concentration, before anaerobic treatment started.

VC concentrations remained undetectable or less than 1 μM in the site until day 280 (Figure 3). Larger concentrations approximating 5 μM (310 ppb) were then observed in Wells 2 and 3. The increase in VC formation coincided with the decrease in DCE concentration.

We concluded that the groundwater concentrations of PCE and TCE in the pilot site were converted to DCE and VC by microbial reductive dehalogenation. Also, we concluded that PCE could be biodegraded to below detection limits.

Phase II: PCE Daughter Product Biodegradation

The well field schematic for Phase II is shown in Figure 4. To test the effects of circulation on the volatile chlorinated concentrations, Well 99 was circulated to Well 174. Figure 4 displays the concentrations of PCE in Wells 2, 3, 7N, 7, 7S, and 99. Circulation of Well 99 to Well 174 continued from day 0 to day 110, approximately 4 months. During this period, groundwater PCE concentration increased at Well 99 from 15 to 24 μM (2,475 to 3,960 ppb). This compares to groundwater PCE concentrations of nondetectable for Wells 2 and 3 (<0.1 μM; <15 ppb) in the pilot site. This demonstrates that simple circulation of water did not cause the disappearance of the PCE from the site. Similar results were obtained for TCE (data not shown).

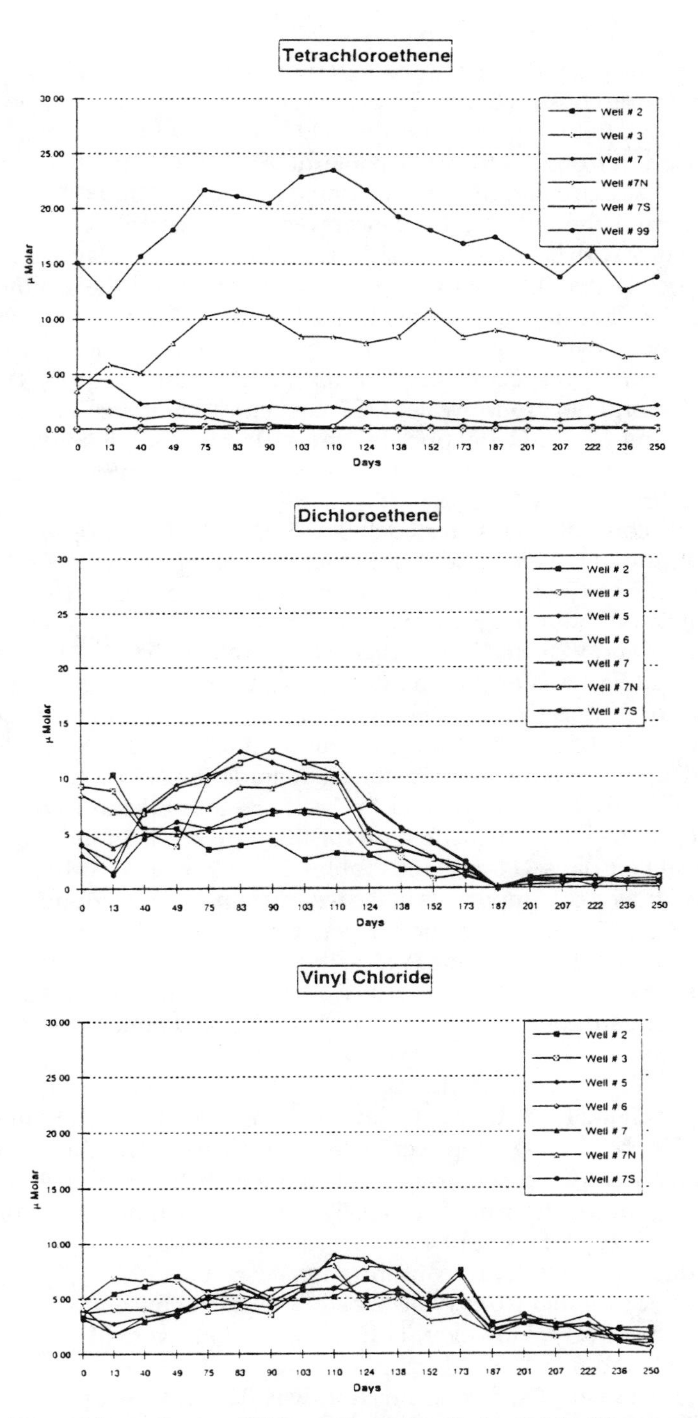

FIGURE 4. Tetrachloroethene, dichloroethene, and vinyl chloride concentrations during Phase II. Note that the time (days) restarts with zero for Phase II.

Both recovery and circulatory systems were stopped for approximately 3 months during construction of the phase II system. Upon completion and restart of the new well field (Figure 4, day 0), no PCE was observed in well 3, indicating that desorption from the formally contaminated solids had not occurred.

Phase II site circulation began on day 0 for Figure 4 and lasted for 250 days. During Phase II, the sulfate anion concentration was kept above 0.6 mM in all the pilot site wells, with the exception of the results obtained on day 426, when sulfate anion approximated 0.1 mM (data not shown). As before, the groundwater PCE and TCE were transformed rapidly into DCE and VC in the pilot site (Figure 4; TCE data not shown). The PCE and TCE concentrations typically were less than 0.03 and 0.04 μM, respectively, in Wells 2, 3, 5, and 6 (<5 ppb; TCE data not shown). In contrast, groundwater DCE concentration increased to a maximum approximating 12 μM (1,150 ppb) by day 110 in Wells 3, 5, and 6. Thereafter, DCE declined until day 187 in all the wells, nearing 1 μM (96 ppb), where it remained through day 250.

The VC concentrations started at 3 to 5 μM (186 to 310 ppb) and increased to a maximum between 5 and 9 μM (190 and 560 ppb) for Wells 2, 3, 5, and 6 (Figure 4). The VC concentration started to decline on day 110 and continued to decline throughout the remainder of the year.

The DCE and VC could be biotransformed to ethene gas, but previous research indicated that biotransformation rates from VC to ethene gas were slow under anaerobic conditions (Major & Hodgins 1991). Also, a comparison of feed and daughter product formation in the site needed to be performed.

The daughter production from the biodegradation of PCE has been identified as TCE, DCE, VC, DCA, CA, and both ethene and ethane (Sims et al. 1991). Therefore, we began an investigation for these products in the groundwater at our pilot site on days 314 and 426 (Tables 1 and 2, respectively).

Concentration of components in the feed stream were determined by analyzing Wells 7N, 7, and 7S for the specific halogenated compounds. These concentrations were multiplied by the fractional contribution for each well to define the total feed stream concentration. The results were then summed for Wells 7N, 7 and 7S as the feed stream concentration. Flow from the feed tank into this stream was ignored.

Table 1 displays the specific components in groundwater from each well, plus the calculated feed stream on day 314. The feed stream contained 1.42 μM of PCE (235 ppb; Table 1). However, as the flow moved downgradient past Well 2, the groundwater PCE concentration was reduced by more than 90% to 0.06 μM (10 ppb). Groundwater from Wells 3, 5, and 6 revealed no detectable PCE concentration (<0.03 μM; <5 ppb).

The feed stream TCE concentration was 0.72 μM (95 ppb; Table 1). Groundwater sampled from Wells 2 and 3 revealed concentrations of 0.09 and 0.19 μM, respectively. However, TCE concentrations were BDL in groundwater from Wells 5 and 6 (<0.03 μM; <5 ppb).

The feed stream DCE concentration was 0.85 μM (83 ppb; Table 1). This decreased in the groundwater samples taken from downgradient wells to 0.25 μM in Well 6, a 71% decrease across the pilot site.

TABLE 1. Groundwater concentrations (μM) of PCE, TCE, DCE, VC, CA, ethane, and ethene in the Wells 1N-1S, 2, 3, 5, and 6 and control well 15 on day 314.

	(μM)					
Component/Well	Feed to 1N-1S	2	3	5	6	15
PCE	1.42	0.06	ND	ND	ND	NS
TCE	0.72	0.09	0.19	ND	ND	NS
DCE (*cis*- and *trans*-)	0.85	0.40	0.37	0.16	0.25	NS
VC	0.48	1.07	0.37	0.29	0.58	NS
CA	0.24	0.31	0.34	0.57	0.50	NS
Ethene	0.87	NS	NS	0.71	1.75	ND
Ethane	ND	NS	NS	ND	ND	ND
Summation	4.58	–	–	1.73	3.08	-

ND = Not detected. For PCE <0.03 μM; 5 ppb. For TCE <0.04 μM; 5ppb. For DCE <0.05 μM; 5 ppb. For VC <0.16 μM; 10 ppb. For CA <0.16 μM; 5 ppb. For ethene <0.35 μM; 10 ppb. For ethane <0.33 μM; 10 ppb.

NS = Not sampled.

The feed stream VC concentration was 0.48 μM (30 ppb; Table 1). The VC concentration increased in the groundwater taken from Well 2 to 1.07 μM (67 ppb) and then decreased to 0.58 (36 ppb) in groundwater taken downgradient in Well 6. The increase in VC is consistent with the biotransformation of PCE to VC. Using the VC concentration in the groundwater from Well 2, the decrease across the pilot site to Well 6 was 46%.

The feed stream CA concentration was 0.24 μM (15 ppb; Table 1). This concentration increased in the groundwater across the site to 0.50 μM (32 ppb) in Well 6. This is a 2-fold increase across the pilot site.

For ethene, the feed stream concentration was 0.87 μM (25 ppb; Table 1). No groundwater samples were taken from Wells 2 and 3 for gas analysis. The ethene concentration increased to 1.75 μM (52 ppb) in groundwater in Well 6. Ethane was not detected in the feed stream or groundwater from the monitoring wells (<0.16 μM; <10 ppb). Ethene was not detected in Well 15, which is outside the pilot site. Therefore, the ethene present in the site was due to the biodegradation of PCE and its daughter products to ethene.

The data shown in Table 1 indicate that PCE entering the pilot site in the feed stream was being biotransformed into VC in the groundwater by Well 2. This was further biotransformed into ethene in groundwater by Well 6. It was concluded that some of PCE entering the site appeared to be biotransformed into ethene.

TABLE 2. Concentrations of PCE, TCE, DCE, VC, CA, ethene, and ethane in Wells 1N-1S, 2, 3, 4N, 4, 4S, 5, and 6 and control wells 99 and 15 on day 426.

	(µM)									
Component/Well	Feed to 1N-1S	2	3	4N	4	4S	5	6	99	15
PCE	0.86	ND	ND	ND	ND	ND	ND	ND	3.98	NT
TCE	0.79	ND	0.06	ND	ND	ND	ND	ND	0.31	NT
DCE (*cis-,trans-*)	0.66	0.06	0.19	ND	ND	ND	ND	ND	ND[a]	NT
VC	ND	ND	ND	ND	ND	ND	ND	ND	ND[a]	NT
CA	ND	ND	ND	ND	ND	ND	ND	0.16	ND[a]	NT
Ethene	0.75	0.46	0.57	1.46	1.43	2.11	0.50	1.71	ND	ND
Ethane	ND	ND	0.30	0.30	ND	ND	ND	ND	ND	ND
Summation	3.06	0.52	1.12	1.76	1.43	2.11	0.50	1.87	–	–

(a) Higher detection limits because a 1:4 dilution of sample was made. Detection limits are 4 times that of ND values.

ND = Not detected. For PCE <0.03 µM; 5 ppb. For TCE <0.04 µM; 5 ppb. For DCE <0.05 µM; 5 ppb. For VC <0.16 µM; 10 ppb. For CA <0.16 µM; 5 ppb. For ethene <0.35 µM; 10 ppb. For ethane <0.33 µM; 10 ppb.

NT = Not tested.

Other daughter products were being formed and degraded in situ. With the exception of CA, these daughter products should also form ethene. To estimate the mass balance in the pilot site, individual halogenated concentrations in the feed stream were summed and compared to the products found in groundwater from Well 6. Concentrations BDL were ignored in these summations. The summation of the components in the feed stream equaled 4.58 µM (Table 1). Summation of the components in Well 6 produced 3.08 µM. Comparison of the products in the groundwater from Well 6 to the reactants in the feed stream approximated 67% recovery.

To interpret the product recovery in the pilot site, a comparison was made to a conservative tracer. This would account for dilution and dispersion with water outside the site. A bromide anion tracer demonstrated that 74% of the bromide placed into the 1N-1S wells was recovered in the 7N-7S wells. Comparison between the recovery of the end products from PCE degradation, 67%, to the recovery of the bromide tracer, 74%, indicated that the major end products from PCE biodegradation in the aquifer were reconciled.

Table 2 displays the specific components in groundwater from each well, plus the calculated feed stream on day 426. PCE was present in the feed stream at 0.86 µM (143 ppb). The PCE concentration was BDL in all of the remaining monitoring wells.

TCE was also present in the feed stream at 0.79 µM (105 ppb; Table 2). It was BDL in Wells 2, 4N, 4, 4S, 5, and 6. The only detectable concentration of TCE was in groundwater taken from Well 3, just above the detection limit, at 0.06 µM (8 ppb).

For DCE, the concentration present in the feed stream was 0.66 µM (64 ppb; Table 2). Groundwater DCE concentrations were detected from Wells 2 and 3 at 0.06 and 0.19 µM (6 and 18 ppb), respectively. The groundwater concentration of DCE in Wells 4N, 4, 4S, 5, and 6 were BDL.

VC was not present in the feed streams, although it was detected on day 314 (Tables 1 and 2). Nor was it present in groundwater from the remaining monitoring wells (<0.16 µM; 10 ppm, Table 2) on the day 426 sampling. CA also was not detected in the pilot site on day 426, with the exception of Well 6 where the concentration of chloroethane was 0.16 µM (10 ppb).

Ethene was detected in the feed stream at 0.75 µM (Table 2). In groundwater taken from Wells 4N, 4, 4S, and 6, the concentration of ethene had increased to 1.46, 1.43, 2.11, and 1.71 µM (47, 46, 68, and 54 ppm), respectively. Ethene was BDL in groundwater taken from Wells 15 and 99 (<0.16 µM; <10 ppb). Because wells outside the pilot site (15 and 99) had no detectable ethene or ethane, then ethene or ethane in the pilot site must have come from the biodegradation of halogenated hydrocarbons, notably PCE biodegradation. Ethane was present only in groundwater taken from Wells 4N and 4 at 0.29 µM (10 ppm). Because ethane was not a detectable background gas, the ethane in the pilot site must have come from the biodegradation of chloroethane or from the reduction of ethene. Neither DCA nor 1,1 DCE was detected in the pilot groundwater and the major DCE isomer was *cis*-DCE. Methane was observed repeatedly in the groundwater from well 15 at concentrations approximating 20 ppb; no significant elevation in methane was noted, during Phase II, in the pilot site monitoring wells.

Recovery of feed stream reactants as products in groundwater from Well 6 reveals approximately 62% recovery on day 426 sampling. This compares favorably to the 74% recovery for the bromide tracer. The ethene found in Well 6 approximated 55% of the chlorinated hydrocarbons in the feed stream. Groundwater taken from Wells 4N, 4, 4S, and 5 were free of demonstrable chlorinated hydrocarbons at the detection limits of EPA Method 8240. Also, groundwaters taken from Wells 2, 3, and 6 reveal low concentrations of PCE daughter products ranging between 0.06 and 0.19 µM for DCE, TCE, or CA.

In summary, PCE and its daughter products, TCE, DCE, VC, and CA, have been anaerobically biodegraded to BDL in wells 4N, 4S, and 5. In some monitoring wells, ethene and ethane, the dechlorinated products from microbial PCE degradation, are detected at near-stoichiometric concentrations based on the chlorinated hydrocarbons in the feed stream. Variations in ethene and mass balance concentrations may be related to differential groundwater movements in the pilot site. We conclude that microbial reductive dehalogenation of PCE can be used to remove chlorinated ethenes from some subsurface groundwater aquifers.

DISCUSSION

Previous studies have demonstrated the anaerobic dechlorination of PCE using aquifer solids and water in the laboratory (Parsons et al. 1984, Schulz-Muramatsu et al. 1990, Suflita et al. 1988, Wilson et al. 1983, Wilson et al. 1986). To be a successful remediation method, PCE and its daughter products must be completely dechlorinated in a contaminated site (Major & Hodgins 1991). Laboratory studies have indicated complete PCE degradation was possible using microbial reductive dehalogenation; however, many other studies have indicated only limited success with complete PCE biotransformation (de Bruin et al. 1992, DiStefano et al. 1991, Freedom & Gossett 1989, Suflita et al. 1988). This work has demonstrated that PCE and its daughter products can be biodegraded to BDL concentrations in situ, using EPA Method 8240. Therefore, microbial reductive dehalogenation is a potential remedial mechanism for halogenated compounds in groundwater aquifers and deserves further investigation.

Ethene was the major metabolic product detected in these wells. To our knowledge, this is the first demonstration that PCE can be degraded to ethene under sulfate-reducing conditions, although PCE has previously been shown to biodegrade to TCE and DCE under sulfate-reducing conditions (Bagley & Gossett 1990, Suflita et al. 1988). The produced ethene is considered to be environmentally acceptable, because ethene has not been associated with long-term toxicological problems and is a natural occurring plant hormone (Sims et al. 1991). Furthermore, ethene is known to further biodegrade to carbon dioxide under aerobic environmental conditions.

VC was thought to persist in anaerobic environments and to be more toxic to bacteria than the parent compounds (Major & Hodgins 1991). Our work does not support either theory. In this work, after about 6 months of operation in Phase II, both VC and DCE appeared to have biodegraded to ethene and ethane. The site was operated under sulfate-reducing conditions to try and control the amount of VC production, because operation under methanogenic conditions might have led to the production of large amounts of VC, a known human carcinogen. The pattern of increase and disappearance of DCE and VC from Phase II is suggestive of microbial succession. We speculate that microbial succession may be a mechanism which completes reductive dehalogenation of DCE and VC to ethene and ethane. From this assumption, it follows that bacterial consortia, rather than a single species, would be required to completely dehalogenate PCE to ethene and ethane.

ACKNOWLEDGMENTS

We thank Clifford Moczygemba, Mary Norvell, Bill Muldoon, John Coleman, Dr. Charles Bleckmann, the entire Du Pont and Conoco organizations, and Nancy Frank of the Texas Water Commission, for their dedicated efforts in support of this project.

REFERENCES

Bagley, David M., and James M. Gossett. 1990. "Tetrachloroethene Transformation to Trichloroethylene and *cis*-1,2 Dichloroethylene by Sulfate-Reducing Enrichment Cultures." *Appl. Environ. Microbiol. 56*: 2511-2516.

De Bruin, Wil P., Michiel J. J. Kotterman, Maarten A. Posthumus, Gosse Schraa, and Alexander J. B. Zehnder. 1992. "Complete Biological Reductive Transformation of Tetrachloroethene to Ethane." *Appl. Environ. Microbiol. 58*: 1996-2000.

DiStefano, Thomas D., James M. Gossett, and Stephen H. Zinder. 1991. "Reductive Dechlorination of High Concentrations of Tetrachloroethene to Ethene by an Anaerobic Enrichment Culture in the Absence of Methanogenesis." *Appl. Environ. Microbiol. 57*: 2287-2292.

Freedman, D. L., and J. M. Gossett. 1989. "Biological Reductive Dechlorination of Tetrachloroethylene and Trichloroethylene to Ethylene Under Methanogenic Conditions." *Appl. Environ. Microbiol. 55*: 2144-2151.

Major, D. W., and E. W. Hodgins. 1991. "Field and Laboratory Evidence of In Situ Biotransformation of Tetrachloroethene to Ethene at a Chemical Transfer Facility in North Toronto." In R. E. Hinchee and R. F. Olfenbuttel (Eds.), *On-Site Bioreclamation: Process for Xenobiotic and Hydrocarbon Treatment*, pp. 147-178. Butterworth-Heinemann, Stoneham, MA.

Parsons, F., P. R. Wood, and J. DeMarco. 1984. "Transformations of Tetrachloroethene and Trichloroethene in Microcosms and Groundwater." *J. Am. Water Works Assoc. 76*: 56-59.

Schloz-Muramatsu, Heidrum, Regine Szewzyk, Ulrich Szewayk, and Suse Gaiser. 1990. "Tetrachloroethylene as Electron Acceptor for the Anaerobic Degradation of Benzoate." *FEMS Microbiol. Letters 66*: 81-86.

Sims, J. L., J. M. Suflita, and H. H. Russell. 1991. "Reductive Dehalogenation of Organic Contaminants in Soils and Ground Waters." *EPA Ground Water Issue*, EPA/540/4-90/054.

Suflita, J. M., S. A. Gibson, and R. E. Beeman. 1988. "Anaerobic Biotransformation of Pollutant Chemicals in Aquifers." *J. Indust. Microbiol. 3*: 179-194.

Wilson B. H., G. B. Smith, and J. F. Rees. 1986. "Biotransformation of Selected Alkylbenzenes and Halogenated Aliphatic Hydrocarbons in Methanogenic Aquifer Material: A Microcosm Study." *Environ. Sci. Technol. 20*: 997-1002.

Wilson, J. T, J. F. McNabb, B. H. Wilson, and M. J. Noonan. 1983. "Biotransformation of Selected Organic Pollutants in Groundwater." *Dev. Ind. Microbiol. 24*: 225-233.

BIOREMEDIATION OF TRICHLOROETHYLENE-CONTAMINATED SOILS BY A METHANE-UTILIZING BACTERIUM *METHYLOCYSTIS* SP. M

O. Yagi, H. Uchiyama, K. Iwasaki, M. Kikuma, and K. Ishizuka

ABSTRACT

A methane-utilizing bacterium, *Methylocystis* sp. M (strain M), was isolated from soil. This strain degrades trichloroethylene (TCE). Strain M can degrade TCE at a relatively high concentration of 35 mg/L, as well as various halogenated aliphatic compounds. Bioremediation of TCE-contaminated soil by strain M was investigated in soil microcosms, using 70-mL serum bottles and 450-mL glass columns as the soil microcosms. Strain M was added to the microcosms, which were filled with TCE-contaminated soil and water mixtures having TCE concentrations ranging from 0.1 to 1 ppm. More than 95% of the TCE in the 0.1 ppm concentration was degraded by the addition of strain M, and degradation of 1,1-, *cis*- and *trans*-dichloroethylene (DCE) was accelerated by the strain. Adding H_2O_2 with strain M significantly promoted TCE degradation. Strain M proved to be very effective in cleaning up TCE-contaminated soil.

INTRODUCTION

Volatile chlorinated aliphatic compounds, such as tetrachloroethylene (PCE), trichloroethylene, and 1,1,1-trichloroethane have been detected in groundwater. A survey of wells by the Japan Environment Agency found more than one-third of the wells had water contaminated by these compounds (Kawasaki 1985). The groundwater contamination is thought to be related to the widespread use of these compounds as cleaning solvents in dry-cleaning operations and in semiconductor manufacturing plants, and leakage from hazardous waste sites. PCE and TCE are considered carcinogenic and are resistant to biodegradation in the environment. As a result, the Japan Ministry of Health and Welfare established drinking water standards of 10, 30, and 300 µg/L for PCE, TCE, and 1,1,1-trichloroethane, respectively.

Various methods have been used to clean up the groundwater pollution, such as air stripping and carbon adsorption, vacuum extraction, and excavation.

But these processes do not eliminate the contamination; they only transfer it to another medium. It is said that bioremediation is effective to clean up soil pollution and is a clean technology. However, little information exists on bioremediation technologies.

Recently, the microbial degradation of trichloroethylene under anaerobic conditions has been studied (Bouwer & McCarty 1983, Vogel & McCarty 1985, Fathepure et al. 1987). Under aerobic conditions, chlorinated alkenes have been reported to resist biodegradation. Wilson and Wilson (1985) reported the existence of TCE-degrading soil columns amended with methane and air. Fogel et al. (1986) reported that a mixed culture containing methanotrophs aerobically degraded TCE to carbon dioxide. Little et al. (1988) also reported aerobic TCE degradation. Nelson et al. (1987) reported that toluene dioxygenase was involved in the TCE-degradative ability of *Pseudomonas putida*. The bioremediation of groundwater contaminated with volatile organic compounds has been reported. However, in situ bioremediation technologies are not well developed (Alan et al. 1989, Boyer et al. 1988, Nelson et al. 1990, Sims et al. 1990, Wilson et al. 1986).

For this study, we constructed soil microcosms that simulated the groundwater environment. We isolated strain M from soil in our laboratory and determined the effect of *Methylocystis* sp. M on the decontamination of TCE-contaminated soil.

MATERIALS AND METHODS

Isolation and Identification of TCE-Degrading Bacteria

In 1988, we collected many soil and activated sludge samples from various parts of Japan and screened them for their ability to degrade TCE. Soil or activated sludge was added to a 155-mL serum bottle containing 30 mL of mineral salt medium, as shown in Table 1. To this, 30 µg TCE and 20 mL methane were added as carbon sources. After shaking at 30°C, the enrichment culture was spread onto

TABLE 1. Composition of medium.

Ingredients	Concentration
KH_2PO_4	0.45 g/L
K_2HPO_4	1.17 g/L
NH_4Cl	214 g/L
$Ca(NO_3)_2 \cdot 2H_2O$	4.8 mg/L
$MgSO_4 \cdot 7H_2O$	121 mg/L
$FeSO_4 \cdot 7H_2O$	28 mg/L
Trace metals	
Distilled water	pH 7.2
CH_4	20 mL/bottle

mineral salt agarose plates. The agarose plates were incubated in a desiccator in an atmosphere of methane:air (1:1, v/v) at 30°C. Isolated colonies were picked up and respread. This procedure was repeated at least five times (Uchiyama et al. 1989a).

We isolated a pure TCE-degrading bacterium, strain M, from a soil sample. This bacterium was rod-shaped, nonmotile, Gram-negative, catarase (–), and oxidase (–), and it grew on methane and methanol. The predominant cellular fatty acid was $C_{18:1}$. Its quinone type was Q-8, and it contained 64.5% guanine and cytosine. From these results, strain M was determined to be a new species and was identified as *Methylocystis* sp. M. Strain M degraded 35 mg/L of TCE, as well as 1,1-dichloroethylene; both *cis*- and *trans*-1,2-dichloroethylene; 1,2-dibromoethylene; 1,1,2-trichloroethane; 1,2-dichloroethane; and chloroform. However, it could not degrade PCE, 1,1,1-trichloroethane, carbon tetrachloride, or aromatic compounds. A hypothetical pathway of TCE degradation by strain M is shown in Figure 1 (Nakajima et al. 1992a; Uchiyama et al. 1989b, 1992a). The immobilized cells with Ca-alginate also could degrade TCE in aqueous and gaseous phase (Uchiyama et al. 1992b).

Strain M was cultivated in a 5-L jar fermentor containing 3 L of mineral salt medium at 30°C. Methane and air were sparged into the fermentor at a rate of 100 mL/min. Cells were harvested with a centrifuge and washed twice with 10 mM phosphate buffer (pH 7.2) when the optical density (OD) at 580 nm reached 1.2. Cells were kept at –80°C while awaiting use (Nakajima et al. 1992b).

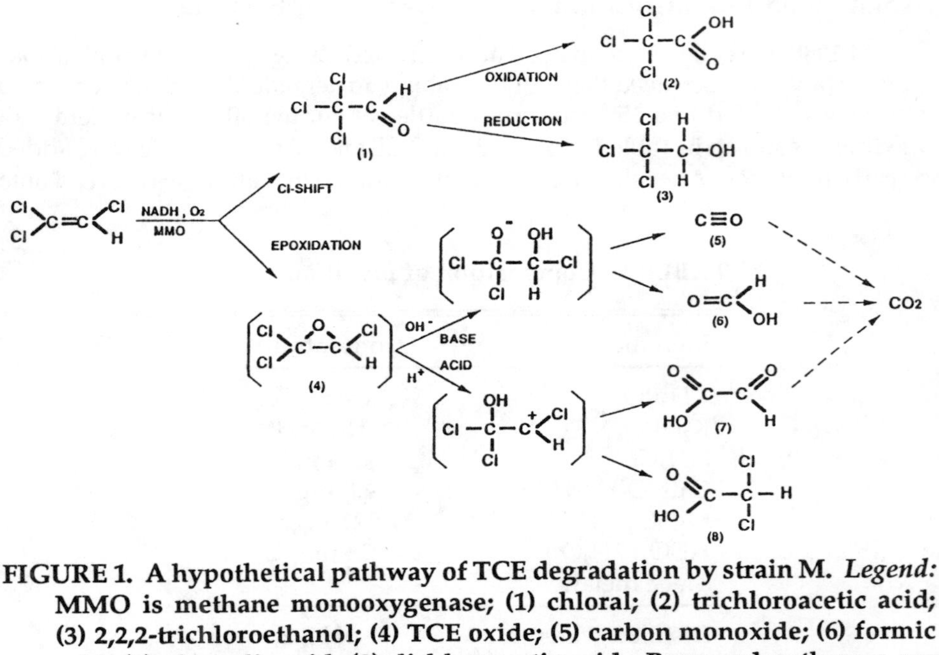

FIGURE 1. A hypothetical pathway of TCE degradation by strain M. ***Legend:*** **MMO is methane monooxygenase; (1) chloral; (2) trichloroacetic acid; (3) 2,2,2-trichloroethanol; (4) TCE oxide; (5) carbon monoxide; (6) formic acid; (7) glyoxylic acid; (8) dichloroacetic acid. Proposed pathways are indicated by dashed lines.**

Soil Microcosms

Biodegradation rates were determined for two types of soil microcosms. In the first microcosm, 30 g of dry soil was put into a 70-mL serum bottle and TCE-contaminated water was filled to 50 mL with a water/soil mixture (1:1, v:v). In the second, into a 450-mL cylindrical glass column (40 mm I.D. by 300 mm high) was put 350 g of dry soil, and TCE-contaminated water was added to fill the column. The TCE-contaminated water was prepared by diluting TCE-saturated water to 1,100 mg/L. Andosol, collected from a vegetable field, was used as the soil sample. It contained 4.3% carbon and 0.35% nitrogen. After the equilibrium of TCE adsorption to soil was established, strain M was added to the soil microcosms. The soil microcosms were incubated in the dark at 20°C. Sterile controls were prepared by autoclaving the soil and water mixture for 90 min. Duplicate microcosms were constructed for each condition. To study the effect of strain M concentrations on TCE degradation, a range of 0.01 to 0.1 at OD 660 nm of strain M was adopted; and for the TCE concentration effect, a range of 5 to 50 µg of TCE was added to the serum bottles.

Determination of TCE Degradation

TCE concentration was determined periodically by gas chromatographic (GC) analysis of a 100-µL headspace gas sample from the serum bottles or from 1 mL of surface water from the glass column, using a ^{63}Ni electron capture detector (ECD) or a flame ionization detector (FID) with a glass column (3 mm by 3 m) and with 15% Silicon DC 550 on 60/80 mesh Uniport B. The injection, oven, and detector temperatures were 300, 120, and 300°C, respectively.

RESULTS AND DISCUSSION

Effect of the Strain M Concentration on TCE Degradation

Figure 2 shows the effect of the strain M concentrations on TCE degradation in the serum bottle microcosms. Cell densities changed to between OD 0.01 and OD 0.1 at 660 nm. The OD 0.01 corresponded to 10^6 cells/mL. TCE was added at 5 µg to each bottle. The andosol had no ability to degrade TCE in 14 days. Figure 2 shows that 25% of TCE was degraded in 1 day and 50% was degraded in 3 days at OD 0.01. In the case of OD 0.02, 40% and 70% of TCE were degraded in 1 and 3 days, respectively. More than 80% of TCE was degraded in 1 day at OD 0.1, and TCE was completely degraded in 3 days. This result shows that strain M is able to degrade TCE in the soil environment, and the high density of cells indicates the high TCE degradation rate.

Effect of TCE Concentration

Figure 3 shows TCE degradation at high and low TCE concentrations in the serum bottle microcosms. Low and high concentrations corresponded to 5 and

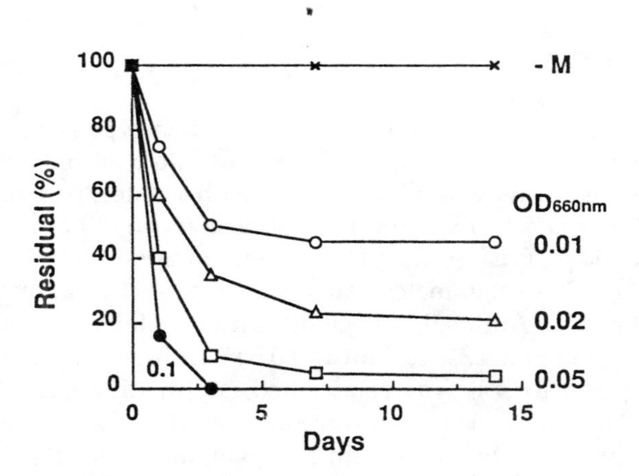

FIGURE 2. Effect of strain M concentration on TCE degradation in serum bottle microcosms.

50 µg of TCE in each bottle, respectively. About 20 and 5% of TCE were degraded at the low and high TCE concentrations in 18 days without strain M. However, TCE degradation was significantly accelerated by the addition of strain M of OD 0.05. TCE was almost completely decomposed in 1 day at low concentration, whereas 15% and 25% of TCE were degraded in 1 and 7 days, respectively, with higher concentrations. The TCE degradation rate decreased with the higher concentrations.

Dichloroethylene Degradation by Strain M

Figure 4 shows the degradation of 1,1-, *cis*-, and *trans*-DCE in serum bottle microcosms. The DCE concentration was 5 µg in each bottle; 1,1-DCE was not degraded, but *cis*- and *trans*-DCE were somewhat degraded in 3 days without strain M. About 20% of 1,1-DCE was degraded in 1 day by the addition of strain M,

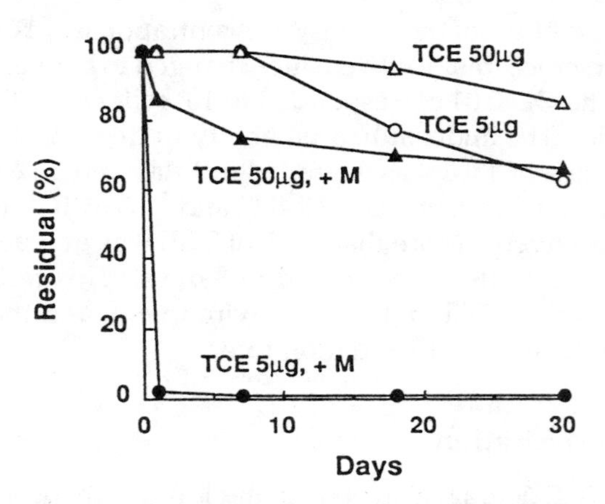

FIGURE 3. Effect of TCE concentration in serum bottle microcosms.

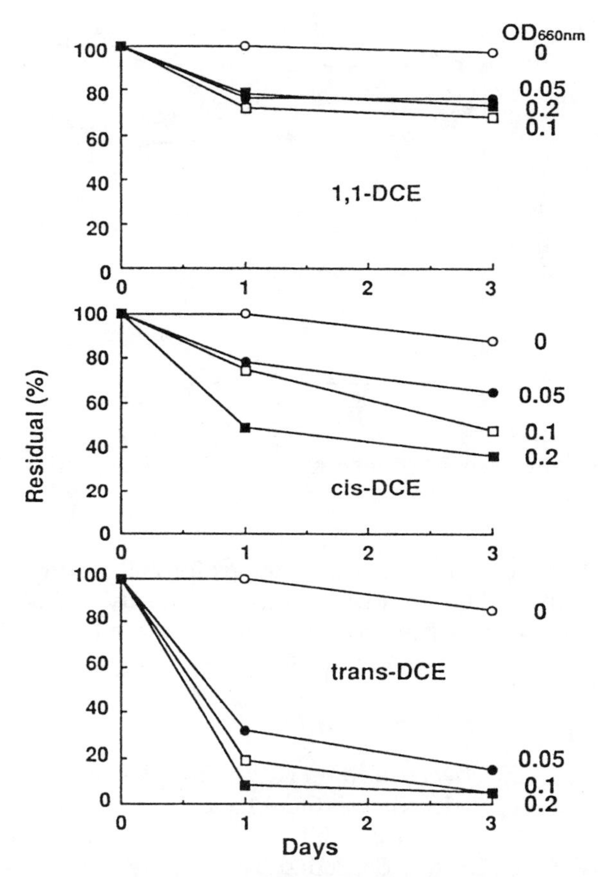

FIGURE 4. Effect of strain M concentration on DCE degradation in serum bottle microcosms.

although cell density did not affect the degradation rate. About 20% of *cis*-DCE was degraded in 1 day at OD 0.05, and 50% was degraded at OD 0.2. More than 70% of *trans*-DCE was degraded in 1 day at OD 0.05. All three DCE degradations were accelerated by adding strain M. For *cis*- and *trans*-DCE, the high cell density showed the high degradation ability. DCEs were degraded at the high rate in 1 day, and at a lower rate after 1 day. It was reported that DCEs were metabolites of TCE and were detected in the soil environment. It is very important to know the biodegradabilities of DCEs. Among DCEs, 1,1-DCE is the most toxic substance. For this reason 1,1-DCE is more difficult to degrade than *cis*- and *trans*-DCE.

TCE Degradation in Column Microcosms

Figure 5 shows the effect of cell density on TCE degradation in glass column microcosms. About 20% of TCE was degraded by the addition of strain M in 1 day,

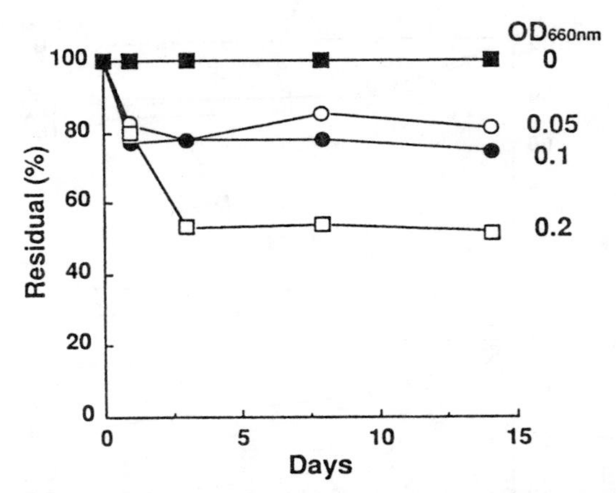

FIGURE 5. Effect of strain M concentration on TCE degradation in glass column microcosms.

and 50% of TCE was degraded in 3 days at the high cell density. However, TCE degradation stopped after 1 day at the low cell density and after 3 days at the high cell density. Degradation may have stopped due to the lack of oxygen.

Effect of H_2O_2 on TCE Degradation

Figure 6 shows the effect of adding H_2O_2 to the glass column microcosms on TCE degradation. The cell density was OD 0.1 and TCE concentration was 0.1 ppm. H_2O_2 was added at a range of 0 to 50 ppm. TCE was not degraded without strain M, but 30% of TCE was degraded by adding strain M. About 40% and 60% of TCE were degraded in 2 days by the addition of 5 ppm and 10 ppm of H_2O_2, respectively. The addition of H_2O_2 accelerated TCE degradation (Berwanger

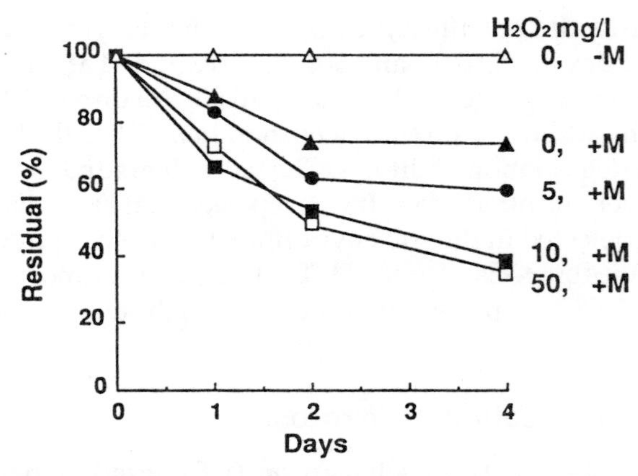

FIGURE 6. Effect of H_2O_2 concentration with strain M on TCE degradation in glass column microcosms.

& Barker 1989). No degradation was observed when H_2O_2 alone was added. H_2O_2 seemed effective in increasing TCE degradation with strain M. It is well known that H_2O_2 is an oxidant and, in high concentrations, is toxic to microorganisms. The sterile control did not degrade TCE. TCE degradation clearly was caused by biological activities. Studies on the behavior of strain M in the soil environment and on the method of applying H_2O_2 are required for effective use of strain M to clean up contaminated soil.

REFERENCES

Alan, T. M., V. Alex, and S. Fogel. 1989. "Biodegradation of *trans*-1,2-Dichloroethylene by Methane-Utilizing Bacteria in an Aquifer Simulator." *Environ. Sci. Technol. 23*:403-406.

Berwanger, D. J., and J. F. Barker. 1989. "Aerobic Biodegradation of Aromatic and Chlorinated Hydrocarbons Commonly Detected in Landfill Leachates." *Water Poll. Res. J. Canada 23*:460-475.

Bouwer, E. J., and P. L. McCarty. 1983. "Transformation of 1-and 2-Carbon Halogenated Aliphatic Organic Compounds under Methanogenic Conditions." *Appl. Environ. Microbiol. 45*:1286-1294.

Boyer, J. D., R. C. Ahlert, and D. S. Kosson. 1988. "Pilot Plant Demonstration of In Situ Biodegradation of 1,1,1-Trichloroethane." *J. Water Pollut. Control Fed. 60*:1843-1849.

Fathepure, B. Z., J. P. Nengu, and S. A. Void. 1987. "Anaerobic Bacteria that Dechlorinate Perchloroethene." *Appl. Environ. Microbiol. 53*:2671-2674.

Fogel, M. M., A. R. Taddeo, and S. Fogel. 1986. "Biodegradation of Chlorinated Ethenes by a Methane-Utilizing Mixed Culture." *Appl. Environ. Microbiol. 51*:720-724.

Kawasaki, T. 1985. "Present Condition and Countermeasure with Groundwater Pollution." *Jpn. J. Water Pollut. Res. 8*:264-268.

Little, C. D., A. V. Palumbo, S. E. Herbes, M. E. Lidstrom, R. L. Tyndall, and P. J. Gilmer. 1988. "Trichloroethylene Biodegradation by a Methane-Oxidizing Bacterium." *Appl. Environ. Microbiol. 54*:951-956.

Nakajima, T., H. Uchiyama, O. Yagi, and T. Nakahara. 1992a. "Novel Metabolite of Trichloroethylene in a Methanotrophic Bacterium, *Methylocystis* sp. M and Hypothetical Degradation Pathway." *Biosci. Biotech. Biochem. 56*:486-489.

Nakajima, T., H. Uchiyama, O. Yagi, and T. Nakahara. 1992b. "Purification and Properties of a Soluble Methane Monooxygenase from *Methylocystis* sp. M." *Biosci. Biotech. Biochem. 56*:736-740.

Nelson, M. J., S. O. Montgomery, W. R. Mahaffey, and P. H. Pritchard. 1987. "Biodegradation of Trichloroethylene and Involvement of an Aromatic Biodegradative Pathway." *Appl. Environ. Microbiol. 53*:949-954.

Nelson, M. J., J. V. Kinsella, and T. Montoya. 1990. "In Situ Biodegradation of TCE Contaminated Groundwater." *Environ. Prog. 9*:190-196.

Sims, J. L., R. C. Sims, and J. E. Matthews. 1990. "Approach to Bioremediation of Contaminated Soil." *Hazardous Waste & Hazardous Materials 7*:117-149.

Uchiyama, H., T. Nakajima, O. Yagi, and T. Tabuchi. 1989a. "Aerobic Degradation of Trichloroethylene in High Concentration by a Methane-Utilizing Mixed Culture." *Agric. Biol. Chem. 53*:1019-1024.

Uchiyama, H., T. Nakajima, O. Yagi, and T. Tabuchi. 1989b. "Aerobic Degradation of Trichloroethylene by a New Methane-Utilizing Bacterium Strain M, Type 2." *Agric. Biol. Chem. 53*:2903-2907.

Uchiyama, H., T. Nakajima, O. Yagi, and T. Nakahara. 1992a. "Role of Heterotrophic Bacteria in Complete Mineralization of Trichloroethylene by *Methylocystis* sp. M." *Appl. Environ. Microbiol. 58*:3067-3071.

Uchiyama, H., K. Oguri, O. Yagi, and E. Kokufuta. 1992b. "Trichloroethylene Degradation by Immobilized Resting Cells of *Methylocystis* sp. M in a Gas-Solid Bioreactor." *Biotechnology Letters 14*:619-622.

Vogel, T. M., and P. L. McCarty. 1985. "Biotransformation of Tetrachloroethylene, Dichloroethylene, Vinyl Chloride and Carbon Dioxide under Methanogenic Conditions." *Appl. Environ. Microbiol. 49*:1080-1083.

Wilson, J. T., and B. H. Wilson. 1985. "Biotransformation of Trichloroethylene in Soil." *Appl. Environ. Microbiol. 49*:242-243.

Wilson, J. T., L. E. Leach, H. Henson, and J. N. Jones. 1986. "In Situ Biorestoration as a Ground Water Remediation Technique." *GWMR Fall*:56-64.

EVALUATING TRICHLOROETHENE BIODEGRADATION BY MEASURING THE IN SITU STATUS AND ACTIVITIES OF MICROBIAL POPULATIONS

E. E. Cox, D. W. Major, D. W. Acton,
T. J. Phelps, and D. C. White

ABSTRACT

Trichloroethene (TCE) released from an unlined waste pond has contaminated the sediments and groundwater beneath and downgradient of a manufacturing facility. Characterization of the site microbiology revealed that dense and diverse microbial populations exist in the sediments and groundwater beneath the facility. The microbial biomass, diversity, growth, catabolism, and anabolism were greater in the sand units than in the silt or clay units. Microbial stress indicators were most prevalent in TCE-contaminated sediments within the unsaturated zone. Aerobic TCE biotransformation was confined mainly to sand units in the vicinity of the former waste pond. Radiolabeled carbon dioxide ($^{14}CO_2$) generation from [1,2-^{14}C]-TCE and [1,2-^{14}C]-vinyl chloride (VC) was also confined mainly to sand units in the vicinity of the former waste pond. A number of anaerobic dechlorination products (*cis*-1,2-dichloroethene, *trans*-1,2-dichloroethene, and VC) were detected during the biodegradation of [1,2-^{14}C]-TCE, but $^{14}CO_2$ was the dominant product in all cases. Overall, the water availability, the silt and clay content, and the distribution of TCE in the subsurface appear to be controlling microbial activity, growth, and TCE biodegradation.

INTRODUCTION

Trichloroethene (TCE) historically was used as a degreasing agent at a manufacturing facility in Merced, California, and was released to the subsurface through seepage from an unlined waste pond. Previous subsurface investigations found that TCE was present in sediments beneath the former waste pond, and in groundwater samples obtained from two of the three aquifers beneath the site, to a depth of approximately 60 m. TCE also was detected in groundwater samples obtained downgradient of the site (Bechtel Environmental Inc. 1988). Interestingly however, the breakdown products, i.e., *trans*- and *cis*-1,2-dichloroethene (*t* and *c*DCE, respectively), and vinyl chloride (VC), which typically are associated with TCE-contaminated sediments and groundwater, were not detected. As a result, we conducted a subsurface investigation to determine if the requisite microorganisms capable of biodegrading TCE were absent or if a microbial population existed that could completely mineralize TCE without the production of VC or other intermediate products.

This paper presents the characterization of the microbial community and an assessment of microbial growth and activity (catabolic and anabolic) in sediment and groundwater samples from two exploratory boreholes (designated MB-1 and MB-2) and three existing extraction wells (designated MW-39, MW-40, and MW-44). The locations of the boreholes and wells, and TCE concentrations detected in sediments in the vicinity of the former waste pond are shown in the geologic cross section presented in Figure 1. TCE and VC biotransformation by the indigenous microorganisms in the sediment and groundwater samples is presented.

SAMPLING AND ANALYSIS

Summaries of sediment and groundwater collection, microbial community analysis, and TCE biotransformation are presented below. Details of the site geology and geochemistry are presented elsewhere (Major et al. in submission).

Sediment Collection

Two exploratory borings were cored using hollow-stem augers. No drilling fluids were used in this investigation. A decontaminated California-modified split-spoon sampler fitted with three sterile brass liners was driven into the ground in advance of the lead augers, and retrieved samples were aseptically processed in an anaerobic glove bag in the field laboratory. Samples were frozen on dry ice and were shipped by courier to the University of Tennessee for initiation of phospholipid fatty acid (PLFA) analysis.

Groundwater Collection

Temporary monitoring wells were installed in each of the exploratory borings. A sand pack was placed in the boring annulus around the well screen, and the well screen was isolated from the overlying formations. The monitoring wells were developed using a decontaminated submersible pump. Groundwater for microbial activity and mass loss experiments was collected using dedicated Teflon™ bailers attached to stainless-steel wires. Samples were transferred directly from the bailer into sample bottles through a Teflon™ stopcock. Sample bottles were filled without headspace and were shipped on ice by courier to the University of Tennessee. Groundwater samples collected for lipid analysis were filtered through 0.2 µm polycarbonate filters to trap the microorganisms. The filters were frozen on dry ice, shipped by courier to the University of Tennessee, and stored until the iipids were extracted and analyzed.

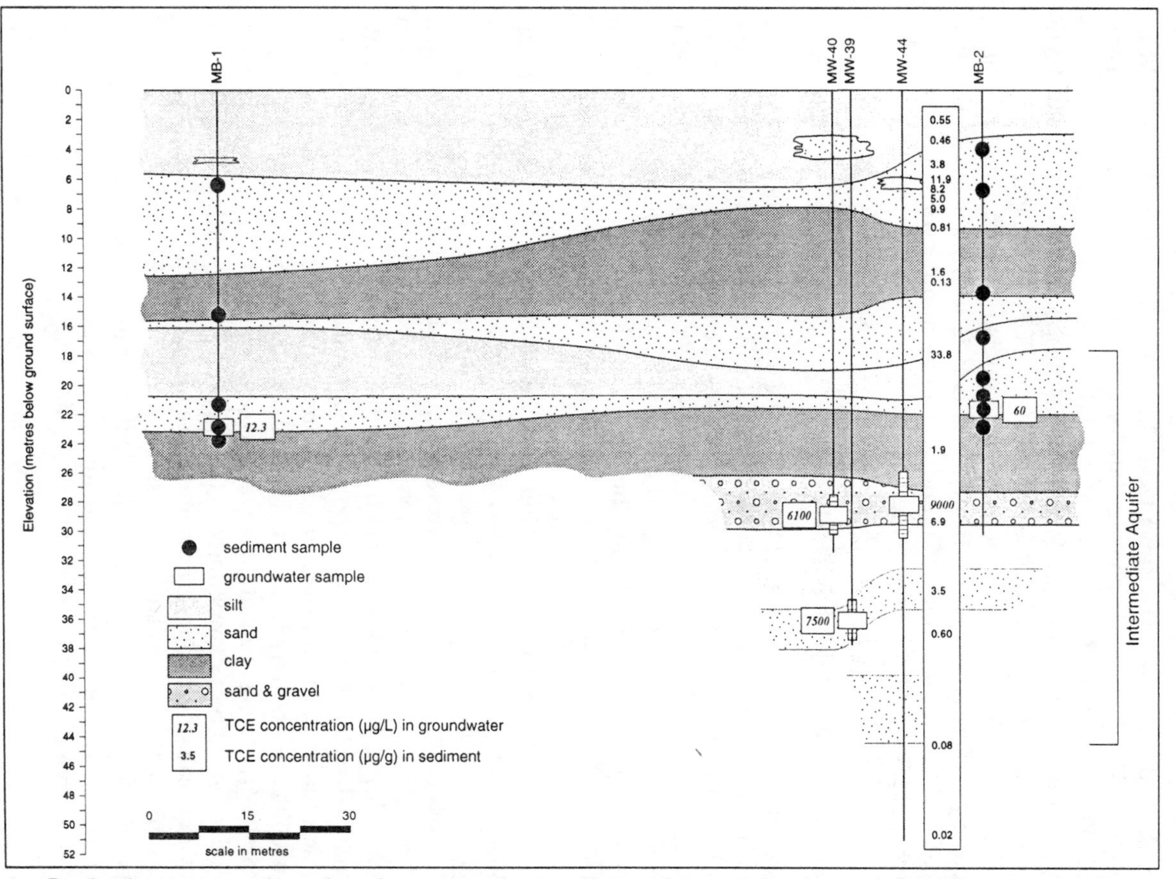

FIGURE 1. Geologic cross section showing extraction wells, exploratory borings, and TCE concentrations in sediments and groundwater in the vicinity of the former waste pond.

Phosphlipid Analyses

PLFA analysis was used to measure the total viable biomass, the community composition, and the nutritional and physiological status of the microbial populations. Total phospholipids were extracted from 110 g (dry weight) of frozen and lyophilized (dehydrated) sediment samples or filters by a modification (White et al. 1979, 1983) of the single-phase chloroform-methanol method of Bligh and Dyer (1959). Fatty acids were designated by the total number of carbon atoms and by the number of double bonds followed by the position of the double bond closest to the ω (aliphatic) end. Geometry of the PLFA is designated by '*c*' for *cis* or '*t*' for *trans*; 'i', 'a', and 'br' refer to iso-, anteiso-, and methyl-branching, respectively. Methyl-branching from the Δ end is indicated by its position followed by the designation 'Me' and the total number of carbon atoms. Cyclopropyl fatty acids are designated as 'cy'.

Assessment of Microbial Activity

Microbial growth was assessed by measuring the rate of [*methyl*-^{3}H]thymidine (76.0 mCi/mmole) incorporation into macromolecules of microorganisms in the sampled sediments. Catabolic potential was assessed by measuring the rate of [1,2-^{14}C]acetate (56 mCi/mmole) or [U-^{14}C]glucose (2.8 mCi/mmole) mineralization by the microorganisms in the sampled sediments and groundwater. Anabolic potential was assessed by measuring the rate of [1,2-^{14}C]acetate (56 mCi/mmole) or [^{3}H]acetate (3.3 mCi/mmole) incorporation by the microorganisms in the sampled sediments and groundwater.

All activity experiments were initiated within 72 hours of sediment and groundwater collection. Experiments were conducted under both aerobic and anaerobic conditions in either sterile polypropylene centrifuge tubes or crimp-top tubes containing sediment or groundwater and sterile distilled water. Radiolabeled substrate transfers were made using gastight syringes (Hamilton Co., Reno, Nevada). The amount of radiotracer added during activity experiments was small so that the electron donor pool size was not appreciably increased (i.e., from micromolar to millimolar concentrations). Time frames were minimized to ensure linearity of the results (Phelps et al. 1989). Additionally, radiotracers were carrier-free to ensure that there were no competing substrates (e.g., ethanol as a carrier would be a competing electron donor with the radiotracer acetate). Tubes were sealed with butyl septa and were incubated in the dark at ambient temperatures that were similar to the in situ temperature of 21 to 24°C. At selected time points, duplicate tubes were inhibited and the appropriate activity was analyzed. Results are expressed as disintegrations per minute/day/cell based on the corresponding PLFA biomass measurements.

Microcosm Biotransformation Experiments

Microcosm enrichment studies were conducted to examine the loss of TCE mass over time under various nutrient conditions. The microcosms consisted of 58-mL serum vials containing either 10 g of sediment or 10 mL of groundwater, plus 5 mL of distilled water or medium (as indicated below). Serum vials were sealed with Teflon™ septa. Atmospheric gas made up the headspace in the microcosm, unless otherwise indicated in the treatments listed below. All experiments were performed in duplicate. Four experimental treatments were used:

Treatment 1: 5 mL distilled water (equivalent treatment with groundwater was not conducted).
Treatment 2: 5 mL mineral salts medium plus 5 mM phosphate.
Treatment 3: 5 mL distilled water plus 5% methane headspace (vol:vol).
Treatment 4: 5 mL of mineral salts medium plus 3% each of methane and propane headspace (vol:vol).

Treatment 1 examined the effect of water on releasing bound nutrients in the sediment. Treatment 2 attempted to determine if phosphate was limiting activity. Treatment 3 attempted to induce methanotrophic bacteria to degrade TCE, and Treatment 4 attempted to induce other alkane-alkene oxidizers to degrade TCE. The TCE was added to each vial at approximately 20 mg/L. Incubations were in the dark, at ambient temperature, and in an inverted position (to minimize TCE loss by volatilization). After 0, 2, 4, and 12 weeks of incubation, duplicate vials were sacrificed and the TCE mass loss was determined; 1 hour prior to analysis, each vial was acidified with 0.5 mL of 6 M HCl to stop microbial activity.

Microcosm controls consisted of (1) autoclaved site groundwater and sediment, (2) autoclaved site groundwater and sediment plus respective nutrient amendments, and (3) autoclaved site groundwater and sediment plus respective nutrient amendments, and 0.5 mL of 2 M NaOH (inhibitor).

Gas chromatography (GC) was used to measure the concentrations of TCE, *c*DCE, *t*DCE, VC, propane, methane, and CO_2 in the headspace of the microcosms. The limits of detection for sampled compounds were as follows: TCE, 0.1 µg/L; *c*DCE, 10 µg/L; *t*DCE, 10 µg/L; VC, 10 µg/L; and propane, methane, and CO_2, 0.05% (vol:vol). The detection limits of the volatile organic compounds (VOCs) are dissolved concentrations as determined by the Ideal Gas Law, Henry's law, and solubility coefficients.

Radiolabeled Mineralization Experiments

Aerobic and anaerobic mineralization experiments were conducted in 25-mL crimp top tubes containing either 2.0 mL of groundwater, or 2.0 g of sediment and 1.0 mL of sterile water, and carrier-free isotope. Either [1,2-^{14}C]VC (0.53 mCi/mmole), or [1,2-^{14}C]TCE (10 mCi/mmole) was added at 0.66 and 0.59 µCi per tube, respectively, using gastight syringes (Hamilton Co., Reno, Nevada). Time course experiments were performed in duplicate with multiple time points ranging from t_0 to 1 month with a minimum of 4 time points per isotope examined. All tubes had Teflon™ septa and were incubated in the dark at ambient temperatures similar to the *in situ* temperature of 21 to 24°C. At selected time points, duplicate tubes were inhibited with 0.5 mL of 2.0 M sodium hydroxide. Each tube was acidified with 0.5 mL of 6M HCl 1 hour prior to analysis to inhibit microbial activity.

Radioactive $^{14}CO_2$, $^{14}CH_4$, and daughter products released during biodegradation of the radiolabeled parent compounds were analyzed by GC-gas proportional counting (Fliermans et al. 1988; Nelson & Zeikus 1974). Results were calculated based on the initial slope of product evolution and are reported as disintegrations per minute/day/cell based on the corresponding PLFA biomass measurements.

RESULTS AND DISCUSSION

Microbial Biomass

Table 1 presents the PLFA biomass results for the sampled sediments and groundwater. The total viable biomass was calculated assuming a conversion factor of $4x10^4$ colony-forming units (CFU) per pmole of PLFA (White et al. 1979). Microbial biomass was highest in the unsaturated sand units of both MB-1 and MB-2, and decreased with depth. In the saturated sediments, microbial biomass was generally an order of magnitude lower in MB-2 than in MB-1. The higher concentration of TCE in MB-2 may be inhibiting microbial growth. The smallest population densities were consistently found in the silt and clay units within the unsaturated and saturated sediments. Previous studies by White et al. (1983) confirm that decreased biomass generally is detected in sediments with high clay content. Microbial biomass estimates were 4 to 5 orders of magnitude higher for groundwater samples from MB-1 and MB-2 compared to groundwater from extraction wells MW-39, MW-40, and MW-44. The higher concentration of TCE in the extraction well groundwater may be inhibiting microbial growth.

Microbial Community Composition

Table 1 summarizes the PLFA profiles for the sampled sediments and groundwater. Significant microbial population densities and complex community structures were detected in the unsaturated sand units and groundwater of MB-1 and MB-2, as evidenced by the number of different PLFA detected in these samples. Community diversity and complexity generally decreased with depth, and was lower in silt and clay units than in sand units.

Uncontaminated subsurface sediments generally contain high proportions of Gram-positive microorganisms, however, shifts in the proportions of Gram-positive to Gram-negative microbiota have been observed in contaminated sediments (Smith et al. 1985). Samples collected from MB-1 and MB-2 consistently had higher percentages of monounsaturated PLFA (usually Gram-negative) than terminally branched, saturated PLFA (usually Gram-positive) (Table 1). The predominance of monounsaturated PLFA as compared to terminally branched, saturated PLFA in saturated sediment and groundwater samples suggests that there has been a shift of the microbial community in response to the TCE contamination.

TABLE 1. Summary of phospholipid fatty acid (PLFA) and polyhydroxyalkonate (PHA) profiles for sampled sediments and groundwater.

	Unsaturated Sediments						Saturated Sediments							Groundwater				
Sample Location	MB-1		MB-2				MB-1			MB-2				MB-1	MB-2	MW-39	MW-40	MW-44
Depth (m bgs)	6.7	15.2	4.0	6.7	13.7	16.8	21.3	22.9	23.8	19.5	20.7	21.6	22.9	21.6	21.6	45.1	40.8	46.0
Sample Description	sand	clay	sand	sand	clay	silt	sand	sand	silt/clay	sand	sand	sand	silt/clay					
Number PLFA dectected	28	11	30	30	13	5	12	12	6	8	4	1	4	20	18	4	4	5
PLFA (mole %)																		
Saturates	20.2	81.7	31.8	40.6	33.4	63.2	64.8	32.5	61.7	38.0	67.8	100	55.8	23.7	19.4	40.5	40.4	26.7
Terminally branched saturates	24.3	5.42	27.4	16.9	2.67	-	7.68	1.10	-	6.71	-	-	-	2.55	2.16	-	-	28.2
Mid-chain branched saturates	6.60	-	9.35	5.70	-	-	-	4.07	-	9.26	-	-	-	-	-	-	-	-
monounsaturates	41.1	7.77	29.9	34.0	39.3	16.6	25.1	25.2	31.1	37.0	17.2	-	25.3	69.1	72.5	59.5	59.6	45.1
polyunsaturates	7.76	5.15	1.06	2.86	23.8	20.2	2.48	37.1	7.24	8.98	15.0	-	18.9	4.59	5.97	-	-	-
Total	100	100	100	100	99	100	100	100	100	100	100	100	100	100	100	100	100	100
pmole PLFA/g or mL	62.2	5.14	374	124	13.3	1.35	4.68	15.9	1.16	1.96	0.64	0.18	1.63	423	271	0.10	0.02	0.03
Biomass (CFU/g or mL)	2.5E+6	2.1E+5	1.5E+7	4.9E+6	5.3E+5	5.4E+4	1.9E+5	6.4E+5	4.6E+4	7.8E+4	2.6E+4	7.2E+3	6.5E+4	1.7E+7	1.1E+7	4.0E+3	8.0E+2	1.2E+3
PHA/PLFA ratio	2.09	1.95	45.0	19.9	15.8	-	38.5	-	-	-	-	-	-	0.14	-	-	-	-
Trans/cis ratio of 16:1w7	0.04	-	0.66	0.43	-	-	-	-	-	-	-	-	-	0.17	0.07	-	-	-
Trans/cis ratio of 18:1w7	-	-	0.23	0.17	-	-	-	-	-	-	-	-	-	0.13	0.10	-	-	-
Cyclo/mono ratio of cy17	0.28	-	0.57	0.76	-	-	0.99	-	-	**	-	-	-	-	-	-	-	-
Cyclo/mono ratio of cy19	0.17	-	0.49	0.30	-	-	-	-	-	-	-	-	-	-	-	-	-	-

Notes:
bgs = below ground surface
- = Not Detected
Saturates = Sum of 12:0, 14:0, 15:0, 16:0, 17:0, 18:0, 20:0, and 22:0
Terminally branched saturates = Sum of i14:0, i15:0, a15:0, i16:0, i17:1w7c, a17:1w8c, i17:0, a17:0/1w8
Mid-chain branched saturates = Sum of Sat17:0, Sat18:0, 10Me16:0, 10Me18:0, Br16:1, Br18:1, Br19:1
Monounsaturates = Sum of 15:1w6, 16:1w5c, 16:1w7c, 16:1w7t, i16:1w9c, Cy17/17:1, 18:1w7c, 18:1w7t, 18:1w9c, Cy19, 19:1w12c
Polyunsaturates = Sum of 18:2w6, 18:3w3, 20:3w6
g or mL = gram dry weight of soil or millilitre of groundwater
Biomass calculated using a conversion factor of 4e+4 CFUs per pmole of PLFA
** = ratio approaches infinity

The PLFAs detected in groundwater samples from wells MW-39, MW-40, and MW-44 were considerably different than those obtained from the groundwater of MB-1 and MB-2 (Table 1). A maximum of 5 PLFAs were detected in MW-39, MW-40, and MW-44, whereas more than 18 PLFAs were detected in the groundwater samples from MB-1 and MB-2. The decreased community diversity in groundwater from MW-39, MW-40, and MW-44 may be attributable to the elevated TCE concentrations in the groundwater in the deeper sand units.

A large percentage (4.5 to 58.9 mole %) of the PLFAs detected in the sediment and groundwater samples was characteristic of eukaryotes (sum of the polyunsaturates, the saturates 20:0 and 22:0, and the monounsaturates i16:1w9c and 18:1w9c). Of particular note was the dominance of 18:2w6 in sediment and groundwater samples from MB-1 and MB-2. PLFA 18:2w6 is a product of aerobic desaturation activity (carried out by eukaryotic microorganisms such as protozoans) and is a known fungal fatty acid. The high percentage of this fatty acid and the presence of 18:1w9c (a precursor to polyunsaturate synthesis) suggest a strong presence of protozoans and fungi. Madsen et al. (1991) have shown that protozoan activity (e.g., grazing on bacteria) can be correlated with biodegradation activities of microorganisms and, as such, the high proportion of eukaryotic PLFAs should not be unexpected given the elevated microbial biomass of these borings.

Microbial Community Status

The polyhydroxyalkonate (PHA):PLFA ratio, the *trans:cis* ratio of 16:1w7 and 18:1w7, and the ratio of the cyclopropyl fatty acids (cy19 and cy17) to their precursor monounsaturates (18:1w7c and 16:1w7c) can be used to determine the nutritional and physiological status of microbial populations (Guckert et al. 1986). Nonideal growth conditions (ratio of cy17/16:1w7c and cy19/18:1w7c greater than 0.1) were evident in sand samples from the unsaturated zone in both MB-1 and MB-2, and in two sand samples from the saturated zone (MB-1, 21.3 m, and MB-2, 19.5 m). Physiological stress (starved) conditions (*trans/cis* ratio of 16:1w7 and 18:1w7 greater than 0.1) were evident in sand samples from the unsaturated zone and from the groundwater of both MB-1 and MB-2. PHA accumulation was evident in the unsaturated sediment samples from both wells, and in MB-1, 21.3 m, and MB-1, H_2O. The presence of PHA in these samples suggests that a carbon source (dissolved organic carbon or TCE) is available that the microorganisms cannot use for growth but can accumulate as storage lipid.

The highest stress levels were detected in microbial populations in MB-2, 4.0 m, and MB-2, 6.7 m. Limited water availability and residual TCE in the sediments and in the soil gas at these locations is likely affecting the microbial populations in these sediments.

Microbial Activity

Figure 2 presents the rates of thymidine incorporation and the rates of catabolic and anabolic activity in the sampled sediments and groundwater. Microbial growth, mineralization, and anabolism generally were higher in sand

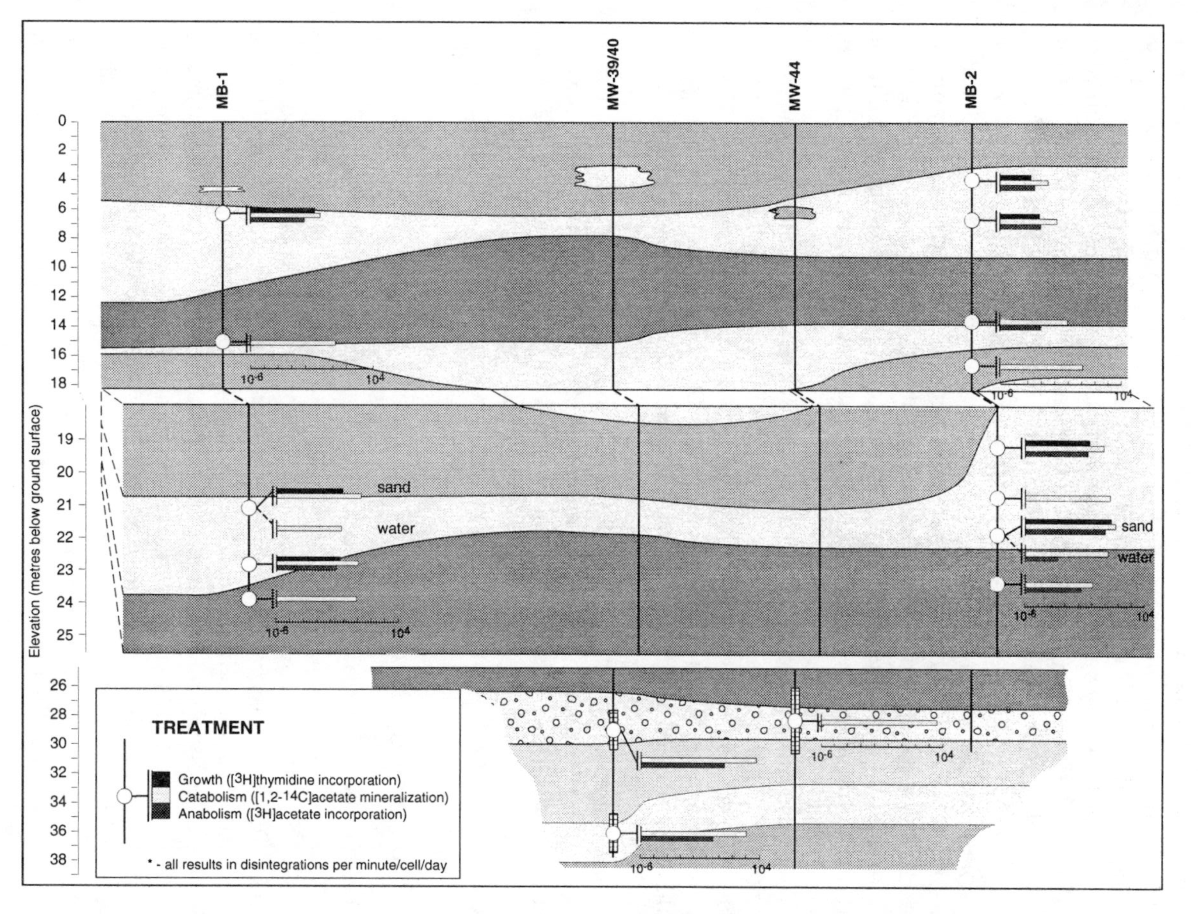

FIGURE 2. Microbial catabolism, anabolism, and growth in sampled sediments and groundwater.

samples than in silt and clay samples. These results are consistent with the lower population densities generally detected in the silt and clay units. The unsaturated sediment samples from MB-2 showed incorporation and mineralization rates several orders of magnitude lower than in comparable sediment samples from MB-1 and saturated sediment samples from both borings. Nutritional and physiological stress attributable to low water availability and exposure to elevated concentrations of TCE may be inhibiting the microbial populations in these samples.

Mineralization rates (glucose and acetate) and rates of acetate incorporation detected in the groundwater samples obtained from the extraction wells MW-39, MW-40, and MW-44 were 3 to 4 orders of magnitude higher than those detected in samples obtained from MB-1 and MB-2. However, population densities detected in groundwater samples from the extraction wells were 4 to 5 orders of magnitude lower than those observed in MB-1 and MB-2 (Table 1). The absence of detectable nutritional and physiological stress indicators (e.g., PHA accumulation) in the microbial communities in the extraction wells may explain the increased per cell activity observed.

The predominant production of CO_2 rather than CH_4 indicates that anaerobic metabolism at this site is not directed toward methanogenesis. The only samples in which methanogenesis was detected were MB-1, 21.3 m, and MB-1, 22.9 m. Had methanogenesis been significant, more than 85% of the acetate methyl groups would have formed methane, whereas less than 2% methane formation was observed in these two samples.

Enrichment with carbon-labeled glucose and acetate was detected during the activity experiments. Tenfold stimulations of aerobic catabolic activity were detected in a number of sediment samples and in groundwater samples from the extraction wells. In comparison, the [^{3}H]acetate experiments showed no increase in activity over time and thus provide a better snapshot of in situ activity.

TCE Biotransformation

Biotransformation of nonradiolabeled TCE was assessed under various aerobic nutrient conditions, due to the predominantly oxidizing geochemical conditions at the site. Figure 3 presents the percentage loss of TCE as compared to the sterile controls. TCE mass loss values less than 20% were considered to be experimental error. In samples that exhibited TCE mass loss, the percentage loss was similar between treatments. The similarity of the results from each treatment suggests that sufficient nutrients are present in the sediments but their accessibility may be limited under in situ conditions. Soil structure was probably disrupted during experimental setup, possibly leading to a release of previously inaccessible nutrients. This is particularly evident in the samples that received water. The addition of water to the disrupted sediments likely served to both relieve water stress in unsaturated sediment samples, and to release bound nutrients.

Sediment samples from MB-2 generally showed greater percentage decrease in TCE concentration than did equivalent samples from MB-1. Interestingly, the groundwater samples from MW-39, MW-40, and MW-44, all of which had low microbial population densities, showed TCE-biotransformation activity similar

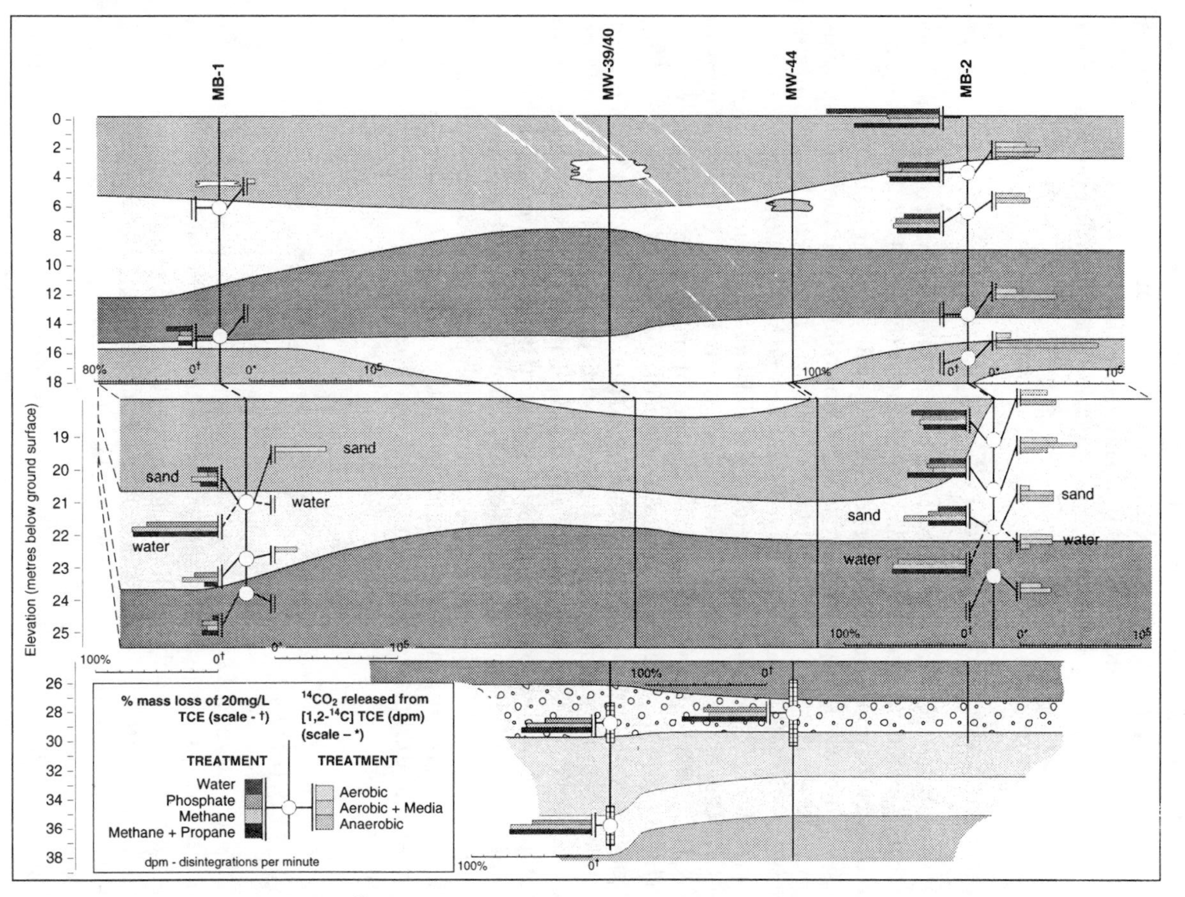

FIGURE 3. TCE biotransformation and $^{14}CO_2$ generation in sampled sediments and groundwater.

to the groundwater samples from MB-1 and MB-2. This activity indicates good biotransformation capacity by the existing microbial populations in the second and third sand units of the intermediate aquifer. No TCE biotransformation was observed in samples from the silt and/or clay units. Anaerobic dechlorination products *c*DCE and *t*DCE were detected at trace concentrations in several samples. Many of the groundwater samples that showed TCE mass loss became turbid with microbial growth, indicating good activity and growth potential in the presence of TCE.

Mineralization of Radiolabeled TCE and VC

Radiolabeled CO_2 was generated from at least one sediment or groundwater sample under all treatment conditions investigated (Figure 3). All sediment and groundwater samples from MB-2 showed mineralization of radiolabeled TCE to $^{14}CO_2$, whereas $^{14}CO_2$ production in MB-1 was confined to the sand samples. The production of $^{14}CO_2$ from radiolabeled VC was limited to the unsaturated sand samples in MB-2, and the saturated sand sample MB-2, 21.6 m. In addition to the production of $^{14}CO_2$, several anaerobic dechlorination products ([1,2-^{14}C]*c*DCE, [1,2-^{14}C]*t*DCE, and [1,2-^{14}C]VC) were detected during the biodegradation of [1,2-^{14}C]TCE. However, $^{14}CO_2$ was the dominant product in all cases.

The maximum extent of mineralization of TCE to CO_2 was 5%. This percentage would appear to be low in comparison to the amount of TCE mass loss observed in the nonradiolabeled experiments (average of 35%). However, typical catabolic activities of a carbon compound result in approximately one-third of the total carbon flow to cellular carbon, one-third to other water-soluble carbon compounds, and the remaining third to CO_2. Thus, a reasonable estimate of the expected amount of CO_2 over a 30-day incubation period would be 10% (based on the 35% average), which is in the range of the observed CO_2 levels. Given the amount of CO_2 produced, the time required for turnover of TCE would be months to years.

CONCLUSIONS

The collective PLFA data and the measured microbial activities (growth, catabolism, and anabolism) indicate that significant microbial biomass exists in the subsurface beneath the facility that is capable of TCE mineralization to CO_2, both with and without the production of dechlorination intermediates (*c*DCE, *t*DCE, and VC). The biomass and microbial activities are confined mainly to the sand units. The biomass is generally poised for aerobic activity as a result of the highly oxidizing conditions at the site. Anaerobic mineralization of TCE, although not directed toward methanogenesis, also can be enhanced or induced. The water availability, the silt and clay content, the abundance of eukaryotic organisms, and the distribution of TCE in the subsurface appear to be the controlling factors of microbial activity, growth, and TCE biotransformation. The ability to biotransform TCE both aerobically and anaerobically in the subsurface sediments at this site provides a promising and cost-effective remedial alternative to the current pump-and-treat remedial strategy.

ACKNOWLEDGMENTS

We thank the General Electric Company for funding this research, Dr. Deborah Hankins for her support, and Scott Neville, Robert Mackowski, and David Ringelberg for their assistance.

REFERENCES

Bechtel Environmental, Inc. 1988. *Site Investigation and Remedial Actions, Merced, California.*

Bligh, E. G., and W. J. Dyer. 1959. "A rapid method of lipid extraction and purification." *Can. J. Biochem. Physiol. 35*:911-917.

Fliermans, C. B., T. J. Phelps, D. Ringelberg, A. T. Mikell, and D. C. White. 1988. "Mineralization of trichloroethylene by heterotrophic enrichment cultures." *Appl. Environ. Microbiol. 54*:1709-1714.

Guckert, J. B., M. A. Hood, and D. C. White. 1986. "Phospholipid ester-linked fatty acid profile changes during nutrient deprivation of *Vibrio cholerae*: Increase in the *trans/cis* ratio and proportions of cyclopropyl fatty acid." *Appl. Environ. Microbiol. 52*:794-801.

Madsen, E. L., J. L. Sinclair, and W. C. Ghiorse. 1991. "*In situ* biodegradation: Microbiological patterns in a contaminated aquifer." *Science 252*:830-833.

Major, D. W., E. E. Cox, D. W. Acton, T. J. Phelps, and D. C. White. In Submission. "Evaluating trichloroethene biotransformation in subsurface sediments and groundwater by measuring the *in situ* status and activities of microbial populations."

Nelson, D. R., and J. G. Ziekus. 1974. "Rapid method for the radioisotopic analysis of gaseous end products of anaerobic metabolism." *Appl. Microbiol. 28*:258-261.

Phelps, T. J., D. B. Hendrick, D. Ringelberg, C. B. Fliermans, and D. C. White. 1989. "Utility of radiotracer activity measurements for subsurface microbiology studies." *J. Microbiol. Methods 9*:15-27.

Smith, G. A., J. S. Nickels, J. D. Davis, R. H. Findlay, P. S. Vashio, J. T. Wilson, and D. C. White. 1985. "Indices identifying subsurface microbial communities that are adapted to organic pollution." In: *Second International Conference on Ground Water Quality Research Proceedings*, pp. 212-213. Oklahoma State University, Stillwater, OK.

White, D. C., R. J. Bobbie, J. D. King, J. Nickels, and P. Amoe. 1979. "Lipid analysis of sediments for microbial biomass and community structure." In: Litchfield, C. D. & P. L. Seyfried (Eds.), *Methodology for Biomass Determinations and Microbial Activities in Sediments*, pp. 87-103. ASTM STP 673. American Society for Testing and Materials.

White, D. C., G. A. Smith, M. J. Gehron, J. H. Parker, R. H. Findlay, R. F. Martz, and H. L. Fredrickson. 1983. "The ground water aquifer microbiota: Biomass, community structure and nutritional status." *Develop. Indust. Microbiol. 24*:201-211.

CONSTITUTIVE DEGRADATION OF TRICHLOROETHYLENE BY AN ALTERED BACTERIUM IN A GAS-PHASE BIOREACTOR

M. S. Shields, M. J. Reagin, R. R. Gerger, C. Somerville, R. Schaubhut, R. Campbell, and J. Hu-Primmer

ABSTRACT

Pseudomonas cepacia G4 expresses a unique toluene *ortho*-monooxygenase (Tom) that enables it to degrade toluene and trichloroethylene (TCE). Transposon mutants of G4 have been isolated that constitutively express Tom. Two fixed-film bioreactor designs were investigated for the exploitation of one such constitutive strain (G4 PR1) in the degradation of vapor-phase TCE. One received air-entrained TCE along with a continuous inflow of nutrient medium. Strain G4 was unable to degrade TCE in such a vapor-phase bioreactor (100% re-feed), whereas PR1, present at $\geq 1.4 \times 10^8$ bacteria gram^{-1} of the support material, completely removed the recirculating 80 µM (10 mg TCE/L air) TCE. Tests of strain PR1 in a continuous-flow vapor-phase reactor resulted in an average of 92.1% removal of TCE at an average input concentration of 12.6 µM (2 mg/L) TCE (at 0% re-feed) over a 72-h test. The genes responsible for this and the oxidative cleavage of catechol have been cloned into pGEM3Z as an 11 kb *Eco* RI fragment (pMS64) and expressed in *Escherichia coli*. *E. coli* (pMS64) is capable of converting toluene to *ortho*-cresol, *ortho*-cresol to 3-methylcatechol, and phenol to catechol, and degrading TCE in a batch liquid reactor. DNA probe analysis indicates that the genes responsible for this toluene catabolic pathway are located on a large plasmid of G4 (≥150 kb) termed pG4L. We propose that pG4L will serve as the archetype for a new class of catabolic plasmid known as Tom, which encodes an *ortho*-hydroxylation pathway for the degradation of benzene, toluene, *o*-xylene, cresols, and phenol.

INTRODUCTION

The largest category of groundwater pollutants includes the chemicals classified as volatile organics. A portion of this classification is made up of the chloroaliphatics, which include the chloroethylenes: trichloroethylene (TCE);

tetrachloroethylene; *trans*-1,2-dichloroethylene (DCE); 1,1-DCE; and vinyl chloride (Rajagopal 1986). Despite the demonstrated environmental hazard of these pollutants, their industrial use continues because few alternatives exist (Love & Eilers 1982, Vogel et al. 1987). The environmental behavior of TCE is due in part to its physical properties (i.e., high density, high water solubility, and low chemical reactivity), and in part to its biological recalcitrance. All these factors contribute to TCE's notoriety as a persistent point-source pollutant. This is the case despite reports of bacterial transformation capabilities. Anaerobic bacterial degradation is a very slow process that frequently results in the production of vinyl chloride (Barrio-Lage et al. 1988, Bouwer & McCarty 1983, Bouwer et al. 1981, Freedman & Gossett 1989, Kleopfer et al. 1985, Vogel & McCarty 1985). Aerobic bacteria can metabolize TCE more rapidly, but do so only in a cometabolic fashion. TCE serves as a cooxidative substrate for various oxygenases of these aerobic bacteria, but not as an inducer of these enzymes. These bacteria require cosubstrates that include toluene (Kaphammer et al. 1990, Nelson et al. 1986, Nelson et al. 1987, Nelson et al. 1988, Wackett & Gibson 1988, Winter et al. 1989); phenol (Harker & Kim 1990, Montgomery et al. 1989, Nelson et al. 1986), methane (Fox et al. 1990, Henry & Grbić-Galić 1991, Little et al. 1988, Oldenhuis et al. 1989, Tsein et al. 1989, Wackett & Householder 1989); ammonia (Arciero et al. 1989, Vannelli et al. 1990); isoprene (Ewers et al. 1990); or 2,4-dichlorophenoxyacetic acid (2,4-D) (Harker & Kim 1990).

Our research has centered on the toluene- and phenol-utilizing bacterium *P. cepacia* G4 (Shields et al. 1989, 1991). We have sought to use a nonrecombinant revertant of a *Tn*5-induced mutant that no longer requires induction of the enzyme required for TCE degradation (i.e., toluene *ortho*-monooxygenase [Tom]). Primary emphasis has been for the development of an oyster shell associated biofilm capable of constitutive TCE degradation from an airstream.

MATERIALS AND METHODS

Bacterial Strains, Plasmids, and Culture Conditions

Bacterial strains and plasmids used in this study included *P. cepacia* strains G4 (Phe^+,TCE^+) (Shields et al. 1991); G4 5223 (G4:*Tn*5) (Phe^-,TCE^-) revertible to phenol utilization; PR1, a phenol-utilizing (Tom constitutive) revertant of G4 5223; *E. coli* C600 (pRZ102), a ColE1:*Tn*5 (mob^+, kanamycin sulfate [Km] resistant to 50 μg mL^{-1}) suicide vector for delivery of *Tn*5 into *Pseudomonas* sp. (Jorgensen et al. 1979); *E. coli* HB101 (pRK2013), a ColE1:RK2 Tra, Km^r mobilization vector for pRZ102 (Figurski & Helinski 1979); and *Alcaligenes eutrophus* AEO106 (pRO101), a pJP4 derivative containing *Tn* 1721 (encoding tetracycline resistance) (Harker et al. 1989).

Column nutrient additives consisted of a basal salts medium (BSM) (Shields et al. 1991) containing sodium lactate as the sole carbon source (16 g/L) at pH 7.0. *E. coli* strains were grown on Luria-Bertani (LB) medium (Maniatis et al. 1982) with the appropriate selective antibiotics; pGEM4Z was obtained from Promega Biotech, Inc., Madison, Wisconsin.

Microbiological Methods

Tn5 mutagenesis was carried out via triparental matings between G4, *E. coli* C600 (pRZ102), and *E. coli* HB101 (pRK2013). The G4:*Tn5* insertion derivatives were selected by growth on BSM-lactate with Km (50 µg/mL). Such matings are possible because *E. coli* cannot utilize lactate as a sole source of carbon, and strain G4 is sensitive to Km at these levels.

Detection of Toluene *ortho*-monooxygenase Mutants

Tn5-induced mutants of G4, defective in the toluene degradative pathway, were detected by their inability to oxidize *meta*-trifluoromethylphenol (TFMP) (Aldrich Chemical Co., Milwaukee, Wisconsin) to trifluoroheptadienoic acid (TFHA [the yellow transformation product]) (Engesser et al. 1988) as previously described (Shields et al. 1991). This was accomplished by lifting bacterial colonies to nitrocellulose filters impregnated with TFMP, where such mutants remained colorless. Likewise, colonies constitutive for its oxidation turned yellow within 10 minutes in the absence of an aromatic inducer.

Chemical Analysis and Degradation Assays

TCE analysis was carried out by gas chromatography of pentane extracts of culture medium as previously described (Shields et al. 1991). Overnight TCE degradation assays of cell preparations were performed as previously described (Shields et al. 1989). TCE degradation rates were determined using cell suspensions in a 50-mL glass syringe with a Teflon™ plunger without an air headspace as previously described (Folsom et al. 1990). This assay allowed multiple nondestructive sampling (1-mL each) without the introduction of an air headspace. Residual concentrations of 2,4-D and recognizable metabolites in BSM were determined through high-performance liquid chromatography (HPLC) analyses of culture supernatants.

Bioreactor Design

The bioreactor was designed around the concept of providing a biofilm of active PR1 bacteria attached to oyster shell. This active biofilm could then be used as a biofilter for air-entrained TCE. The 26-L column reactor described in this communication consists of 8-inch process pipe with pressure ring seals. The seals contain both the beaded glass end of the pipe and a machined ½-inch 316 stainless steel endplate. The two are held together by a compressible rubber outer seal and a Teflon™ inner seal. The steel endplates were tapped for Swagelok™ fittings to stainless steel tubing. Recycle was achieved via peristaltic pumps with Viton™ tubing. Nutrients were delivered via peristaltic pumps as well. The TCE was delivered as a custom gas mixture (2,000 psi cylinder) by Holox Corp., Pensacola, Florida. Oyster shell was autoclaved for 2 hours and loaded into the reactor which had been sterilized by addition of ~10 g ethylene oxide (Fluka) and recirculated overnight. Inoculation of the reactor was via an attached chemostat.

RESULTS

Creation of PR1

*Tn*5-induced mutants of G4 were selected as km^r lactate using colonies from the triparental mating of *E. coli* C600 (pRZ102) × *E. coli* HB101(pRK2013) × *P. cepacia* G4. Those mutants specifically lacking Tom activity were detected by their failure to oxidize TFMP to TFHA. Four putative *Tn*5-induced mutants lacking Tom activity were identified in this manner. One of these, G4 5233, proved to be unstable, reverting to growth on minimal media containing 2 mM phenol as the sole carbon source at approximately 1 × 10^{-4} per generation. These revertants were then grown on minimal media containing 20 mM lactate as the sole carbon source and lifted onto 7-cm nitrocellulose discs impregnated with TFMP and compared to wild type *P. cepacia* G4 similarly grown on this noninducing medium. All colonies that turned yellow due to TFHA production were found to be constitutive for their expression of toluene *ortho*-monooxygenase and catechol-2,3-dioxygenase (C23O). One such revertant of G4 5223, designated strain PR1 (for phenol revertant number one) was the subject of all investigations reported herein. Because all mutants were isolated from a single *Tn*5-mutagenized culture, the distinct possibility existed that there could be some relationship between them (i.e., siblings). Cell lysates of PR1 grown without exposure to phenol were found to oxidize catechol and 3-methylcatechol at 156 and 50 nmoles/min per milligram protein, respectively. Neither G4 nor G4 5223 was found to express detectable C23O activity toward either of these substrates under these growth conditions. Kanamycin resistance has been stably maintained in all strains investigated so far. No Km^s revertants of PR1 have ever been found despite attempts to find such revertants following 100 generations of nonselective transfer.

Constitutive TCE Metabolism by PR1

Strains G4 5223 and PR1 were grown on lactate, pelleted, and suspended in BSM and sealed in a glass vial (Teflon™-lined butyl rubber septum with a crimp cap). TCE was added to these vials to ~85 µM, and they were incubated inverted, with shaking at 30°C overnight. Only PR1 was capable of degrading the TCE (Table 1) due to constitutive expression of this phenotype.

The potential for cellular toxicity due to reactive intermediates of TCE following oxidative metabolism has been amply demonstrated (Ewers et al. 1990, Tsein et al. 1989, Wackett et al. 1989). The fundamental difference between G4 and PR1 is the metabolism of TCE. G4 will not metabolize TCE without prior induction by aromatics. This allows separation of direct toxicity by TCE versus that caused by an oxidative TCE metabolite.

TABLE 1. TCE degradation by selected strains.

Strain	TCE Remaining (μM)[(a)]
Uninoculated	83.3 ± 7.6
G4 5223	86.2 ± 3.1
G4 5223 PR1	1.4 ± 1.4

(a) Mean TCE concentration of triplicate samples ± standard deviation after an overnight incubation with the indicated strains, grown with lactate, no inducer present.

G4 and PR1 were grown in BSM containing 0.05% yeast extract, 0.5% glucose, and TCE at 35, 210, 530, and 3,260 μM (measured by direct extraction of culture liquor). Growth was inhibited substantially in both strains at 3,260 μM TCE. This was interpreted to be due to a direct toxicity of TCE. At 35 and 210 μM, there was no detectable depression of growth rate in either strain compared to that seen without any added TCE. G4 attained the same growth rate in the presence of 530 μM TCE as that seen for G4 or PR1 in the absence of TCE (doubling time ~3.2 hr). The doubling time of PR1 at 530 μM TCE, however, was depressed to ~5.6 hr. One possible cause of this would be the production of a TCE metabolite that is more toxic than TCE.

Effects of Physical Variables

The effects of oxygen, temperature, and pH on the rate of TCE degradation by PR1 were measured in sealed reaction vessels (no air headspace), and the results are presented in Figure 1. They indicate no substantial change in the degradation rate of TCE over expected aquifer pH ranges, or at oxygen concentrations above 2 mg/L. Somewhat surprisingly, a specific activity of approximately 30% of that measured at 30°C was monitored at 4°C. This could be very important for considerations of in situ applications of PR1, where aquifer temperatures are often below 15°C.

Molecular Biology

Expansion of Substrate Range. TCE in aquifers is frequently found in combination with other pollutants, including chloroaromatics. These represent a pollutant class of considerable concern to the use of PR1 due to their metabolism through *meta*-fission (i.e., C23O) to toxic nonmetabolizable chlorocatechols. In an effort to extend the range of pollutants degraded by PR1 and offer some protection to metabolic suicide with chloroaromatics, plasmid pRO101 (specifying the degradation of 2,4-D) was introduced into PR1. The transconjugant, PR1 (pRO101), expresses enzymes for *ortho*-fission (i.e., catechol-1,2-dioxygenase [C12O]) of the

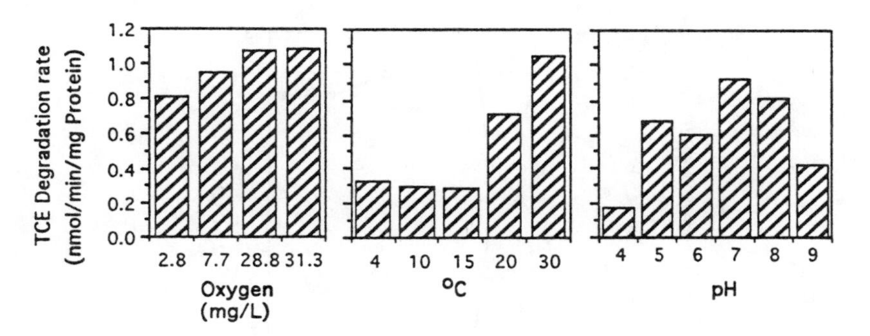

FIGURE 1. Range of TCE degradative activity under various ranges of temperature, pH, and oxygen concentration.

resultant 3-chlorocatechol. The C23O of PR1 will accept 3-chlorocatechol as a substrate in a nonproductive *meta*-cleavage mechanism that inactivates itself (Figure 2). This results in chlorocatechol accumulation that is highly toxic and consequently inhibitory of cellular metabolism. This necessarily limits the ability of the bacterium to successfully degrade many compounds including TCE. PR1 (pRO101) is capable of growth on 0.05% 2,4-D (Figure 3), chlorobenzene, or 2-chlorophenol as sole carbon sources in BSM in shake flasks at 30°C. In addition it is also effective in degrading TCE at approximately 20 µM, following a 24-hour pre-exposure to 2-chlorophenol at 1 mM (Table 2), thus demonstrating its capacity to avoid chronic toxic effects. The established range of Tom activity now encompasses 10 aromatic substrates and 5 aliphatics (Table 3).

Toluene/TCE Degradative Genes Reside on a New Toluene Catabolic Plasmid. A derivative of PR1 (pRO101) was isolated that was unable to grow on phenol as the sole carbon source and was unable to degrade TCE. Plasmid analysis of G4 and its derivatives revealed the presence of two plasmids in each strain of

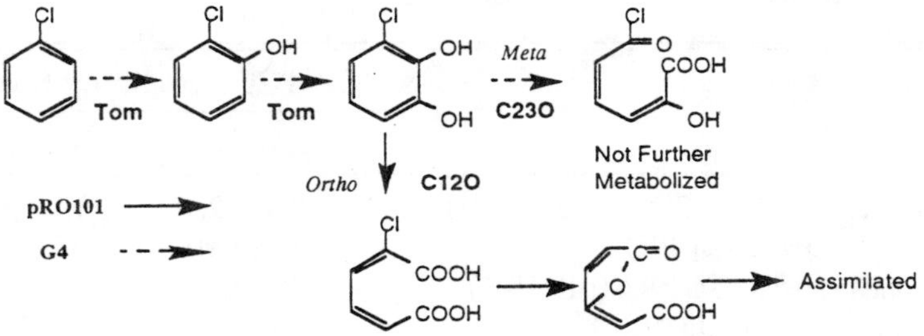

FIGURE 2. Pathway of chloroaromatic catabolism by *Pseudomonas cepacia* G4. Reactions catalyzed by the G4 encoded *meta*-cleavage pathway enzymes are shown as broken lines; the *ortho*-cleavage enzymes encoded by pRO101 are shown as solid lines.

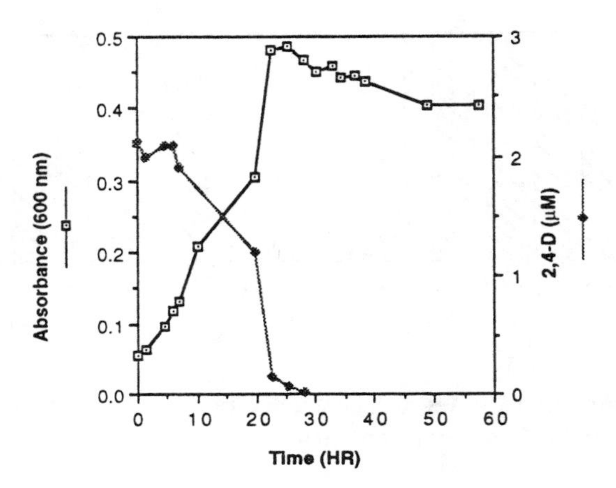

FIGURE 3. Pseudomonas cepacia G4 PR1 (pRO101) grown with 2,4-D as a sole source of carbon and energy. PR1 was unable to grown in this medium or affect 2,4-D concentration.

approximately 50 and 110 kb (designated pG4S and pG4L, respectively, for those found in strain G4 and likewise pPR1S and pPR1L, respectively, for those in strain PR1). Following 100 generations of nonselective growth, PR1(pRO101) isolates were discovered that could no longer grow on phenol as a sole carbon source. These isolates were found to have retained plasmid DNA corresponding to pRO101 and pPR1S but not to pPR1L.

Strain G4 (Km^s and phenol inducible for TCE degradation) (Nelson et al. 1987) was mated to PR1(Km^r) (pRO101, pPR1L$^-$). A Km^r transconjugant was isolated that carried pG4L, pPR1S, and pRO101. This strain was found to have regained the ability to degrade phenol and TCE following phenol induction. This strain was not capable of degrading TCE constitutively (i.e., without phenol induction).

TABLE 2. Inhibition of TCE degradation.

	TCE Remaining (μM)[a]	
	No Addition	1 mM 2CP
Alcaligenes eutrophus AEO106 (pRO101)	18.4 ± 2.1	13.8 ± 1.5[b]
Pseudomonas cepacia G4 PR1	1.0 ± 0.01	18.9 ± 0.50
Pseudomonas cepacia G4 PR1 (pRO101)	1.0 ± 0.02	1.0 ± 0.04
No Cell Control (media only)	ND	20.2 ± 1.6

(a) TCE remaining following an overnight sealed bottle degradation assay at 23°C containing cells grown with lactate (no addition) or lactate with 1 mM 2-chlorophenol (2CP). ND, not determined.

(b) Harker and Kim (1990).

TABLE 3. Susbstrates of the *P. cepacia* G4 PR1 Tom pathway.

Substrate	Metabolite Detected
TCE	One unidentified[a]
cis-1,2-Dichloroethylene	One unidentified[a]
trans-1,2-Dichloroethylene	One unidentified[a]
1,1-Dichloroethylene	None detected
Vinyl chloride	None detected
Toluene	3-Methylcatechol[b]
ortho- and *meta*-cresol	3-Methylcatechol[b]
ortho-xylene	2,3-Dimethylphenol[b]
2,3-Dimethylphenol	2,3-Dimethylcatechol[b]
Phenol	Catechol[b]
Benzene	Phenol[b]
Naphthalene	1-Hydroxynaphthalene[c]
Chlorobenzene[e]	2-Chlorophenol[d]
2-Chlorophenol[e]	3-Chlorocatechol[c]

(a) Transiently detected during intial stages of vapor-phase treatment by gas chromatography, each with a unique retention time dependent upon the chloroaliphatic substrate.
(b) First metabolic product only, substrate is completely metabolized.
(c) Terminal oxidation product.
(d) First metabolic product only, substrate is not completely metabolized.
(e) Completely metabolizable in G4 PR1 (pRO101).

The same PR1 (pRO101, pPR1S) strain was rendered resistant to nalidixic acid (Nal) and rifampicin (Rif) and mated to PR1 (pPR1S, pPR1L) (Nal^s, Rif^s). The 110 kb pPR1L was transferred from the Nal^s, Rif^s strain to the phenol-, Nal^r, Rif^r PR1 (pRO101, pPR1S) strain to create the Nal^r, Rif^r strain PR1 (pRO101, pPR1S, pPR1L). PR1 (pRO101, pPR1S, pPR1L) also was able to degrade phenol and TCE. However, PR1 (pRO101, pPR1S, pPR1L) degraded TCE constitutively. This demonstrates conclusively that not only are the phenol and TCE degradative enzymes located on the large indigenous plasmid, but the genetic determinant responsible for their constitutive expression is as well.

Cloning

An 11 kb *Eco* RI fragment of plasmid material isolated from PR1 was cloned into the vector pGEM4Z (Promega Biotech) to create a recombinant plasmid: pMS64. *E. coli* JM109 (pMS64) constitutively degraded 20 μM TCE, and produced *o*-cresol from toluene and 3-methylcatechol from *m*-cresol and *o*-cresol (all phenotypes associated with Tom activity in G4) (Shields et al. 1989, 1991). In addition, C23O activity (encoded by the Tom pathway in G4) was detectable in this recombinant *E. coli*. Hybridization of these cloned genes with total DNA isolated from G4 (pG4S, pG4L) and its derived strains, PR1 (pRO101, pPR1S), PR1 (pRO101, pPR1S, pPR1L), took place only with the large ~110 kb plasmid (i.e., pG4L or pPR1L) in each case.

A PR1 Biofilter for the Vapor-Phase Degradation of TCE

A biofilter was used to apply the axenic cometabolic process to the degradation of air-entrained TCE. A number of materials were tested for their ability to support growth of a PR1 biofilm that was capable of TCE degradation (Figure 4). The material exhibiting the greatest degree of TCE removal per gram of colonized material was crushed oyster shell (patent applied for). In addition to performance criteria, the cost and buffering capacity of the oyster shell matrix contributed to its selection for further vapor-phase column research.

All column designs tested thus far involved the metered application of nutrients (yeast extract, peptone, and glucose or minimal medium with lactic acid) to a glass or stainless steel column, packed with oyster shell and colonized with PR1. Air-entrained TCE was passed through the column (bottom to top). Initially, a 3-L column assembled in this manner was capable of the stable removal of 7 to 10 μM air-entrained TCE introduced at a rate of 4 mL per minute over a 96-hour period (Figure 5).

This reactor was scaled up to a 26-L laboratory column reactor (Figure 6) with several design changes. The most important among these were to allow for nutrient liquid recycle and a continuous input of freshly grown cells during the inoculation phase. This reactor received a continuous input of PR1 cells

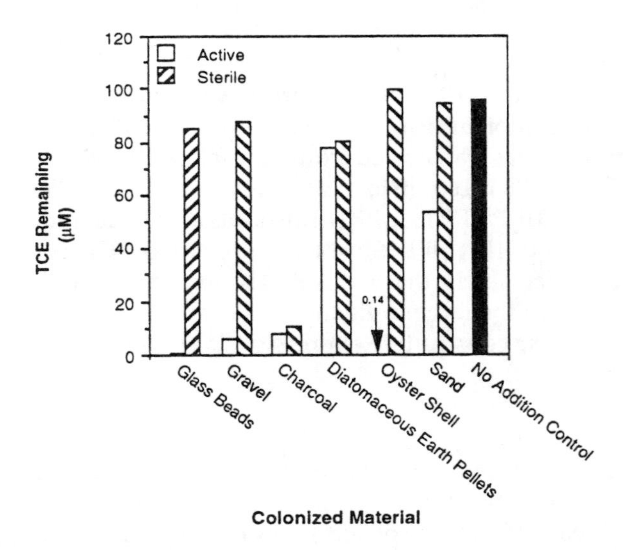

FIGURE 4. Materials tested for their ability to support a TCE-degrading biofilm of *Pseudomonas cepacia* G4 PR1. TCE analyses were performed following overnight incubations at 30°C. The "no addition control" consisted solely of BSM and TCE incubated overnight in the same manner.

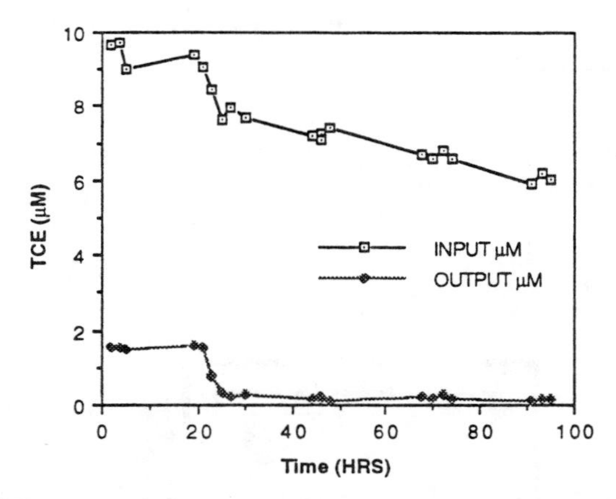

FIGURE 5. TCE removal from an airstream passed through a 3-L column supporting a biofilm of *Pseudomonas cepacia* G4 PR1 on crushed oyster shell. TCE concentration in the output line equaled the input line after 2 hours in an uninoculated or G4-inoculated column.

produced from a chemostat. After a 48-hour inoculation period, air-entrained TCE (~ 1μM [131 μg/L]) was introduced at 100 mL per min total flowrate (ca 58% oxygen). Under these conditions ~90% of the TCE present in the gas phase was continuously removed during a 4-day test period (Figure 7).

DISCUSSION

Physiology

The cometabolism of TCE by bacteria is well documented (Arciero et al. 1989, Ensley 1991, Ewers et al. 1990, Henry & Grbić-Galić 1991, Montgomery et al. 1989, Nelson et al. 1986, Wackett et al. 1989, Wackett & Householder 1989, Wilson & Wilson 1985, Zylstra et al. 1988, Zylstra et al. 1989). Several of these bacteria have been, or are in the process of being, investigated for their potential role in TCE degradation in bioreactors or for in situ applications. It has been the purpose of the work reported here to present preliminary findings on the use of a particular *P. cepacia* strain rendered constitutive for the production of *ortho*-toluene monooxygenase, in a vapor-phase bioreactor. Although several toluene oxygenases capable of TCE oxidation have been cloned to *E. coli* and thus no longer require aromatic induction (Kaphammer et al. 1990, Winter et al. 1989, Zylstra et al. 1989), this is the first reported bioreactor application of a nonrecombinant bacterium capable of doing so.

Constitutive expression of Tom by liquid cultures of PR1 was detectable by the rate of TFMP conversion to TFHA (a reaction requiring both Tom and C23O), as well as by TCE degradation. The 8 nmol TFHA produced per minute per milligram of protein by uninduced PR1 was similar to the 6.5 nmol per minute per milligram protein produced by phenol-induced G4. Likewise an ~1 nmol per

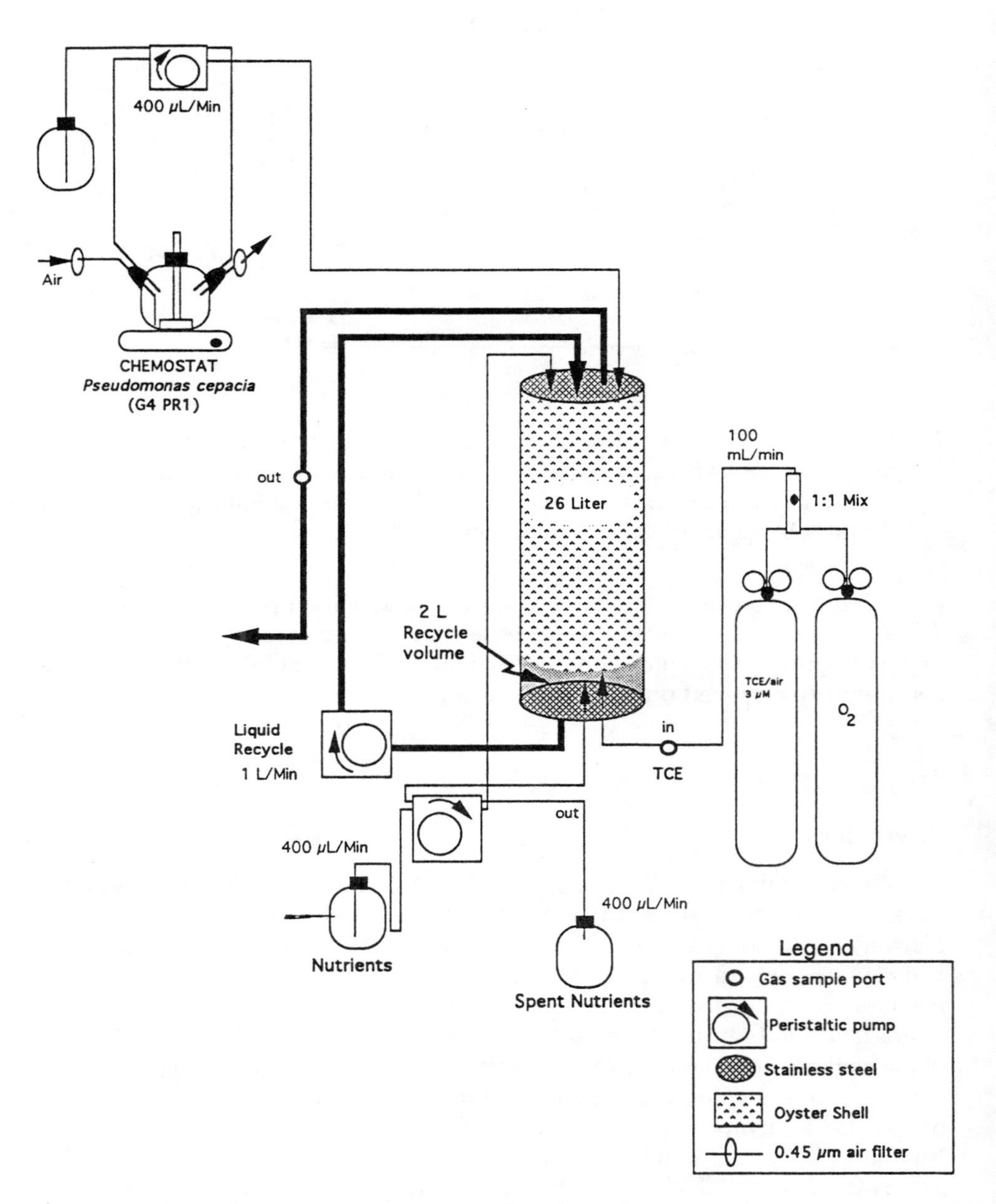

FIGURE 6. A 26-L biofilter containing crushed oyster shell and colonized with *Pseudomonas cepacia* G4 PR1. Air-entrained TCE was delivered from a compressed gas cylinder and mixed with oxygen to balance nutrient requirements during inoculation-colonization.

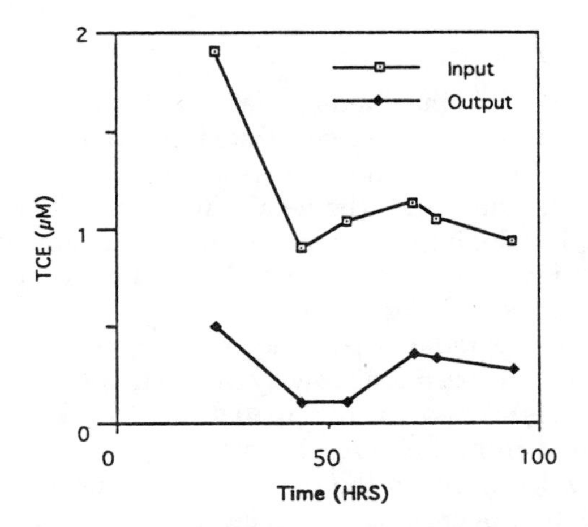

FIGURE 7. TCE removal from an airstream passed through a 26-L column supporting a biofilm of *Pseudomonas cepacia* G4 PR1 on crushed oyster shell. TCE concentration in the output line equaled that of the input line after 20 hours in an uninoculated column.

minute per mg protein rate of uninduced TCE removal by liquid cultured PR1 is approximately one-fourth of the average daily TCE degradation rate attained with phenol-induced (chemostat grown) G4 (Folsom & Chapman 1991). The effect of varying physical conditions on the ability of PR1 to degrade TCE indicated a fairly broad range through which appreciable activity was maintained: pH 5 to 9, 4 to 30°C, 0 to 20 parts per thousand saline, and oxygen concentrations from 2.8 to 31 mg/L. The most unexpected finding was the essentially unchanged rates of degradation between 4 and 15°C.

Toxic end products from bacterial oxygenase action on TCE have been shown to be detrimental to many bacteria. *P. putida* F1, for instance, has been shown to experience such detrimental effects due to the action of toluene dioxygenase (Wackett & Householder 1989). The Tom system of *P. cepacia* G4, however, is only moderately affected. The depression in growth rate of PR1 at 473 μM TCE (without a similar growth rate effect detectable in G4 cultures at 554 μM TCE) indicates a probable TCE metabolite effect, because PR1 is responsive to TCE without induction and G4 is not. This observation is similar to effects reported from exposure of two isoprene metabolizers to TCE (Ewers et al. 1990).

The ability to constitutively cometabolize several other substrates by virtue of the Tom pathway also has been established. The substrate include vinyl chloride; *cis-* and *trans*-1,2-dichloroethylene; 1,1-dichloroethylene; and toluene, phenol, xylenes, and cresols. *P. cepacia* PR1 (pRO101), the nonrecombinant derivative, is capable of using these same substrates in addition to growth on 2,4-D, chlorobenzene, and 2-chlorophenol. The efficacy for their treatment in a bioreactor or in situ system remains to be demonstrated.

Molecular Biology

The introduction of another biodegradative plasmid (pRO101) into *P. cepacia* 5223 PR1 has already enhanced the substrate range of this organism to include chlorobenzene and 2-chlorophenol (determination of the utilization of other chloroaromatics is pending) and is better able to degrade TCE during prolonged exposure to them. The cloning and expression of these degradative genes in *E. coli* now allow us to study the effects of expression of these genes in a well-defined genetic and physiological background.

Genes of the Tom degradative pathway necessary to degrade the compounds listed in Table 3 have now been conclusively shown to reside on a newly described catabolic plasmid. It is necessary to differentiate this ~110 kb *Pseudomonas* toluene degradative plasmid from the TOL plasmid (archetype: pWWO) (Worsey & Williams 1975), which encodes the oxidation of toluene through a series of oxygenases and dehydrogenases to benzoate and then to catechol. The toluene *ortho*-monooxygenase is (so far) unique to G4 (Shields et al. 1989) and is its largest indigenous plasmid. For this reason we have proposed the name Tom for the ~110 kb toluene degradative plasmid described here, for which pG4L serves as the archetype. Its existence now poses some provocative possibilities for environmental remediation of several aromatic and aliphatic priority pollutants. The precise nature of the original *Tn*5 mutation that allowed the large plasmid of G4 to become constitutive for TCE degradation remains unknown. However, due to our analysis of similar mutants (Shields et al. 1991), it seems reasonable to conclude that the simplest explanation for the genesis of PR1 is a *Tn*5 insertion in pG4L that subsequently underwent a genetic exchange with the chromosome, transferring the Km^r genes of *Tn*5 but not the Tom degradative pathway genes. This latter transfer of *Tn*5 would appear to be independent of the genetic alteration leading to consititutivity of the Tom pathway.

Future Research Goals

Laboratory. Clearly TCE degradation in the bioreactor columns described above merely reflects a window of activity and does not necessarily represent optimized conditions for TCE treatment. Current investigations center on nutrient/oxygen/substrate (i.e., TCE) balances, selective growth substrates, antibiotic resistances to limit invasion to aid in maintaining the axenic culture, and column operational parameters including nutrient and gas recycle.

One of the more challenging aspects of this application is the maintenance of a bioreactor carrying on a cometabolic transformation. Unlike other biodegradative systems where the degradation of the pollutant enhances the survival of the biodegradative organism, there is no direct selective advantage for *P. cepacia* in the metabolism of haloaliphatics. One approach will employ the cycling of selective nutrients to maintain a population enriched for bacteria best able to use each sequentially. Such an approach is best evaluated through laboratory studies first.

Field. Current plans include for field testing the bioreactor/biofilter column concept using PR1 will involve a much larger version (two ~100L stainless steel reactors) developed by the U.S. Air Force (USAF), at the Civil Engineering Support Laboratory, Tyndall Air Force Base in Panama City, Florida. These reactors are to be employed in a test of the system under field conditions at a USAF site to be determined. Field research in conjunction with SBP Technologies Inc., the University of Waterloo, Waterloo Ontario, Canada, and the U.S. Environmental Protection Agency (EPA), Gulf Breeze, Florida, is also planned to test PR1 in situ bioreactor technology under controlled field conditions.

ACKNOWLEDGMENTS

Portions of this work was supported through a research grant from SBP Technologies Inc.; Contract No. 68-03-3479 between Technical Resources Inc. and the EPA Research Laboratory, Gulf Breeze, Florida; and EPA Grant CR 820704-01-0 and USAF Contract F 08635-92-C-0103 to the University of West Florida.

REFERENCES

Arciero, D. T., M. Vannelli, M. Logan, and A. B. Hooper. 1989. "Degradation of trichloroethylene by the ammonia-oxidizing bacterium *Nitrosomonas europaea*." *Biochem. Biophys. Res. Commun. 159*: 640-643.

Barrio-Lage, G. A., F. Z. Parsons, and P. A. Lorenzo. 1988. "Inhibition and stimulation of trichloroethylene biodegradation in microaerophilic microcosms." *Environ. Toxicol. Chem. 7*: 889-895.

Bouwer, E. J., and P. L. McCarty. 1983. "Transformations of 1- and 2-carbon halogenated aliphatic compounds under methanogenic conditions." *Appl. Environ. Microbiol. 45*: 1286-1294.

Bouwer, E. J., B. J. Rittmann, and P. L. McCarty. 1981. "Anaerobic degradation of halogenated 1-and 2-carbon organic compounds." *Environ. Sci. Technol. 15*: 596-599.

Engesser, K. H., R. B. Cain, and H. J. Knackmuss. 1988. "Bacterial metabolism of side chain fluorinated aromatics: Cometabolism of 3- trifluoromethyl (TFM)-benzoate by *Pseudomonas putida* (arvilla) mt-2 and *Rhodococcus rubropertinctus* N657." *Arch. Microbiol. 149*: 188-197.

Ensley, B. D. 1991. "Biochemical Diversity of Trichloroethylene Metabolism." *Annu. Rev. Microbiol. 45*: 283-299.

Ewers, J., D. Freier-Schroder, and H-J. Knackmuss. 1990. "Selection of trichloroethene (TCE) degrading bacteria that resist inactivation by TCE." *Arch. Microbiol. 154*: 410-413.

Figurski, D. H., and D. R. Helinski. 1979. "Replication of an origin-containing derivative of plasmid RK2 dependent on a plasmid function provided in trans." *Proc. Natl. Acad. Sci. USA 76*: 1648-1652.

Folsom, B. R., P. J. Chapman, and P. H. Pritchard. 1990. "Phenol and trichloroethylene degradation by *Pseudomonas cepacia* G4: Kinetics and interactions between substrates." *Appl. Environ. Microbiol. 56*: 1279-1285.

Folsom, B. R., and P. J. Chapman. 1991. "Performance characterization of a model bioreactor for the biodegradation of trichloroethylene by *Pseudomonas cepacia* G4." *Appl. Environ. Microbiol. 57*: 1602-1608.

Fox, B. G., J. G. Borneman, L. P. Wackett, and J. D. Lipscomb. 1990. "Haloalkane oxidation by the soluble methane monooxygenase from *Methylosinus trichosporium* OB3b: Mechanistic and environmental implications." *Biochemistry 29*: 6419-6427.

Freedman, D. L., and J. M. Gossett. 1989. "Biological reductive dechlorination of tetrachloroethylene and trichloroethylene to ethylene under methanogenic conditions." *Appl. Environ. Microbiol. 55*: 2144-2151.

Harker, A. R., and Y. Kim. 1990. "Trichloroethylene degradation by two independent aromatic-degrading pathways in *Alcaligenes eutrophus* JMP134." *Appl. Environ Microbiol. 56*: 1179-1181.

Harker, A. R., R. H. Olsen, and R. J. Seidler. 1989. "Phenoxyacetic acid degradation by the 2,4-dichlorophenoxyaceticacid (TFD) pathway f plasmid pJP4: Mapping and characterization of the TFD regulatory gene, *tfd* R." *J. Bacteriol. 171*: 314-320.

Henry, S. M., and D. Grbić-Galić. 1991. "Influence of endogenous and exogenous electron donors and trichloroethylene oxidation toxicity on trichloethylene oxidation by methanotrophic cultures from a groundwater aquifer." *Appl. Environ. Microbiol. 57*: 236-244.

Jorgensen, R. A., S. J. Rothstein, and W. S. Reznikoff. 1979. "A restriction enzyme cleavage map of *Tn*5 and location of a region encoding neomycin resistance." *Molec. Gen. Genet. 177*: 65-72.

Kaphammer, B., J. J. Kukor, and R. H. Olsen. 1990. "Cloning and characterization of a novel toluene degradative pathway from *Pseudomonas pickettii* PKO1." Abstr K-145, p. 243. *Abstr. 90th Annu. Meet. A. Soc. Microbiol.*

Kleopfer, R. D., D. M. Easley, B. B. Haas Jr., and T. G. Deihl. 1985. "Anaerobic degradation of trichloroethylene in soil." *Environ. Sci. Technol. 19*: 277-280.

Little, C. D., A. V. Palumbo, S. E. Herbes, M. E. Lidstrom, R. L. Tyndall and P. J. Gilmer. 1988. "Trichloroethylene biodegradation by a methane-oxidizing bacterium." *Appl. Environ. Microbiol. 54*: 951-956.

Love, O. T. Jr., and R. G. Eilers. 1982. Treatment of drinking water containing trichloroethylene and related industrial solvents. *J. Am. Waterworks Assn. 80*: 413-425.

Maniatis, T., E. F. Fritsch, and J. Sambrook. 1982. *Molecular cloning: A laboratory manual.* Cold Spring Harbor Laboratory, Cold Spring Harbor, NY.

Montgomery, S. O., M. S. Shields, P. J. Chapman, and P. H. Pritchard. 1989. "Identification and characterization of trichloroethylene-degrading bacteria" Abstr. K-68, p. 256. *Abstr. 89th Annu. Meet. Am. Soc. Microbiol.* 1989.

Nelson, M. J. K., S. O. Montgomery, W. R. Mahaffey, and P. H. Pritchard. 1987. "Biodegradation of trichloroethylene and involvement of an aromatic biodegradative pathway." *Appl. Environ. Microbiol. 53*: 949-954.

Nelson, M. J. K., S. O. Montgomery, E. J. O'Neill, and P. H. Pritchard. 1986. "Aerobic metabolism of trichloroethylene by a bacterial isolate." *Appl. Environ. Microbiol. 52*: 383-384.

Nelson, M. J. K., S. O. Montgomery, and P. H. Pritchard. 1988. "Trichloroethylene metabolism by microorganisms that degrade aromatic compounds." *Appl. Environ. Microbiol. 54*: 604-606.

Oldenhuis, R., R. L. J. M. Vink, D. B. Janssen, and B. Witholt. 1989. "Degradation of chlorinated aliphatic hydrocarbons by *Methylosinus trichosporium* OB3b expressing soluble methane monooxygenase." *Appl. Environ. Microbiol. 55*: 2819-2826.

Rajagopal, R. 1986. "Conceptual design for a groundwater quality monitoring strategy." *The Environ. Professional. 8*: 244-264.

Shields, M. S., S. O. Montgomery, S. M. Cuskey, P. J. Chapman, and P. H. Pritchard. 1991. "Mutants of *Pseudomonas cepacia* strain G4 defective in catabolism of aromatic compounds and trichloroethylene." *Appl. Environ. Microbiol. 57*: 1935-1941.

Shields, M. S., S. O. Montgomery, P. J. Chapman, S. M. Cuskey, and P. H. Pritchard. 1989. "Novel pathway of toluene catabolism in the trichloroethylene-degrading bacterium G4." *Appl. Environ. Microbiol. 55*: 1624-1629.

Tsein, H-C., G. A. Brusseau, R. S. Hanson, and L. P. Wackett. 1989. "Biodegradation of trichloroethylene by *Methylosinus trichosporium* OB3b." *Appl. Environ. Microbiol. 55*: 3155-3161.

Vannelli, T., M. Logan, D. M. Arciero, and A. B. Hooper. 1990. "Degradation of halogenated aliphatic compounds by the ammonia-oxidizing bacterium *Nitrosomonas europaea*." *Appl. Environ. Microbiol. 56*: 1169-1171.

Vogel, T. M., and P. L. McCarty. 1985. "Biotransformation of tetrachloroethylene to trichloroethylene, dichloroethylene, vinyl chloride, and carbon dioxide under methanogenic conditions." *Appl. Environ. Microbiol. 49*: 1080-1083.

Vogel, T. M., C. S. Criddle, and P. L. McCarty. 1987. "Transformations of haologenated aliphatic compounds." *Environ. Sci. Technol. 21*: 722-736.

Wackett, L. P., G. A. Brusseau, S. R. Householder, and R. S. Hanson. 1989. "A survey of microbial oxygenases: Trichloroethylene degradation by propane-oxidizing bacteria." *Appl. Environ. Microbiol. 55:* 2960-2964.

Wackett, L. P., and D. T. Gibson. 1988. "Degradation of trichloroethylene by toluene dioxygenase in whole-cell studies with *Pseudomonas putida* F1." *Appl. Environ. Microbiol. 54*: 1703-1708.

Wackett, L. P., and S. R. Householder. 1989. "Toxicity of trichloroethylene to *Pseudomonas putida* F1 is mediated by toluene dioxygenase." *Appl. Environ. Microbiol. 55*: 2723-2725.

Wilson, J. T., and B. H. Wilson. 1985. Biotransformation of trichloroethylene in soil. *Appl. and Environ. Microbiol. 49*: 242-243.

Winter, R. B., K.-M. Yen, and B. D. Ensley. 1989. "Efficient degradation of trichloroethylene by a recombinant *Escherichia coli*." *Bio/Technology. 7*: 282-285.

Worsey, M. J., and P. A. Williams. 1975. "Metabolism of Toluene and xylenes by *Pseudomonas putida (arvilla)* mt-2: evidence for a new function of the TOL plasmid." *J. Bacteriol. 124*: 7-13.

Zylstra, G. J., W. R. McCombie, D. T. Gibson, and B. A. Finette. 1988. "Toluene degradation by *Pseudomonas putida* F1: genetic organization of the *tod* operon." *Appl. Environ. Microbiol. 54*: 1498-1503.

Zylstra, G. J., L. P. Wackett, and D. T. Gibson. 1989. "Trichlorotethylene degradtion by *Escherichia coli* containing the cloned *Pseudomonas putida* F1 toluene dioxygenase genes." *Appl. and Environ. Microbiol. 55*:3162-3166.

BIODEGRADATION OF CARBON TETRACHLORIDE UNDER ANOXIC CONDITIONS

H. D. Stensel and L. J. DeJong

ABSTRACT

The biological degradation kinetics of carbon tetrachloride (CT) was studied in a laboratory anoxic fluidized-bed reactor over a period in excess of 400 days. The original culture was isolated from groundwater contaminated with CT and nitrate. The addition of acetate stimulated both denitrification activity and CT degradation activity, but CT degradation did not occur under endogenous conditions with excess nitrate present. Typical operating conditions for the fluidized-bed reactor were a 7.0-hour detention time, a chemical oxygen demand (COD) loading of 1.2 g/L-d, and an influent CT concentration of 800 μg/L. The specific CT degradation rates (SCTDRs) ranged from 200 to 1,200 μg CT/g volatile suspended solids (VSS)-d, and a relationship between SCTDR, COD loading, and CT concentration was developed. Operation at zero nitrate concentration resulted in SCTDRs 7 to 12 times higher than with nitrate present. However, the higher SCTDRs at zero nitrate concentration were associated with higher chloroform (CF) production. The CT degradation capability was greatly reduced after 220 days of operation following a change in the method used to remove biomass growth. A reactor design evaluation showed that the total reactor detention time, COD requirements, and solids production are a function of the number of reactor stages and the required effluent CT concentration.

INTRODUCTION

One of the most common categories of hazardous wastes found in contaminated groundwaters is organic solvents (Westrick et al. 1984), which includes CT, also listed as a priority pollutant by the U.S. Environmental Protection Agency (EPA). Significant levels of CT (ranging from 0.5 to 2.0 mg/L) and nitrate exist in the groundwater at Hanford's U.S. Department of Energy nuclear reservation as a result of disposal of liquid wastes for more than 40 years (Last et al. 1991). A biological culture found by Brouns et al.(1990) in the groundwater at the site has been able to degrade CT during nitrate reduction with the addition of acetate.

They also found that CT degradation did not occur under endogenous conditions with nitrate present.

In situ degradation of CT may be promoted by appropriate additions of acetate, nitrate, and perhaps other nutrients to the groundwater. Another alternative is to pump and treat groundwater in an aboveground reactor under more controlled conditions. In either case, rates of CT degradation and factors controlling these rates must be understood. The purpose of this study was to determine if the CT degradation rate is related to CT concentration and acetate feed level in an anoxic laboratory fluidized-bed reactor. A fluidized bed-reactor is a feasible aboveground treatment system with advantages that include the ability to maintain a high biomass concentration with minimal detention times.

BACKGROUND

One of the first investigations on CT degradation using denitrifying bacteria was by Bouwer and McCarty (1983b). Using radiolabeled CT, they found that after 13 weeks of incubation 23% of the CT was converted to CF and the rest was found in carbon dioxide and cell material. Although the pathway is not fully known, CT degradation under anoxic conditions appears to occur by a different pathway than by reductive dehalogenation observed for CT degradation under sulfate reduction (Egli et al. 1987) and methanogenic conditions (Bouwer & McCarty 1983a, Egli et al. 1988, Mikesell & Boyd 1990).

Limited work has been done to isolate specific cultures capable of CT degradation under anoxic conditions. A pure denitrifying culture, *Pseudomonas* sp. Strain KC, was isolated from a contaminated site by Criddle et al. (1990a) and was found to degrade CT to carbon dioxide with little or no production of CF. The addition of reduced iron at a 5 μm concentration and cobalt seemed to inhibit the degradation of CT, but did improve the growth of cells. Criddle et al. (1990b) found that *Escherichia coli* K-12, an organism capable of diverse respiratory processes, did not degrade CT when oxygen or nitrate was used as the electron acceptor. Stuart (1990) found, under acetate feed conditions, similar degradative capabilities between a mixed denitrifying culture derived from a municipal activated sludge facility and that for an enrichment derived from the Hanford groundwater. SCTDRs ranged from 20 to 80 μg CT/g VSS-hr and varied with initial COD levels in serum bottle batch tests, but no correlation between COD and SCTDR was possible.

Additional studies on CT degradation under anoxic conditions have been done in biofilm columns and soil columns (Bouwer & Wright 1988, Cobb & Bouwer 1991). Bae and Rittmann (1990) studied the effect of acetate and nitrate additions on CT degradation in a soil column. With acetate and no nitrate added the CT removal was 28%, without acetate and with excess nitrate the removal was only 3%, and with both acetate and nitrate the removal was 9%. About 5 to 13% of the CT degraded was found as CF. They concluded that CT was removed by both reductive dehalogenation and another pathway related to nitrate reduction. Semprini et al. (1991) studied CT transformation under denitrifying conditions in laboratory columns from a site at Moffett Field, California. They showed 30 to 40% conversion of CT to CF. One explanation presented for the high CF production was that a sulfate-reducing population may have been present that competed for the acetate fed to stimulate the denitrifying culture activity.

EXPERIMENTAL PLAN AND METHODS

A bench-scale fluidized-bed reactor was operated for more than 1 year to study CT removal under different acetate, CT, and nitrate feed conditions. The reactor was seeded with the denitrifying, CT-degrading culture derived from contaminated groundwater at the Hanford site. The laboratory reactor was a 1.3-L glass column with a diameter and length of 5.1 and 65 cm, respectively. It contained about 370 ml of 14 × 30 mesh silica media (Celite™) that had a highly porous structure and a specific density of about 1.4. The expanded fluidized bed depth was maintained at 27 to 28 cm by providing effluent recycle at a rate in excess of 20 times the influent flow rate. Acetate and nitrate-inorganic feed solutions were delivered separately by peristaltic pumps. A saturated CT water solution was delivered by a syringe pump to a mixing/sampling chamber in the nitrate-inorganic line to provide controlled amounts of CT to the reactor. The reactor effluent was sampled from a port in the recycle line. Gas production was monitored by a Precision Scientific wet test gas meter connected by Viton™ tubing to the reactor headspace. A gas sample port was located in the gas line.

The acetate and nitrate-inorganic feed solutions were prepared with tap water. Acetate was added as sodium acetate to yield reactor influent COD concentrations of about 300 mg/L. Nitrate was added as sodium nitrate in the feed and influent concentrations were generally about 90 mg/L as N. The influent CT concentration was about 800 μg/L except when transient testing was in progress. An inorganic feed was selected to simulate Hanford groundwater conditions and to provide nutrients and pH buffer. The inorganic compounds and concentrations in the combined reactor influent are as follows: $Na_2SiO_3 \cdot 9H_2O$, 10 mg/L; $NaHCO_3$, 300 mg/L; $NaSO_4$, 6 mg/L; NaF, 44 mg/L; $CaCl_2 \cdot 2H_2O$, 81 mg/L; $MgCl_2 \cdot 6H_2O$, 117 mg/L; KH_2PO_4, 860 mg/L; and NH_4Cl, 106 mg/L.

Table 1 summarizes the normal operating conditions for the fluidized bed operation. Changes in the COD loading and influent CT concentration were made during transient testing to determine the effect of COD loading and CT concentration on CT removal rates. A target influent COD/N ratio of 3.4 was selected to make sure that excess nitrate would be present in the reactor. This ratio was calculated based on a net solids yield of 0.13 g VSS/g COD removed (Randall et al. 1992). A 7-hour hydraulic detention time, based on the expanded-media bed volume, was arbitrarily selected based on earlier work by Brouns et al. (1990). These requirements limited the normal organic loading to about 1.2 g COD/L-d, which is a relatively low loading for a fluidized-bed operation. Organic loadings of 4 to 6 g COD/L-d have been used for nitrate reduction, with acetate addition, in a fluidized-bed reactor treating municipal wastewaters (Gommers et al. 1988). The reactors were operated at room temperature that varied between 25 and 30°C

TABLE 1. Summary of fluidized-bed experimental reactor normal operating conditions.

Parameter	Value
Flow	2.0 L/d
Hydraulic Retention Time	7.0 hr
Organic Loading	1.2 g COD/L-d
Influent COD/N ratio	3.4
Reactor Temperature	25 to 30°C
Reactor pH	7.5 to 7.8
Expanded-Media Depth	28 cm
Reactor VSS Conc.	1,900 to 3,100 mg/L
Effluent Soluble COD Conc.	20 to 50 mg/L
Effluent NO_3-N Conc.	2 to 8 mg/L
Effluent NO_2-N Conc.	0.2 to 0.3 mg/L

over the course of the experiments. The reactor VSS concentration based on the expanded-media bed volume ranged from about 1,900 mg/L during the early phase of the operation to 2,200 mg/L at day 200 and 3,100 mg/L at day 400 of the operation. Effluent COD concentrations were independent of the reactor COD loading and appeared to a function of by-product production from substrate utilization and endogenous decay.

Excess biomass growth formed a sticky mass with the media at the top of the reactor and, without regular removal, bed fluidization was impaired. During the first 150 days of operation the biomass control method consisted of stirring the upper portion of the bed about every 2 weeks and allowing the excess growth to discharge in the effluent. Between 150 and 180 days of operation, the sticky biomass growth became more troublesome and an additional step was used that involved removing about 15% of the media every 2 weeks and replacing it with fresh media. After 180 days a more vigorous cleaning method was employed on a weekly basis. The top of the reactor was removed, the entire media was vigorously stirred with a steel rod, and the detached biomass was flushed out with 2 L of tap water containing 100 mg/L of NO_3-N.

Determining CT Degradation Kinetics

To determine the effect of CT concentration, the influent CT concentration was varied for different experiments lasting approximately 5 days, while maintaining a constant COD loading and excess nitrate in the reactor. Three to four CT loading tests were done at three different COD loading levels. After each CT loading change, an equilibrium time of 3 hydraulic detention times was allowed before data collection. Samples were then taken at three different times over a 24-hour period. A similar procedure was followed to observe the effect of COD loading on CT degradation kinetics. COD loadings were varied by changing the influent COD concentration, and the CT feed concentration was held constant. Three COD loadings changes were made for two different influent CT levels. The influent nitrate concentration was changed with influent COD concentration changes to ensure excess nitrate in the reactor. On a few occasions, the nitrate level was not adjusted sufficiently and CT degradation was observed in the reactor at a zero nitrate concentration. This test and data collection condition generally occurred over a 1- to 2-day period.

Data were obtained to provide a CT mass balance based on the flowrate, CT feed and effluent liquid concentrations, effluent CT gas concentration, and gas production rate. Sorption of CT was not considered in this balance because the system was subjected to long-term CT exposure to allow sorption equilibrium conditions. The influent, reactor liquid effluent, and gas effluent were sampled with a 5-mL disposable plastic syringe with a 22.5-gauge, 1.5-inch (3.8 cm) disposable needle. The CT degradation rate in µg/d was based on the difference in the mass of CT applied in the influent and the measured mass of CT lost in the liquid and gas effluents and corrected for a reactor CT loss rate. The SCTDR in µg CT/g VSS-d was calculated by dividing the CT degradation rate by the expanded-bed volume and the expanded-bed VSS concentration.

Leak tests were performed to determine a reactor CT loss rate. CT loss was observed and was assumed to have occurred due to diffusion through tubing and connections and possibly undetected small gas leaks. The CT leak test was performed by periodically monitoring the change in CT mass in the reactor under endogenous conditions with excess nitrate available. We have shown in previous work that CT degradation does not occur under endogenous conditions with nitrate present and lack of acetate addition (Stuart 1990). Figure 1 shows the relationship found between the reactor CT loss rate and reactor CT concentration. This relationship was used to correct the CT mass balance data to calculate a CT degradation rate. The CT loss rate accounted for 30 to 50% of the total reactor CT removal rate for the different operating conditions studied. During the later part of the study when the CT degradation capability was lost, the calculated CT degradation rate was negative, indicating that this approach may yield a somewhat lower CT degradation rate.

Analytical Methods

Influent COD concentrations were calculated based on feeding a known concentration of acetate and measuring the flowrates of the acetate and nitrate-inorganic feeds. Effluent COD concentrations were measured by the spectrophotometric method using prepared dichromate ampules from Hach Chemical. Low range (0 to 150 mg/L) ampules received 2 mL of sample after filtration through a 0.45-µ filter. Samples were analyzed in duplicate and were generally within 5% of each other.

Influent nitrate-N concentrations were also calculated based on feeding a known concentration of nitrate in the nitrate-inorganic feed and measuring the two

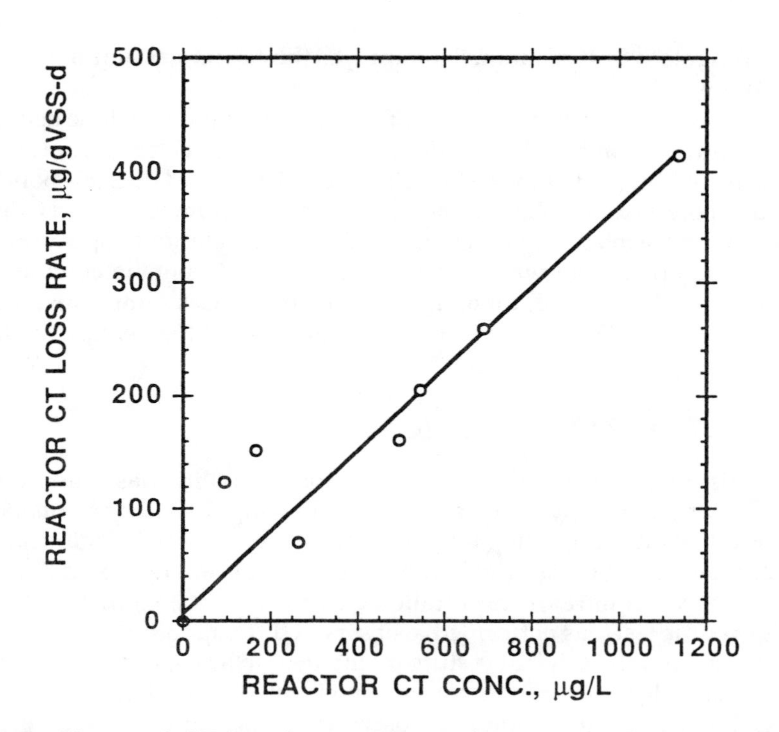

FIGURE 1. Observed CT loss from reactor versus reactor CT concentration at endogenous conditions with excess nitrate.

feed flowrates. Effluent nitrate and nitrite concentrations were measure by a spectrophotometric method using reagent powder pillows from Hach Chemical. NitraVer 5 reagent was used with 10-mL samples for nitrate analysis within a 0- to 4.5-mg/L range. NitraVer 3 pillows were used with 10-mL samples for nitrite analysis within a 0- to 0.2-mg/L range. Duplicate analyses were done, and the variation between samples was within 5%.

Total suspended solids (TSS) and VSS analysis were analyzed as described in Standard Methods (American Public Health Association 1985) using a glass fiber filter pad for filtration of liquid samples. To determine the VSS content of the fluidized-bed media, a known volume of media was dried at 103°C and ignited in the 550°C oven. The difference in the dried and ignition weights was assumed to be the VSS content of biological solids. No loss of fresh media mass was observed under similar treatment.

Influent and effluent 5-mL samples were analyzed for CT and CF during the first 170 days of the reactor operation using a Perkin Elmer 8700 purge-and-trap gas chromatograph with a Tekmar model ALS automatic sampler and a model 1000 Hall Detector. A 105-m, 0.53-mm inner diameter (ID) wide bore capillary, 3-μm cross-bonded phenyl/methyl polysiloxane column was used. Oven temperatures of 40°C and 150°C were used at isocratic times of 10 and 5 min, respectively.

The ramp rate was 10°C/min. Helium was the carrier gas and the total sample run time was 26 min.

After 170 days of operation the CT and CF analyses were done using a Perkin Elmer Auto System gas chromatograph (GC) with an electron capture detector following a 30-m, 0.32-mm-ID wide bore capillary, RTX-1, cross-bonded 100% dimethyl polysiloxane column. The temperature program was 80°C isothermal for 4.2 min, with the injection temperature at 250°C and detector temperature at 380°C. A 2-mL liquid sample was extracted into 2-mL of pentane containing 0.05% 1,2-dibromoethane (EDB) for an internal standard. Gas samples were similarly prepared, except 4 mL of gas were used with 2 mL of the pentane/EDB carrier.

RESULTS AND DISCUSSION

A distinct change in the CT degradation capability was found after about the 220th operating day. On many of the following days, the calculated SCTDR was negative after correcting for reactor CT losses. The only significant SCTDR values after the 220th day were where the CT application rate was very high or where the reactor nitrate concentration was zero. On the 420th day of operation, a serum bottle test was performed to observe CT degradation with the fluidized-bed culture and the original culture maintained in a suspended growth reactor. The suspended-growth culture was operated with daily addition of acetate, nitrate and CT. No CT degradation was observed for the fluidized-bed culture in the presence of excess nitrate with or without acetate addition, but the suspended-growth culture had maintained its CT degradation ability.

Thus it appeared that the fluidized-bed reactor had lost its original capacity for CT degradation due to either the long-term CT feeding or some operating condition. The only changes in the reactor operation were the biomass control method and the transient testing with wide variations in CT and COD loadings. After day 180, a more vigorous cleaning method was employed and with greater frequency. About 40 days later, the CT degradation ability appeared to decrease. Although not conclusive, a proposed explanation is that the active CT-degrading bacteria were on the surface of the media and the vigorous cleaning method selectively removed them. The bacteria remaining in the pores of the media would be older and may have lost their CT-degrading capability with age. Another possibility is that another denitrifying bacteria with little CT degradation capability may have developed in the inner layers of the biofilm. The inner layers of biofilm growth would likely have had sufficient nitrate, but substrate would have been limited. Perhaps a different culture was more competitive and predominated under the low carbon levels. Based on these observations, the data collected before the 220th operating day were used to evaluate the CT degradation kinetics.

During the cleaning procedures, excess solids were collected and their mass measured. These results were used with the long-term COD removal data to calculate an observed solids yield for the system. The average observed solids yield was 0.13 g VSS/g COD removed. The average solids retention time (SRT) for the reactor, based on the total average reactor mass and solids removal rate, was 30 days. Stuart (1990) showed a higher average observed solids yield of 0.18 g VSS/g COD removed, but her suspended-growth reactor had an operating SRT of about 9 days.

CT Degradation Kinetics

CT degradation rates are evaluated as a function of COD loading and reactor CT concentration. Kinetic relationships developed are considered as only estimates, because of the apparent inaccuracies for reactor CT loss estimates, which were used to determine the biological degradation rates. The effect of COD loading on CT degradation rates was found by observing SCTDR data at COD loadings ranging from 0.1 to 1.2 g COD/g VSS-d within a narrow range of reactor CT concentrations, from 450 to 675 µg/L. A similar approach was used to evaluate the effect of CT concentration on SCTDRs, using data within a narrow range of COD loadings, from 0.9 to 1.2 g COD/g VSS-d. Figure 2 shows that higher SCTDRs were obtained as the COD loading to the reactor was increased. The range of

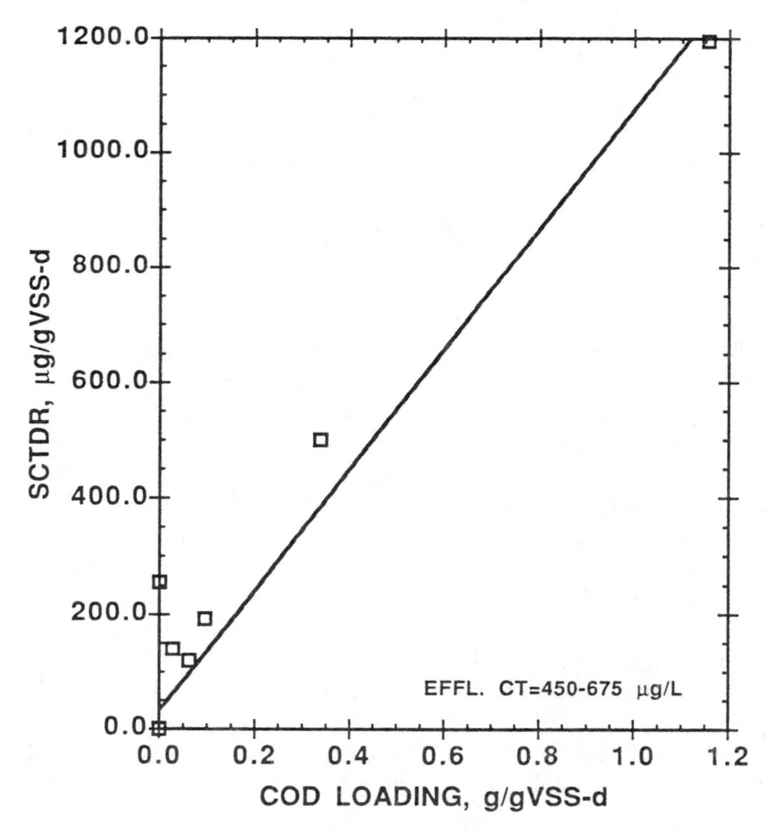

FIGURE 2. Specific CT degradation rate versus reactor-specific COD loading (reactor CT concentration, 450 to 675 µg/L).

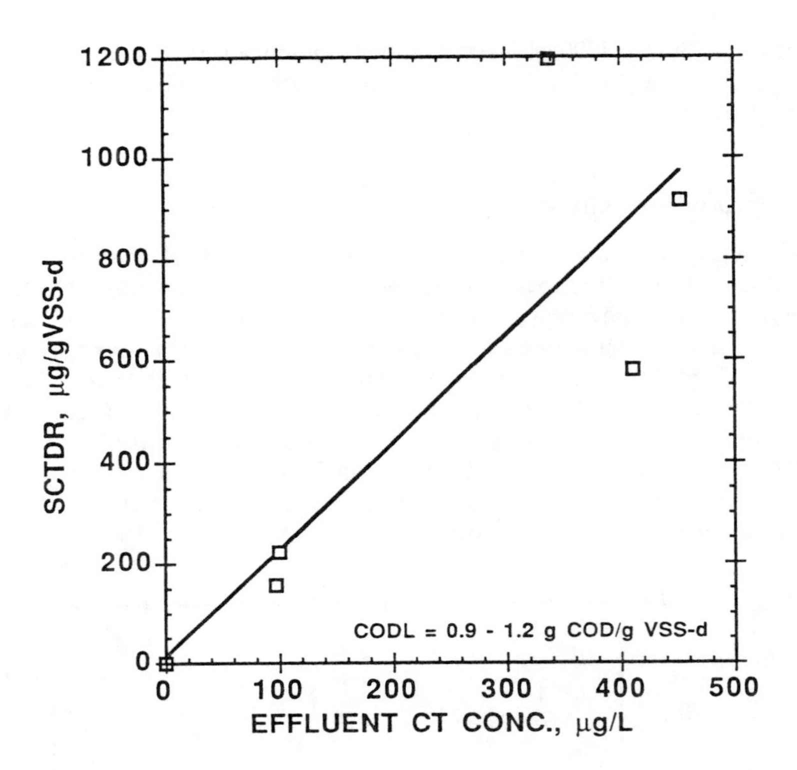

FIGURE 3. Specific CT degradation rate versus reactor CT concentration (reactor-specific COD loading, 0.9 to 1.2 g COD/g VSS-d).

SCTDRs shown encompasses the range of SCTDRs reported previously by Brouns et al. (1990) and Stuart (1990). SCTDRs also increased with increasing reactor CT concentrations (Figure 3) to indicate that both COD loading and CT concentration affect the CT degradation kinetics.

The relationship between SCTDRs, COD loadings, and CT concentrations was evaluated using least squares regressions with a statistical program, Systat (version 5.1) on a Macintosh computer. Linear and nonlinear models were evaluated. The relationships are considered significant for the fluidized-bed reactor operation for the first 180 days, prior to the change in the excess biomass removal method. The biomass removal method for the first 180 days, which consisted of removing and replacing media, is more typical of fluidized-bed operating procedures than the method used that resulted in a decrease in the CT degradation ability.

The linear model relates the SCTDR to the reactor CT concentration and the specific COD loading as follows:

$$SCTDR = K(CT)(CODL) \tag{1}$$

where SCTDR = specific CT degradation rate, μg/g VSS-d
K = model coefficient, L/g
CT = reactor CT concentration, μg/L
CODL = specific COD loading, g COD/g VSS-d

A value for K of 2.308 L/g was produced from the Systat program with an r^2 of 0.902. The nonlinear model assumes that there is a maximum COD loading where the effect of COD loading on the SCTDR is maximized and additional increases in COD loading do not affect the SCTDR. This is similar to the effect of substrate concentration on substrate removal rates described in the Michaelis-Menten equation. This model is described as follows:

$$SCTDR = K1(CT)(CODL)/(CODL + K2) \quad (2)$$

where K1 =model coefficient, L/g-d
K2 =half-velocity coefficient, g COD/g VSS-d

The least squares regression yielded a value for K1 of 31.17 L/g-d and a value for K2 of 11.8 g COD/g VSS-d with an r^2 of 0.904. The high value for K2 suggest that below specific COD loadings of about 10.0 g COD/g VSS-d the effect of CODL could be adequately described by the linear model. COD loadings well below the K2 value would be preferred to minimize the release of acetate in the reactor effluent. Thus, the linear model is adequate for estimating CT removal rates within expected operating ranges of the fluidized-bed reactor.

Effect of Limited Nitrate Conditions

On some days changes in operating conditions occurred, to result in insufficient nitrate addition relative to the COD added, and CT degradation continued at zero nitrate concentrations. Table 2 summarizes the SCTDRs for these events along with the COD loadings and CT concentrations. Days 141 and 160 are compared because the same COD loading occurred on both days and the reactor CT concentrations were close. At zero nitrate concentration the SCTDR was 12.6 times higher than with excess nitrate present. Comparison of days 165 and 179 show that the SCTDR was 7.2 times higher for the zero nitrate concentration operation when only about a 40% increase should have been expected based on changes in COD loading. The reactor CT concentration at day 180 is twice that at day 165 and the COD loading is 30% higher, but the SCTDR is 9.2 times higher. The data shown for days 220 and beyond indicate that even when no CT degradation was taking place under excess nitrate conditions, significant CT degradation occurred at zero nitrate concentrations, although not as high a rate as for the earlier SCTDRs.

The higher SCTDRs may have been related to the more reduced conditions that existed in the reactor at zero nitrate concentration. The basis for this explanation is the assumption that CT competes with nitrate for electron donors.

TABLE 2. Effect of nitrate-limited conditions on SCTDR.

Oper. Day	Reactor Nitrate Conc., mg/L	COD Loading g COD/L-d	Reactor CT Conc., µg/L	SCTDR µg/g VSS-d
141	38.0	1.9	100	150
160	0.0	1.9	60	1900
165	9.0	1.3	100	250
179	0.0	1.8	100	1800
180	0.0	1.7	200	2300
220	8.0	1.4	150	0
221	0.0	1.8	80	300
222	0.0	0.5	300	500
233	8.0	1.4	300	0
234	0.0	0.6	60	300
238	0.0	1.5	40	250

With nitrate being limited, a greater amount of electron donor is available to increase the amount of CT reduction and the more reduced conditions increase the rate of CT reduction. Other investigators (Bae & Rittmann 1990, Bouwer & Wright 1988) have shown that CT degradation rates increased with the establishment of more reduced conditions, but the change in rates were not as dramatic as these.

The amount of CF production increased at zero nitrate concentration conditions. The moles of CF produced per mole of CT degraded was about 0.20 at zero nitrate compared to typical values of less than 0.10 during the other conditions.

Reactor Design Implications

The linear model was used to evaluate a fluidized-bed reactor design. The form of the equation suggests that staged designs would provide CT degradation at reduced total reactor volumes. Other design considerations include the amount of acetate that must be added and the excess sludge production. The latter two parameters would also be important considerations for in situ treatment, because acetate and possibly nitrate would have to be added to promote CT degradation. Significant levels of solids production could plug pores and affect flow distribution in the subsurface.

The following equation was developed to describe the detention time needed to treat an influent CT concentration to a desired effluent concentration. The mass balance to develop the equation assumed the linear model to describe the specific rate of CT degradation and considers the CT lost to the gas phase during gas production as a result of nitrate reduction. Half a mole of nitrogen gas is assumed produced per mole of nitrate nitrogen reduced, and the CT is assumed to quickly reach equilibrium between the gas and liquid phase. A dimensionless Henry's constant of 1.0 is assumed for CT and a COD/N ratio of 4.0 was assumed, which was needed to determine the amount of nitrate consumed for a given reactor COD loading. The detention time is then calculated as:

$$HRT = V/Q = (CTo\text{-}CTe)/[2.308*CT*CODL*X + 0.2*CT*CODL*X] \quad (3)$$

where HRT = hydraulic detention time, d
Q = influent flow, L/d
V = reactor volume, L
CTo = influent CT concentration, µg/L
CTe = effluent CT concentration, µg/L
X = reactor VSS concentration, g/L

The second portion of the equation in the denominator represents the CT loss from the reactor due to stripping into the nitrogen gas produced. This represents 0.2/2.308 or 8.6% of the CT removed.

A reactor design evaluation using the linear kinetic model showed that four-stage system requires a total detention time of 10 hours to treat an influent CT concentration from 1.0 mg/L to an effluent CT concentration of 100 µg/L at a CODL of 0.6 g COD/g VSS-d. This compares to about 29 hours for a single-stage system at the same CODL. A lower influent concentration would result in lower required reactor detention times. If the CODL to each reactor were doubled, the required detention time would be halved.

The reactor design evaluation included an evaluation of the amount of COD consumed and solids produced per unit of CT removed. Table 3 shows that the amount of COD required and amount of solids produced per unit of CT removed can be very significant, especially to achieve low CT concentrations. Further advantages for a stage system are demonstrated, as less COD was required and less sludge production thus occurs. This analysis assumed that a constant supply of COD is needed. CT degradation with intermittent additions of acetate should

TABLE 3. COD requirements and solids production for CT degradation.

Effluent CT Conc., µg/L	mg COD/mg CD removed 1 Stage	mg COD/mg CD removed 4 Stage	mg VSS/mg CT removed 1 Stage	mg VSS/mg CT removed 4 Stage
25	26,800	4,200	3,500	550
50	13,400	3,200	1,740	420
100	6,700	2,300	870	300
200	3,400	1,700	440	220
400	1,700	1,200	220	160

be studied to determine if the CT degradation can continue after substrate addition is stopped. This would be an important consideration for in situ bioremediation.

CONCLUSIONS

A laboratory fluidized-bed reactor containing Celite media was operated at about a 7-hour detention time for a number of months under denitrifying conditions with acetate addition to observe CT degradation kinetics as a function of the reactor COD loading, reactor CT concentration, and nitrate level. CT degradation kinetics were studied under transient conditions by varying the COD loading or influent CT concentration. The following conclusions can be made from this study.

1. SCTDRs varied from 100 to 1,200 μg/g VSS-d as the CODL was increased from 0.2 to 2.0 g COD as acetate/g VSS-d.
2. A linear increase in SCTDR was also observed as the reactor CT concentration was increased.
3. The SCTDR in the fluidized-bed reactor can be predicted by a product of a constant equal to 2.308, the reactor CT concentration (μg/L), and the CODL (g/g VSS-d).
4. Operation of the fluidized-bed at zero nitrate concentrations with acetate and CT addition resulted in increases in SCTDRs that were 7 to 12 times higher than SCTDRs with similar CT concentrations and with excess nitrate available.
5. After about 220 days of operation the CT degradation capacity of the reactor was greatly reduced and may have been a result of the method used after 180 days to remove excess biomass.
6. Chloroform production was normally less than 10% of the CT degraded on a molar basis with excess nitrate, but was close to 20% under nitrate-limiting conditions.
7. Application of the CT degradation kinetic model showed that a multiple-staged system greatly reduces total reactor detention time, the amount of COD needed per unit of CT degraded, and the amount of solids produced compared to a single-stage reactor.
8. The amount of COD required per unit of CT degraded can range from 1,200 to 26,800 mg/mg with higher amounts required at low effluent CT concentrations. Similarly, the sludge production may vary from 160 to 3,500 mg/mg.

ACKNOWLEDGMENTS

The laboratory experimental work was supported by the Battelle Pacific Northwest Laboratories and partial graduate student support for this work was received by the University of Washington Valle Scandinavian Scholarship Program.

Special acknowledgement is given to Tom Brouns and Sheryl Stuart for their technical assistance on the initial phase of this research.

REFERENCES

American Public Health Association. 1985. *Standard Methods for the Examination of Water and Wastewater,* 15th ed. APHA, New York, NY.

Bae, W., and B. E. Rittmann. 1990. "Effects of Electron Acceptor and Electron Donor on Biodegradation of CCl_4 by Biofilms." In *Proc. of the 1990 American Society of Civil Engineers National Conference on Environmental Engineering,* pp. 53-68. ASCE, New York, NY.

Bouwer, E. J., and P. L. McCarty. 1983a. "Transformation of 1- and 2-Carbon Halogenated Aliphatic Organic Compounds under Methanogenic Conditions." *Applied Environ. Microbiol. 5*: 1286-1294.

Bouwer, E. J., and P. L. McCarty. 1983b. "Transformations of Halogenated Organic Compounds Under Denitrification Conditions." *Applied Environ. Microbiol. 45*: 1295-1299.

Bouwer, E. J., and J. P. Wright. 1988. "Transformations of Trace Halogenated Aliphatics in Anoxic Biofilm Columns." *J. Cont. Hydrol.* 2: 155-169.

Brouns, T. M., S. S. Koegler, W. O. Heath, J. K. Fredrickson, H. D. Stensel, D. L. Johnstone, and T. L. Donaldson. 1990. *Development of a Biological Treatment System for Hanford Groundwater Remediation FY 1989 Status Report.* Battelle Pacific Northwest Laboratories, Richland, WA.

Cobb G. D., and E. J. Bouwer. 1991. "Effects of Electron acceptors on Halogenated Organic Compound Biotransformation in a Biofilm Column." *Environ. Sci. Technol. 25:* 1068-1074.

Criddle, C. S., J. T. DeWitt, D. Grbić-Galić, and P. L. McCarty. 1990a. "Transformation of Carbon Tetrachloride by *Pseudomonas* sp. Strain KC Under Denitrification Conditions." *Applied Environ. Microbiol. 56:* 3240-3246.

Criddle, C. S., J. T. DeWitt, and P. L. McCarty. 1990b. "Reductive Dehalogenation of Carbon Tetrachloride by *Escherichia coli* K-12." *Applied Environ. Microbiol. 56*: 3247-3254.

Egli, C., R. Scholtz, A. M. Cook, and T. Leisinger. 1987. "Anaerobic Dechlorination of Tetrachloromethane and 1,2-Dichloroethane to Degradable Products by Pure Cultures of *Desulfobacterium* sp. and *Methanobacterium* sp." *FEMS Microbiology Letters 43*: 257-261.

Egli, C., T. Tschan, R. Scholtz, A. M. Cook, and T. Leisinger. 1988. "Transformation of Tetrachloromethane and Carbon Dioxide by *Acetobacterium woodii.*" *Applied Environ. Microbiol. 54*: 2819-2824.

Gommers, P. J., W. Bijleveld, and J. G. Kuenen. 1988. "Simultaneous Sulfide and Acetate Oxidation in a Denitrifying Fluidized Bed Reactor-I." *Water Research* 22: 1075-1083.

Last, G. V., R. J. Lenhard, B. N. Bjornstad, J. C. Evans, K. R. Roberson, F. A. Spanea, J. E. Amonette, and M. L. Rochhold. 1991. *Characteristics of the Volatile Organic Compound-Arid Integrated Demonstration Site.* Battelle Pacific Northwest Laboratories, Richland, WA.

Mikesell, M. D., and S. A. Boyd. 1990. "Dechlorination of Chloroform by *Methanosarcina* Strains." *Applied Environ. Microbiol. 56*: 1198-1201.

Randall, C. W., J. L. Barnard, and H. D. Stensel. 1992. *Design and Retrofit of Wastewater Treatment Plants for Biological Nutrient Removal.* Technomic Publishing Co., Inc., Lancaster, PA.

Semprini, L., G. D. Hopkins, D. B. Janssen, M. Lang, P. V. Roberts, and P. L. McCarty. 1991. *In-Situ Biotransformation of Carbon Tetrachloride Under Anoxic Conditions.* Final Report CR 815816. Robert S. Kerr Environmental Research Laboratory, U.S. Environmental Protection Agency, Ada, OK.

Stuart, S. L. 1990. "Biodegradation of Carbon Tetrachloride by Two Mixed Denitrifying Cultures." Masters Thesis, School of Engineering, University of Washington, Seattle, WA.

Westrick, J. J., W. Mello, and R. F. Thomas. 1984. "The Groundwater Supply Survey." *Jour. Amer. Water Works Assoc.* May, pp. 52-59.

COMPARATIVE EFFICIENCY OF MICROBIAL SYSTEMS FOR DESTROYING CARBON TETRACHLORIDE CONTAMINATION IN HANFORD GROUNDWATER

M. J. Truex, R. S. Skeen, S. M. Caley, and D. J. Workman

ABSTRACT

Past waste disposal practices at the U.S. Department of Energy's (DOE) Hanford site have resulted in carbon tetrachloride and nitrate contamination in the groundwater. In situ bioremediation is currently being investigated as a cost-effective means to destroy these groundwater contaminants. A key factor in the cost effectiveness of the process is the nutrient amendments required to stimulate microbial destruction of the contaminants. Current and previous research has focused on determining the reaction kinetics and microbial processes for carbon tetrachloride destruction using acetate as the electron donor and the indigenous microbes. In this study, kinetic experiments were conducted and a first-order model was used to compare the rate of carbon tetrachloride destruction and biomass production stimulated using acetate, ethanol, glycerol, and methanol as substrates under denitrification conditions. All of the substrates stimulated comparable carbon tetrachloride destruction and biomass production. Glycerol treatments exhibited a unique biphasic pattern of carbon tetrachloride destruction. In addition, it was determined that the rate of carbon tetrachloride destruction was first order with respect to carbon tetrachloride concentration.

INTRODUCTION

Past waste disposal practices at the U.S. Department of Energy's (DOE) Hanford site have resulted in carbon tetrachloride (CCl_4) and nitrate contamination in the groundwater. In situ bioremediation is currently being investigated as a cost-effective means to destroy these groundwater contaminants. The cost effectiveness of bioremediation is significantly influenced by the nutrient amendments required to sustain the contaminant destruction reactions (Skeen et al. 1992b, Truex et al. 1992). This is particularly important for bioremediation of CCl_4 because its biodestruction is the result of a cometabolic process. Nutrient amendments are also important in controlling the growth characteristics of the bacteria to prevent biofouling. Current and previous research has focused on determining the reaction kinetics and microbial processes for CCl_4 destruction using acetate as the electron donor for indigenous microbes (Brouns et al. 1990, Skeen et al. 1992a, Semprini et al. 1991). This study was conducted to determine if electron donors other than acetate may be more cost effective, or may provide a better means of process control, for in situ bioremediation of CCl_4 contamination. Three alternative electron donors – glycerol, methanol, and ethanol – were screened for their ability to stimulate CCl_4 destruction. Detailed reaction kinetic experiments using an indigenous microbial consortium with these substrates and with acetate were conducted to determine the efficiency of each in destroying CCl_4.

MATERIALS AND METHODS

A bacterial consortium isolated from aquifer sediments at the DOE's Hanford site was stored in 50% glycerol at −70°C as a stock cell culture. The growth medium used for all cell culturing and kinetic experiments was a simulated groundwater (SGW) (Skeen et al. 1992a) amended with the appropriate electron donor and acceptor for the specific experiment. Cell culture and inoculation protocol were as described by Skeen et al. (1992a). Kinetic experiments (duplicated under similar conditions) were conducted using acetate, methanol, ethanol, and glycerol as electron donors at an initial concentration of 0.1 mole of carbon/L. Nitrate was provided at an initial concentration that was not limiting for the duration of the experiments. Abiotic controls were performed to assay for loss of CCl_4 due to the experimental procedure.

Experiments were conducted in a gastight, 1-L reaction vessel designed to allow samples to be removed from the reactor with negligible losses of CCl_4 (Skeen et al. 1992c). Experimental procedures were identical to those used by Skeen et al. (1992a). Briefly, the vessel was sterilized in an autoclave; pressure tested to 10 psi to ensure there were no gas leaks; and then charged with SGW, the specific electron donor, and CCl_4. Carbon tetrachloride was added in a saturated water solution to produce a nominal aqueous concentration between 300 µg/L and 4,500 µg/L. Based on preliminary experiments, a 3- to 4-hour lag time before injection of the bacterial inoculum was used to allow the CCl_4 to reach equilibrium partitioning between the gas and aqueous phases. Bacteria were inoculated at a nominal biomass concentration of 25 mg-dry weight/L as measured by total suspended solids. The starting aqueous volume in the reactor was 750 mL, and the pH of the reactor was buffered at 7.0. The reactor was operated as a mixed batch reactor in a constant temperature bath (17°C) under denitrification conditions.

Periodic samples were withdrawn to determine the aqueous-phase concentration of the solution anions, biomass, and CCl_4. Carbon tetrachloride was extracted into hexane and measured using gas chromatography. Solution anion concentrations were measured from filtered aqueous samples using ion chromatography. Biomass was measured spectrophotometrically and correlated to total suspended solids (mg-dry weight/L) using a standard curve. Values for kinetic parameters that could not be estimated directly from experimental data were numerically determined using the software Simusolv (Dow Chemical Company, Midland, Michigan).

RESULTS AND DISCUSSION

Denitrification rates were not significantly different for any of the treatments. In addition, each treatment demonstrated greater destruction of CCl_4 than the losses measured in abiotic controls. For comparison purposes, the biomass production and CCl_4 destruction rates for each treatment were quantified by numerically fitting the experimental data to the rate expressions described by equations 1 and 2. In these equations, X is the biomass concentration (mg-dry weight/L), CT is the CCl_4 concentration (µg/L), and K_{CT} (hr^{-1} (mg-dry weight/L)$^{-1}$) and K_X (hr^{-1}) are rate constants.

$$d(X)/dt = K_x(X) \quad (1)$$

$$d(CT)/dt = -K_{CT}(X)(CT) \quad (2)$$

During the fitting procedure, the amount of CCl_4 that partitioned into the headspace of the reactor was calculated using a Henry's law constant of 0.856 (mg/L gas phase)/(mg/L aqueous phase) (Gossett 1987) to account for changes in the headspace volume due to sampling and assuming instantaneous equilibrium.

Cell growth rate was assumed to be first order in cell numbers (biomass). For the CCl_4 destruction reaction, the rate of CCl_4 destruction was estimated to be first order with respect to CCl_4 concentration based on an initial rate analysis of the two experiments using acetate as the electron donor, and two additional experiments using acetate performed at different initial concentrations of CCl_4. These experiments provided CCl_4 data over a range from 300 µg/L to 4,500 µg/L. In this analysis, the natural logarithm of the initial rate of change in CCl_4 concentration, normalized by the initial biomass concentration, is plotted versus the natural logarithm of the initial CCl_4 concentration. As shown in equation 3, the slope of the resulting line indicates the order of the CCl_4 destruction rate.

$$\ln[-d(CT)/dt(1/X)] = \ln(K_{CT}) + n[\ln(CT)] \quad (3)$$

Equation 3 was obtained by evaluating equation 2 at a specific instant in time and by assuming that the reaction depends on CCl_4 concentration to the nth power. This method for determining the power dependency can be applied with the assumption that the biomass and CCl_4 concentration are constant over the time period used to estimate the initial rate. The initial CCl_4 destruction rate was estimated by linear regression of the data for the first day of incubation. During this time period, the biomass and CCl_4 concentrations changed an average of 35% and 13%, respectively. The small variation in CCl_4 concentration was acceptable for the assumption of constant CCl_4 concentration. Although the assumption of constant biomass does not appear to be entirely valid, the initial rate analysis results were relatively insensitive to changes in biomass concentration over the observed range. This is indicated by the similar power dependencies calculated when the minimum (n=1.07), average (n=1.18), and maximum (n=1.25) values of the first-day biomass concentrations were used in the initial rate analysis. The r^2 values for the linear regression of data for equation 3 used to generate these n values were 0.86, 0.87, and 0.92, respectively. From these three estimates of the power dependency, the assumption that the rate of CCl_4 destruction is first order with respect to CCl_4 concentration is justified.

Reaction constants for CCl_4 destruction were comparable for the acetate, methanol, and ethanol treatments (Table 1). The biomass production constant, which is a measure of the exponential growth rate, was also similar for acetate, ethanol, and methanol treatments. The time-phase pattern of contaminant destruction for the duration of the experiment was similar for all treatments except for glycerol. For this reason, equations 1 and 2 did not accurately describe the experimental data for glycerol and the results are not shown in Table 1. Two additional experiments were performed with glycerol as the substrate and similar results were obtained. Three of the four glycerol treatments exhibited a biphasic pattern where greater than 80% of the CCl_4 destruction occurred in the first 30 to 40 hours of the experiment (Figure 1A). Denitrification and biomass production, however, exhibited the opposite trend (Figure 1B). In treatments with the other substrates, the time-phase pattern of CCl_4 destruction followed a first-order decrease and the lag phase before exponential microbial growth was much shorter than the lag phase for glycerol treatments.

Because glycerol can be used as a substrate for fermentation, the initial CCl_4 destruction may have occurred during fermentation reactions that took place before vigorous denitrification began. Potentially fumarate, a TCA cycle intermediate, may have been the electron acceptor supporting the fermentation of glycerol. Significant CCl_4 destruction previously has been demonstrated in batch experiments amended with glycerol and fumarate, while treatments conducted with glucose fermentation (glycolysis) did not promote significant CCl_4 destruction (J. K. Fredrickson, unpublished data). Other mechanisms may have been responsible for the biphasic pattern of CCl_4 destruction in the glycerol treatments. However, these mechanisms have not been fully investigated.

TABLE 1. Reaction constants for a first-order model of CCl_4 destruction (K_{CT}) and biomass production (K_X) in kinetic experiments.

Treatment	K_{CT} ($\times 10^{-4}$)	%VE	K_X ($\times 10^{-3}$)	%VE
Acetate	1.2 ± 0.4	57, 90	9.6 ± 6.1	89, 99
Methanol	1.3 ± 0.8	79, 45	7.4 ± 3.1	96, 71
Ethanol	1.0 ± 0.7	66, 66	10 ± 11	76, 98

%VE = the percent of data variation explained by the first-order model (first experiment, second experiment).

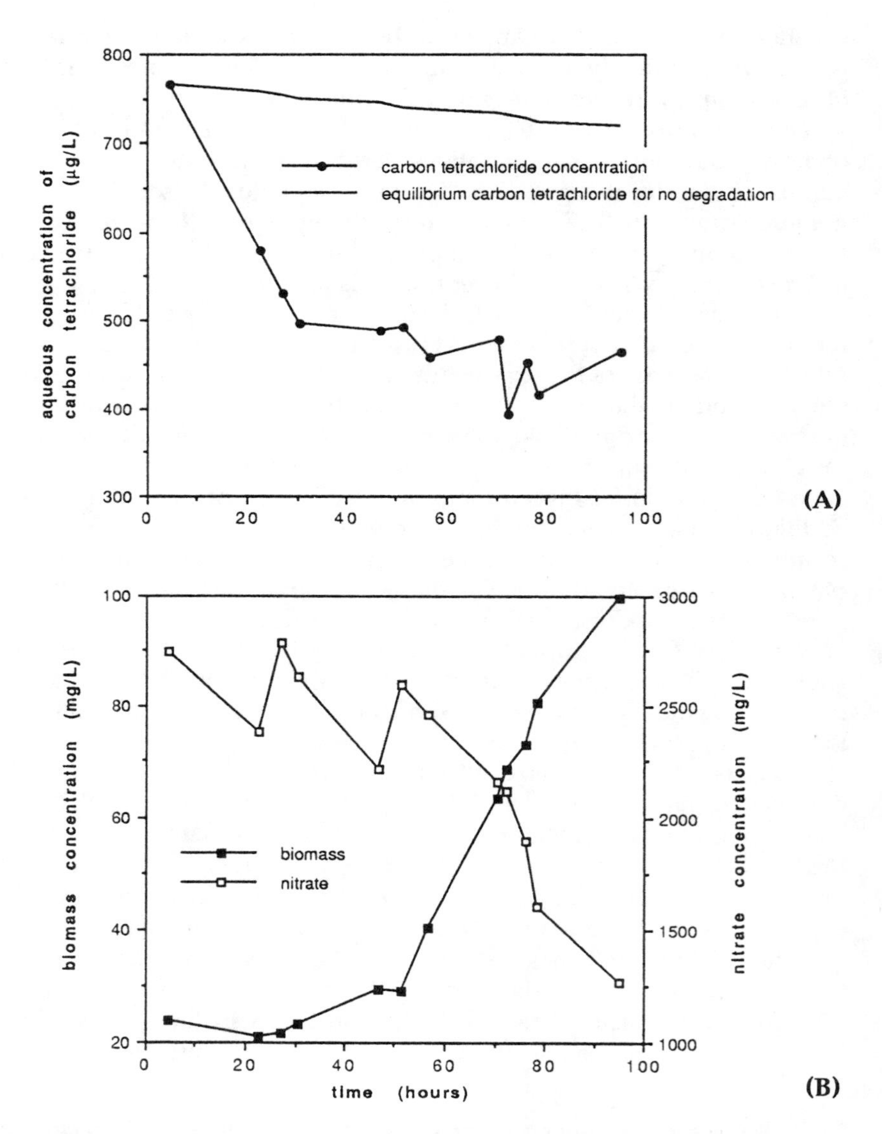

FIGURE 1. Typical time-phase pattern of CCl_4 destruction (A), denitrification, and biomass production (B) in batch kinetic experiments using glycerol as the electron donor.

The impact of these results for in situ bioremediation of CCl_4 may be important relative to cost effectiveness, optimization, and control of the process. Each of these compounds may be potentially useful for stimulating in situ bioremediation of CCl_4 because each induced similar rates of CCl_4 destruction and microbial growth. Additional information on regulatory constraints, relative cost per unit mass of CCl_4 destroyed, and the microbial growth response in a porous medium will determine which substrate will be most cost effective for application at the Hanford site. Experiments to evaluate these additional cost and application issues are currently under way. Glycerol, in particular, may offer significant advantages because nitrate may not be required to stimulate CCl_4 destruction. Use of a substrate such as glycerol, where both denitrification and fermentation reactions can support CCl_4 destruction, would greatly reduce the amount of nitrate required for contaminant remediation at the Hanford site, and consequently, may result in an improved cost effectiveness over systems using substrates that require denitrification conditions. Optimization of nutrient conditions may result in relatively high CCl_4 destruction rates and low denitrification rates using glycerol as an electron donor.

ACKNOWLEDGMENTS

This work was supported by the U.S. Department of Energy, Office of Technology Development under Contract DE-AC06-76RLO 1830 as part of the VOC-Arid Site Integrated Demonstration. Pacific Northwest Laboratory is operated by Battelle Memorial Institute for the U.S. Department of Energy under contract DE-AC06-76RLO 1830.

REFERENCES

Brouns, T. M., S. S. Koegler, W. O. Heath, J. K. Fredrickson, H. D. Stensel, D. L. Johnstone, and T. L. Donaldson. 1990. *Development of a Biological Treatment System for Hanford Groundwater Remediation: FY 1989 Status Report*. PNL-7290, Pacific Northwest Laboratory, Richland, WA.

Gossett, J. M. 1987. "Measurement of Henry's Law Constants for C_1 and C_2 Chlorinated Hydrocarbons." *Environ. Sci. Technol.* 21:202-208.

Semprini, L., G. D. Hopkins, D. B. Janssen, M. Lang, P. V. Roberts, and P. L. McCarty. 1991. *In-Situ Biotransformation of Carbon Tetrachloride Under Anoxic Conditions*. EPA/2-90/060. U.S. Environmental Protection Agency, Ada, OK.

Skeen, R. S., K. M. Amos, M. Shouche, and J. N. Petersen. 1992a. "Degradation of Carbon Tetrachloride by a Denitrifying Microbial Consortium." *Appl. Environ. Microbiol.* (submitted for publication).

Skeen, R. S., S. P. Lutrell, T. M. Brouns, B. S. Hooker, and J. N. Petersen. 1992b. "In Situ Bioremediation of Hanford Groundwater." *Remediation* (accepted for publication).

Skeen, R. S., J. N. Petersen, and M. J. Truex. 1992c. "A Batch Reactor for Monitoring Biodegradation of Volatile Organics." *Appl. Environ. Microbiol.* (submitted for publication).

Truex, M. J., D. R. Brown, and D. B. Elliott. 1992. *Cost/Benefit Analysis Comparing Ex Situ Treatment Technologies for Removing Carbon Tetrachloride from Hanford Ground Water.* PNL 8334, Pacific Northwest Laboratory, Richland, WA.

ANAEROBIC BIOTRANSFORMATION AND TRANSPORT OF CHLORINATED HYDROCARBONS IN GROUNDWATER

M. A. Hossain and M. Y. Corapcioglu

ABSTRACT

In situ biorestoration is a remediation technique in which the indigenous aquifer bacteria are stimulated by injecting compounds to provide food and energy. Stimulated bacteria transform the target contaminants such as tetrachloroethylene (PCE) and trichloroethylene (TCE) into secondary substrates. In this study, we develop a model to simulate the substrate-limited biotransformation of the halogenated solvents present in anoxic groundwater by sequential reductive dehalogenation under methanogenic conditions. The model consists of conservation of mass equations for the primary substrate; immobile indigenous biomass; organic solvents such as PCE and TCE; and their intermediate products trichloroethylene, dichloroethylene (DCE), and vinyl chloride (VC). The limiting factor on bacterial growth is assumed to be the primary substrate. The microbial yield coefficient is determined from the stoichiometric equation describing the anaerobic process. Numerical solutions are presented for a practical problem. The general model is then simplified by assuming an abundance of primary substrate and constant population of biomass. The simplified model is applied to simulate the biotransformation of PCE and TCE along an anaerobic fixed-film column. The numerical results show a favorable match with experimental data reported in the literature.

INTRODUCTION

Chlorinated hydrocarbons such as PCE, TCE, and 1,1,1-trichloroethane (TCA) enter soils and groundwater through accidents and leaks at chemical disposal sites. These chlorinated hydrocarbons, which are halogenated aliphatic compounds, are commonly found in solvents, cleaning agents, polyvinyl chloride (PVC) pipes, and dry-cleaning substances. Among them, TCE is the most frequently detected compound in contaminated groundwaters and superfund sites. It is estimated that by 1989, 22 million kilograms of TCE had entered the environment in the United States. Contamination of groundwater by chlorinated solvents is one of the most difficult problems to treat. Within the last decade, researchers have shown that bioremediation can be used as a decontamination technique at the site (Wilson et al. 1986, Criddle et al. 1991, McCarty et al. 1991).

In situ biotransformation with subsequent mineralization is a potentially promising method due to cost effectiveness and minimum disturbance of the aquifer. Indigenous aquifer bacteria are stimulated to biodegrade dissolved hydrocarbon contaminants in groundwater. Stimulation can be achieved by injecting compounds to provide food and energy to microorganisms under either aerobic or anaerobic conditions. Stimulated bacteria would then transform the target contaminants such as PCE and TCE as secondary substrates. Under aerobic conditions, bacteria known as methanotrophs provided with methane and oxygen transform organic solvents through cometabolic processes (McCarty et al. 1991).

Cometabolic transformation of a nongrowth substrate by methanotrophs in the presence of an electron donor, methane, and an electron acceptor, oxygen, has been simulated by Semprini and McCarty (1992). Halogenated solvents present in oxygenated groundwater also can be transformed by sequential reductive dehalogenation under methanogenic conditions. For example, studies by Parsons and Barrio-Lage (1985) and Vogel and McCarty (1985) indicate the biotransformation pathway of PCE to VC. Furthermore, Vogel and McCarty (1985) have demonstrated the partial mineralization of VC. As noted by Bouwer and Cobb (1986), anoxic conditions prevail when the aerobic conditions cease due to depletion of oxygen. Under anoxic conditions, denitrification by using nitrate, sulfate restoration by using sulfate, and methanogenesis by using CO_2 as electron acceptors sequentially follow. The primary substrate to be provided as an electron donor is acetate or some other nontoxic growth compound.

Estimating the strength of primary substrate, as well as concentration of secondary substrate and intermediates, is an important aspect of efficient aquifer recovery operations that require the ability to predict the transport of contaminants under a variety of conditions. Predictions of this type frequently are facilitated by using a mathematical model designed to represent the physical system under consideration in a simplified, but meaningful, function. Simulation of a site-specific bioremediation effort provides a relatively inexpensive and convenient tool to test the efficiency of a biorestoration strategy. The result of such an effort would assist in the design and evaluation various remedial schemes and the estimation of contaminant plume migration (Corapcioglu & Hossain 1990a).

Corapcioglu and Hossain (1990b) reviewed various modeling techniques employed to quantify anaerobic and aerobic biodegradation processes. Later, Corapcioglu et al. (1991) developed a mathematical model for transport and methanogenic biotransformation of chlorinated hydrocarbons in groundwater. In this study, we develop a model to simulate the substrate-limited biotransformation of the halogenated solvents present in anoxic groundwater by sequential reductive dehalogenation under methanogenic conditions.

MODEL DEVELOPMENT

The system under consideration consists of (1) one or more dissolved solvents that are transformed as secondary substrates by sequential reductive dehalogenation, (2) stimulated indigenous biomass to transform secondary substrates, and (3) primary substrate injected to stimulate the growth of these natural bacteria. A modeling effort should include conservation of mass equations of the primary substrate, biomass, secondary substrate(s), and intermediates.

Our starting point is the conservation of mass equation for the primary substrate in a rigid saturated porous medium. In a one-dimensional space, this equation can be stated as

$$n\frac{\partial P}{\partial t} + q_w\frac{\partial P}{\partial x} - D\frac{\partial^2 P}{\partial x^2} = \frac{P}{K_s^P + P}\mu_{max}\, nB \quad (1)$$

where P is the molar concentration of the primary substrate (moles per unit volume of water), n is the porosity, q_w is the volumetric flow rate of groundwater, D is the hydrodynamic dispersion coefficient, μ_{max} is the maximum rate of primary substrate use (moles of primary substrate per mole of bacteria per unit time), K_s^P is the half-saturation constant of the primary substrate, B is the molar concentration of biomass (moles of bacteria per unit volume of water), t is the time, and x is the space variable. In Eq. (1), we assume that the rate of microbial growth is a function of the growth substrate. Then, for an immobile biomass, the mass balance equation can be stated as (Corapcioglu & Haridas 1985)

$$n\frac{\partial B}{\partial t} = Y\frac{P}{K_s^P + P}\mu_{max}nB - K_dBn + K_cY_cC_dn \quad (2)$$

where Y is the microbial yield per unit mole of substrate consumed (moles of bacteria per unit mole of primary substrate), K_d is the first-order microbial decay coefficient, K_c is the first-order decay rate of natural organic carbon, Y_c is the microbial yield per unit mole of organic carbon, and C_d is the molar concentration of natural organic carbon. The last term in Eq. (2) represents the maintenance of biomass at survival mode under undisturbed natural conditions. In this case, the growth of bacteria by using natural organic carbon would be equal to the decay of biomass. Eq. (2) assumes that the bacteria grow on both natural organic carbon and alien primary substrate injected for stimulation.

In this study, the mobility of bacteria is neglected. Corapcioglu and Haridas (1984, 1985) proposed a mathematical model for a mobile biomass stimulated by a single substrate. They considered mobile bacteria suspended in pore water as well as captured on soil grains. The microbial yield coefficient Y is determined from the stoichiometric equation describing anaerobic substrate use. Assuming a constant cellular composition of bacteria, a generalized stoichiometric equation for primary substrate use under methanogenic conditions is of the form

$$\begin{aligned} n_1CH_3COO^- + n_2NH_4^+ + n_3H_2O + n_4CO_2 \rightarrow \\ m_1C_5H_7O_2N + m_2CH_4 + m_3HCO_3 \end{aligned} \quad (3)$$

where CH_3COO^- is acetate, NH_4^+ is ammonium, $C_5H_7O_2N$ is the microbial cell, and HCO_3 is bicarbonate. Bouwer and Cobb (1986) balanced Eq. (3) and obtained

n_1=1.0, n_2=0.019, n_3=0.952, n_4=0.024, m_1=0.019, m_2=0.954, and m_3=0.984. Based on these values the value of Y (=m_1/n_1) is equal to 0.019 mole of bacteria per unit mole of acetate.

Thus, the conservation of mass equation for an organic solvent is represented by (Corapcioglu et al. 1991)

$$n\frac{\partial C_1}{\partial t} + q_w\frac{\partial C_1}{\partial x} - D_1\frac{\partial^2 C_1}{\partial x^2} = -\frac{\mu_{max}}{K_s^1 + C_1} nC_1 \frac{P}{K_s^P + P} B\alpha_1 \tag{4}$$

where C_1 is the molar concentration of organic solvent, D_1 is the hydrodynamic dispersion coefficient, K_s^1 is the half-saturation constant for the organic solvent, and α_1 is the molar ratio of parent compound biotransformed to primary substrate consumed. One can define α_1 as the ratio of maximum organic biotransformation rate to maximum rate of primary substrate use. Intermediate products biotransformed from the parent solvent can be formulated sequentially by respective mass balance equations. At this point, let us assume that the reductive dehalogenation pathway includes three intermediate compounds. An example of this type of one-parent-product, three-intermediate-product pathway would be the biotransformation of PCE (PCE → TCE → DCE → VC). In this case, the mass balance equations for the intermediate compounds can be stated as

$$n\frac{\partial C_2}{\partial t} + q_w\frac{\partial C_2}{\partial x} - D_2\frac{\partial^2 C_2}{\partial x^2} = \left(-\frac{\alpha_2}{K_s^2 + C_2}C_2 + \frac{\alpha_1}{K_s^1 + C_1}C_1\right)\mu_{max} nB\frac{P}{K_s^P + P} \tag{5}$$

$$n\frac{\partial C_3}{\partial t} + q_w\frac{\partial C_3}{\partial x} - D_3\frac{\partial^2 C_3}{\partial x^2} = \left(-\frac{\alpha_3}{K_s^3 + C_3}C_3 + \frac{\alpha_2}{K_s^2 + C_2}C_2\right)\mu_{max} nB\frac{P}{K_s^P + P} \tag{6}$$

$$n\frac{\partial C_4}{\partial t} + q_w\frac{\partial C_4}{\partial x} - D_4\frac{\partial^2 C_4}{\partial x^2} = \left(-\frac{\alpha_4}{K_s^4 + C_4}C_4 + \frac{\alpha_3}{K_s^3 + C_3}C_3\right)\mu_{max} nB\frac{P}{K_s^P + P} \tag{7}$$

where C_2, C_3, and C_4 are the respective molar concentrations of the intermediate products; D_2, D_3, and D_4 are the respective hydrodynamic dispersion coefficients; and α_2, α_3, and α_4 are respective ratios of product biotransformed to primary substrate used. As noted earlier, these ratios can be defined as the ratios of the respective maximum rate of primary substrate use. K^1_s, K^2_s, K^3_s, and K^4_s are the half-saturation constants for parent and intermediate products. Corapcioglu and Hossain (1991) presented a general technique to estimate the biokinetic parameters of sequential multi-intermediate systems from linear plots of batch data and verified their technique by using the biotransformation data of carbon tetrachloride.

APPLICATION OF THE MODEL

The model has been solved by a finite difference solution technique and applied to study in situ transformation of target compounds by indigenous soil microorganisms. A scenario has been considered to illustrate the applicability of the proposed model. The primary substrate such as acetate is injected into an aquifer contaminated by two target compounds to provide food and energy to microorganisms. We assume that halogenated solvents such as PCE and TCE present in anoxic groundwater will be transformed by sequential reductive dehalogenation under methanogenic conditions.

We investigate a scenario in which the aquifer is contaminated by the organic solvents, PCE and TCE. Initial biomass concentration is estimated as 7.29×10^{-7} mole/L. A primary substrate, i.e., acetate, is injected at x=0 at a concentration of 0.166 mole/L as a single-pulse input at t=0. Because it is extremely difficult to reach the solubility level concentrations in groundwater due to various factors such as hardness and pH, this value was taken as one-third of the solubility of acetate in pure water under ideal conditions. The biotransformation pathway can be written as

$$\underset{(CCl_2 = CCl_2)}{\text{PCE}} \rightarrow \underset{(CHCl{=}CCl_2)}{\text{TCE}} \rightarrow \underset{(CHCl{=}CHCl)}{\text{DCE}} \rightarrow \underset{(CH_2{=}CHCl)}{\text{VC}}$$

In this case, subscripts 1, 2, 3, and 4 in Eqs. (4) through (7) denote the PCE, TCE, DCE, and VC, respectively. Model parameters and other constants are given in Table 1. A constant flow velocity is assumed throughout a 3-m saturated column. Parameters employed in the simulations are within realizable limits and are consistent with laboratory (Vogel & McCarty 1985) and experimental (Semprini et al. 1990) values reported in the literature. The initial and boundary conditions of the problem can be stated as

$$\begin{aligned} &\text{at } t = 0, \quad B = 7.29 \times 10^{-7} \text{ mole/L;} \\ &\text{at } P = 0.0, \quad C_1 = 90 \times 10^{-6} \text{ mole/L;} \\ &\text{at } C_2 = 8{,}300 \times 10^{-6} \text{ mole/L} \end{aligned} \tag{8}$$

$$\text{at } x = 0 \text{ and } t = 0, \quad P = 0.166 \text{ mole/L} \tag{9}$$

The migration of primary substrate peak and the growth of biomass along the column are shown in Figures 1 and 2, respectively. Figure 1 illustrates that the peak of primary substrate gradually decreases from 48 hours to 144 hours. The change in porosity due to microbial growth at 144 hours is calculated as 6×10^{-4} cm^3/cm^3 $\sigma = Bn/\rho$, where σ is the volume of biomass per unit total volume and ρ is the density of the biomass which can be taken as 1 gm/cm^3. This indicates minimal clogging of the pores due to microbial growth. Similar to Eq. (2), this equation also neglects microbial mobility and assumes that biomass is totally attached to the soil grains without any microbial cells in suspension.

Biodegradation of PCE is illustrated in Figure 3. Disappearance of PCE is faster than that of TCE due to the initial presence of TCE in addition to the generated compound. Figure 4 indicates the accumulation of DCE at a much lower concentration level than the parent compound(s). However, as noted by various

TABLE 1. Model parameters.

Molar Ratio of Organic Biotransformed to Primary Substrate	Half-Saturation Constants
$\alpha_1 = 0.045$	$K_s^1 = 0.00015$ mole/L
$\alpha_2 = 0.06$	$K_s^2 = 0.00022$ mole/L
$\alpha_3 = 0.1$	$K_s^3 = 0.00460$ mole/L
$\alpha_4 = 2.55 \times 10^{-4}$	$K_s^4 = 0.00110$ mole/L
$\mu_{max} = 35.96$ mole/L/d	$K_s^P = 0.00041$ mole/L

Mechanical dispersion coefficient, D=0.033 cm^2/d
Flowrate, $q_w = 3.0$ cm/d
Time step, $\Delta t = 0.05$ d
Space interval, $\Delta x = 10$ cm
Porosity, n = 0.19
Column length, l = 300 cm
Injected acetate concentration = 0.166 mole/L
Initial concentration of PCE = 90×10^{-6} mole/L
Initial concentration of TCE = $8{,}300 \times 10^{-6}$ mole/L
Molecular weight of PCE = 166.0 g
Molecular weight of TCE = 131.5 g
Molecular weight of DCE = 97.0 g
Molecular weight of VC = 62.5 g
Molecular weight of acetate = 60.0 g
Molecular weight of bacteria = 137.0 g
Initial biomass concentration = 7.29×10^{-7} mole/L
Yield coefficient = Y = 0.019 mole of bacteria/mole of acetate
$Y_c C_d = 1{,}000$ mg/L
Decay rate = $K_d = 1 \times 10^{-2}$/d

researchers (e.g., Wilson et al. 1986), DCE and VC accumulation is an undesirable end-product of such an operation. Despite its occurrence at low levels, VC is known to be more toxic and carcinogenic than either PCE or TCE. This is one aspect that requires further investigation for successful mineralization of VC. One possibility is the further reductive dehalogenation of VC to ethylene ($CH_2=CH_2$) and then to ethane (C_2H_6) (De Bruin et al. 1991). Another alternative is to prevent the formation of VC by sulfate inhibition as reported by Kästner (1991).

The effect of primary substrate concentration on biodegradation of target compounds is investigated by assigning different magnitudes of initial pulse injection. The original value P=0.166 mole/L was increased to 0.298 mole/L and then decreased to 0.083 mole/L. The remaining percentage of total initial mass of PCE plus TCE is plotted in Figure 5 as a function of time for different primary substrate concentration values. The difference is around 3% each way 140 hours after the injection of acetate. However, the difference is slightly higher at earlier times.

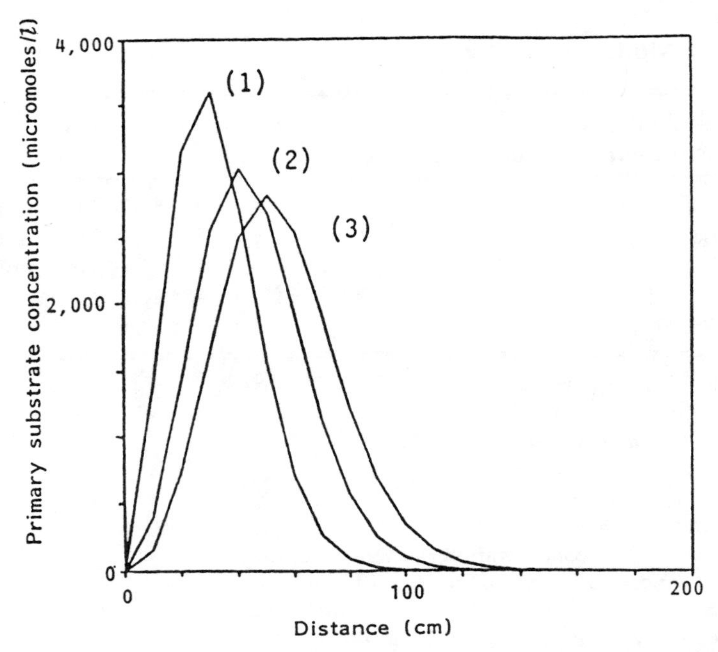

FIGURE 1. Spatial variation of primary substrate at different times: (1) at 48 hrs, (2) at 96 hrs, (3) at 144 hrs.

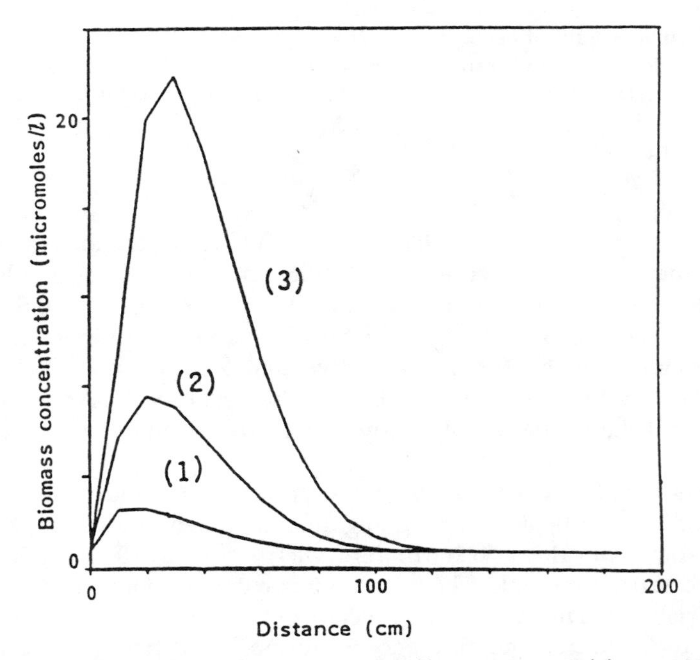

FIGURE 2. Spatial variation of biomass at different times: (1) at 48 hrs, (2) at 96 hrs, (3) at 144 hrs.

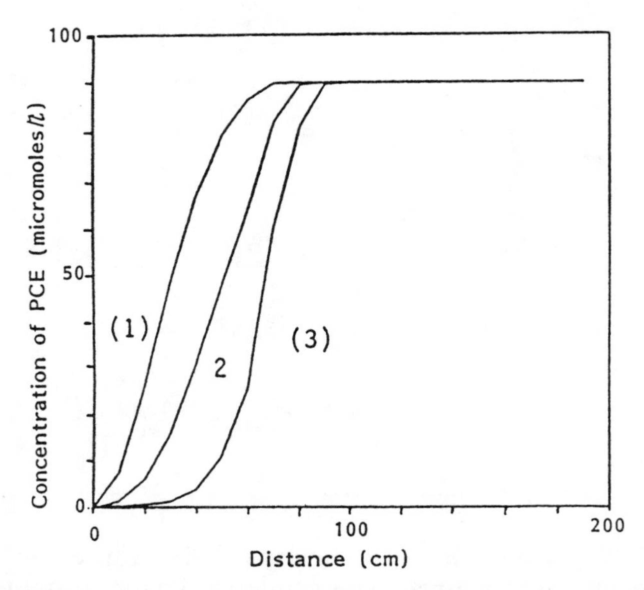

FIGURE 3. Spatial variation of PCE at different times: (1) at 48 hrs, (2) at 96 hrs, (3) at 144 hrs.

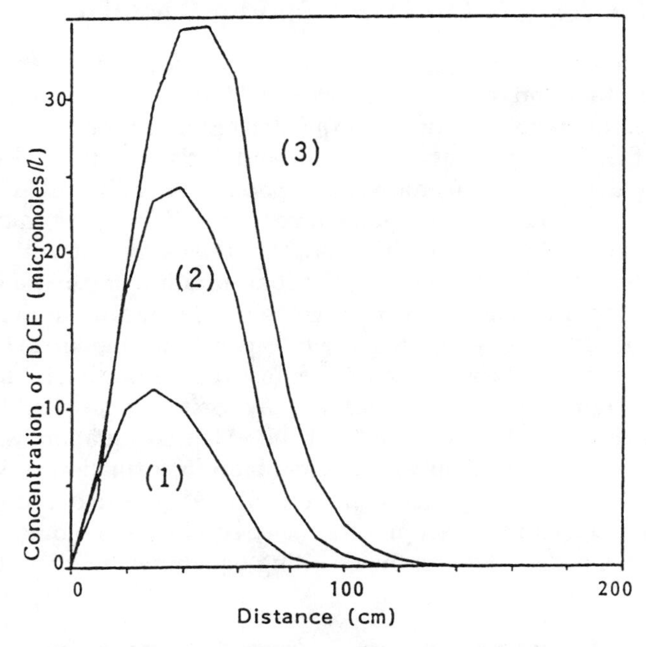

FIGURE 4. Spatial variation of DCE at different times: (1) at 48 hrs, (2) at 96 hrs, (3) at 144 hrs.

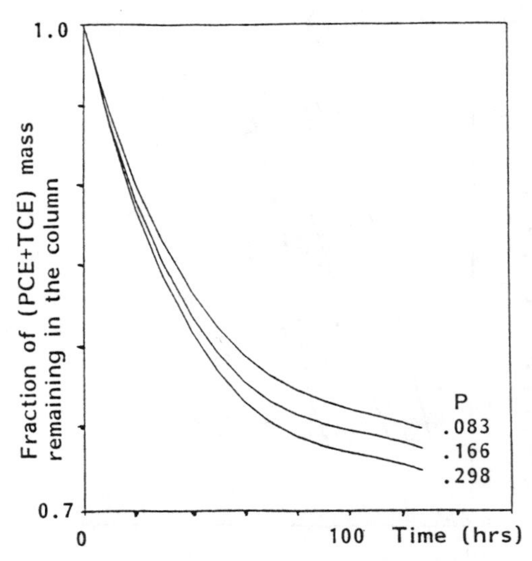

FIGURE 5. The effect of injected primary substrate concentration on the mass of contaminant remained in the column (primary substrate concentration P is given in moles/L).

SIMPLIFICATION OF THE GENERAL MODEL

Corapcioglu et al. (1991) presented a model by assuming the abundance of primary substrate and constant population of biomass. They employed Michaelis-Menten rate equations to model the sequential reductive dehalogenation reactions. Based on the measurements of Wilson and McNabb (1983), McCarty (1984) commented that a microbial groundwater population of 10^6/mL to 10^7/mL (which corresponds to about 1 microorganism per grain of fine sand) can biotransform secondary substrates such as PCE and TCE relatively fast with respect to the groundwater flowrate. In this case, the concentration of substrate surrounding the microorganisms attached to grains is essentially the same as in the bulk liquid. This corresponds to the assumption of a fully penetrated biofilm in biofilm models.

For cases where the active microorganism concentration in the aquifer does not change significantly with time, we may assume a constant biomass. This assumption justifies the use of the Michaelis-Menten equation, which describes the depletion of bacterial substrate when biotransformation processes are not related to the growth of biomass. If there is a net growth of indigenous micro-organisms in the aquifer, then mass balance equations of biomass and primary substrate are needed. Hence, the resulting equations are given by

$$\frac{\partial(C_1 n)}{\partial t} + q_w \frac{\partial C_1}{\partial x} - D\frac{\partial^2 C_1}{\partial x^2} = -\frac{\mu^1_{max}}{K^1_s + C_1} n C_1 \qquad (10)$$

$$\frac{\partial(C_2 n)}{\partial t} + q_w \frac{\partial C_2}{\partial x} - D\frac{\partial^2 C_2}{\partial x^2} = -\frac{\mu^2_{max}}{K^2_s + C_2} nC_2 + \frac{\mu^1_{max}}{K^1_s + C_1} nC_1 \tag{11}$$

$$\frac{\partial(C_3 n)}{\partial t} + q_w \frac{\partial C_3}{\partial x} - D\frac{\partial^2 C_3}{\partial x^2} = -\frac{\mu^3_{max}}{K^3_s + C_3} nC_3 + \frac{\mu^2_{max}}{K^2_s + C_2} nC_2 \tag{12}$$

$$\frac{\partial(C_4 n)}{\partial t} + q_w \frac{\partial C_4}{\partial x} - D\frac{\partial^2 C_4}{\partial x^2} = -\frac{\mu^4_{max}}{K^4_s + C_4} nC_4 + \frac{\mu^3_{max}}{K^3_s + C_3} nC_3 \tag{13}$$

where μ^k_{max} is the maximum rate of kth substrate depletion; and K_s^k is the half-saturation constant. In Eqs. (10) through (13), we assume that hydrodynamic dispersion coefficients are the same, i.e., $D_1=D_2=D_3=D_4=D$. The μ_s^k and K_s^k are affected by conditions such as temperature and pH at which the biotransformation takes place. The values of μ^k_{max} and K_s^k can be estimated from a plot of data — dC_k/dt versus C_k. After obtaining the constants of the rectangular hyperbola from a linear plot of the data, they can be employed in a numerical solution of the governing equations (Corapcioglu & Hossain 1991).

The proposed biotransformation model has been applied to simulate the biotransformation of PCE and TCE along a large, anaerobic, fixed-film column as reported by Vogel and McCarty (1985). A 2-m-long upflow plastic column with a 20-cm internal diameter was filled with smooth quartzite rocks containing an active methanogenic bacterial population. The porosity of the column was 0.19. An influent solution of PCE, TCE, nutrients, and primary substrates was continuously fed with a velocity of 33 cm/day. Samples were removed through ports at different heights. Concentrations of PCE, TCE, total DCE, and VC were presented as percentages of PCE plus TCE molar concentrations. As noted earlier, maximum rate of substrate depletion and half-saturation constants were determined from a linear plot of data. Model parameters used in the numerical simulation are $\mu^1_{max} = 0.117$ mole/L/day, $\mu^2_{max} = 0.98$ mole/L/day, $\mu^3_{max} = 0.191$ mole/L/day, $\mu^4_{max} = 0.05$ mole/L/day, and $K_s^1 = 0.06$ mole/L, $K_s^2 = 0.007$ mole/L, $K_s^3 = 0.008$ mole/L, $K_s^4 = 0.05$ mole/L. These values are consistent with those reported in the literature (Barrio-Lage et al. 1987).

Other model parameters are mechanical dispersion coefficient $D = 0.033$ cm^2/day; flow velocity $V = q_w/n = 33$cm/day; critical DCE concentration = 0.285 micromole/L; time step $\Delta t = 0.05$ day; and space interval $\Delta x = 10$ cm. The dispersion coefficient was calculated by multiplying the flow velocity by the dispersivity, which was taken as 0.01 m. This is a realizable dispersivity value for a laboratory column. The numerical results obtained after 10 days of column operation are plotted in Figure 6 along with the experimental data of Vogel and McCarty (1985).

Numerical results show a favorable match with the data. Figure 6 shows that, despite a relatively high flowrate, PCE and TCE were almost totally biotransformed to intermediate compounds within the first fourth of the column. The major discrepancy between the experimental and numerical results was a

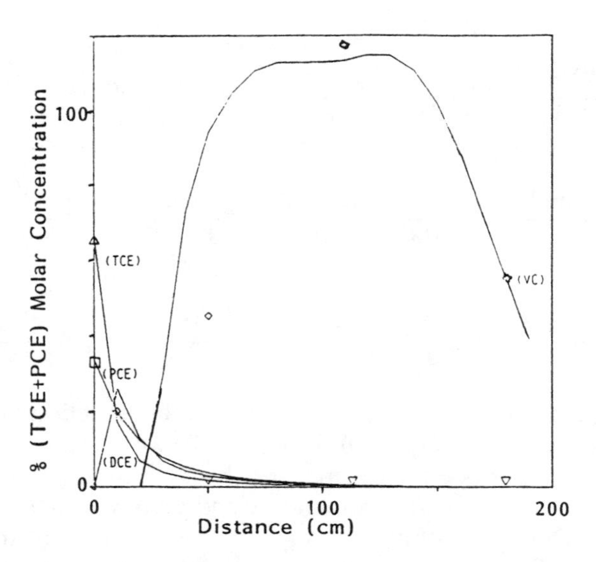

FIGURE 6. Comparison of numerical results with experimental data of Vogel and McCarty (1985). (□=PCE, ▲=TCE, ▼=DCE, and ◇=VC).

simulated DCE peak, although the DCE concentrations remained quite low throughout the column during the experiment. However, both experimental and numerical results indicate that VC increases sharply when DCE concentrations are low, and PCE and TCE disappear. The deviations from the predicted values can be attributed to various factors such as fluctuations in concentrations of nutrients, primary substrates, and biomass along the column during the course of the experiment. Information on these variables is not available.

Another possible reason for apparent discrepancies between predicted and measured concentrations is the deviation from the suggested biotransformation pathway due to other biotic and/or abiotic conversions. A direct stoichiometric relationship between the parent compounds and reaction products as discussed earlier is not always observed. Sometimes there is a compound mass unaccounted for in mass balances and neglected in models due to lack of information, thus causing deviations.

CONCLUSIONS

Biotransformation and migration of chlorinated aliphatic compounds in a saturated column have been simulated by employing a mathematical model. The model predicts the primary substrate-controlled sequential reductive dehalogenation of dissolved compounds under methanogenic conditions. The model consists of mass balance equations of a nontoxic primary growth substrate, secondary nongrowth substrates and byproducts, and biomass of indigenous bacteria. Governing equations are coupled through cometabolic biotransformation bacteria. Governing equations are coupled through cometabolic biotransformation kinetics represented by a Monod equation. It has been assumed that the rate of microbial growth is a function of growth substrate. In a cometabolic biotransformation process, the competition is between the primary substrate and secondary substrate(s) and byproducts for enzyme-catalyzed reactions. The numerical results of the simplified model under the assumption of abundance of primary substrate and constant population of biomass have been compared with experimental data of Vogel and McCarty (1981). The comparison showed a favorable match.

REFERENCES

Barrio-Lage, G., F. Z. Parsons, and R. S. Nassar. 1987. "Kinetics of the Depletion of Trichloroethene." *Environmental Science and Technology 21*(4): 366-370.

Bouwer, E. J., and G. D. Cobb. 1986. "Modeling of Biological Processes in the Subsurface." *Water Science and Technology 19*(3): 769-779.

Corapcioglu, M. Y., and A. Haridas. 1984. "Transport and Fate of Microorganisms in Porous Media: A Theoretical Investigation." *Journal of Hydrology 72*: 149-162.

Corapcioglu, M. Y., and A. Haridas. 1985. "Microbial Transport in Soils and Groundwater: A Numerical Model." *Advances in Water Resources 8*(6): 188-200.

Corapcioglu, M. Y., and M. A. Hossain. 1990a. "Groundwater Contamination by High-density Immiscible Hydrocarbon Slugs in Gravity-driven Aquifers." *Ground Water 28*(3): 403-410.

Corapcioglu, M. Y., and M. A. Hossain. 1990b. "Theoretical Modeling of Biodegradation and Biotransformation of Hydrocarbons in Subsurface Environments." *J. Theoretical Biol. 142*: 503-516.

Corapcioglu, M. Y., and M. A. Hossain. 1991. "Estimating Biotransformation Rate Constants for Sequential Reductive Dehalogenation Reactions." *Journal of Environmental Engineering (ASCE) 117*(5): 631-639.

Corapcioglu, M. Y., M. A. Hossain, and M. A. Hossain. 1991. "Methanogenic Biotransformation of Chlorinated Hydrocarbons in Groundwater." *Journal of Environmental Engineering (ASCE) 117*(1): 47-65.

Criddle, C. S., L. A. Alvarez, and P. L. McCarty. 1991. "Microbial Processes in Porous Media." In J. Bear and M. Y. Corapcioglu (Eds.), *Transport Processes in Porous Media*, pp. 639-691. Kluwer, Dordrecht, The Netherlands.

De Bruin, W. P., M. J. J. Kotterman, M. A. Posthumus, G. Shraa, and A. J. B. Zelunder. 1991. "Complete Anaerobic Reductive Dechlorination of Tetrachloroethene in a Bioreactor." *Appl. Environ. Microbiol. 58*: 1996-2000.

Kästner, M. 1991. "Reductive Dechlorination of Tri- and Tetrachloroethylene by Nonmethanogenic Enrichment Cultures." In R. E. Hinchee and R. F. Olfenbuttel (Eds.), *On-Site Bioreclamation*, pp. 134-146. Butterworth-Heinemann, Stoneham, MA.

McCarty, P. L. 1984. "Application of Biological Transformations in Groundwater." *Proceedings of the Second International Conference on Groundwater Quality Research*, pp. 6-11. March 26-29, Tulsa, University Center for Water Research, Stillwater, OK.

McCarty, P. L., L. Semprini, M. E. Dolan, T. C. Harman, C. Tiedeman, and S. M. Gorelick. 1991. "In Situ Methanotrophic Bioremediation for Contaminated Groundwater at St. Joseph, Michigan." In R. E. Hinchee and R. F. Olfenbuttel (Eds.), *On-Site Bioreclamation*, pp. 16-40, Butterworth Heinemann, Stoneham, MA.

Parsons, F., and G. Barrio-Lage. 1985. "Chlorinated Organics in Simulated Groundwater Environments." *Journal of American Water Works Association 77*(5): 52-59.

Semprini, L., P. V. Roberts, G. D. Hopkins, and P. L. McCarty. 1990. "A Field Evaluation of In-Situ Biodegradation of Chlorinated Ethenes: Part 2. Results of Biostimulation and Biotransformation Experiments." *Ground Water 28*(5): 715-727.

Semprini, L., and P. L. McCarty. 1992. "Comparison Between Model Simulations and Field Results for In-Situ Biorestoration of Chlorinated Aliphatics: Part 2. Cometabolic Transformation." *Ground Water 30*(1): 37-44.

Vogel, T. M., and P. L. McCarty. 1985. "Biotransformation of Tetrachloroethylene to Trichloroethylene, Dichlorethylene, Vinyl Chloride, and Carbon Dioxide under Methanogenic Conditions." *Applied and Environmental Microbiology 49*(5): 1080-1083.

Wilson J. T., and J. F. McNabb. 1983. "Biological Transformation of Organic Pollutants in Groundwater." *Trans. American Geophysical Union 64*(33): 505.

Wilson, J. T., L. E. Leach, M. Henson, and J. N. Jones. 1986. "In Situ Biorestoration as a Groundwater Remediation Technique." *Groundwater Monitoring Review, Fall*: 56-64.

IN SITU BIOREMEDIATION AT A WOOD-PRESERVING SITE IN A COLD, SEMI-ARID CLIMATE: FEASIBILITY AND FIELD PILOT DESIGN

M. R. Trudell, J. M. Marowitch, D. G. Thomson, C. W. Fulton, and R. E. Hoffmann

ABSTRACT

This study evaluated the feasibility of in situ bioremediation and design of a field pilot study for an abandoned pentachlorophenol (PCP) site south of Calgary, Alberta. A laboratory treatability study was carried out for 35 weeks at temperatures of 5, 15, and 25°C, with initial PCP concentrations of 300, 500, and 900 mg/kg. Substantial degradation of PCP and oil and grease (O&G) was observed in all treatments. Residual PCP levels after 35 weeks were below 10 mg/kg in four of the nine treatments, and below 200 mg/kg in all treatments. Evidence from the laboratory treatability study was the basis for the design of a field pilot study begun in fall, 1992. For a 2,500-m^2 treatment plot, contaminated material was excavated to 2.5 m depth, fertilizer added and homogenized, and the material replaced in the excavation. A drainage blanket and leachate collection system were installed at the base of the excavation to collect any infiltration and serve as a secondary front for the diffusion of oxygen. The design features a plastic surface covering to intercept precipitation and control both infiltration and evaporation. A drip irrigation system under the plastic cover is used to maintain optimum moisture.

INTRODUCTION

The Peerless Wood Preservers site, located near Cayley, 60 km south of Calgary, Alberta, was a pressure treatment wood-preserving facility that used pentachlorophenol (PCP) applied with a fuel-oil carrier. The company was in operation from 1958 to 1982. The site, now abandoned, is contaminated with the process residues of the operation. Previous studies have found soil PCP concentrations up to 850 mg/kg and groundwater PCP concentrations up to 33,000 mg/L (Monenco 1990).

The study involved two major components: a laboratory treatability study to determine the effectiveness of degrading the contaminants biologically, and an investigation of hydrogeologic conditions to evaluate the feasibility of implementing bioremediation in the field under the geologic and climatic conditions prevalent at the study site and given the distribution of contaminants.

The mean annual temperature at the study site is 3.4°C. The mean annual precipitation is 400 mm, with mean annual evaporation of 620 mm, giving an annual moisture deficit of 220 mm. The site is underlain by lacustrine deposits of primarily sand and silt to a depth of 1.5 to 2.5 m, beneath which is fractured clay-loam till down to bedrock at a depth of 7 to 8 m. The upper 0.5 to 1.5 m of bedrock is a fine-grained, fractured sandstone unit that is a very poor aquifer. Below the sandstone is a predominantly shale unit underlain by an aquifer used for domestic water supply at a depth of approximately 30 m.

TREATABILITY STUDY

The biodegradability of PCP is well established as summarized, for example, in the review paper by Boyd et al. (1989). In addition to biodegradation, loss of PCP in soils can be attributed to sorption and to incorporation of PCP into soil organic matter by oxidative coupling (Boyd et al. 1989). Under alkaline soil conditions such as those at the study site, PCP will be almost entirely in its dissociated (phenolate anion) form. Sorption of dissociated PCP is minimal compared with the undissociated form (Callahan et al. 1979), and sorption that does occur is expected to be largely reversible (Bellin et al. 1990). Mass balance studies of PCP degradation in aerobic, alkaline soils have found no evidence of PCP irreversibly bound in soil organic matter (Bellin et al. 1990; Mueller et al. 1991a, b). Based on these studies, as well as the work of Seech et al. (1991), the treatability study designed was to provide indirect evidence of PCP and hydrocarbon biodegradation, without relying on the more rigorous balance method used to document PCP biodegradation in previous studies. Because abiotic losses are not accounted for in this study, the results cannot be taken as a conclusive demonstration of biodegradation and the degradation rates and residual concentrations must be viewed as optimistic. The lines of indirect evidence used to infer biodegradation from the treatability study are:

1. Decrease in PCP and hydrocarbon concentrations over time.
2. Decrease in soil toxicity (Microtox) over time.
3. Increase in microbial population (both heterotrophic and PCP-acclimated microorganisms) over time.
4. Production of CO_2 over time.
5. Preferential removal of C17 and C18 *n*-alkanes relative to pristane and phytane.

Taken together, these lines of indirect evidence were considered adequate to indicate that biodegradation was a significant component of observed contaminant reduction over time. Although analysis of chloride may have provided more direct evidence of biodegradation, chloride analyses were not done in this study.

A backhoe was used to collect contaminated soil samples of approximately 60 L from 10 to 30 cm depth. In the laboratory, the soils were air-dried for approximately 48 hours, sieved through a 2-mm mesh screen, and then homogenized. Samples were analyzed for O&G content and PCP concentration. PCP was determined by Soxhlet extraction with methylene chloride, analyzed by high-performance liquid chromatography (HPLC) using a C-18 column, a flowrate of 1.6 mL/min, and an ultraviolet (UV) detector at a wavelength of 300 nm. The O&G content was determined gravimetrically by Soxhlet extraction with methylene chloride. Based on these analyses, a soil sample with 900 mg/kg PCP and 1.8% O&G was selected to represent soil with a high (H) PCP concentration. A soil sample with 300 mg/kg PCP and 1.3% O&G was selected to represent soil with a low (L)-PCP-concentration soil. A 50:50 mixture of the two soils with 500 mg/kg PCP and 1.6% O&G was prepared as a soil with a moderate (M) PCP concentration. The treatability study design is given in Table 1. The three concentrations of contaminants were evaluated at three temperatures, although only five of the nine treatments were sampled in detail.

Treatment consisted of placing 250 g dry wt. of soil in 2-L glass jars. The soil moisture contents were adjusted to 70% of field capacity, and nutrients (162 mg of NH_4NO_3 and 48 mg NH4H_2PO_4) were added. The soils were incubated at constant temperature and aerated by mixing once a week. Each trial was conducted in triplicate.

The initial (week 0) soil samples were analyzed for soil texture, cation exchange capacity and exchangeable cations, electrical conductivity and soluble ions, total nitrogen, available macronutrients, and total elements. Initial samples were characterized using gas chromatography/mass spectroscopy (GC/MS) and analyzed for dioxins and furans. The GC/MS characterization was repeated at the end of the experiment. CO_2 measurements were made weekly on all jars by collecting gas samples in a syringe, through a septum; CO_2 concentration was determined using fast gas chromatography. The initial soil samples and the soil samples obtained throughout the experiment were analyzed for PCP concentration, O&G content, relative toxicity (Microtox), pH, and total carbon and total organic

TABLE 1. Treatability study design.

	Temperature		
Concentration	5°C	15°C	25°
Low	O[(a)]	X[(b)]	X
Moderate	X	X	X
High	O	X	O

(a) O — Sampled at 0, 35 weeks (weekly CO_2).
(b) X — Sampled at 0, 2, 4, 12, 20, and 35 weeks.

carbon (TOC). Initial soil pH ranged from 7.6 to 8.1, and soil pH remained between 7.5 and 8.2 over the course of treatment.

Plate counts of colony-forming units (CFUs) were made on trypticase soy agar (TSA) to determine the total number of microorganisms, and on TSA amended with 200 and 500 mg/kg PCP to determine the number of microorganisms acclimated to elevated levels of PCP.

PCP Degradation

Figure 1 shows that reduction in PCP concentrations, evidently by biodegradation, occurred at all temperatures, and for all initial PCP concentrations. After 35 weeks (Table 2), soils with the low level of PCP that incubated at 5, 15, or 25°C, and soils with the moderate level of PCP that incubated at 15 or 25°C, had PCP levels below 25 mg/kg. Highly contaminated soils, and the soil with a moderate level of PCP incubated at 5°C, all showed PCP levels below 200 mg/kg.

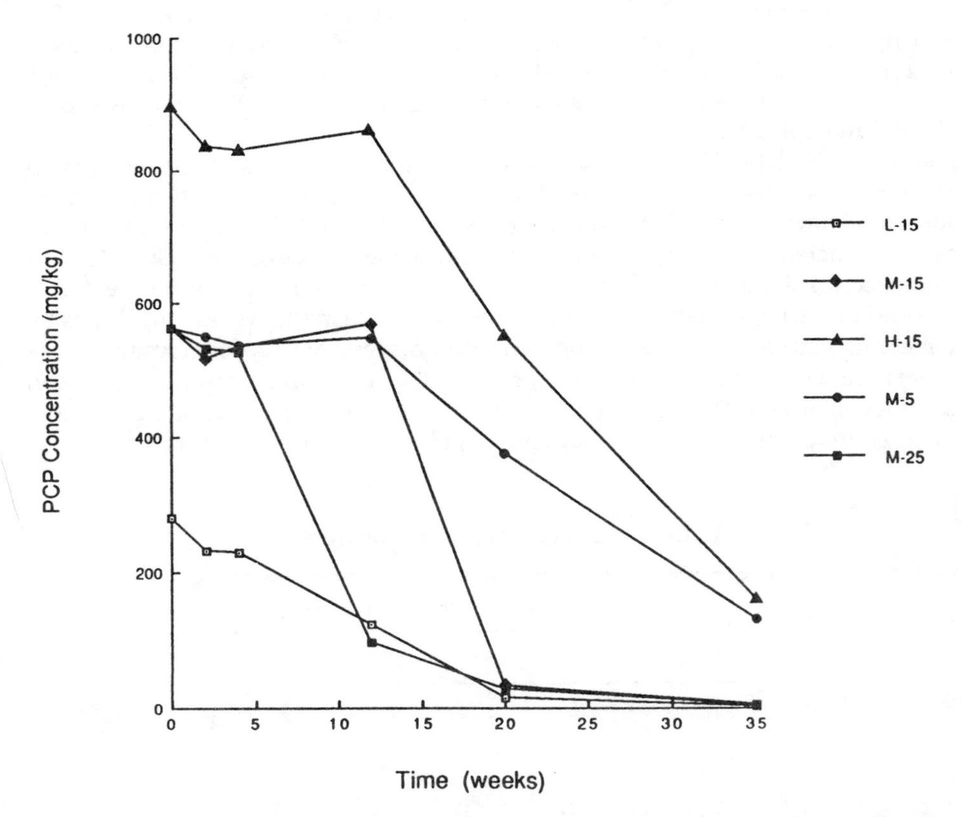

FIGURE 1. Effect of treatments on PCP concentrations. Legend scheme: Treatment concentration (Low, Moderate, High); Treatment temperature in °C (5, 15, or 25).

TABLE 2. Residual concentration of PCP (mg/kg) after 35 weeks.

Treatment	Temperature		
	5°C	15°C	25°C
Low (300 mg/kg)	22	2	1
Moderate (500 mg/kg)	130	6	4
High (900 mg/kg)	185	160	150

The initial PCP concentration and incubation temperature also affected the time required for degradation to begin. At 15°C, after 2 weeks PCP in soil with a low level of contamination showed a decrease, whereas degradation in soil with a moderate or high level of contamination occurred only after 12 weeks. The lag period of soil with the moderate level of concentration was reduced to 4 weeks when incubated at 25°C.

The stoichiometry of PCP degradation is expressed as

$$C_6Cl_5OH + 17/4\ O_2 + 5/2\ H_2O = 6\ CO_2 + 5\ H^+ + 5\ Cl^- \qquad (1)$$

According to the equation, 0.5 kg of oxygen is required to degrade 1 kg of PCP, assuming that all of the carbon in PCP is converted to CO_2. The PCP degradation rate, assuming first-order kinetics, can be expressed in terms of the half-life ($T_{1/2}$). Estimates of $T_{1/2}$ for PCP degradation from the treatability study are summarized in Table 3. Individual values range from 60 to 179 days, with the longer half-life values associated with lower temperature treatments or higher initial PCP concentration. These $T_{1/2}$ values are within the range of 23 to 178 days summarized by Howard et al. (1991, p. 242) for the degradation of PCP in soils.

TABLE 3. Estimated PCP degradation rates: PCP half-life (days).

Treatment	Temperature		
	5°C	15°C	25°C
Low (300 ppm)	—[(a)]	74	—
Moderate (500 ppm)	179	113	60
High (900 ppm)	—	165	—

(a) Not determined.

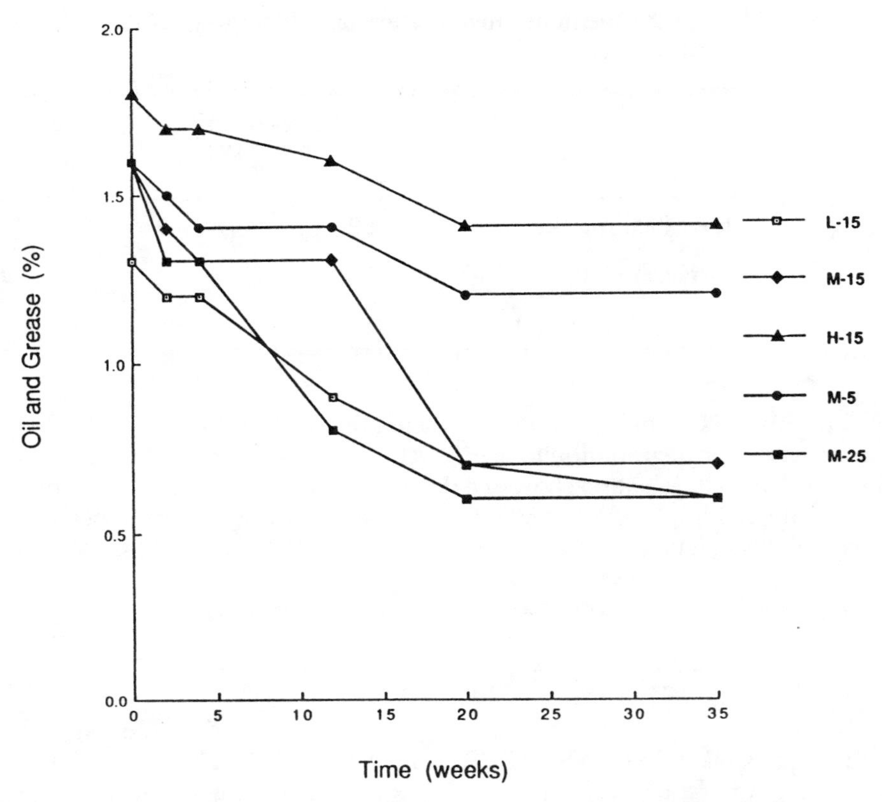

FIGURE 2. Effect of treatments on oil and grease concentrations. Legend scheme: Treatment concentration (Low, Moderate, High); Treatment temperature in °C (5, 15, or 25).

Oil and Grease Degradation

Figure 2 shows that, as with the PCP, O&G levels decreased in all treatments. Table 4 gives the O&G concentration in all treatments after 35 weeks. After 35 weeks, the soils with a low level of O&G incubated at 5, 15, or 25°C, and soils with a moderate level of O&G incubated at 15 or 25°C showed O&G levels below 1%, representing a reduction in O&G of 38 to 62%. Soil with a moderate level of contamination incubated at 5°C, and soil with a high level of contamination incubated at 15 or 25°C, had O&G contents below 1.5% representing a reduction in O&G of 22 to 33%. Highly contaminated soil incubated at 5°C showed only a slight (11%) reduction in O&G content to 1.6%. Unlike PCP degradation, O&G degradation was evident after 2 weeks in all of the treatments shown in Figure 2.

The percent reduction in O&G content is much lower than the percent reduction in PCP concentration. However, the amount of O&G degraded is much greater, because the amount initially present is much greater (13,000 to 18,000 mg/kg).

TABLE 4. Residual concentration of O&G (%) after 35 weeks.

	Temperature		
Treatment	5°C	15°C	25°C
Low (1.3%)	0.8	0.6	0.8
Moderate (1.6%)	1.2	0.7	0.6
High (1.8%)	1.6	1.4	1.2

The rate of degradation of O&G was substantially reduced after 20 weeks, most likely due to the change in the distribution of the residual compounds. The GC/MS analysis suggested that, by 20 weeks, the amount of the more biodegradable compounds had been reduced and the residual compounds were more recalcitrant.

Carbon Dioxide Production

Figure 3 shows typical average cumulative CO_2 evolution with incubation time. The decrease in the rate of CO_2 evolution with time correlates with the

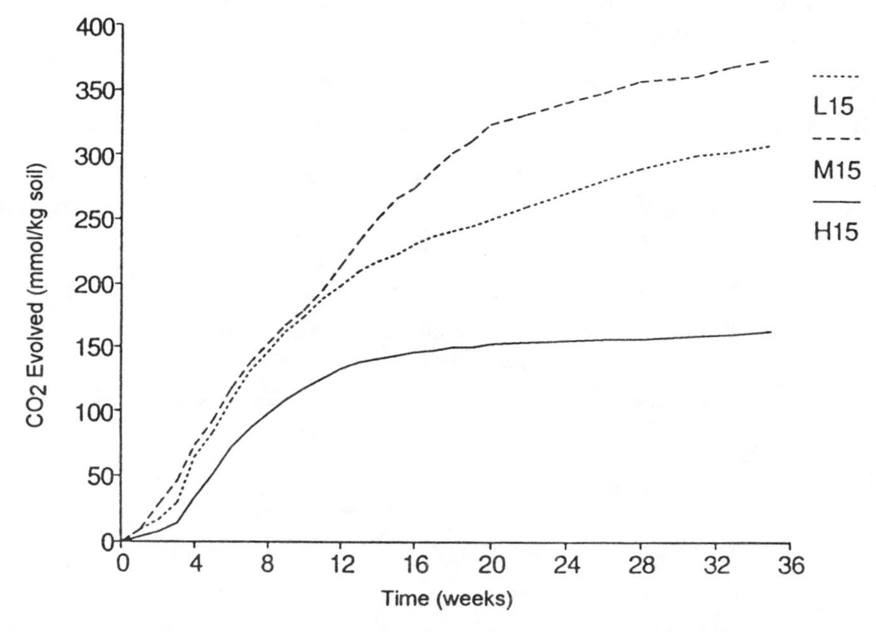

FIGURE 3. Evolution of carbon dioxide from treatments at 15°C. Treatments at other temperatures were similar.

decrease in the rate of degradation of O&G. Temperature, as expected, had a pronounced effect on CO_2 evolution, with higher temperatures resulting in higher rates of evolution. The concentrations of PCP and O&G also had a pronounced effect, with high levels of contamination resulting in decreased CO_2 evolution. The CO_2 evolution rate gives a measure of microbiological activity. The higher respiration rates observed with the moderately contaminated soils compared to the soil with a low level of contamination likely are due to the greater availability of substrate with the higher level of contamination.

The O&G degradation stoichiometry can be estimated empirically from the measurements of CO_2 production, because each mole of CO_2 produced from hydrocarbon degradation requires 1 mole of oxygen. Table 5 gives the amount of CO_2 produced, in moles, per kilogram of O&G degraded. Values range from 33.7 to 50.8 mol CO_2 per kilogram of O&G, with a mean of 41.36 mol CO_2 per kilogram of O&G. Expressed in terms of O_2 required, this corresponds to 1.32 kg O_2 per kilogram of O&G.

Oxygen Utilization Rate. The data on the rate of CO_2 production also can be used to estimate the O_2 utilization rate. Assuming first-order kinetics for CO_2 production and O_2 utilization, the estimated half-lives for CO_2 production are given in Table 6. From the nine treatments, $T_{½}$ for CO_2 production ranges from 32.2 to 123.2 days, with a mean $T_{½}$ of 72.9 days. Half-lives of about 100 days correspond to the treatments at 5°C; at higher temperatures, $T_{½}$ is between 32 and 73 days. Assuming a CO_2:O_2 ratio of 1:1, then the rate of O_2 utilization is equal to the rate of CO_2 production, and the half-lives given in Table 6 can be considered as $T_{½}$ values for oxygen utilization.

Soil Toxicity

Figure 4 shows the effect of the treatments on the toxicity of the soil, as measured by Microtox assay. Results for all treatments after 35 weeks are given in Table 7. The results are 5-min EC_{50}s, which represent the concentration of the

TABLE 5. Total CO_2 production (moles) per kilogram O&G degraded.

	Temperature		
Treatment	5°C	15°C	25°C
Low	—[a]	42	—
Moderate	34	41	51
High (1.8%)	—	40	—

(a) Not determined.

TABLE 6. Rate of CO_2 production, and inferred O_2 consumption: CO_2 half-life.

Treatment	Temperature		
	5°C	15°C	25°C
Low	97	60	32
Moderate	99	72	71
High	123	46	57

eluted soil solution that causes a 50% decrease in fluorescence after 5 min exposure. All three soils used in the experiment initially were rated as extremely toxic with EC_{50}s at less than 3%. Most treatments produced in an initial increase in toxicity. As with the PCP and O&G concentrations, soil toxicity eventually was reduced in all treatments. However after 35 weeks of incubation, all the highly contaminated soils and the moderately contaminated soil incubated at 5°C remained extremely toxic (EC_{50} less than 2%), and the soil with a low level of contamination incubated at 5°C was very toxic (EC_{50} at 29%). The moderately contaminated soil incubated at 15°C was moderately toxic (EC_{50} at 85%). Only the soils with a low level of contamination incubated at 15 and 25°C, and the moderately contaminated soil incubated at 25°C, showed no toxic effect (EC_{50} greater than or equal to 100%).

With further incubation, the toxicity levels of the other soils likely could be further reduced. The indigenous organisms were able to remediate soils with a moderate level of contamination (i.e., mixtures with low and high levels of contamination) to nontoxic levels. This could indicate that the compounds making the soil toxic are biodegradable.

Microbiological Analyses

Figure 5 shows the results of the microbiological analyses. All three soils initially showed similar heterotrophic counts. After 2 weeks of incubation, all treatments except at 5°C showed a 100-fold increase in heterotrophic counts. The 5°C treatment showed only a 10-fold increase, but by 4 weeks, the counts were similar to those of the other treatments. The number of heterotrophic organisms remained approximately the same during the remainder of the experiment. Counts of organisms acclimated at 200 mg/kg PCP were similar to counts of total organisms. At 500 mg/kg PCP, the soil with the initially low PCP concentration showed acclimated counts 100-fold less than total counts, and the soils with the initially moderate and high levels of PCP showed acclimated counts 10-fold less than total counts. After 4 weeks of incubation, the number of acclimated organisms declined, most significantly in the treatments that were most effective in reducing the PCP concentration.

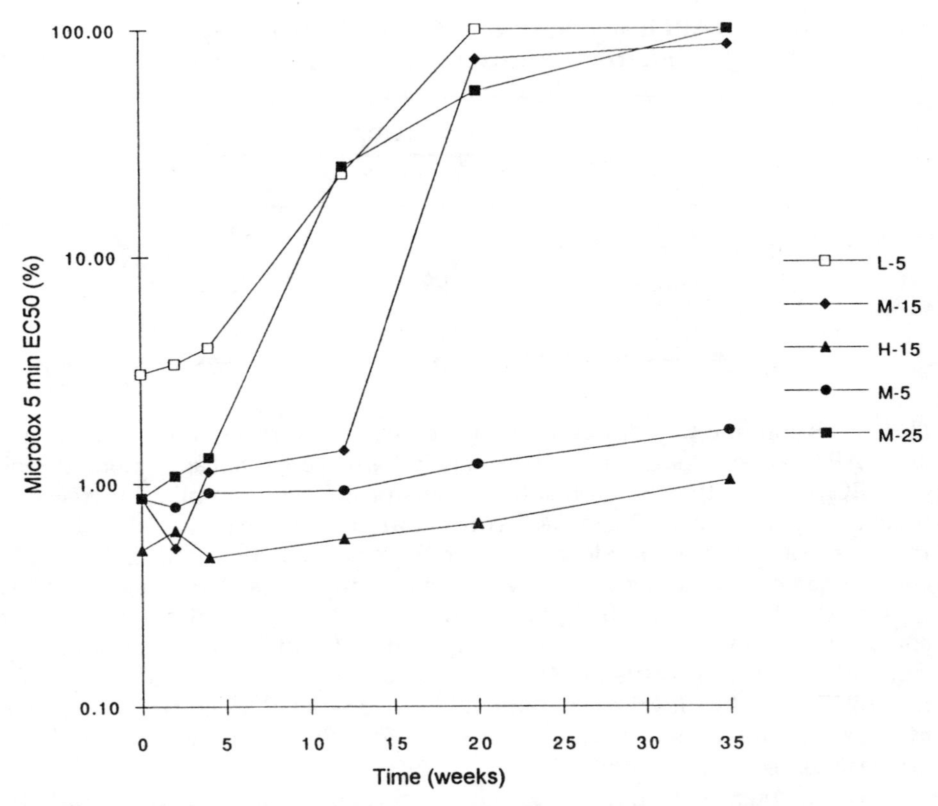

FIGURE 4. Effect of treatments on soil toxicity (Microtox); 5-minute EC_{50} concentration (% soil solution extract). Legend scheme: Treatment concentration (Low, Moderate, High); Treatment temperature in °C (5, 15, or 25).

GC/MS Analyses

From the GC/MS analyses, both the saturate and aromatic fractions showed distributions of compounds typical to those of petroleum distillates. The predominant aromatics in the soils are naphthalenes, dibenzothiophenes, phenanthracenes, and fluorenes. Selected polycyclic aromatic hydrocarbons were quantitatively analyzed, and all were less than 10 mg/kg in the soil. The third fraction, extracted with methylene chloride, showed the presence of chlorinated compounds dominated by PCP, but including dimers of polychlorinated phenols. Dioxins and furans also were present.

A GC/MS scan was done for the soil with a moderate level of contamination after 12 weeks of incubation at 25°C, and indicated that biodegradation of hydrocarbons was occurring. The saturate fraction showed an increase in pristane and phytane, relative to C-17 and C-18 *n*-alkanes. The aromatic fraction showed a relative decrease of the lighter aromatic compounds.

TABLE 7. Microtox toxicity of treated soil after 35 weeks: 5-minute EC_{50} concentration (% soil solution) that produces 50% toxicity response after 5 minutes of exposure.

Treatment	Temperature 5°C	15°C	25°C
Low	29	NTE[a]	NTE
Moderate	1.7	85	NRE
High	0.5	1.0	1.2

(a) NTE – No Toxic Effect, EC_{50} at greater than 100%.

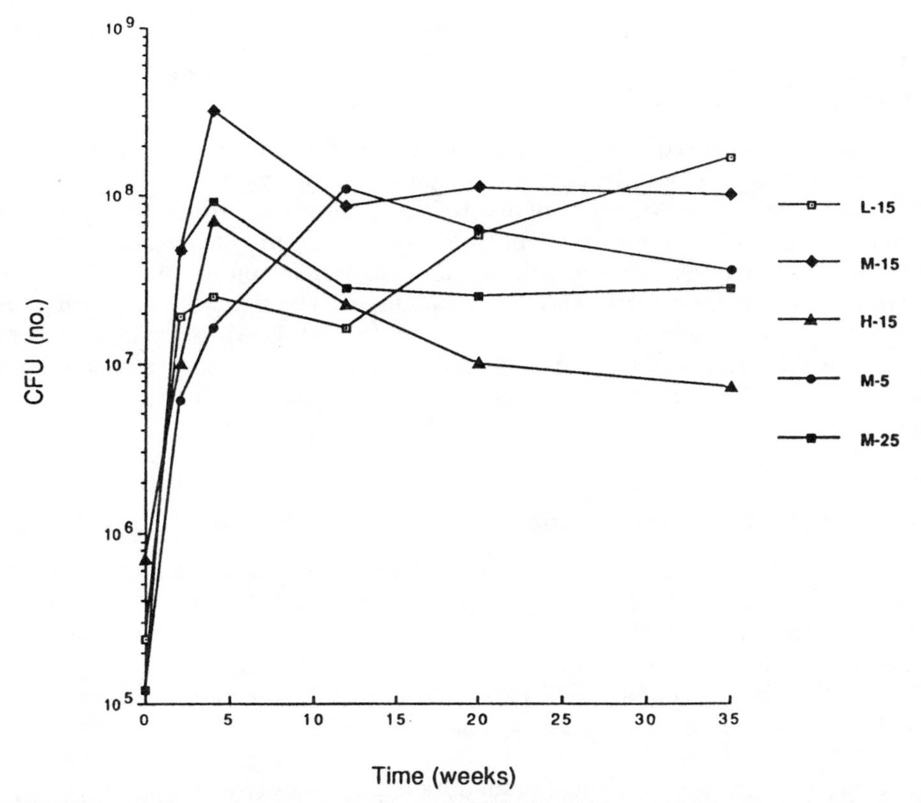

FIGURE 5. Effect of treatments on microbiological population, as number of colony-forming units (CFU). Legend scheme: Treatment concentration (Low, Moderate, High); Treatment temperature in °C (5, 15, or 25).

The GC/MS analysis was repeated after 20 weeks of incubation at 15°C. Changes in the distribution of saturated hydrocarbons occurred in soils with low, moderate, and high levels of contamination. The soils with moderate and high levels showed a selective loss of C-17 and C-18 *n*-alkanes relative to pristane and phytane. The *n*-alkane/isoprenoid ratio is used to indicate biodegradation because isoprenoids are considered recalcitrant to biodegradation and usually degrade at much lower rates.

Dioxin/Furan Analysis

The three initial soil samples were found to be very highly contaminated with dioxins and furans, but none showed detectable levels of 2,3,7,8-T4CDD. The GC/MS analysis found dioxins and furans in all soils after 20 weeks of incubation at 25°C. However, the method can detect only dioxins with more than 4 chlorine atoms.

FEASIBILITY OF BIOREMEDIATION UNDER FIELD CONDITIONS

The treatability study was conducted under optimum conditions of moisture content and oxygen supply, and under uniform temperature and contaminant concentrations. The degree to which field conditions approach the optimum conditions of the laboratory treatability study will certainly influence the transferability of laboratory observations to the field. If field conditions deviate substantially from optimum, the rate of degradation could be significantly reduced, or degradation may be inhibited to a great extent. In the following sections the field conditions that may influence the effectiveness of biodegradation are examined relative to the results of the treatability study and design of the field pilot bioremediation study.

Vertical Distribution of Contaminants

Continuous core samples were collected in Shelby tubes from each lithologic unit in 40-cm sampling intervals and were analyzed for PCP and O&G (Figure 6). Peak concentrations of both occur at a depth of about 1.6 m, near the interface between the lacustrine sediment and the underlying lower permeability till. More than 90% of the total mass of O&G is within the top 2.25 m of geologic material. Within this same interval is 74% of the total mass of PCP, and PCP concentrations above 200 mg/kg are restricted to depths of less than 5 m. Because of the concentration of contaminant mass at shallow depths, it is reasonable to consider focusing on treatment of the upper 2 to 2.5 m, with the potential for enhanced in situ degradation of contaminants to about 4 to 5 m depth.

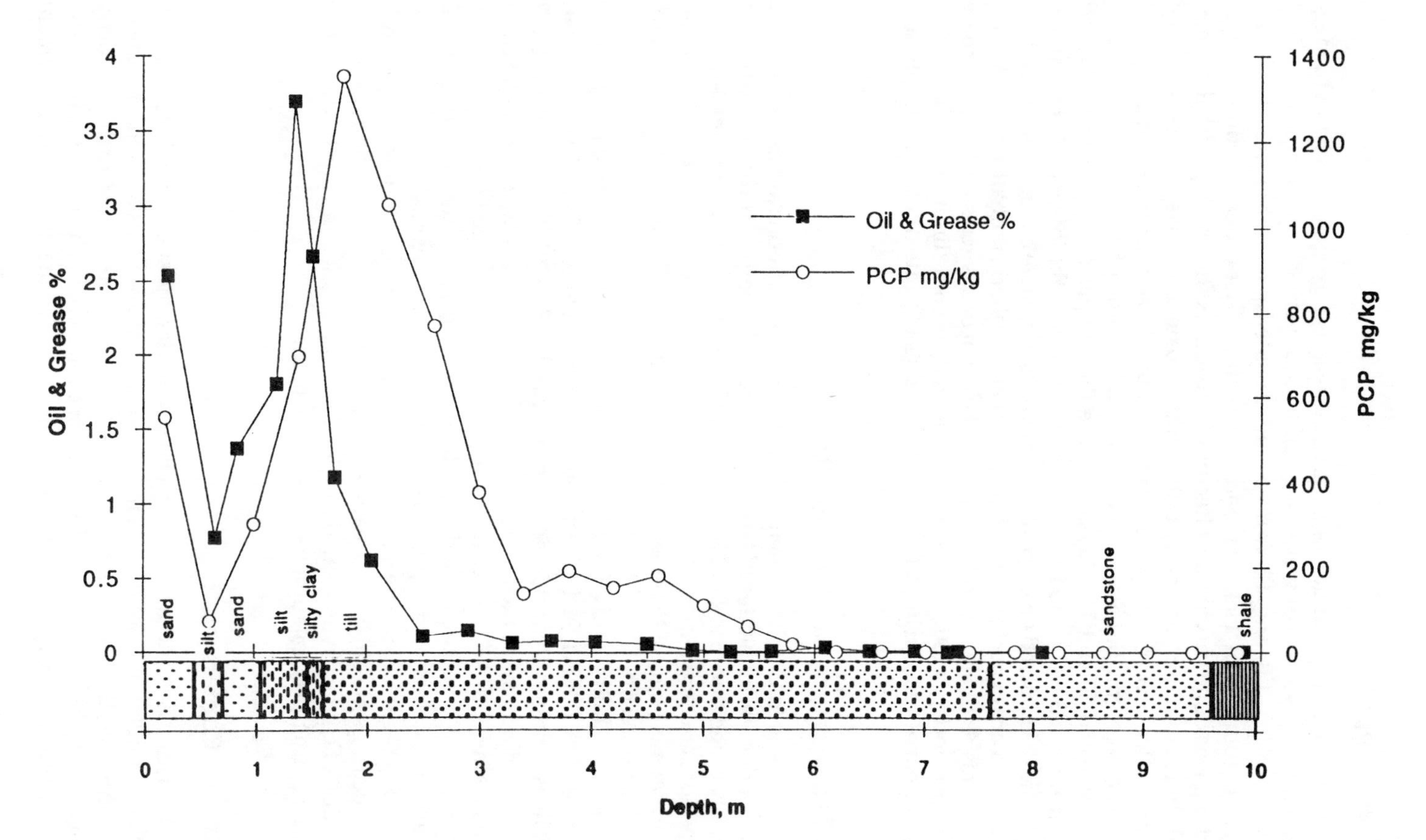

FIGURE 6. Profiles of PCP and O&G with depth. Interface between lacustrine sediment and till is at 1.6 m.

Oxygen Supply

To evaluate the potential for passive diffusion of oxygen to support biodegradation, soil gas samples were collected from 2.5-cm-diameter pipes installed at depths of 0.5, 1.0, and 1.5 from a contaminated part of the site. Using an analytical solution of Fick's second law (with a first-order reaction), curve-matching was done varying the effective diffusion coefficient (D_e) with the range of half-lives from the treatability testing until the field data were bracketed. The field and calculated results are shown in Figure 7. A reasonable bracketing of the field oxygen profile was obtained with a D_e value of 1×10^{-3} m^2/h, and $T_{1/2}$ values between 60 and 100 days. Because the soil gas samples were collected after an extended wet period, these probably represent the lower end of the range of D_e in the field. Based on these observations, steady-state oxygen concentrations at 2 m depth are expected to be 25 to 35% of atmospheric for half-lives of 60 and 100 days, respectively. Therefore, passive diffusion of oxygen should be adequate to support biodegradation, although oxygen may become limiting if other conditions are enhanced during treatment, increasing the rate of O_2 consumption.

Soil Temperature and Moisture Regime

From May to October, when soil temperature conditions are expected to be most conducive to biodegradation, average soil temperatures near the site generally range from 6 to 11°C (MacMillan, 1987). In the warmest months during this period, average temperatures at 10 cm depth range from 15 to 18°C. These measurements suggest that the results of the 5 and 15°C trials of the treatability study best reflect field soil temperature conditions.

The treatability study for this site was conducted at a constant moisture content of 70% field capacity. A range of moisture content from 60 to 80% field capacity is commonly considered as optimum (U.S. EPA 1989). MacMillan (1987) reported soil moisture measurements for two sites near Cayley, monitored at depths from 0.2 to 1.8 m over a 2-year period. The average water content was generally at or below 60% field capacity, suggesting that augmentation of soil moisture may be required. It is not known what effect cycles of alternative wetting and drying might have on the process of biodegradation. Infiltration and groundwater recharge through the treatment area, which can cause the downward migration of contaminants, can be avoided by either collecting infiltration water or intercepting precipitation on the site, or both.

Hydraulic Containment

The hydraulic conductivity of the till underlying the site is low (9×10^{-8} m/s), which will impede the downward movement of contaminants from the treatment area. However, the till also is fractured, and the estimated fracture flow velocity is relatively high (4×10^{-5} m/s). Diffusion of contaminants between the till matrix and the fractures can operate in two ways. It can retard the downward movement

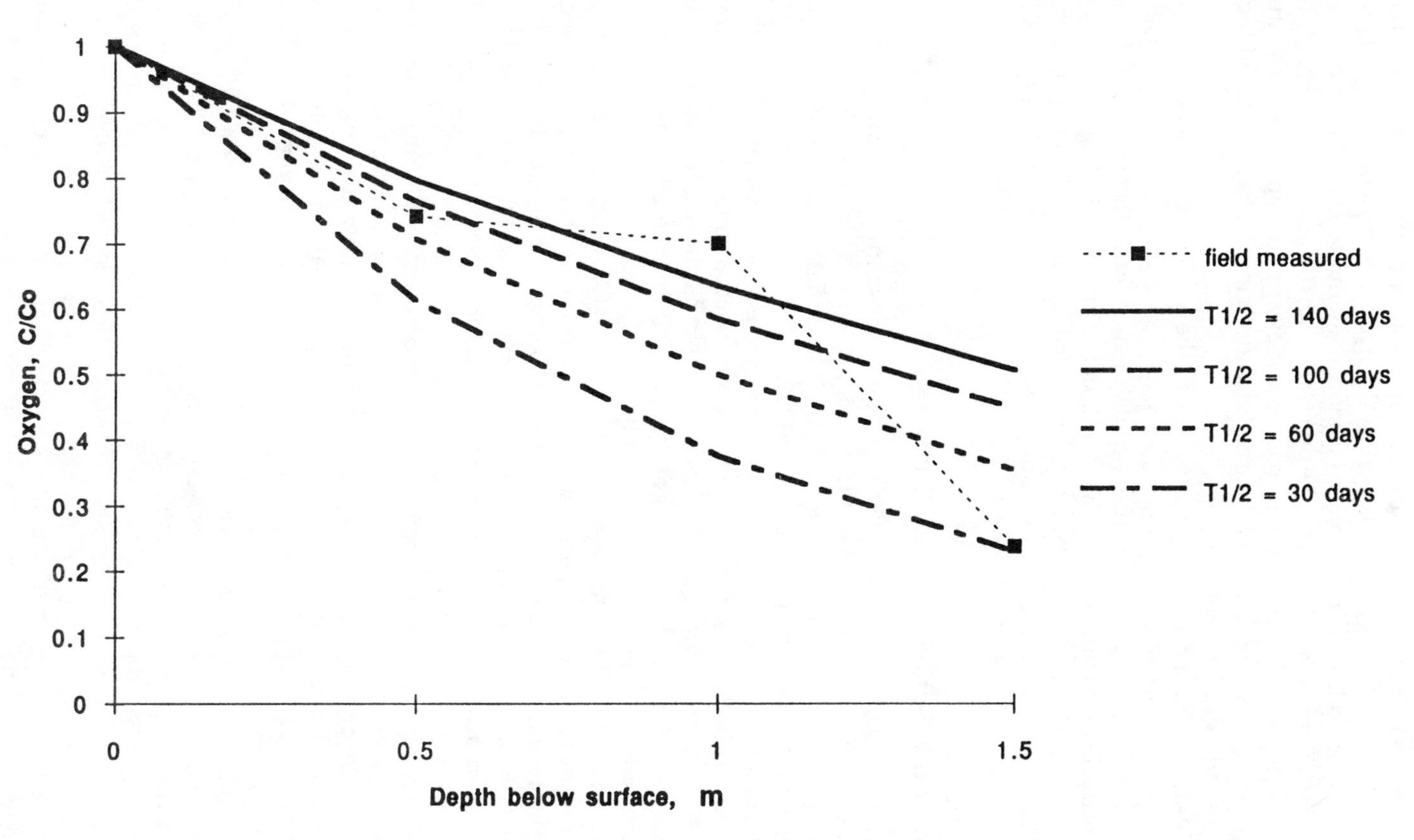

FIGURE 7. Oxygen profile measured in soil gas, with calculated steady state oxygen diffusion profiles, for $D_e = 1 \times 10^{-3}$ m^2/h.

of contaminants in groundwater by diffusion of contaminants from the fractures into the matrix. This mechanism is part of the hydraulic containment scheme for the treatment cell. Where the matrix is more contaminated than the groundwater in the fractures the process can be reversed, with diffusion from the matrix into the fractures. This mechanism has potential application for the in situ remediation of the contaminated fractured till underlying the treatment cell and will be the subject of a future paper.

A 14-day pumping test was conducted in the shallow sandstone aquifer underlying the till to evaluate if pumping this aquifer would hydraulically isolate the site and provide containment for any contaminated fluids that might infiltrate vertically during bioremediation. The results of the pumping test indicated that a radius of influence of about 20 m could be expected for this aquifer at a pumping rate of 11.2 m^3/d. Consequently, hydraulic containment of much of the site could be achieved with three production wells.

Design of the Field-Scale Pilot Study

Because of the limitations imposed by the cold, dry climate and the generally fine-textured soils, the field pilot study design (Figure 8) includes features to provide as close to optimal conditions as possible for field bioremediation. The field pilot study was begun in the fall of 1992.

A test plot 35 m wide by 60 m long was excavated to a depth of 2.5 m. The soil was stockpiled as it was excavated and processed through an agricultural manure spreader to reduce particle size. The mixing diluted the highly contaminated zone at 1.6 m depth and brought this material into more intimate contact with both nutrients and native soil microbes.

Fertilizer was added to supplement N and P_2O_5 at the rate of 0.2 and 0.09 kg/m^3, respectively. Sawdust and wood chips from the previous pole peeling, cutting, and sharpening operations were added as bulking agents to half of the treatment plot at the rate of one part sawdust/wood chips to three parts soil (by volume) to improve the soil texture, permeability, and water-holding properties. The backfill was sloped at the surface to promote runoff.

A 10-cm depth of 2 cm washed gravel was placed in the bottom of the excavated test cell as a subsurface drainage layer. Slotted drainage pipe was placed in the gravel layer on 3-m centers to collect leachate or shallow groundwater from the bottom of the excavation. The underdrain piping also was vented to the surface through risers, providing a zone for the introduction of air at the base of the treatment cell.

A drip irrigation system was installed on the surface of the test cell to add moisture to the soil. Tensiometers were installed within the test cell to prevent overwatering by providing an indirect measurement of soil moisture and overriding the time-based irrigation control system, if water content within the bed should exceed 80% of field capacity. The drip irrigation system also provides the ability to add nutrients, if required at a later date.

An evaporation/infiltration barrier consisting of 0.5-mm-thick, black woven polyethylene covers the treatment cell to intercept precipitation, reduce infiltration,

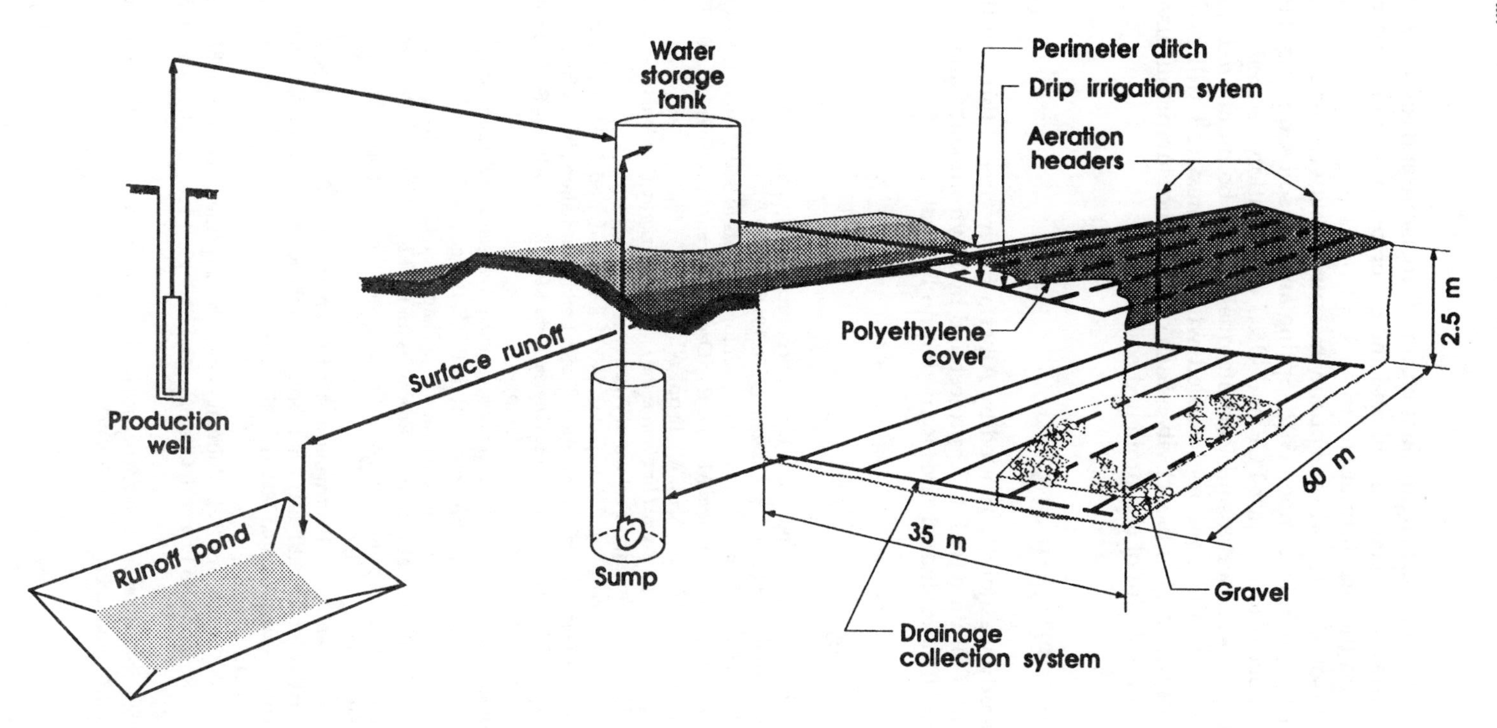

FIGURE 8. Schematic diagram showing design of the treatment cell for the field pilot. The production well is used to provide hydraulic containment of the site.

and reduce evaporation, to maintain uniform soil moisture and conserve moisture. A layer of straw between the soil and the cover provides both space for air circulation and thermal insulation.

Baseline soil and groundwater quality measurements were made immediately after construction in December 1992. Spring and fall soil sampling campaigns will be conducted to monitor PCP and O&G degradation, as well as nutrient availability. Groundwater and leachate quality will be monitored quarterly. Temperature and O_2 monitors installed within the test cell will provide an indication of the level of bioactivity. Each of these measurements also will be made in a control plot of contaminated soil.

ACKNOWLEDGMENTS

This study was jointly funded by Alberta Environment through the Help End Landfill Pollution (H.E.L.P.) project and by Environment Canada through the National Contaminated Sites Remediation Program.

REFERENCES

Bellin, C. A., G. A. O'Connor, and Y. Jin. 1990. "Sorption and Degradation of Pentachlorophenol in Sludge Amended Soils." *J. Environ. Qual. 19*: 603-608.

Boyd, S. A., M. J. Mikesell, and J. F. Lee. 1989. "Chlorophenols in Soils." In B. L. Sawhney and K. Brown (Eds.), *Reactions and Movement of Organic Chemicals in Soils*, pp. 209-229. SSSA Spec. Publ. 22, SSSA and ASA, Madison, WI.

Callahan, M. A. 1979. *Water-Related Environmental Fate of 129 Priority Pollutants.* Report 440/4-79-029a. U.S. Environmental Protection Agency.

Howard, P. H., R. S., Boethling, W. F. Jarvis, W. M. Meyland, and E. M. Michalenko. 1991. *Handbook of Environmental Degradation Rates.* Lewis Publishers, Inc., Chelsea, MI.

MacMillan, R. A. 1987. *Soil Survey of the Calgary Urban Perimeter.* Alberta Soil Survey Report No. 45, Alberta Research Council, 244p.

Monenco Consultants Ltd. 1990. *Phase II Environmental Assessment of the Peerless Wood Preservers' Site.* Prepared for Alberta Environment (H.E.L.P. Project), ADE 8598-8/6053, 1990-05-25.

Mueller, J. G., S. E. Lantz, B. O. Blattmann, and P. J. Chapman. 1991(a). "Bench-Scale Evaluation of Alternative Biological Treatment Processes for the Remediation of Pentachlorophenol- and Creosote-Contaminated Materials: Solid-Phase Bioremediation." *Environ. Sci. Technol. 25*:1045-1055.

Mueller, J. G., S. E. Lantz, B. O. Blattmann, and P. J. Chapman. 1991(b). "Bench-Scale Evaluation of Alternative Biological Treatment Processes for the Remediation of Pentachlorophenol- and Creosote-Contaminated Materials: Slurry-Phase Bioremediation." *Environ. Sci. Technol. 25*:1055-1061.

Seech, A. G., J. T. Trevors, and T. L. Bulman. 1991. "Biodegradation of Pentachlorophenol in Soil: The Response to Physical, Chemical and Biological Treatments." *Can. J. Microbiol. 37*:440-444.

U.S. EPA. 1989. *Seminar on Site Characterization for Subsurface Remediations.* Report CERI-89-224. U.S. Environmental Protection Agency, September.

PRACTICABILITY OF IN SITU BIOREMEDIATION AT A WOOD-PRESERVING SITE

K. R. Piontek and T. J. Simpkin

ABSTRACT

The results of large-scale in situ bioremediation pilot tests at a former wood-preserving facility are described. A 3-year testing program was undertaken to develop and test site-specific applications of several in situ treatment techniques. The testing program culminated in field pilot tests conducted in test cells up to approximately 1,600 square meters in size. In situ bioremediation techniques evaluated in field pilot tests were bioremediation combined with waterflood oil recovery and chemically enhanced soil flushing, use of nitrate as an alternative electron acceptor, and bioventing. The pilot test results were thoroughly evaluated to determine the applicability of full-scale in situ bioremediation at the site in terms of cleanup level achievable, duration, cost, and technical practicability. The tests and evaluation constitute a thorough assessment of the applicability of in situ bioremediation for cleanup of immiscible wood-preserving contamination in the subsurface. An overview of the tests and evaluations is presented. Test/evaluation methodology, data interpretation, and findings are described in detail in other technical reports and articles.

PROJECT BACKGROUND

CH2M HILL has been assisting Union Pacific Railroad (UPRR) in a comprehensive program to assess and address the contamination at the Laramie Tie Plant site in Laramie, Wyoming. For decades, railroad tie treating operations were conducted at the site. Creosote was the primary wood-preserving agent used and comprises the majority of the contamination now present at the site. Pentachlorophenol (PCP) also was used, but in much smaller quantities.

The remedial investigation of the site revealed widespread contamination by wood-preserving wastes consisting largely of an immiscible, heavier-than-water mixture of creosote and PCP in carrier oil. This mixture is slightly heavier than water, and as such is classified as dense, nonaqueous-phase liquid (DNAPL). Contamination at the site ranges from oil-saturated sands and gravels to groundwater containing microgram per liter levels of dissolved DNAPL constituents.

By far, most of the contamination at the site exists in a DNAPL pool that has accumulated at the base of a highly permeable alluvial deposit, at an average depth of approximately 3 meters. It is estimated that this alluvial deposit contains approximately 19 to 30 million liters of DNAPL over an area of approximately 36 hectares (90 acres). Fortunately, DNAPL migration into the underlying bedrock has been limited generally by the fine-grained character of the bedrock and the naturally upward groundwater flow.

While the site investigation was still under way, UPRR began implementing a series of measures to address the potential risks to human health and the environment posed by the site contamination. UPRR voluntarily ceased wood-preserving operations in 1983. In 1984, UPRR demolished the wood-preserving facility and remediated surface impoundments containing sludges and oily wastewaters.

The most significant potential risks posed by the site were associated with the intermittent seepage of DNAPL into the Laramie River as well as the more constant flux of contaminated alluvial groundwater into the river. A system to prevent further contaminant migration from the site, called the Contaminant Isolation System (CIS), was installed and began operating in 1986. This system consists of both a physical and a hydraulic barrier to contaminant migration. The physical barrier is a 3,000-meter-long soil-bentonite cutoff wall that surrounds the alluvial contamination. The hydraulic barrier is provided by approximately 5,000 meters of horizontal drainline that sustain inward groundwater flow to the site. The contaminated groundwater generated in the hydraulic containment system is treated in a system comprised of oil/water separation and activated carbon treatment.

The conclusion reached after approximately 5 years of operating and monitoring this system is that the actual and most imminent risks formerly posed by site contamination have been addressed by these remedial actions and through the other site management practices that are currently being employed.

TESTING PROGRAM OVERVIEW

In 1986, the U.S. Environmental Protection Agency (EPA) approved the CIS as an interim source control remedy, but stated that there was insufficient information to support selection of a final remedy for the site. Two issues that prevented selection of a final remedy were controversy over the long-term effectiveness of the CIS as a component of a final remedy and the unknown effectiveness of certain in situ treatment techniques that were viewed as having potential applicability.

UPRR implemented a testing program to evaluate several in situ treatment methods for remediation of the subsurface contamination at the site. The objectives of the 3-year program of bench-scale and field pilot tests were (1) to develop site-specific applications of these techniques, (2) to allow a more definitive evaluation of the techniques, and (3) to obtain the data needed for their potential application at the Laramie Tie Plant site.

At most sites with readily degradable contaminants, maintaining sufficient oxygen will be the rate-limiting factor for in situ bioremediation. Wilson and Sims (1989) state that the rate of biodegradation generally is not limited by the metabolic capability of microorganisms, but rather by the stoichiometry and mass transport limitations on the supply of oxygen. The premise used in structuring the testing program on the Laramie Tie Plant Site project was that the schedule for and the cost of full-scale in situ bioremediation would be driven by meeting the oxygen demand posed during the process.

As previously mentioned, it is estimated that as much as 30 million liters of DNAPL is present at the Laramie Tie Plant site. Early in the project, it was recognized that the DNAPL pool would exert a very large oxygen demand, even after DNAPL recovery operations, and that the technical practicability of in situ bioremediation at the site hinged on cost-effectively meeting this oxygen demand. Therefore, there were three key elements of the in situ bioremediation testing program: (1) quantifying the oxygen demand that would have to be met, (2) developing and evaluating methods for reducing the oxygen demand, and (3) evaluating options for meeting the oxygen demand.

Oxygen Demand

The oxygen demand is the mass of oxygen required to achieve the desired degree of contaminant biodegradation in the subsurface. For very preliminary evaluations, the stoichiometric oxygen requirement provides an appropriate estimate. However, given the importance of the oxygen demand on the practicability of in situ bioremediation at the Laramie Tie Plant site, this factor was determined experimentally in a column study (CH2M HILL 1990). The actual oxygen demand coefficient calculated from the test data ranged from 2.3 to 2.8 kg of oxygen per kg of organic removed.

Methods for Reducing the Oxygen Demand

Two pretreatment methods for reducing the subsurface oxygen demand that would be posed during subsurface bioremediation were examined: a waterflood oil recovery approach, and a chemically enhanced in situ soil washing technique. Reduction of subsurface contaminant levels through contaminant leaching during in situ bioremediation also was examined.

Waterflood Oil Recovery. The waterflood oil recovery approach that was developed for and pilot-tested at the Laramie Tie Plant site is based on the use of dual, horizontal recovery drainlines and parallel delivery drainlines. Production of water from the upper water recovery drainline and DNAPL from the lower drainlines enhances conditions for DNAPL recovery. Delivery of water to the subsurface in a parallel drainline is used to enhance the DNAPL flow gradient to the recovery system (Kuhn et al. 1989).

The success of this oil recovery method was illustrated in a 1989 pilot test in which nearly 900,000 liters of reusable oil was recovered from a subsurface area of approximately 1 hectare (2 acres) in a 90-day test period. Through removal of mobile DNAPL, the waterflood oil recovery approach is capable of significantly reducing the oxygen demand that would be exerted in subsurface bioremediation. During the pilot testing program, soil cores were collected before and after the waterflood oil recovery test, and the volume of oil produced during the test was tracked. It was determined that the total extractable organic (TEO) contaminant concentrations in alluvial materials initially containing mobile DNAPL could be reduced from an average of approximately 96,000 to an average of 25,000 mg/kg (dry weight basis). Assuming the same degree of contaminant reduction could be achieved in all areas of the site containing mobile DNAPL, the total mass of oily contamination in the alluvium could theoretically be reduced by approximately 50%. Assuming an initial oil volume of 25 million liters, and using the oxygen demand coefficient discussed earlier, this remaining contaminant mass corresponds to an oxygen demand of approximately 27 to 36 million kilograms of oxygen.

In Situ Soil Flushing. Chemically enhanced in situ soil flushing also was evaluated as a method that could potentially be used to further reduce the oxygen demand, thus increasing the technical practicality of subsurface bioremediation. A technique adapted from enhanced oil recovery techniques was first evaluated in bench-scale tests, and then in a small-scale field pilot test. Based on the results obtained in the initial pilot test, a larger field pilot test was conducted (Pitts et al. 1993).

Sheet-pile walls were driven to divide the waterflood oil recovery test area into three test cells. In 1989, an in situ soil flushing pilot test was performed in the middle cell, in an area of approximately 1,600 square meters. Chemical solutions were delivered to a central 37-meter delivery drainline. Through the delivery of chemicals and subsequent flushing of mobilized oil and residual chemicals, an equivalent of approximately 90,000 liters of oil were produced. The corresponding reduction in soil contaminant concentrations was from an initial concentration of approximately 25,000 mg oil/kg dry soil to a final concentration of 4,000 mg oil/kg dry soil. Final concentrations ranged from 290 to 11,000 mg oil/kg dry soil, reflecting the impact of preferred flow paths within the contaminated interval.

Based on the pilot test results, the combination of oil recovery and in situ soil flushing is capable of significantly reducing the subsurface oxygen demand in alluvial sands and gravels. The contaminant mass in the subsurface, expressed as the oxygen demand that would have to be met during subsequent in situ bioremediation, would be approximately 6 to 9 million kilograms. This oxygen demand projection considers both residual contaminants and residual polymer/surfactant from the soil flushing process.

This oxygen demand projection assumes that the oxygen demand exerted by residual polymer/surfactant is not significant. The flushing sequence used in the pilot test (including a polymer taper and multipore volume water flush) were effective in flushing out the polymer and surfactant. Additionally, given the solubility of these chemicals, their primary fate in full-scale bioremediation applications would be leaching from the subsurface and treatment in an aboveground process versus in situ biodegradation.

While the degree of contaminant removal achieved with this in situ soil flushing technique was encouraging, significant technical limitations of the technique were identified in the testing program (CH2M HILL 1990). The chemical solution used in the process resulted in the formation of a very stable oil emulsion that (1) requires extensive treatment, (2) limits the ability to recover the creosote from the spent formulation and reuse the soil washing formulation, and (3) creates significant amounts of process residuals that must be managed. Given these technical challenges, the scale and complexity of operations that would be involved in fullscale in situ soil flushing at the site, and associated costs, the technical practicability of this technique is questionable.

Contaminant Leaching. The third process examined for reducing the oxygen demand was contaminant leaching during subsurface bioremediation. The primary contaminant removal processes in subsurface bioremediation involving soil flushing with oxygen-enriched water are (1) contaminant biodegradation in the subsurface contaminated zone, and (2) extraction of aqueous-phase contamination from the subsurface, followed by contaminant removal in an aboveground treatment process. Prior to the pilot tests, it was recognized that high rates of contaminant leaching would lower the subsurface oxygen demand that would have to be met, and would therefore increase the technical practicality of subsurface bioremediation. Based on reports in the literature, it was hoped that stimulation of subsurface microbial activity would result in enhanced rates of contaminant leaching through microbial processes (e.g., biosurfactant production).

The pilot tests showed that some contaminant leaching does occur, and that biological water treatment can be employed to treat the water before it is reoxygenated and recycled. The rate of contaminant leaching that was observed, however, was lower than hoped. The leachate chemical oxygen demand (COD) dropped fairly rapidly with time (see Figure 1), approximating a first-order decay model. If the first-order decay is projected for more than 2 years, leaching with aboveground contaminant removal would account for less than 20% of the total subsurface oxygen demand that would have to be met in subsurface bioremediation directly following oil recovery at the Laramie Tie Plant site.

Options for Meeting the Oxygen Demand

In addition to examination of processes that would reduce the oxygen demand, the testing program included examination of three methods for meeting the subsurface oxygen demand that would be posed during subsurface bioremediation. These were (1) the use of pure oxygen, (2) the use of nitrate instead of oxygen, and (3) bioventing. Hydrogen peroxide was not used in the pilot tests. Based on the high subsurface oxygen demand, resulting H_2O_2 requirements, H_2O_2 costs, likely inefficiencies in H_2O_2 resulting from premature H_2O_2 decomposition (Huling et al. 1990), it was concluded prior to the pilot tests that meeting the subsurface oxygen demand with H_2O_2 would not be practical in this application.

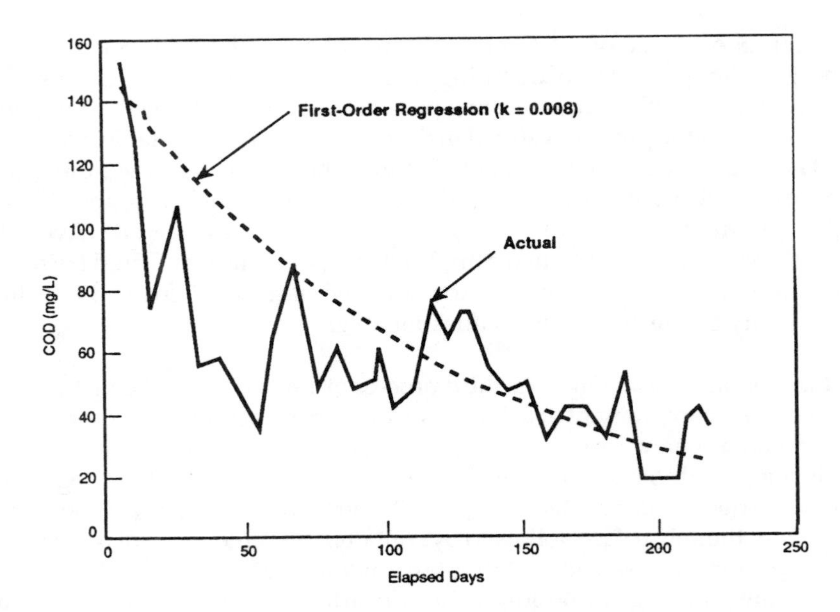

FIGURE 1. Contaminant leaching during bioremediation pilot study.

Use of Pure Oxygen. Use of pure oxygen was the "conventional" in situ bioremediation method used in the pilot tests. In this method, oxygenated, nutrient-enriched water was delivered to the subsurface contaminated zone via injection drainlines. Using pure oxygen and pressurized mixing conditions, oxygen concentrations of 30 to 40 mg/L typically were achieved. Two separate field pilot tests of this method were conducted in 1989. A test was performed in an approximate 60-square-meter test cell to examine the efficacy of using in situ bioremediation as a polishing step following chemically enhanced in situ soil washing. In situ bioremediation directly following waterflood oil recovery was tested in a larger, approximate 1,600-square-meter test cell.

In both tests, stimulation of subsurface microbial contaminant biodegradation was evident. One indication was the rapid depletion of oxygen from the injected water. Figure 2 presents the dissolved oxygen breakthrough curves for monitoring wells located approximately 1.5 meters downgradient of the injection point. In the pilot test of bioremediation directly following oil recovery, oxygen breakthrough occurred after 21 weeks. In the pilot test of bioremediation following both oil recovery and soil flushing, oxygen breakthrough occurred after 7 weeks, due to the lower residual oxygen demand.

The combination of oxygen/contaminant mass balance data and results of associated tests (e.g., microbial enumeration) established that the contaminant biodegradation activity of indigenous microorganisms could be stimulated in both the heavily contaminated soils that are present after waterflood oil recovery, and in the soils that had been subjected to the harsh conditions associated with

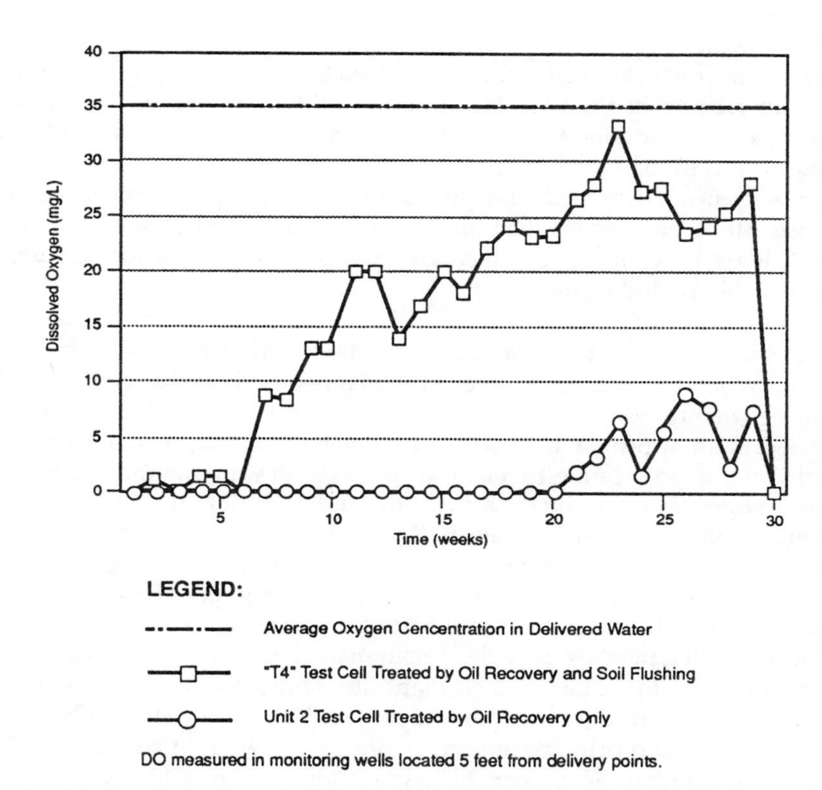

FIGURE 2. Oxygen breakthrough in subsurface bioremediation pilot study.

the in situ soil flushing process. The tests also confirmed the conceptual model of in situ bioremediation that had been developed for the site, i.e., the concept that the schedule for and cost of the process would be driven by meeting the oxygen demand posed by the subsurface contamination.

Use of Nitrate. In the bioremediation field, biodegradation of certain contaminants under denitrifying conditions is considered a promising technique. Under denitrifying conditions, nitrate is used instead of oxygen as the electron acceptor for microbial metabolism and contaminant breakdown. In planning the testing program, it was thought that the relatively high solubility and lower cost of nitrate could potentially provide significant advantages over oxygen in terms of cost and schedule. A pilot test was performed to determine if the indigenous soil microorganisms could use nitrate instead of oxygen to biodegrade site contaminants and, if so, to determine the rate of nitrate reduction.

A pilot test involving injection of nitrate- and nutrient-enriched water was performed. A key element of the test protocol was determination of dissolved oxygen and nitrate in the injected water and in a network of monitoring wells.

The monitoring data obtained in the test showed the oxygen was consumed in the immediate vicinity of the injection drainline. Some nitrate reduction in the downgradient anoxic zone was observed. However, the rate was rather low and decreased with time.

This subsurface bioremediation pilot test showed that the soil microorganisms could use nitrate but also showed that the sustainable nitrate use rate would be relatively low and would not significantly reduce the subsurface oxygen demand in a reasonable period of time.

Bioventing. Bioventing was also evaluated during the field pilot testing program. When the pilot tests were being planned and conducted (1986-1989), bioventing was still considered a rather innovative approach. The pilot test was conducted in an approximate 1,600-square-meter test cell. The application of this technique at the Laramie Tie Plant site involved dewatering the contaminated subsurface interval and inducing airflow through the interval. In this approach, air, rather than oxygenated water, is the medium that carries oxygen to the biodegradation zone. Relative to water, air has a higher oxygen content and a lower viscosity, and the permeability of soil to air is higher. Thus, oxygen mass transport to the subsurface can more cost-effectively be achieved using air.

The bioventing pilot test provided preliminary data on the potential applicability of the approach to the Laramie Tie Plant site. Due to the overall schedule and regulatory/administrative constraints on the project, a test lasting only 1 month was conducted. The primary objectives of the test were to confirm the basic process feasibility (that contaminant biodegradation could be achieved by coupling dewatering and air venting), to obtain some information on the process rate, and to verify that the entire subsurface contaminated zone could be oxygenated at reasonable airflow rates.

The primary monitoring during the test was subsurface gas composition to track microbial contaminant biodegradation activity as evidenced by oxygen depletion and carbon dioxide generation. Test data provided strong evidence that the approach was effective in stimulating the contaminant biodegradation of the native microorganisms. The oxygen use rate determined in the test was 50 to 140 kilograms of oxygen per day (15 to 45 mg oxygen/kilogram soil per day). A higher oxygen delivery and use rate was achieved in this bioventing pilot test than the maximum rate achieved in the comparable pilot test involving injection of oxygenated water (approximately 7 kg oxygen per day).

The pilot test confirmed the basic feasibility of a bioventing approach and illustrated the potential advantages of such an approach. The rate-limiting factor in the conventual bioremediation approach involving delivery of oxygenated water, electron acceptor mass transfer to the contaminated zone, was overcome. However, given the limited duration of the pilot test, there are still uncertainties with application of this technique to the site. These include the biodegradation rate that could be sustained over time as the more readily biodegraded contaminants are removed, possible changes in rate resulting from changes in the vadose zone water chemistry, requirements and methods for management of soil moisture and inorganic nutrient levels, and the cleanup level achievable.

POTENTIAL FOR FULL-SCALE IN SITU BIOREMEDIATION AT THE LARAMIE TIE PLANT SITE

This section summarizes an evaluation of the potential application of subsurface bioremediation to the UPRR Laramie Tie Plant site. The discussion is based on an extrapolation of the pilot test results, where possible.

Cleanup Levels Achievable

The pilot test results suggest that although the soil cleanup levels achievable with subsurface bioremediation in the most permeable subsurface soils may be relatively low, the spatial variations in cleanup levels are likely to be significant. In the most permeable soils nearest the injection drainlines, concentrations of total polycyclic aromatic hydrocarbons (PAHs) after bioremediation were as low as 5 mg/kg in the pilot test cell where both oil recovery and in situ soil flushing were employed as "pretreatment" steps, and as low as 45 mg/kg in the test cell where bioremediation directly followed oil recovery.

Subsurface heterogeneities are a significant impediment to bioremediation at the Laramie Tie Plant site. The variation in the cleanup levels achieved in the bioremediation pilot tests was as great as two orders of magnitude. These variations are believed to be due primarily to heterogeneities in the subsurface soil permeability. The importance of relative permeability is illustrated by the soil coring results presented in Figure 3, which shows the contaminant concentrations at the conclusion of the sequence of the oil recovery, in situ soil flushing, and bioremediation pilot tests in the smaller pilot test facility. Soil core samples collected at the conclusion of the pilot tests had highly variable contaminant concentrations. Soil from a coarse sand layer had contaminant levels that were much lower than soil from adjacent layers of fine sand.

Based on these pilot test findings, the following observations can be made regarding the soil cleanup levels achievable in a full-scale application of subsurface bioremediation: (1) low cleanup levels may be achieved in soils of high permeability in reasonable time periods, (2) extended periods of time will be required to achieve similar levels in soils of lower but still moderate permeability such as fine sands, and (3) there will always be some soils that will receive little or no treatment, due to their very low permeabilities and the resulting lack of oxygen transfer into the soils.

Duration and Magnitude of Operations

In addition to the bulk hydrogeologic characteristics, heterogeneities in subsurface conditions will significantly impact the cost and the effectiveness of in situ bioremediation. At a site with interbedded layers of contaminated soil with varying hydraulic conductivities, the contaminated layer of lowest conductivity often will control the time required for remediation. Consequently, when calculating the remediation period from Darcy's equation, the hydraulic

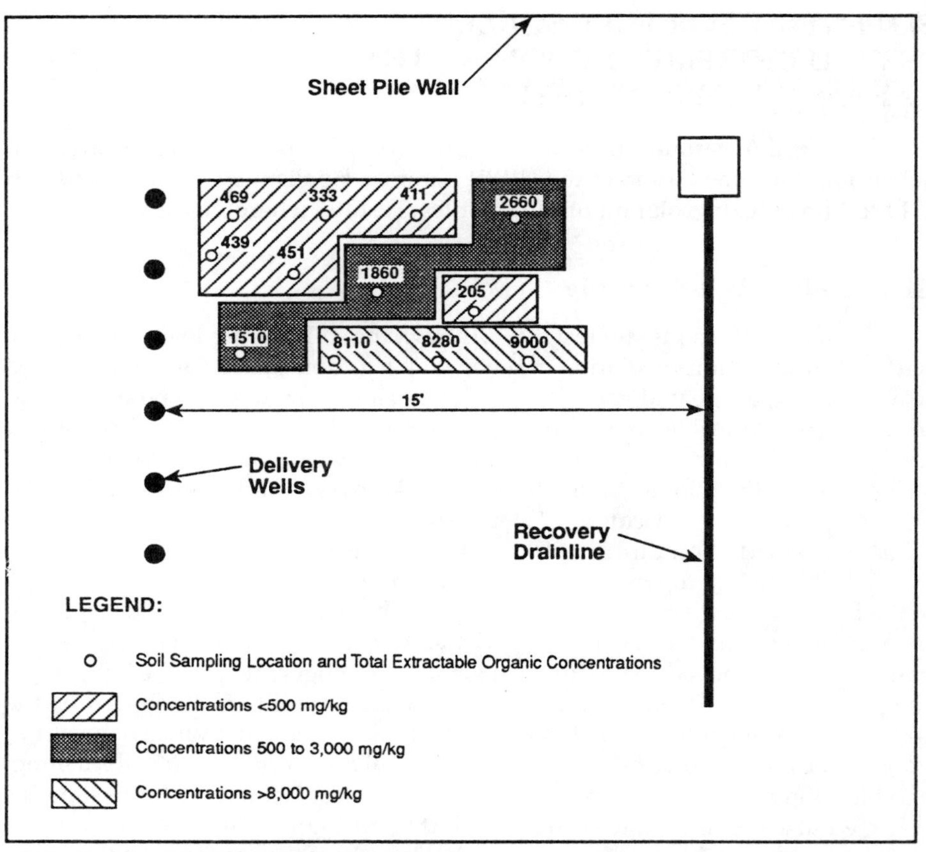

Note: Locations shown are approximate.

FIGURE 3. Total extractable organic soil concentrations after oil recovery, soil flushing, and in situ bioremediation.

conductivity of the least permeable layer that must be remediated should be considered.

A simple model of subsurface bioremediation incorporating the considerations outlined above was used to estimate the duration of subsurface bioremediation at the Laramie Tie Plant site. The model was based on a mass balance of subsurface oxygen demand. For subsurface bioremediation immediately following oil recovery, the model predicts a remediation period ranging from approximately 200 to 350 years with a oxygenated water delivery rate of approximately 60 million liters (15 million gallons) per day. With this scenario, there would be a great deal of uncertainty regarding the ability to sustain this high flowrate for the predicted required duration.

For subsurface bioremediation following both oil recovery and in situ soil flushing, the model predicts a remediation period ranging from approximately

45 to 60 years. This illustrates the desirability of reducing the oxygen demand before commencing bioremediation. However, as discussed earlier, concerns regarding the technical feasibility and practicability of in situ soil flushing were identified in the testing program.

CONCLUSIONS

At the Laramie Tie Plant site, it was recognized that the large DNAPL mass present in the subsurface, and resulting oxygen demand, posed very significant challenges to cost-effective implementation of bioremediation. A strategy for evaluating full-scale bioremediation at the site was developed around reduction of subsurface contaminant levels through pretreatment and effective means of meeting the subsurface oxygen demand. A variety of methods of reducing or meeting the bioremediation oxygen demand were examined in a testing program that culminated in field pilot tests.

Only the waterflood oil recovery method was proven a technically practicable method for significantly reducing the oxygen demand. Full-scale application of the method is projected capable of reducing the bioremediation oxygen demand by about 50%. Methods examined for meeting the oxygen demand were use of pure oxygen in a conventional flooding approach, use of nitrate instead of oxygen, and bioventing. The conventional approach effectively stimulated contaminant biodegradation, but relative permeability effects were found to increase oxygen delivery requirements above the theoretical, stoichiometric requirement due to resulting inefficiencies in oxygen delivery. Based on these findings, the flowrate and duration of full-scale subsurface bioremediation following oil recovery were projected. The results illustrate the challenges posed by the high oxygen demand, and show that a conventional in situ bioremediation approach operation would be very lengthy and expensive.

The pilot tests indicated that use of nitrate would not significantly reduce the subsurface oxygen demand. A preliminary pilot test of short duration indicated that bioventing could be effective for meeting the subsurface oxygen demand. This method would require dewatering of the contaminated zone for an extended period, a factor that significantly limits the cost-effectiveness of this approach.

One factor that significantly impacts the practicability of any in situ treatment technique is the contaminant mass that would remain in the subsurface after treatment. Although in situ treatment culminating with bioremediation would achieve significant contaminant removal from a "percent of initial contamination" perspective, contamination would remain in low permeability strata such as silt lenses in the alluvium and the shallow bedrock. This contamination would be sufficient to act as a lesser, but continuing, source of groundwater contamination. Thus, implementation of full-scale application of subsurface bioremediation at the Laramie Tie Plant would not necessarily eliminate or reduce the need to continue current groundwater management practices, including CIS operations and groundwater monitoring.

UPRR is proceeding with full-scale application of the waterflood DNAPL recovery technique at the Laramie Tie Plant site. This technique is a cost-effective means of removing the mobile, concentrated contamination from the subsurface. The determination of what, if any, additional remedial measures are to be undertaken at the site will be based on a RCRA Corrective Measures Study (CMS) being performed by UPRR. The CMS will include a comparative evaluation of various remedial alternatives, including the in situ treatment methods discussed in this paper. The CMS will examine the benefits, limitations, and costs of in situ bioremediation based on the findings summarized above.

REFERENCES

CH2M HILL. 1990. *Union Pacific Railroad Laramie Tie Plant, Milestone IV Report.* Submitted to U.S. Environmental Protection Agency Region VIII, Denver, CO.

Huling, S.G., B. E. Bledsoe, and M. V. White. 1990. *Enhanced Bioremediation Utilizing Hydrogen Peroxide as a Supplemental Source of Oxygen: A Laboratory and Field Study.* Report EPA/600/2-90/006, U.S. Enviromental Protection Agency, February.

Kuhn, R. C., T. Sale, K. Piontek, and D. Stieb. 1989. "Recovery of Wood-Treating Oils from an Alluvial Aquifer Using Dual Drainlines." *Proceedings of the Conference on Petroleum Hydrocarbons and Organic Chemicals in Groundwater.* National Water Well Association, Dublin, OH.

Pitts, M. J., Kon Wyatt, T. C. Sale, and K. Piontek. 1993. *Utilization of Chemical-Enhanced Oil Recovery Technology to Remove Hazardous Oily Water From Alluvium.* SPE 25153, Society of Professional Engineers, Richardson, TX.

Wilson, J. T., and R. C. Sims. 1989. "In Situ Treatment Design—Surface and Subsurface." *Bioremediation of Hazardous Waste Sites Workshop,* U.S. EPA Technology Transfer CERI-89-11, February 1989.

BIOLOGICAL TREATABILITY STUDIES ON SURFACE IMPOUNDMENT SLUDGE FROM A CHEMICAL MANUFACTURING FACILITY

M. D. Lee, W. A. Butler, T. F. Mistretta, I. J. Zanikos, and R. E. Perkins

ABSTRACT

A series of laboratory treatability studies and a pilot-scale test were performed to evaluate biological treatment of sludge contaminated with a wide range of volatile and semivolatile organic compounds from two surface impoundments at a chemical manufacturing facility. Volatilization played a significant role in the removal of methylene chloride, toluene, tetrachloroethene, chlorobenzene, xylenes, and 1,2-dichlorobenzene in the pilot study. Biological removal of these volatile compounds was promoted with a lower air flowrate in the laboratory studies. Removal efficiencies generally exceeded 50% for the semivolatile organics in the laboratory and pilot studies. Up to 1.5%/day of the added ^{14}C-labeled chlorobenzene, 1,2-dichlorobenzene, naphthalene, and *bis*(2-ethylhexyl) phthalate were converted to carbon dioxide. Numbers of heterotrophic bacteria increased up to 7 orders of magnitude during the studies and specific-contaminant-utilizers increased by up to 4 orders of magnitude. Based on these studies, aerobic biological treatment would be effective in reducing the concentrations of the organic contaminants in the sludge, although volatilization would account for a significant portion of the removal of the volatile contaminants. A less expensive alternative was chosen for full-scale remediation of the basins.

INTRODUCTION

At the chemical manufacturing facility discussed in this investigation, process wastewaters, process sludges, and activated carbon from the on-site wastewater treatment plant were discharged to two surface impoundments. The 300,000 m^3 of sludge within the surface impoundments is contaminated with a number of organic compounds as well as heavy metals. Chlorobenzene (CB), 1,2-dichlorobenzene (*o*DCB), butyl benzyl phthalate (BBP), and *bis*(2-ethylhexyl) phthalate (*b*EHP) were the organic contaminants present at the highest concentrations in the sludge. The sludge has a solids content of 25 to 35% and a pH of about 5.

Slurry-phase biological treatment was one of the remediation alternatives evaluated for treatment of these surface impoundments. Processes that can be operated in situ may avoid the Land Disposal Restrictions under the Resource Conservation and Recovery Act (RCRA) and thus qualify for less stringent cleanup goals (Zanikos et al. 1992). Stabilization of the heavy metals would be required after biological treatment of the organics.

A series of laboratory treatability studies were run to determine if there were active microbes in the sludge capable of biodegrading the organic contaminants and the extent of removal of these organics that could be achieved. A pilot bioremediation system was installed and operated in one of the surface impoundments. This paper compares the removal efficiencies and organic contaminant mass balances seen in the laboratory studies with those of the field pilot study.

METHODS

Aerobic Drum Bioreactor Studies

Four aerobic drum bioreactor studies were run to determine the effectiveness of aerobic treatment, to show the variability in the treatment effectiveness with the different batches of sludge from the impoundments, and to generate biotreated sludge for dewatering and heavy metal stabilization investigations. The drum bioreactor was constructed from a plastic-lined drum and was sealed with a drum lid. The sludge was placed into the 208-L (55-gallon) drum and diluted with tapwater, and inorganic nutrients were added.

Table 1 summarizes the quantities of sludge, nutrients, water, and pH-adjustment agents, as well as operational parameters for each of the drum bioreactor (DB) studies. The sludge within the reactor was aerated with compressed air using a fine-bubble diffuser and stirred with a mixer. The effluent air was treated by passage through a liquid overflow trap to collect foam and a canister containing approximately 5.5 kg of 6 to 14 mesh activated carbon to remove volatiles, and then was discharged to the outside. A vacuum pump was used to create a slight negative pressure in the drum bioreactor to pull the air through the carbon canister. The drum bioreactor studies were operated between 18 and 48 days. The first three drum bioreactor studies (DB-I, DB-II, and DB-III) were operated at summer temperatures; the final drum bioreactor study (DB-IV) was operated at 5 to 8°C to simulate winter conditions.

Field Pilot Study

A 520-m^2 bioremediation pilot cell was constructed within one of the surface impoundments using a custom-designed floating baffle to separate the pilot cell from the surrounding sludge and water (Zanikos et al. 1992). A schematic depicting the pilot cell is presented in Figure 1 and operational parameters for the pilot

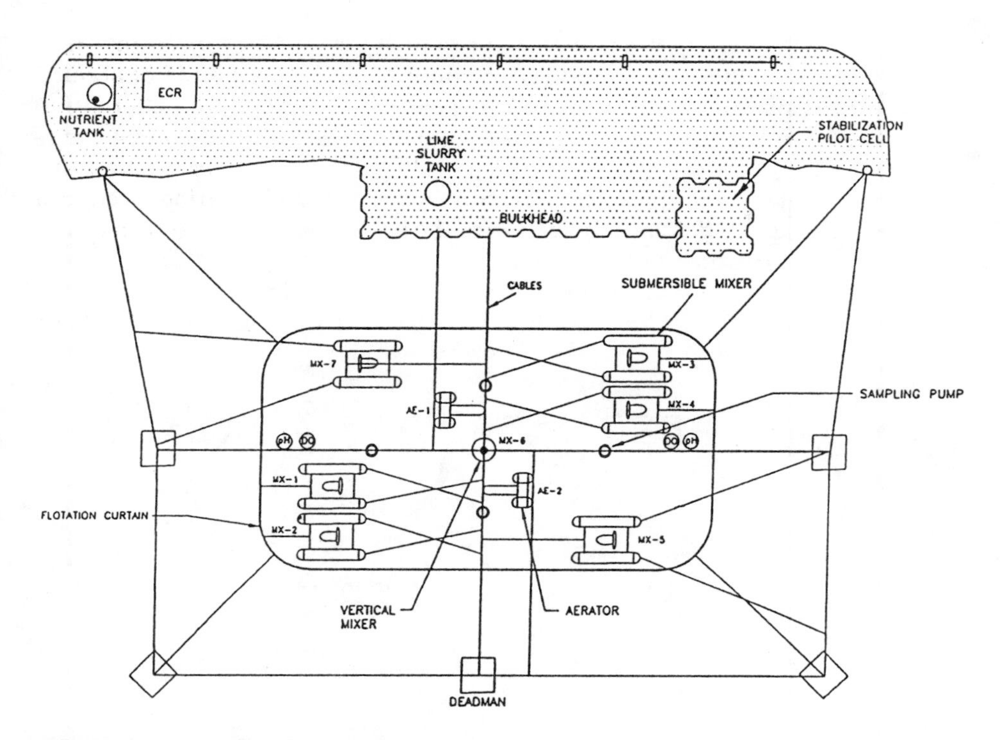

FIGURE 1. Pilot study equipment schematic.

are presented in Table 1. Six submersible mixers and one vertical mixer were used to mix the sludge in the pilot cell. Two aspirating aerators provided up to 17.0 m^3 of air per minute or the equivalent of 33 L/min per square meter of surface area. The mixers and aerators were able to uplift, on average, 1.9 m of the 2.4 m of the sludge in the pilot cell. The solids concentration of the slurry in the cell was maintained at 10%. Feed systems for lime slurry and nutrients also were installed to maintain the pH and to provide the necessary inorganic nutrients.

Enumeration of the Microbial Community

Enumeration of both heterotrophic and specific-contaminant-utilizing microbial populations were made on the sludge samples. The heterotrophic microbial population estimates were made by pour-plating dilutions of the samples on nutrient agar (Bacto) and incubating the plates at room temperature (22°C) for 7 to 21 days under aerobic conditions. The specific-contaminant-utilizing populations were enumerated by spread-plating dilutions of the samples onto triple-washed agar plates prepared with a mineral salts medium (1,000 mg ammonia nitrate, 400 mg monopotassium phosphate, 600 mg disodium phosphate, 10 mg calcium chloride, 100 mg sodium carbonate, 200 mg magnesium sulfate, 20 mg manganese

TABLE 1. Bioreactor and pilot studies operational data.

Study	DB-I	DB-II	DB-III	DB-IV	Pilot
Weight of Sludge (kg)	50	50.6	50.3	64.3	1,400,000
Nutrient Solution					
Tap water (L)	62	62	62	15	9,100
Na_2HPO_4 (g)	500	500	500	600	816,000
NH_4Cl (g)	100	100	100	120	136,000
$CaCO_3$ (g)	500	1,000	0	0	
pH Adjustment and Maintenance					
$CaCO_3$ (g)	0	4,000	17,000	0	
NaOH (g)	14	0	165	0	
$Ca(OH)_2$ (g)	0	0	0	35	530,000
Dry Weight (percent)					
Initial	11.2	4.4	25.0	17.0	18.3
Final	14.9	17.7	23.5	15.9	4.0
Average	12.5	9.8	23.5	16.3	11.7
Air Flowrate (L/min per m^2 surface area for x days)	57 (30 D)	57 (2 D) 24 (4 D) 33 (12 D)	24 (28 D)	8 (48)	33 (42)
Temperature (°C)	22-25	20-30	27-37	5-8	
Total Weight (kg)	113	118	130	80	1,400,000
Final pH	7.1	5.8	7.0	7.2	7.0
Treatment Time (Days)	30	18	28	48	42

sulfate, and 5 mg ferric chloride per L distilled water) and incubating the plates at 22°C for about 28 days in an aerobic atmosphere containing CB, *o*DCB, hexadecane, NAP, or *b*EHP.

Mineralization Studies

Samples of the sludge from the DB-IV and the pilot studies also were used to determine the mineralization of 1-^{14}C-labeled NAP, uniformly ^{14}C-labeled CB, uniformly ^{14}C-labeled *o*DCB, ring ^{14}C-labeled *b*EHP (Sigma Chemicals), and uniformly ^{14}C-labeled glucose (New England Nuclear). Mineralization assays for the DB-IV and pilot studies were conducted using two procedures. For both procedures, the vials were incubated at room temperature (25 ± 2°C) with shaking at 150 rpm. The quantity of carbon dioxide produced in the abiotic controls (poisoned with a final concentration of 3% formaldehyde) was subtracted from the active vials.

For CB and *o*DCB, the sludge was placed in small vials and 10 µL of the ^{14}C-radiolabeled substrate dissolved in methanol was added to triplicate vials. For the DB-IV study Days 0 to 20 samples, the 7-mL vials were completely filled with sludge. Thereafter, 5 mL of sludge was added to the vials, leaving 2 mL of headspace. After incubation for 7 days, the quantity of ^{14}C-radiolabeled carbon dioxide produced was determined using liquid scintillation counting following stripping of aliquots to which acid and base had been added (DB-IV samples from Days 0 to 27) or by collection of the carbon dioxide in a sodium hydroxide trap after acidification of the sludge (DB-IV samples from Days 34 to 48 and the pilot study samples).

The second procedure was used for the glucose, NAP, and *b*EHP. To each of three vials including a center well and 26 mL of headspace was added 5 mL of the sludge. Ten µL of the ^{14}C-radiolabeled substrate dissolved in a solvent of 18% ethanol for glucose, acetone for NAP, or methanol for *b*EHP was added to the sludge. At the end of the incubation period (24 hours for glucose and either 2 or 7 days for the *b*EHP and NAP), the samples were acidified and the $^{14}CO_2$ collected in the sodium hydroxide added to the center well.

Organic Contaminant Analyses

Sludge samples were collected before turning on the air initially and periodically thereafter. The samples were analyzed by purge-and-trap gas chromatography/mass spectrometry (GC/MS) for volatile organic compounds by EPA SW-846 Method 8240 and by solvent extraction and GC/MS quantitation for the semivolatile organic compounds by EPA SW-846 Method 8270 (United States Environmental Protection Agency 1986).

The quantities of volatile organic contaminants in the effluent airstream were determined for the DB-II, DB-IV, and pilot studies. Method TO-14 was used for the DB-II and pilot studies; this method collects a sample of the airstream over an 8-hour period in an evacuated canister and analyses are made on a portion of the collected air by GC/MS (United States Environmental Protection Agency

1991). EPA Method TO-1 was used for the DB-IV study; a portion of the airstream was passed through a charcoal tube for a given period, the tube extracted with methanol, and the extract analyzed by GC/MS (United States Environmental Protection Agency 1991). During the pilot study, air emission rate samples were collected from three emission isolation flux chambers placed over the mixing and aeration zones after 0, 10, and 29 days (Zanikos et al. 1992).

RESULTS

Organic Contaminant Removal From Sludge

The maximum concentrations of the organic contaminants found during each study and the percent of the maximum contaminant concentrations that were removed are reported in Table 2. Where 100% removal is noted, the final concentration was below the analytical method detection limit. Many of the volatiles found in the sludge also were detected in the air emitted from the drum bioreactor and pilot studies. The total volatilized quantity of each compound was calculated by determining the arithmetic mean concentration between air samples and multiplying by the average air flowrate for the period between samples and the number of days in each period. Table 3 presents the initial quantities of the contaminant in the sludge (the maximum measured concentration of the contaminant on a dry weight basis times the mass of sludge solids), the quantity of contaminant remaining in the sludge at the end of the treatment period, and the calculated quantity removed by volatilization. The percent of the initial quantity of each organic contaminant remaining at the end of the study, the percent volatilized, and the percent biodegraded (the fraction not accounted for in the residual or removed by volatilization) also are presented in Table 3.

The concentrations of the volatile organic contaminants in the drum bioreactor and pilot studies were rapidly reduced by the microbial population and by stripping from solution with reductions of greater than 99%, except for methylene chloride, xylenes, and ethylbenzene which were reduced greater than 90%. Very little liquid or foam was found in the liquid trap from the drum bioreactor which would suggest that there was no appreciable removal of the organics by this mechanism. The maximum concentrations of the semivolatiles detected in the analyses for the aerobic drum bioreactor and pilot studies were reduced between 24 and 100% (Table 2). The only semivolatiles detected consistently in the air effluent samples were *o*DCB, 1,4-dichlorobenzene, and 1,2,4-trichlorobenzene (Table 3).

Microbial Enumeration and Mineralization Activity

The heterotrophic microbial population estimates for the aerobic drum bioreactor and pilot studies increased to greater than 10^{10} colony-forming units per g dry weight (CFU/g) with the maximum counts of 3.6×10^{13} CFU/g observed in the DB-IV study at the lower winter temperature (Figure 2). The maximum number of heterotrophic microbes in the pilot study was 1.2×10^{10} CFU/g.

TABLE 2. Contaminant removal in drum bioreactor and pilot studies.

Study	DB-I		DB-II		DB-III		DB-IV		Pilot	
Conc. (mg/kg dry wt)	Max	% Rem	Max	% Rem	Max	% Rem	Max	% Rem	Max	% Rem
Volatiles										
Methylene Chloride	2.4	100	67.0	93	3.4	96	5.8	100		
Chlorobenzene	1.7	100	9.9	100	1.4	100	280.0	97	1100.0	>99
Toluene	1.9	100	9.9	100	0.3	100	59.0	96	36.0	100
Xylenes	1.4	100	8.4	100	0.5	100	120.0	93	110.0	95
Tetrachloroethene			1.7	100	0.3	100	5.4	100	0.4	100
2-Butanone	7.6	100					52.0	100		
Ethylbenzene			1.9	100			9.4	83	23.0	100
Acetone					1.6	100	9.9	100	0.8	100
Chloroform					0.3	100	230.0	100	6.3	100
Semivolatiles										
1,2-Dichlorobenzene	150.0	97	370	92	120.0	81	470.0	79	5900.0	99
1,3-Dichlorobenzene					8.2	76	11.0	100	1.5	100
1,4-Dichlorobenzene	3.4	100	12.0	100	25.0	70	41.0	66	50.0	100
1,2,4-Trichlorobenzene	15.0	81	180.0	70	82.0	73	88.0	66	2400	78
Hexachlorobenzene			11.0	61	3.4	36			6.1	62
Nitrobenzene	8.0	65	48.0	73	160.0	42			170.0	92
2,4-Dinitrotoluene			700.0	77	90.0	84			450.0	87
2,6-Dinitrotoluene			180.0	86					230.0	72

(table continues)

TABLE 2. (continued)

Study	DB-I		DB-II		DB-III		DB-IV		Pilot	
Conc. (mg/kg dry wt)	Max	% Rem	Max	% Rem	Max	% Rem	Max	% Rem	Max	% Rem
Semivolatiles (cont'd)										
Aniline							35.0	100	43.0	58
2-Nitroaniline			74.0	96	18.0	78				
3-Nitroaniline			110.0	71	14.0	77				
4-Nitroaniline			180.0	66	18.0	72			48.0	100
4-Chloroaniline					12.0	100	590.0	90	11.0	100
2,4-Dichlorophenol			37.0	54	7.7	100	8.7	100		
Naphthalene	7.0	73	64.0	67	21.0	81	87.0	79	120.0	98
2-Methylnaphthalene	6.9	64	92.0	63	22.0	82	56.0	70	18.0	67
Phenanthrene	6.6	26	94.0	57	24.0	70	18.0	28	11.0	50
Anthracene					1.0	100	5.6	100		
Dibenzofuran			29.0	59	8.0	66	9.3	100	2.6	100
Fluorene			33.0	64	8.0	72	8.2	24	3.2	100
Fluoranthene			39.0	56	8.0	100			4.3	100
Pyrene			50.0	60	7.0	57			1.6	100
Acenaphthalene			44.0	64	12.0	68	19.0	26		
Butyl Benzyl Phthalate	160.0	50	2900.0	62	400.0	86	720.0	74	210.0	73
Bis(2 ethylhexyl) Phthalate	2700.0	73	7600.0	58	1600.0	88	1900	70	2100.0	68
Di-n-octyl Phthalate					57.0	94	12.0	100	38.0	100
N-Nitrosodiphenylamine					9.7	76			240.0	92

Max = Maximum concentration found during study.
% Rem = % of maximum concentration removed during study.
100% Removal = removal to below the analytical method detection limit.

TABLE 3. Fate of volatile organic contaminants in drum bioreactor and pilot studies.

Compound	Study	mg in Reactor	mg Remaining	% Remaining	mg Volatilized	% Volatilized	% Biodegraded
Methylene Chloride	DB-II	348	94	27	908	>100	0
	DB-IV	78.7	0	0	91.1	>100	0
Toluene	DB-II	51	0	0	928	>100	0
	DB-IV	800	26.7	3	42.8	5	92
	Pilot (kg)	9.2	0	0	43.7	>100	0
Tetrachloroethene	DB-II	8.8	0	0	82	>100	0
	DB-IV	73.3	0	0	6.2	8	92
	Pilot (kg)	0.1	0	0			100
Chloroform	DB-IV	3120	0	0	6.6	0	>99
	Pilot (kg)	1.6	0	0			100
Chlorobenzene	DB-II	51.4	0	0	560	>100	0
	DB-IV	3800	104	3	63.8	2	95
	Pilot (kg)	282	0.1	0	376.8	>100	0
Xylenes	DB-II	43.6	0	0	505	>100	0
	DB-IV	1628	112	7	55.1	3	90
	Pilot (kg)	28.2	0.3	1	24.7	88	11
1,2-Dichlorobenzene	DB-II	1920	627	33	811	42	25
	DB-IV	6380	1284	20	0.7	0	80
	Pilot (kg)	1510	4.4	0	578	38	62
1,4-Dichlorobenzene	DB-II	62.3	0	0	2.3	4	96
	DB-IV	556	178	32			68
	Pilot (kg)	12.8	0	0			100
1,2,4-Trichlorobenzene	DB-II	935	1130	>100	52	6	0
	DB-IV	1194	382	32			68
	Pilot (kg)	615	30.2	5	99.1	16	79

mg in Reactor = initial quantity in sludge. mg Remaining = quantity remaining after treatment. % Remaining = percent remaining after treatment. mg Volatilized = quantity in air emissions. % Volatilized = percent of initial quantity volatilized; >100 % where more volatilized than estimated to have been present initially. % Biodegraded = percent not accounted for in remaining or volatilization fractions.

For the DB-IV study, the numbers of *o*DCB-utilizers increased from 3.4×10^4 CFU/g to a maximum of 2.1×10^8 CFU/g, NAP-utilizers increased from 2.0×10^4 to 2.0×10^8 CFU/g, and CB-utilizers went from 3.6×10^5 to a maximum of 2.2×10^8 CFU/g. In the pilot studies, the maximum numbers of *o*DCB-, CB-, NAP-, and hexadecane-utilizers were 1.5×10^8, 3.0×10^7, 1.0×10^8, and 8.9×10^7 CFU/g, respectively.

Not only did the total numbers of the microbes and populations capable of growth on specific contaminants increase, the microbes were able to mineralize the radiolabeled compounds added to aliquots of the sludge. The extent of mineralization over time for glucose, *b*EHP, NAP, CB, and *o*DCB in the DB-IV study is presented in Figure 3. The microbial population was active as indicated by the rapid mineralization of up to 23.9%/day of the added glucose. NAP, CB, *o*DCB, and *b*EHP were all mineralized to some extent, proving that the microbes were involved in the removal of these compounds.

Mineralization of these compounds was also tested in two samples of sludge collected from the pilot study on Day 41. The average mineralization rates in the pilot study samples for glucose, *b*EHP, NAP, CB, and *o*DCB were 2.89, 1.27, 1.18, 1.23, and 1.84%/day, respectively.

DISCUSSION

Organic Contaminant Removal From Sludge

Volatiles. The concentrations of volatiles in the drum bioreactor and pilot studies were reduced by a combination of biodegradation and volatilization (Tables 2 and 3). All of the methylene chloride was thought to have been volatilized because more was emitted from the drum bioreactor studies than

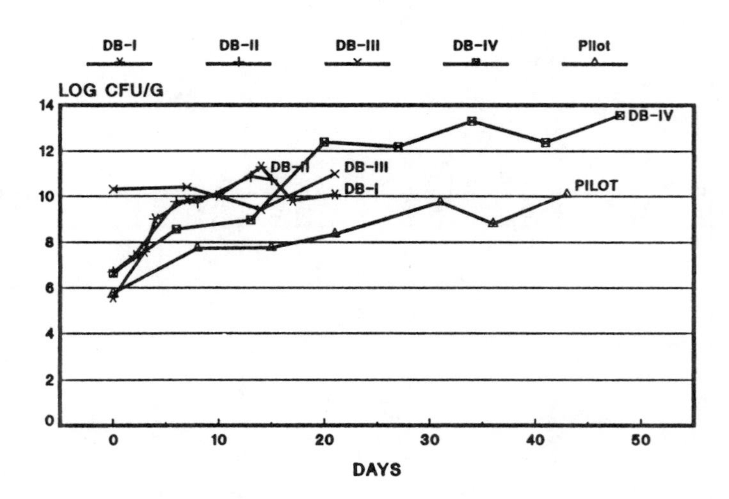

FIGURE 2. Heterotrophic microbial counts in drum bioreactor and pilot studies.

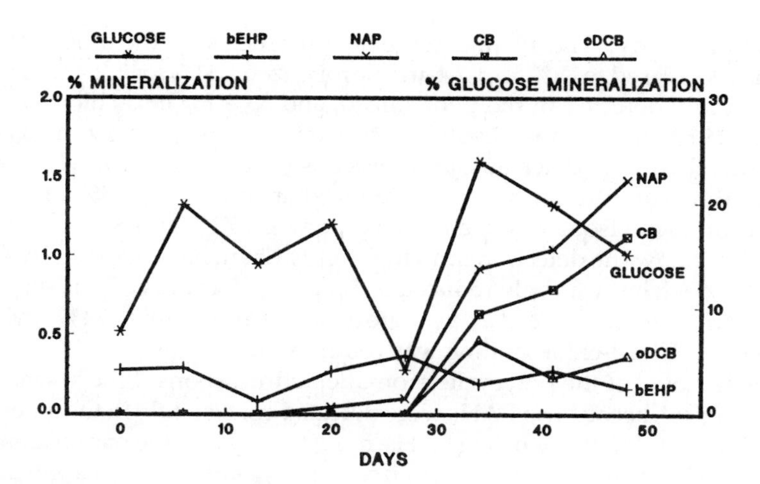

FIGURE 3. Extent of mineralization per day for drum bioreactor IV studies.

estimated to have been present in the initial sludge. Chloroform was not detected in the air samples collected in the pilot study or the DB-IV study after 3 days, which suggested that its removal was a result of biodegradation. All of the tetrachloroethene (PCE) was removed by volatilization in the DB-II study, but less than 10% was stripped in the DB-IV and the pilot studies. PCE may have been cometabolized or present at a concentration below the detection limits for the air samples which would explain the high estimate for biodegradation of this compound which is generally only poorly aerobically biodegradable.

The average CB removal efficiency was greater than 99% in the four drum bioreactor studies and the pilot study. In the DB-II and the pilot studies, volatilization was shown to be have been the principal removal mechanism (Table 3), but not in the DB-IV study where less than 2% of the CB was volatilized. The mineralization studies confirmed that the sludge microbial population was capable of biodegrading CB with as much as 1.12% of the added radiolabel converted to carbon dioxide per day (Figure 3). The mineralization studies were not sparged with air, so removal by volatilization could not be assessed in these assays.

The monoaromatic compounds, including toluene, ethylbenzene, and the xylenes, were reduced rapidly in the sludge with removal efficiencies of greater than 80% (Table 2). Acetone and 2-butanone, the other volatile compounds found in two or more of the studies, were completely removed (Table 2) with no detectable volatilization (Table 3).

Semivolatiles. The less-chlorinated benzene compounds were generally more extensively biodegraded than the more-chlorinated compounds (Table 2). The concentrations of the 1,2-, 1,3-, and 1,4-dichlorobenzene isomers were reduced, on the average, by more than 87% with biodegradation accounting for between 25% and 80% of the removal. Mineralization studies confirmed biodegradation with as much as 0.458% of the added radiolabeled *o*DCB mineralized per day in the DB-IV study. The average removal efficiencies for the other chlorinated benzenes decreased to 78% for 1,2,4-trichlorobenzene, 31% for 1,2,4,5-tetrachlorobenzene (only detected in the pilot study), and 53% for hexachlorobenzene.

The nitro- or amino-substituted benzene compounds were extensively biodegraded with no detectable air emissions (Tables 2 and 3). The maximum concentrations of nitrobenzene were reduced an average of 68%. The 2,4- and 2,6-dinitrotoluene isomers were reduced between 72% and 87%. Aniline was reduced to below the detection limit in the only drum bioreactor study in which it was detected, but was only reduced 58% in the pilot study. The 2-nitroaniline concentrations were reduced an average of 87%, 3-nitroaniline 74%, and 4-nitroaniline 69%. The 4-chloroaniline was reduced an average of 97%.

The removal of the polycyclic aromatic hydrocarbons (PAHs) was thought to be biological, because no PAHs were detected consistently in the air emissions (Tables 2 and 3). For the two-ring PAHs, more than 67% of the naphthalene (NAP) was removed with up to 1.48% mineralized per day, and 2-methylnaphthalene was reduced an average of 69%. PAHs with three rings, including dibenzofuran, phenanthrene, anthracene, fluorene, fluoranthene, and acenaphthalene, were removed from the sludge by overall averages of between 46 and 100%. Pyrene, the only four-ring compound detected consistently, showed reductions averaging 52%.

The phthalates, which were present at the highest concentrations of any of the organic contaminants, also were extensively attacked (Table 2). For example, *b*EHP, found at initial levels of more than 1,000 mg/kg, was reduced between 58 and 88% with as much as 0.368% of the radiolabeled *b*EHP mineralized per day in the DB-IV study. BBP was reduced an average of 69%. The concentrations of the BBP and the *b*EHP varied widely—possibly due to analytical and sampling variability; it was also possible that the plastic liner in the drum bioreactor studies leached some of these phthalates. Di-*n*-octyl phthalate was reduced an average of 98%, and di-*n*-butyl phthalate was reduced to below the detection limit in the one study in which it was detected.

The other semivolatiles detected in two or more studies showed high removal efficiencies (Table 2). The concentrations of 2,4-dichlorophenol, *n*-nitrosodiphenylamine, and benzidine were reduced by averages of 85%, 84%, and 89%, respectively.

Volatilization

The calculated concentrations in the air emissions were in excess of what was estimated to be present initially in the sludges for a number of the organic contaminants. The quantity volatilized was based on only three points for both the DB-II and the pilot studies. More data were collected in the DB-IV study, where 15 samples were collected over the first 6 days and weekly samples for the next 42 days. Another explanation for the higher quantity of material estimated to be present in the air emissions than in the sludge could be analytical problems such as incomplete removal of the volatiles adsorbed to the activated carbon in the sludge during the purge-and-trap analyses.

The volatilization data suggested that volatile emissions would occur during bioremediation of the sludge. The quantity of total volatiles emitted into the air ranged from 0.004 kg/m^3 for the DB-IV study, to 0.036 kg/m^3 for the DB-II study, to 0.88 kg/m^3 for the pilot study. The air flowrate in the DB-IV study was much lower (8 L/min per m^2 surface area) than in the DB-II study where the average air flowrate was 32 L/min per m^2 or the pilot study where the air flowrate was 33 L/min per m^2.

Reducing the air flowrate in the DB-IV study was able to achieve as effective removal efficiencies as seen in the DB-II and pilot studies, but with much lower air emissions. The lower temperature in the DB-IV study probably also contributed to the reduced air emissions. Installation of floating covers to collect the air emissions was successful in this pilot study (Zanikos et al. 1992). Other strategies have been used to control volatilization of organic contaminants and supply oxygen to slurry-phase bioremediation projects. At the French Limited Superfund site outside Houston, Texas, finely dispersed oxygen bubbles were sparged into contaminated sludge that was pressurized to 2 to 4 atmospheres (Bergman et al. 1992). The sludge was pressurized to increase oxygen transfer and to reduce volatilization of organic contaminants. Using oxygen rather than air as the oxygen source also decreases the volume of gas that must be used and consequently the volatilization.

Microbial Enumeration and Mineralization Activity

Substantial microbial activity took place in both the drum bioreactor and pilot studies. The highest heterotrophic microbial numbers of up to 10^{13} CFU/g were seen in the drum bioreactor studies where there was complete mixing and aeration. The heterotrophic microbial numbers reached 10^{10} CFU/g in the pilot studies. The numbers of the specific contaminant-utilizers were up to 10^8 CFU/g in both the DB-IV study and the pilot study. Mineralization rates exceeded 1%/day for glucose, NAP, and CB in the drum bioreactor and the pilot study. Mineralization rates for *o*DCB and *b*EHP were higher in the pilot study than in the DB-IV study.

There was a substantial lag of 20 to 34 days in the production of detectable quantities of CO_2 for *o*DCB, CB, and NAP. This lag may be attributable to the acclimation time required for the microbial population to develop the enzyme systems that allowed them to biodegrade these organic contaminants, sequential substrate use, or the observed lags may be an artifact of the mineralization assay procedures. The quantity of oxygen in the vials may have limited the mineralization of these compounds during the first 34 days of treatment when the vials had no headspace. No activity against CB or *o*DCB was detected until the CO_2 collection procedure was switched from stripping to trapping methods. The *b*EHP was attacked under the conditions found in these drum bioreactor studies, but at a relatively low rate of less than 0.37%/day.

Comparison Between Drum Bioreactor and Pilot Studies

The drum bioreactor studies appeared to be relatively good indicators of the individual organic contaminant treatment efficiencies achieved in the pilot study. Removal efficiencies for the semivolatiles were similar in both the drum bioreactors and the pilot studies, except where compounds were reduced to below the detection limits. Many of the compounds were present at low levels initially and could not be detected in the subsequent analyses.

The drum bioreactor studies overestimated the inorganic nutrients, pH-adjustment agents, and oxygen required for treatment of the sludge compared to the pilot study which would inflate cost projections for the full-scale implementation. The pilot study required 630 g Na_2HPO_4 and 100 g NH_4Cl/m^3 of sludge to maintain the phosphate and ammonia concentrations above 5 mg/L. On the average, 5,600 g Na_2HPO_4 and 1,100 g NH_4Cl/m^3 were supplied in the drum bioreactor studies. The pH-adjustment requirements also were less for the pilot study, i.e., 190 g hydroxide equivalents per cubic meter, compared to the average 470 g of hydroxide equivalents per cubic meter for the drum bioreactor studies. Calcium carbonate was not a strong enough base to be an effective pH-adjustment agent even when used at concentrations as high as 140,000 g/m^3 of sludge in the DB-III study. Based on the average dissolved oxygen uptake rate over time (data not shown), oxygen requirements for the drum bioreactor studies ranged from 11,000 to 35,000 g/m^3 of sludge, whereas the oxygen requirement for the pilot study was 9,200 g/m^3 of sludge.

CONCLUSIONS

These laboratory and pilot treatability studies demonstrated the presence of an active microbial population in the sludge that was capable of degrading the organic contaminants. The microbial populations increased during the studies and were able to biodegrade many of the volatile and semivolatile contaminants as shown by the mass balances around these compounds. The microbial populations were able to mineralize CB, *o*DCB, NAP, and *b*EHP.

The laboratory studies provided reasonable estimates of the extent of removal of the organic contaminants seen in the pilot study. The laboratory studies projected greater requirements for nutrients, oxygen, and pH-adjustment agents than found in the pilot study. Cost projections based on the laboratory studies would overestimate the quantities of these materials required for full-scale implementation of slurry-phase bioremediation.

Based on these findings, in situ slurry-phase bioremediation would be a successful remediation technique for the organic contaminants in the impoundment sludge, even at the lower winter temperatures. Aeration and mixing of the sludge will strip much of the volatile organic contaminants from the sludge. However, volatilization can be controlled by reducing the air flowrates, by installing covers, or by alternative aeration strategies. The semivolatile contaminants also were greatly reduced with biodegradation being the principal removal mechanism. Because the longest incubation period of 48 days in the drum bioreactor studies and 6 weeks in the pilot study were not sufficient to completely remove the semivolatile organic contaminants, a longer treatment time would likely be required in the full-scale remediation to reduce the semivolatiles to below risk-based limits. Full-scale implementation would require the following improvements over the pilot system: better prehomogenization of the sludge before mixing, optimization of the mixer arrangement, and better air emissions control (Zanikos et al. 1992).

Full-scale implementation of in situ bioremediation was estimated to require 1.5 to 2 years, which would include biological treatment in stages to reduce the organic contaminant concentrations, dewatering of the biotreated sludge, stabilization of the heavy metals, and the encapsulation of the stabilized sludge in a vault constructed within the impoundments. For the full-scale remediation program, in situ bioremediation was projected to cost approximately $66 per cubic meter of sludge. Stabilization and encapsulation were projected to cost an additional $71 per cubic meter. With the approval of the state regulatory agency overseeing the remediation program, the less expensive alternative of in situ solidification and encapsulation, without treatment of the organic contaminants, is being implemented at the site.

REFERENCES

Bergman, T. J., Jr., J. M. Greene, and T. R. Davis. 1992. "An In-Situ Slurry-Phase Bioremediation Case With Emphasis on Selection and Design of a Pure Oxygen Dissolution System." In *Proceedings: In Situ Treatment of Contaminated Soil and Water Symposium* (Feb. 1992, Cincinnati, OH), Air and Waste Management Association and Risk Reduction Laboratory, U.S. Environmental Protection Agency, Cincinnati, OH.

United States Environmental Protection Agency. 1986. *Test Methods for Evaluating Solid Wastes.* SW-846. Washington, DC.

United States Environmental Protection Agency. 1991. *40 CFR Appendix A Method 18. Measurement of Gaseous Organic Compound Emission by Gas Chromatography*, pp. 982-1011. Washington, DC.

Zanikos, I. J., T. F. Mistretta, and W. A. Butler. 1992. "In-Situ Bioremediation Pilot-Scale Test of Surface Impoundment Sludge." In *Water Environment Federation 65th Annual Conference and Exposition, Hazardous Waste/Groundwater Symposium Proceedings* (Sept. 20-24, 1992, New Orleans, LA), pp. 293-304. Water Environment Federation, Alexandria, VA.

EFFECTS OF DIFFERENT HYDROPNEUMATIC IN SITU REMEDIATION SYSTEMS

M. Nahold and H. Hötzl

ABSTRACT

Contaminated subsoil and groundwater are more or less successfully treated with different combinations of in situ remediation methods. Techniques such as air injection, two-well forced-gradient flow, one-well circulation, and soil air extraction are tools to improve a remediation. They all need careful maintenance. Special conditions within the subground, such as layers of perched water and both hydraulic and hydrochemical effects, earn more attention. The results in this paper are based on different experiments at sites polluted mainly with volatile chlorinated hydrocarbons and volatile aromatic compounds.

INTRODUCTION

Techniques such as hydraulic decontamination of aquifers and soil cleanup by in situ aeration are applied directly to remobilize and to recover volatile organic compounds (VOCs), but also to support the activity of microbes. A brief overview is given by Nolan and Boardman (1991). The contaminants of interest include the dense (D) and light (L) nonaqueous-phase liquids (NAPLs), including tetrachloroethylene (PCE), trichloroethylene (TCE), *cis*-1,2-dichloroethylene (DCE), and the aromatic compounds benzene, toluene, and xylene.

With groundwater extraction, still the most commonly used remedial method (EPA 1989), the concentrations of contaminants decrease initially but further reductions follow moderately. Pump-and-treat procedures may protect contaminated sites or waterworks from being influenced by the site's runoff. Pumping is an expensive method that, in some cases, can be enhanced or replaced by other technologies.

The group of hydropneumatic in situ remediation systems including soil vapor extraction, air injection, one-well techniques, and combinations of different methods, has been investigated. Hydraulic inhomogeneities within the aquifer and induced secondary effects (such as sintering of wells) reduce the efficiency of various in situ remediation techniques.

HYDRAULIC METHODS

Inverted Wells and Hydraulic Dipoles

Introduction and Theory. Both the direction and quantity of groundwater flow may be influenced by withdrawal and reinjection of water. Accordingly, a symmetric or an unsymmetric steady-state flow field must be established. Consequently, the transport of contaminants increases at first but then decreases again. Contaminants are mobilized first in coarse-grained sediments or in the main fissures between the injection well and the withdrawal well.

Forced-gradient tracer experiments with dye tracers have been carried out by Himmelsbach et al. (1992) and by Veulliet and Hötzl (1992). The results derived from those experiments can be approximated by using a stream tube approach, defined as a division of the natural flow field into a finite number of stream tubes, all contributing to the observed total flux. Maloszewski and Zuber (1990) presented a mathematical solution describing the tracer's concentration induced by a Dirac pulse as a function of time for a given distance. Recovery rates from injected tracers are significantly lower in symmetric dipole situations than during conventional recharge from monopole tests with radial convergent flow.

A forced-gradient tracer experiment in a porous aquifer was described recently by Mas-Pla et al. (1992). The authors intended to predict breakthroughs of tracers and discussed the results critically by referring to one- and two-dimensional mathematical models and their limits. Güven et al. (1992) presented a three-dimensional numerical model and stated good agreement with the results of two-well tracer tests.

Results from Field Experiments in a Fissured Aquifer. Symmetric and asymmetric dipole flows have been induced in a fissured limestone aquifer contaminated with PCE and TCE. The withdrawn water passed through a countercurrent packed tower (air stripper with carbon adsorber) before being reinjected at a distance of 30 m. Hydrogeochemical signals were observed, and contaminant concentrations were measured until they became stable. Within the circulation zone, HCO_3^- decreases while the content of dissolved oxygen increases slowly (Figure 1). Depending on the flow velocity, an estimated 2 metric tons of mainly carbonates per year would generate coatings within the stripping tower or sediment in the fissures. Although water without chlorohydrocarbons (CHCs) is reinjected, PCE and TCE contents increase (Figure 2). This effect may be caused by the mobilization of water in dead-end fissures and perhaps by the solution of the last resources from DNAPL pools.

The changing ratio of TCE:PCE (Figure 3) provides evidence of the existence of a mixture of different contaminant occurrences that are affected by the remediation measure (numbers of samples in the diagram indicate progressing time). Finally, the time of recirculation (Figure 4) was estimated by a tracer test (for basic information concerning tracers refer to Käß 1992). It took about 20 recirculations to get the highest amount of PCE. The balance of the tracer (disregarding sorption) showed that water was withdrawn at about 30% of the injected amount of water.

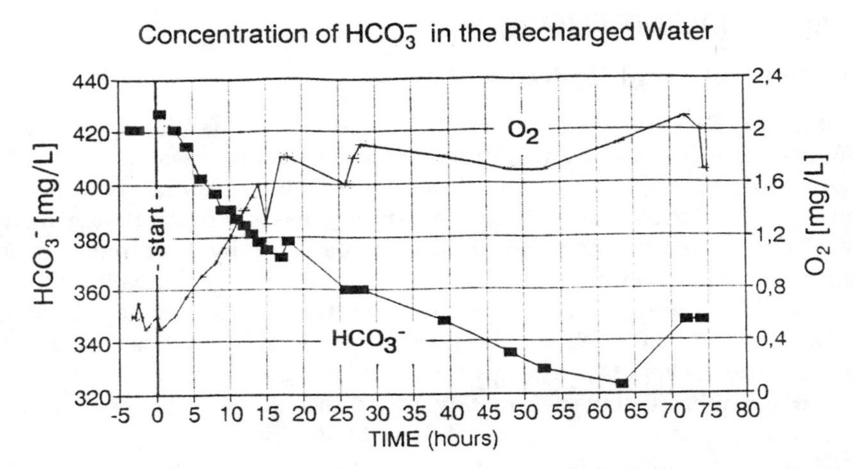

FIGURE 1. Progressive change in groundwater due to mixture with treated water during a dipole flow. HCO_3^- decreases while the content of O_2 increases to a maximum of 2 mg/L.

The efficiency of the dipole flow in this aquifer ends after about 70 recirculations, when infiltration may move to a borehole further away. Tracer tests with asymmetric dipoles (recharge two times injection quantity) could not improve remediation.

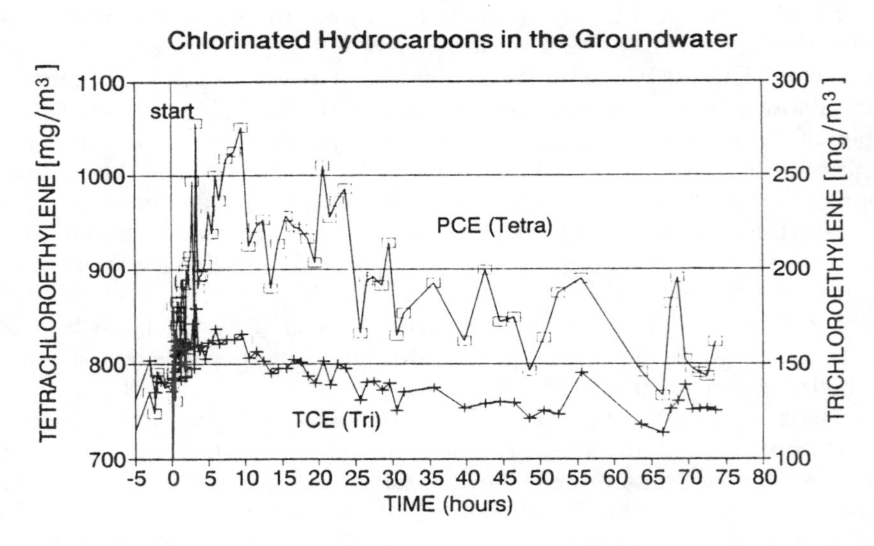

FIGURE 2. Contaminant concentration in the withdrawn water first increases by 2 within the first 10 hours and then goes back to the starting concentration of 750 mg/m³.

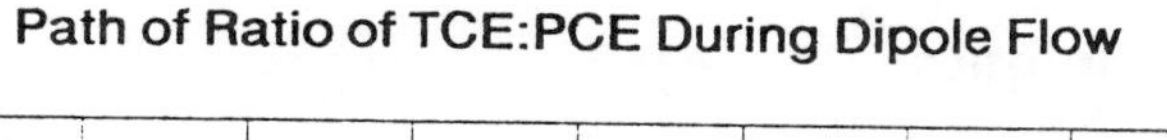

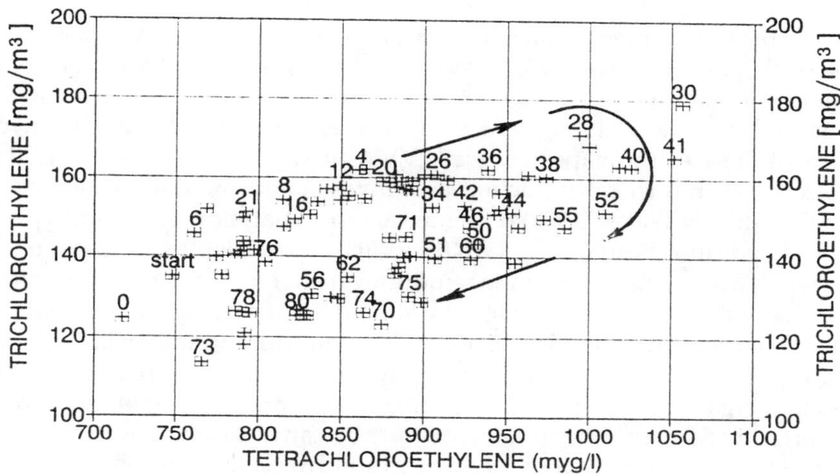

FIGURE 3. The ratio of TCE to PCE changes during cycling of the water. This effect is caused by activated CHC that is older or that is derived from a more bioactive zone.

Information about aquifer parameters, travel times, and the distribution of the contaminants can be estimated with a hydraulic test combined with an analysis of hydrochemical data and a groundwater tracer experiment (Nahold & Hötzl 1992). In this experiment, important information that could be used to optimize the operation of the remediation unit had been gathered within 3 days.

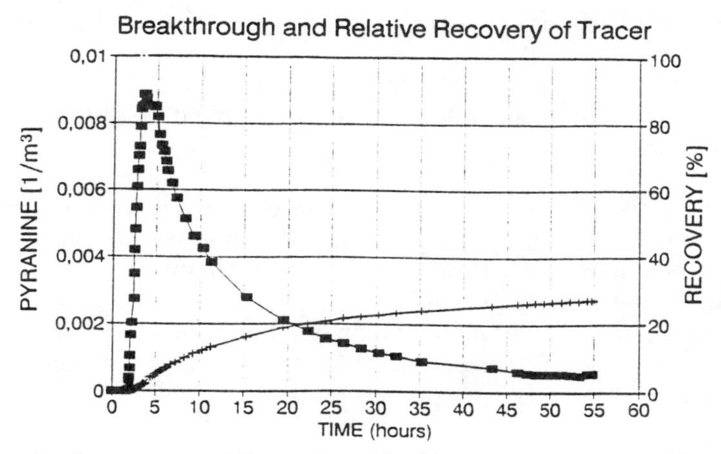

FIGURE 4. Results from a forced-gradient tracer test used to derive the recycling time. Concentrations are divided by the input amount of 5 g Pyranine (color-index 54040).

One-Well Techniques

Different in situ techniques have been developed within the last 5 years emphasizing the benefit of low energy demand. The primary advantage of these techniques is that water need not be elevated to a surface-situated treatment plant. Bürmann (1994) and Herrling et al. (1993) present latest results of their investigations of the vacuum vaporizer well (VVW or UVB). They both carried out analytical and numerical modeling and investigated the effects in the field. Recently, Philip & Walter (1992) presented another analytical approach to predict flow and hydraulic head fields. Two filters at different depths are used to withdraw water from the aquifer and to reinfiltrate it after treatment. If air is injected at the base of the well, it rises inside the well, and the ascending waterflow (within the well) induces a descending recirculation zone around the well. Instead of an air injection pipe, a liner with a pump typically is installed between the two filters to increase discharge and thus enlarge the zone of recirculation. An air stripper, a filter, or any other treatment unit also can be installed within the well. Recent successful applications of such groundwater circulation systems are reported by Herrling et al. (1993).

Supercritical flow in addition to a change in water chemistry by stripping frequently leads to sintering effects. Two important factors must be controlled: (1) fast sintering of filters and treatment units must be avoided by controlling the filter discharge and the hydrochemistry; and (2) the quantities of water recirculating around the well and the water entering or leaving this zone must be measured and calculated. The success of any conventional reinfiltration unit depends on these controls. Groundwater monitoring wells should be installed at different distances and depths to allow in situ measurement and sampling as done by Bürmann (1994) and Wagner (1992). Only small quantities of water must be sampled to avoid misleading data. The filter sections of small-diameter monitoring wells have to be small to prevent short circuits and to ensure that the sample taken is representative of a narrow representative space within the zone of recirculation.

Other one-well techniques use different installations and flow systems within the wells. The following conditions and effects must be differentiated when comparing different systems:

1. The total energy needed to transport or to clean a defined quantity of water (and contaminants) within the processing zone of the system.
2. Zones of dynamic or hydrochemical influence in the sediments and the groundwater around the wells.
3. Times required for decontamination (of 1) or periods until an effect in a defined distance (2) can be detected (distance-dependent recirculation times).

However, in general it is generally not easy to compare one-well systems in the field with ordinary groundwater pumping wells that have a more protective effect but probably a higher demand for energy. Beyond the discussion of critical values, there is a strong need for energy-related descriptions of remediation systems. Although it is not easy to compare in situ technologies directly, some of the systems' effects should be tested at laboratory scale.

Layers of Perched Water

The zones of interest are fine-grained sediments and peat beds contaminated mainly by chlorinated hydrocarbons and gasoline where sorption and anaerobic reactions such as dechlorination take place. Heavily watered layers frequently are penetrated by drill holes, so that more highly contaminated perched water reaches the aquifer. In this scenario, treatment by conventional groundwater extraction followed by on-site treatment is never ending and ineffective.

Organic-enriched clays and peats within the vadose zone cannot be cleaned satisfactorily by aeration. A better method is to drain the perching zones and provide a surface liner to prevent further seepage. An effective drainage system may be situated vertically or as a system of horizontal wells.

PNEUMATIC METHODS

Air Sparging: Aquifer Restoration by Air Injection

Injection of gas into the aquifer to remove volatile NAPLs is a well-known but seldom applied technique. Several attempts have been unsatisfactory because of clogging effects during and after air injection. The injection of CO_2-enriched air into the center of a polluted area should provide better results as a short-term measure within carbonate-rich waters. The dissolved carbonates will be reprecipitated, not within the filter of the well but further away in the aquifer.

A contamination of PCE, TCE, and their decomposition products in a shallow aquifer was treated (Böhler et al. 1990) by injection of 60 to 100 m^3/h of air into the aquifer. Air and contaminants were extracted simultaneously from the unsaturated zone. Böhler's water-tracing test showed that air injection induced circulation flow against the direction of natural groundwater flow. Wehrle and Brauns (1994) found similar effects. Although mobilization shock of DNAPLs was observed in the extracted air, it was not possible to volatilize a satisfactory amount by mere air sparging. High amounts of dissolved contaminants had to be removed by means of an additional water recovery well. Later, the injection well was clogged by precipitation.

Aeration of the Vadose Zone

Zone of Influence. The structure of the sediment in the zone of influence, natural organic matter contents, and porosities are known from exploration of a VOC-contaminated site. However, the need for a more detailed determination of the zone influenced by aeration had been suggested. Field measurements showed that the observation of the areas and depths of low pressure, combined with measurements of VOCs in the off-gas, give sufficient information about remediation progress.

Several details must be taken into account before and during the installation of an air extraction unit. A flow of vadose water toward the extraction well combined with an elevation of the water table with its capillary fringe can be caused initially by air extraction. In wet fine-grained sediments, gentle under-pressures of perhaps −0.1 to −1.0 psi (equals −70 to −700 N/m^2 below the pressure on the surface) can be applied successfully. In coarse-grained sediments, pressures (applied down to −2 psi or −1,400 Pa) and air quantities should be determined by tests or by considering survey data. It is important to find out whether the dominating transport process is determined by desorption or only by partial aeration of bodies of fine-grained sediments within gravels.

A case study was undertaken to determine the effects of vapor extraction. The hydrogeologic setting: the benzene, toluene, and xylene contamination is situated below a fuel storage depot in fluvatile sediments. Sandy gravels are overlain by 1 m silty sand, and the groundwater table is 4 m below the surface. Soil vapor extraction had removed a high amount of the VOCs, but a substantial amount of LNAPLs remained on top of the water table. It was not possible to determine the distribution of the contaminants.

First, the zone affected by aeration had to be determined by measuring air pressures at different depths and at different distances from the extraction well. Air was extracted from eight different wells, one after the other, and finally simultaneously from six wells. Pressure-control measurements were carried out at about 20 of 30 different monitoring wells. Six single situations are shown in Figure 5.

Measurements in the different monitoring wells showed that relative pressures of −100 N/m^2 (equals −0.014 psi or −1 cm of water column) could be detected at distances up to 40 m from the extraction well and at depths of 2.5 to 4 m below the surface.

To explain the results from the field measurements, data were recalculated with the help of a two-dimensional finite-element model (Brauns & Wehrle 1990, Nahold & Gottheil 1991). As it could be proved, the zone of influence of the soil vapor extraction was enlarged due to the silty surface layer becoming wet from rain. Test extractions from six different wells showed an unexpected homogeneity of the zone of influence. Only the air extraction from well no. 3 showed a reduced radius of action, influenced by short circuit (Figure 5).

An adjacent site contaminated with fuel also was found to be affected by far-reaching aeration. An increasing content of CO_2 in the extracted air has been reported. The reduction of benzenes and aliphatic compounds by volatilization seems to be improved by accelerated microbiological in situ decomposition. Finally, the oily layer on top of the groundwater table has to be treated with additional hydraulic measures. As a result of further investigations, it could be shown that a narrow buried channel crosses the clayey base relief of the shallow aquifer. This channel provides discharge pathways in its down-dip direction.

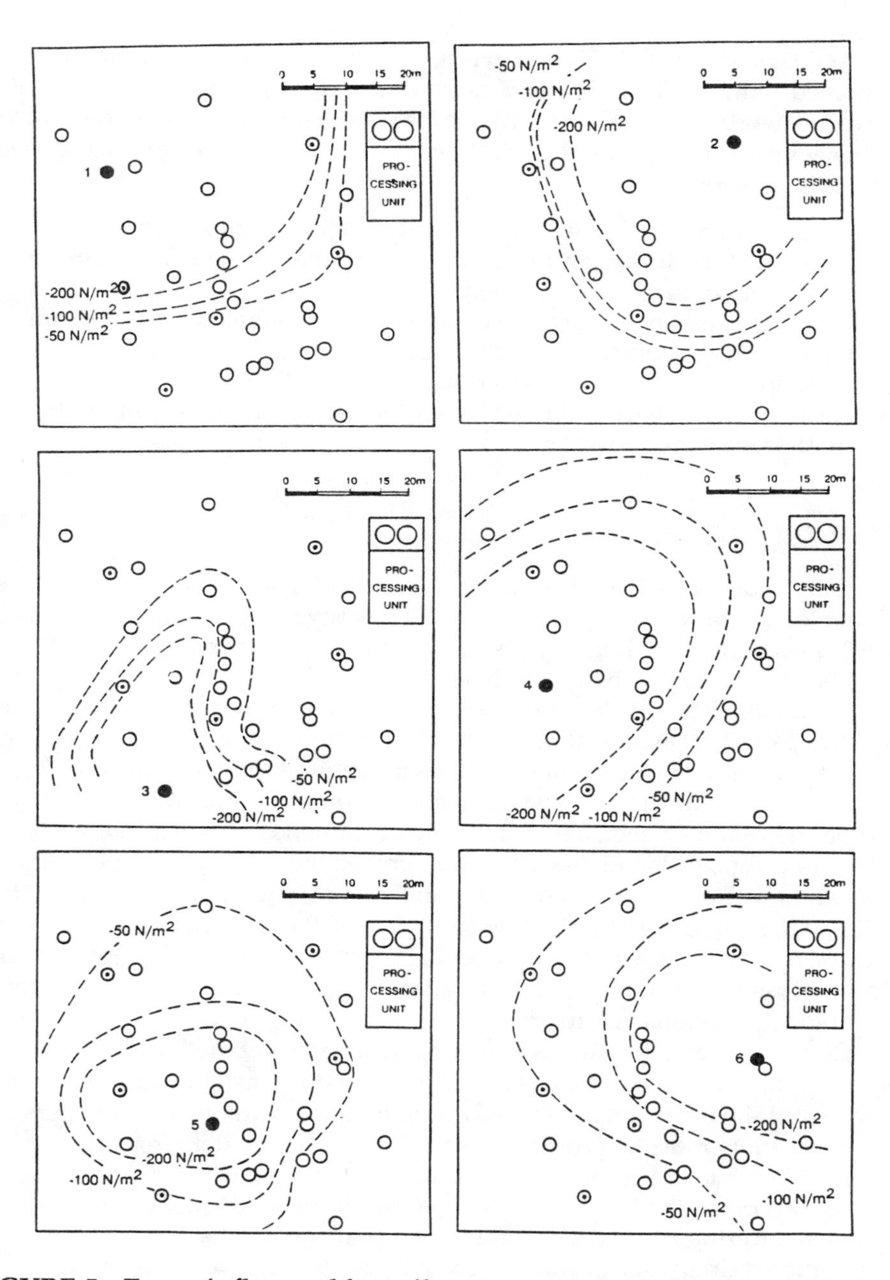

FIGURE 5. Zones influenced by soil vapor extractions showing the results of six individual tests. The relative pressures were measured 2 to 4 m below the surface. A quantity of 70 m³/h air was extracted at a relative pressure of −1 psi (diameter of the extraction well is 0.05 m, diameter of the gravel filter is 0.2 m). Isobars: −50, −100, and −200 N/m²; extraction well (black dot) is numbered.

Aeration of a Depleted Zone in Fissured Limestone. Hydraulic dipole flows combined with air flushing in a fissured aquifer (contaminated with CHCs) increased the efficiency of remediation by tenfold. Figure 6 shows a vertical cross section with schematically indicated flows of water and air. The following effects have been observed:

1. Turbulent flow within fractures enforces vaporization.
2. Air extraction influences the groundwater table, and air pressure can be varied within the unsaturated zone.
3. The flow of highly contaminated vadose water (enforced leakage) from the overlying sediments into the fissured aquifer is increased by low pressure and the lowered water table.
4. Air extraction is accompanied by an increase of contaminants in the (withdrawn) groundwater.

CONCLUSIONS

Only detailed characterization of the subsurface and of chemistry, and best available monitoring allows effective use of energy for remediation. Hoffmann (1993) summarized various aspects as "smart pump and treat." Selected aspects have been investigated by the authors.

Hydraulic tests with dye tracers and contaminants in both porous and fissured aquifers gave information about hydraulic properties of the aquifer and about travel times caused by different techniques for underground remediation. Recovery rates derived from tracer tests and output concentrations of CHCs and other compounds showed the size of the zone of influence. The concentrations of inorganic compounds and dissolved gases proved that the efficiency of waterflow and aeration often suffers from preferential flow. Contaminant biodegradation is limited in large part by the low solubility of the contaminants but mostly because fine-grained sediments are not sufficiently discharged by water (or air). Slow diffusion and desorption from natural organic matter and clay minerals limit biodegradation even more.

Reinjection and in situ recirculating systems combined with aeration of groundwater often suffer from scale deposits and incrustation of the filter and clogging of the formation. To avoid precipitation of carbonates, the filters and discharge units must be properly chosen, and groundwater circulation can be combined with treatment in closed circuits (i.e., no input of atmospheric oxygen during vaporization of CHCs). Nevertheless, the groundwater can be enriched by oxygen during recirculation on top of a phreatic aquifer.

For the recirculation of water, the vacuum vaporizer well or separated withdrawal and injection wells are successfully applied to improve advective transport.

Pneumatic techniques are inexpensive and in some cases excellent tools for remediation of the vadose zone. Their efficiency is, however, limited by porosity and by the content of water and by the amount and quality of organic substances within the sediment. Field measurements of the zone of influence are recommended, with the results explained by parameter studies using models.

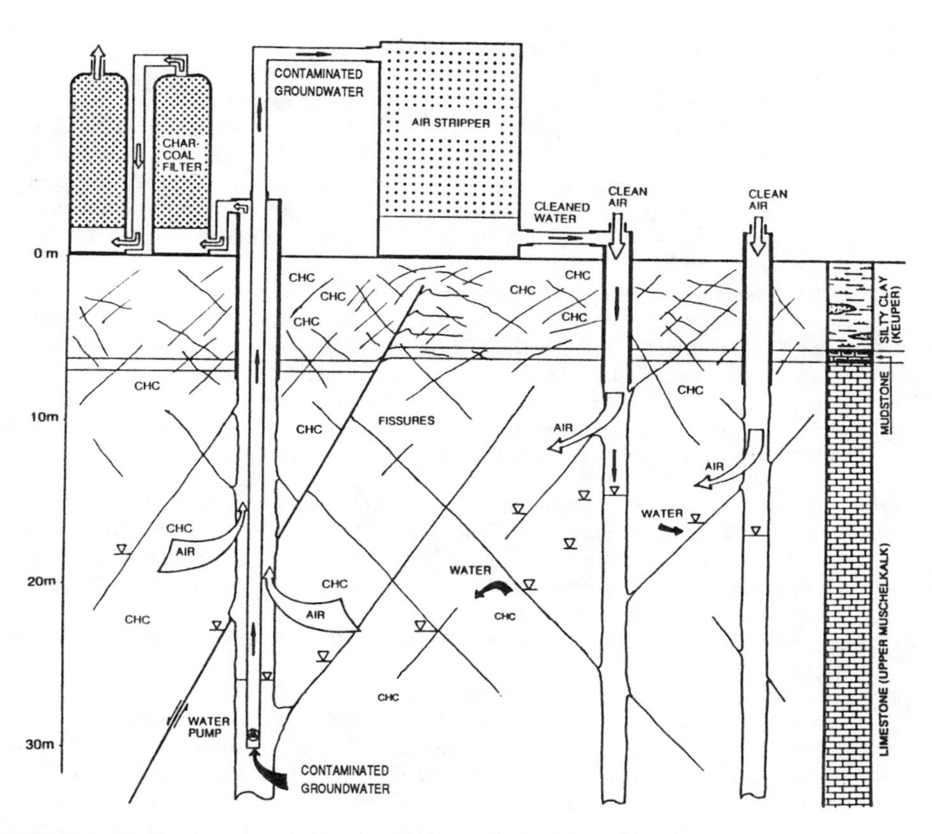

FIGURE 6. Schematic vertical cross section of the contaminated site. The overlying clay is fissured. It contains the center of the contamination and is slowly drained by the fissured limestone. The groundwater is depleted and air is extracted from the unsaturated fissures. Different water table levels in the fissures symbolize different contacts with the wells used in the hydraulic operation.

ACKNOWLEDGMENTS

Results have been gathered from different projects. Research programs have been funded by "Project Water-Waste-Underground" of the state of Baden-Württemberg/Germany and by private industry. Progress has been furthered by co-operation of authorities, private industry, and the owners of contaminated sites with the Department of Applied Geology at Karlsruhe University.

REFERENCES

Böhler, U., J. Brauns, H. Hötzl, and M. Nahold. 1990. "Air injection and soil air extraction as a combined method for cleaning contaminated sites in sediments and solid rocks." In Arendt,

Hinsenveld, and van den Brink (Eds.), *Contaminated Soil '90*: pp. 1097-1104. Kluwer Academic Publishers, The Netherlands.

Brauns, J., and K. Wehrle. 1990. "Zur Dynamik der Bodenluftabsaugung in Lockergestein" (in German). In P. Bock, H. Hötzl, and M. Nahold (Eds.), *Untergrundsanierung mittels Bodenluftabsaugung und In Situ-Strippen, Schr. Angew. Geol. 9*: pp. 123-142. Karlsruhe, Germany.

Bürmann, W. 1994. "Bioremediation by groundwater circulation using the vacuum-vaporizer-well (UVB) technology: Basics and case studies." In R. E. Hinchee, B. C. Alleman, R. E. Hoeppel, and R. N. Miller (Eds.), *Hydrocarbon Bioremediation*. Lewis Publishers, Ann Arbor, MI.

EPA. 1989. *Evaluation of Ground-Water Extraction Remedies. U.S. Environmental Protection Agency Summary Report*. EPA 540/2-89/054. Office of Emergency and Remedial Response.

Güven, O., F. J. Molz, J. G. Melville, S. El Didy, and K. Bomann. 1992. "Three-dimensional modeling of a two-well tracer test." *Ground Water 30*(6): 945-957.

Herrling, B., E. Alesi, G. Bott-Breuning, and S. Diekmann. 1994. "In situ bioremediation of groundwater containing hydrocarbons, pesticides, or nitrate using vertical circulation flows (UVB/GZB technique)." In R. E. Hinchee (Ed.), *Air Sparging*. Lewis Publishers, Ann Arbor, MI.

Himmelsbach T., H. Hötzl, W. Käß, Ch. Leibundgut, P. Maloszewski, T. Meyer, H. Moser, V. Rajner, D. Rank, W. Stichler, P. Trimborn, and E. Veulliet. 1992. "Fractured Rock — Test Site Lindau/Southern Black Forest (Germany)." In Assoc. of Tracer Hydrology (Ed.), *Transport Phenomena in Different Aquifers (Investigations 1987-1992)*, pp. 159-229. *Steir. Beitr. z. Hydrogeologie 43*, Graz, Austria.

Hoffmann, F. 1993. "Ground-water remediation using 'smart pump and treat'." *Ground Water 31*(1): 8-106.

Käß, W. 1992. "Hydrogeologische Markierungstechniken bei der Altlastensanierung" (in German). *Die Geowissenschaften 10*: 199-205, VCH Verlagsgesellschaft, Weinheim, Germany.

Maloszewski, P., and A. Zuber. 1990. "Mathematical modelling of tracer behavior in short-term experiments in fissured rocks." *Water Resour. Res. 26*: 1517-1528, Washington.

Mas-Pla, J., T.-C. Yeh, J. F. McCarthy, and T. M. Williams. 1992. "A forced gradient tracer experiment in a coastal sandy aquifer, Georgetown site, South Carolina." *Ground Water 30*(6): 958-964.

Nahold, M., and K. Gottheil. 1991. "CKW-Schadensfaelle — Die Optimierung von Bodenluftabsaugungen" (in German). *WLB Wasser, Luft und Boden, 11-12/91*: 184-188, Verlag Technik GmbH, Berlin, Germany.

Nahold, M., and H. Hötzl. 1992. "The use of water tracers in order to improve remediation techniques." In H. Hötzl and A. Werner (Eds.), *Tracer Hydrology*, pp. 119-123. Proceedings of the Sixth Symposium of Water Tracing in Karlsruhe. Balkema, Rotterdam, The Netherlands.

Nolan, B.T., and G.D. Boardman. 1991. "Aquifer restoration: Which method?" *Civil Engineering 61*(4): 81-83.

Philip, R. D., and G. R. Walter. 1992. "Prediction of Flow and Hydraulic Head fields for Vertical Circulation Wells." *Ground Water, 30*(5): 765-773.

Veulliet, E., and H. Hötzl. 1992. "Simulation of pollutant migration by in situ step input tracer tests in a jointed aquifer." In H. Hötzl and A. Werner (Eds.), *Tracer Hydrology*, pp. 369-373. Proceedings of the Sixth Symposium of Water Tracing in Karlsruhe, Balkema, Rotterdam, The Netherlands.

Wagner, H. 1992. "Hydraulische, hydrogische und chemische Untersuchungen im Rahmen des UVB-Forschungsvorhabens" (in German). Thesis (Diplomarbeit), Karlsruhe University, Germany.

Wehrle, K., and J. Brauns. 1994. "Induced groundwater circulation due to air injection inside a well: General aspects and experimental results for layered subsoil." In R. E. Hinchee (Ed.), *Air Sparging*. Lewis Publishers, Ann Arbor, MI.

A BIOTREATMENT-TRAIN APPROACH TO A PCP-CONTAMINATED SITE: IN SITU BIOREMEDIATION COUPLED WITH AN ABOVEGROUND BIFAR SYSTEM USING NITRATE AS THE ELECTRON ACCEPTOR

C. D. Litchfield, G. O. Chieruzzi,
D. R. Foster, and D. L. Middleton

ABSTRACT

Pentachlorophenol (PCP) in mineral oil was used for 40 years at a site in the North Central United States and contaminated both the soils and the groundwater. The plant is situated on unconsolidated sand, gravel, clay, and silt with an unsaturated zone of approximately 6.1 to 7.6 m. A light, nonaqueous-phase liquid (LNAPL) layer of mineral spirits and PCP is present in various wells. A leach bed system was installed near the surface in October 1987, and groundwater from the recovery wells was passed through an ultraviolet (UV) light/ozone system. Since then, three sets of borings in the area have demonstrated greater than 90% reduction of the PCP. A second leach bed was constructed under the former drip room in 1989. In 1989 a BIFAR unit, a fluidized bed-activated carbon tower, replaced the UV/ozone system and has been operated with nitrate as the major electron acceptor resulting in an average 93.1% reduction in PCP after the initial startup period. This parallels a decrease in nitrate levels and a concomitant increase in the chloride levels in the recovered groundwaters, indicating that nitrate has been successfully used to stimulate the indigenous microorganisms to degrade PCP both in situ and in the aboveground BIFAR reactor.

INTRODUCTION

Pentachlorophenol (PCP) was used for many years in the wood-treating industry as a preservative. When it was withdrawn in the 1980s it was believed to be recalcitrant to biodegradation. Since then numerous reports have appeared in the literature both in the laboratory and in case studies of field biodegradation of PCP under aerobic conditions (Compeau et al. 1991, Frick et al. 1988, Harmsen 1991, Mueller et al. 1989, and Mueller et al. 1991, to cite a few).

Anaerobic dehalogenation of PCP also has been reported (Hendriksen et al. 1991, Mikesell & Boyd 1988). In the latter study an anaerobic sludge was used as the inoculum for contaminated soil in a system operated under methanogenic conditions. In the laboratory, Mikesell and Boyd (1988) demonstrated that 66% of the radiolabeled PCP went to CO_2 and CH_4. Hendriksen et al. (1991) found similar results in a fixed-film bioreactor inoculated with anaerobic digester sludge. They noted, however, that mineralization was more extensive when glucose was present in the medium. In their study, 72% of the PCP was dehalogenated beyond trichlorophenol (TCP) with glucose, compared to only 27% dehalogenation beyond TCP without glucose in the medium; the electron acceptor was not described in this paper.

Our paper presents a case study of PCP biodegradation in the field by indigenous microorganisms in soil and groundwater operating under denitrifying conditions.

PCP was used for more than 40 years at a site in North Central United States to preserve wooden window frames. The frames were dipped in a mineral spirit solution containing approximately 5% PCP. The treated wood was allowed to dry over an area in the dip room that had a sheet metal collection pan underneath the floor. This pan was on a slant so the drippings could be collected in a tank and reused. With expansion of the business, the drip area became too small and additional space was used that did not contain the drip pan. When the operation ceased in the mid-1980s, soils and groundwater at the site were found to be contaminated with PCP.

PROCESS AND SITE DESCRIPTIONS

Site Geology and Hydrogeology

The site geology is shown in cross section in Figure 1. The plant is situated on unconsolidated sand, gravel, clay, and silt. Depth to groundwater is approximately 6 to 7.6 m below ground surface (bgs). Soils in the unsaturated zone contain up to 1% PCP, and groundwater concentrations range from <13 to 90 mg/L. The saturated aquifer thickness is approximately 3 to 4.6 m and is underlain by a low permeability unit. The hydraulic conductivity of the aquifer ranges between 6.1 m/day and 36.6 m/day. An LNAPL composed of mineral spirits and PCP is present in various wells.

BIFAR Design and Operation

An upflow fluidized-bed granular activated carbon unit was designed for the aboveground treatment of the recovered groundwater. The BIFAR unit came on line in November 1989. A flow diagram for the system is shown in Figure 2. The system includes pretreatment to remove oil and grease and suspended particles,

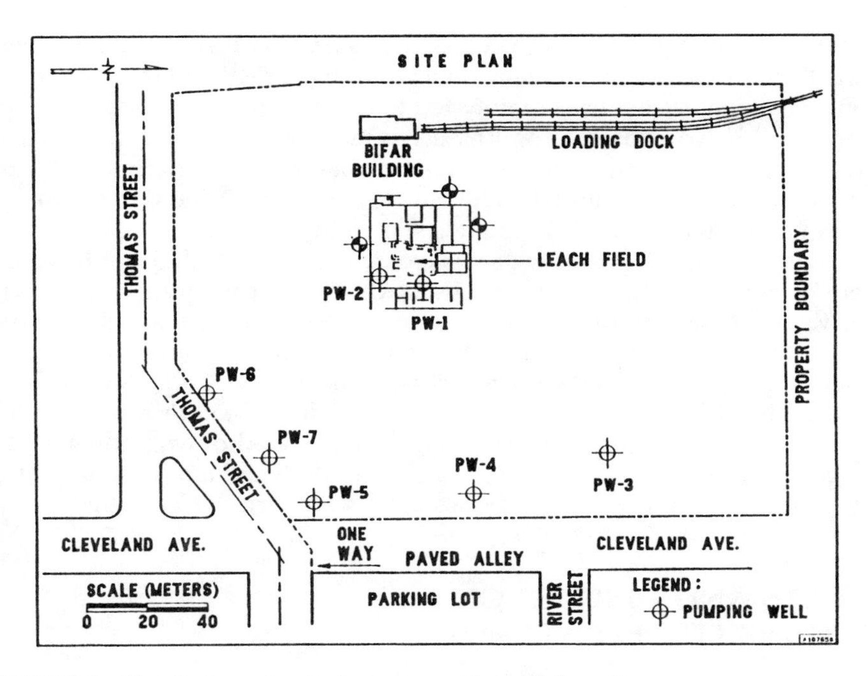

FIGURE 1. Typical geological cross section of the site.

secondary treatment on the BIFAR column, and tertiary treatment components. Pretreatment consists of settling with polymer addition (if required). The secondary treatment is composed of the BIFAR biological reactor and associated recycle system to maintain the desired upflow velocity in the fluidized bed. Tertiary treatment is accomplished using activated carbon in downflow reactors involving a 4,536-kg capacity polishing reactor. The system is designed to treat an average flow of 189.3 L per minute (Lpm) and has been operating at approximately 94.6 Lpm since startup. Activated carbon was replaced in the BIFAR column in April 1991 after 503 days of operation and again with partially spent carbon from the polishing unit in March 1992 after an additional 327 days of operation. Nutrient additions of sodium nitrate and phosphoric acid were made ahead of the BIFAR reactor.

Seepage Beds Design and Operation

Because the vadose zone adjacent to and under the former dip treatment room was impacted by PCP and mineral oil, a shallow subsurface seepage bed system was designed to allow nutrient addition to those soils. The seepage beds were constructed of 10.2-cm-diameter perforated polyvinyl chloride (PVC) pipe laid in a sand and gravel bed at a depth of 1.7 m, below the frost line for the area.

The PVC pipes were covered with geotextile, which was overlain by compacted backfill to grade. Nutrient-supplemented groundwater effluent from the aboveground treatment system was gravity-fed to the seepage beds. The beds were constructed in two phases. The Phase I system was placed in the adjacent courtyard area and became operational in October 1988 to treat directly 200 m^2. The Phase II system was placed on line in November 1989 and directly treated approximately 400 m^2. Their location is shown in Figure 3.

Initially, recovered groundwater was passed through a UV/ozone treatment system before reintroduction to the subsurface. This system was replaced with a BIFAR unit in 1989 because of maintenance problems. A portion of the groundwater that had been treated in the UV/ozone system was supplemented with nitrate and phosphoric acid before recycling to the seepage beds. Groundwater from the BIFAR system was amended with sodium nitrate before recycling. Recovery wells, PW-3 through PW-7 (Figure 2), located downgradient of the seepage beds were used to control groundwater flow and minimize migration of the nutrient-amended water that had passed through the leach field.

RESULTS AND DISCUSSION OF THE FIELD OPERATIONS

In Situ Bioremediation at the Seepage Beds

Because preliminary bench-scale testing had demonstrated slightly better degradation of PCP under nitrate-reducing conditions than under an aerobic treatment system (Table 1), the state approved the use of nitrate for the seepage beds.

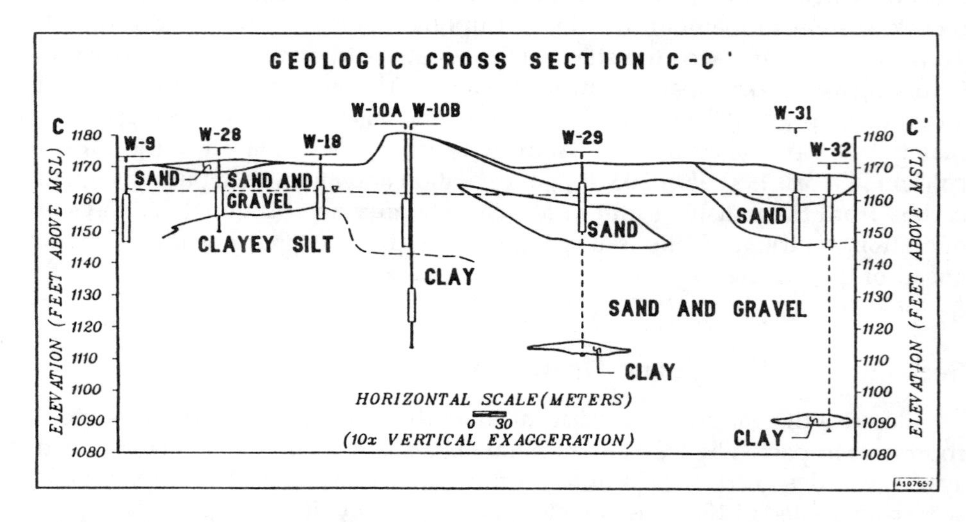

FIGURE 2. Schematic diagram of the BIFAR biological treatment system.

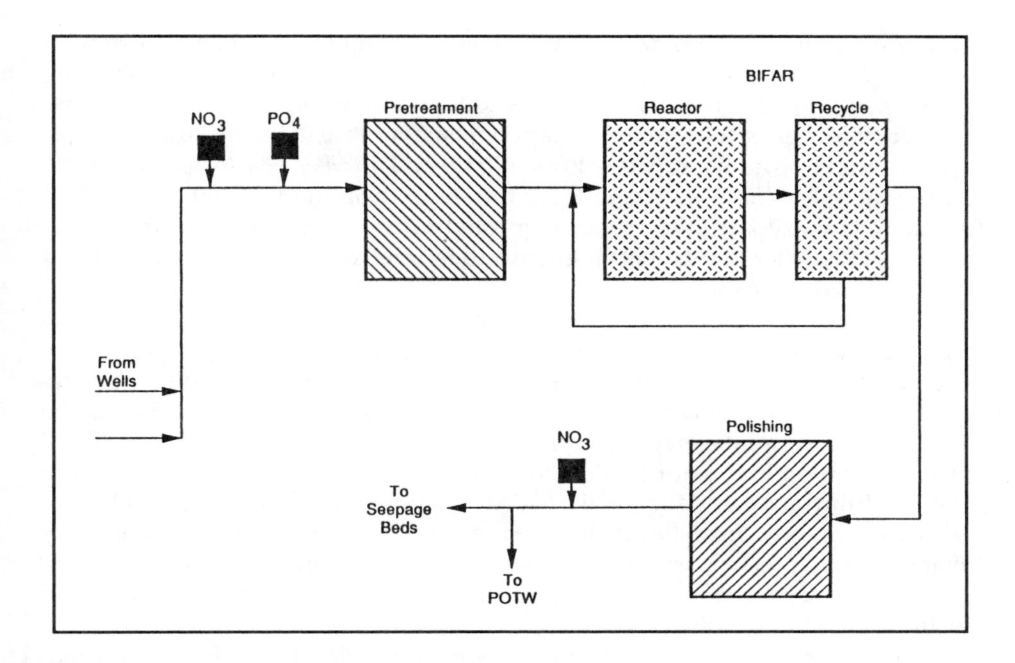

FIGURE 3. Site plan showing locations of wells, seepage beds, and buildings.

Nitrate was added at an initial concentration that maintained a slight excess in the recovery wells. This is shown in Figure 4, which depicts the fate of nitrate in the groundwater during 1989. Throughout the 4 years of operation, nitrate levels in the recovered water have remained in the <0.02 to 5 mg/L range despite fairly high nitrate additions to the infiltration water. Influent nitrate levels were not measured during the first year of operation, although recovery well concentrations

TABLE 1. Summary of the bench-scale biodegradation test for the PCP-contaminated site.

Operating Conditions	PCP Soil Concentration (mg/kg) Initial	After 8 Weeks	Total Mass Removal Leaching	Biodegradation	Total
Aerobic Reactors	13,200	860 - 1,500	5 - 7	76 - 84	83 - 89
Anaerobic Reactors	13,200	640 - 2,100	7 - 10	66 - 85	76 - 92

were determined. These data are not shown on Figure 4 as they also were in the 1 to 5 mg nitrate-N/L range. The results indicate that nitrate was being consumed by the microorganisms in the soils and/or groundwater. In response to agency concerns, the amount of supplemental nitrate to the infiltration water was reduced from approximately 40 mg nitrate-N/L in May 1990 to approximately 24 mg nitrate-N/L. This accounts for the sharp drop in May-June in Figure 4. Again in January 1991, the agency requested a decrease to approximately 5 mg nitrate-N/L. After this date the influent and recovered nitrate levels were almost identical (data not shown).

An examination of the changes in chloride concentrations, Figure 5, shows that during the 4 years of operation there was no significant increase in the chloride levels in an upgradient well, whereas groundwater chloride concentrations in downgradient wells continued to increase, indicating breakdown of the PCP. The most significant portion of Figure 5 is the time period before the BIFAR system went on line, prior to November 1989, as increases in chloride concentrations from biodegradation in the BIFAR system cannot be distinguished from increases resulting from the in situ soil biodegradation. However, it is unlikely that there was much substantial chloride contribution from the in situ biodegradation in the seepage beds after May 1990 because of the extensive biodegradation of the PCP in these soils (Table 2).

The data in Table 2 demonstrate that after approximately 2 years of nutrient enhancement, PCP concentrations in the Phase I seepage bed had been reduced by more than 93%. Soil samples located below the water table and the LNAPL layer also had significant decreases of 97.9 to 99% in the soil PCP levels, reflecting microbial degradation and nutrient enhancement. The water table and the LNAPL layer were at approximately 7 m and 6.1 m, respectively. Infiltration of nutrient-supplemented groundwater into this area continued until 1992 when water flow to the seepage beds was stopped to allow for a more aggressive approach to free product recovery in the capillary fringe. Demolition of the building previously over the Phase II seepage beds has occurred, allowing soil borings to be taken in spring 1993 to evaluate the degree of remediation achieved in this area. At this time it can only be assumed that biodegradation in this area was as effective as in the Phase 1 seepage beds.

BIFAR Operation

The BIFAR system went on line in November 1989 and has operated under nitrate-reducing conditions until June 1992. Recovered groundwater is supplemented with nitrate and phosphoric acid before passage through the unit. The resultant changes in the PCP concentrations are shown in Figure 6. Except for times just before the carbon was changed, effluent concentrations have consistently been below 2 mg/L. At those times when the effluent concentrations have risen, the 4,536-kg polishing unit has ensured that the system effluent has not exceeded the permit limits. The first changeout of the carbon occurred due to increased oil and grease loading fouling the carbon. The second changeout occurred almost

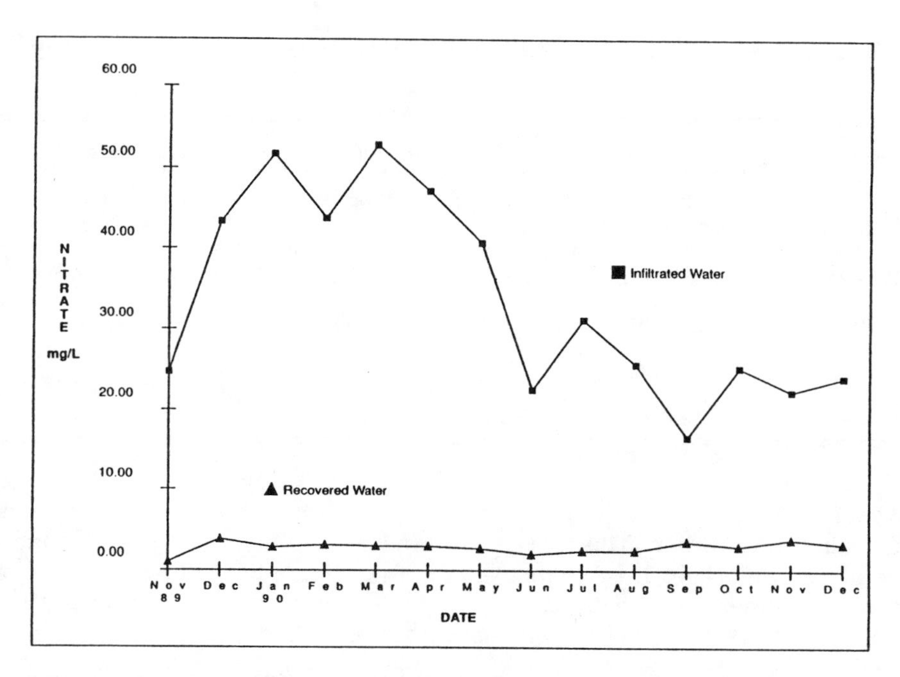

FIGURE 4. Fate of nitrate in the in situ biodegradation of pentachlorophenol.

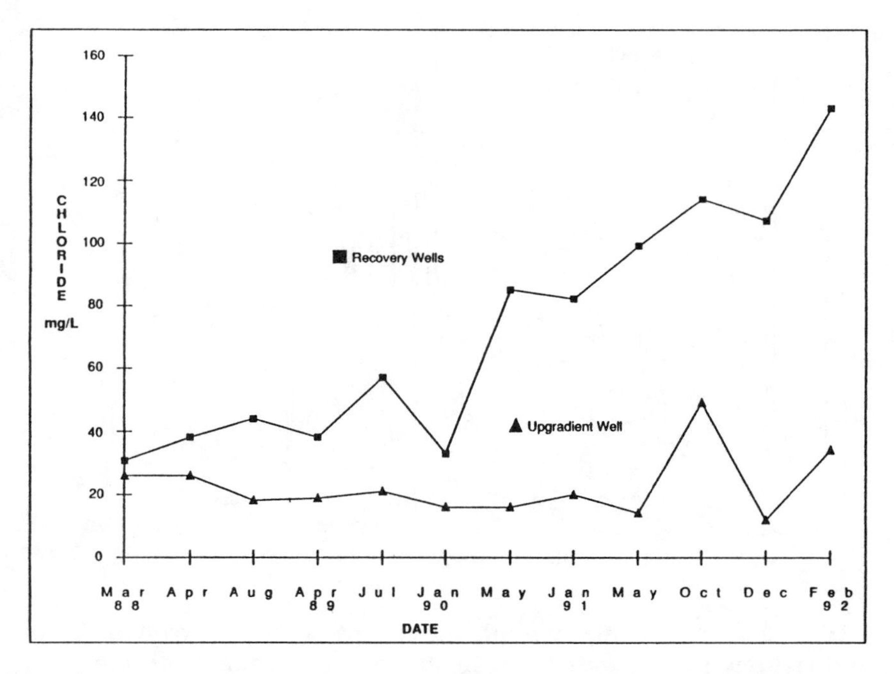

FIGURE 5. Changes in chloride concentration in recovered groundwater.

TABLE 2. Average PCP concentrations in soil borings from the Phase I seepage beds.

	Average PCP Concentrations (µg/kg)				
Depth (m)	October 1987	September 1988	August 1989	May 1990	Percent Reduction
1.2 - 2.4	7,823	1,260	1,099	500	93.6
9.1 - 9.8	27,515	2,024	663	198	99.3
9.8 - 10.4	No Data	1,547	313	602	—
10.4 - 11.9	12,088	2,271	320	249	97.9

a year later when carbon from the partially spent carbon polishing unit replaced the main BIFAR carbon. This reduced the time for biofilm formation and reduced overall carbon usage. Despite the peaks in the effluent level, the BIFAR aboveground treatment system has consistently removed more than 99% of the PCP and the State Water Quality limits have not been exceeded.

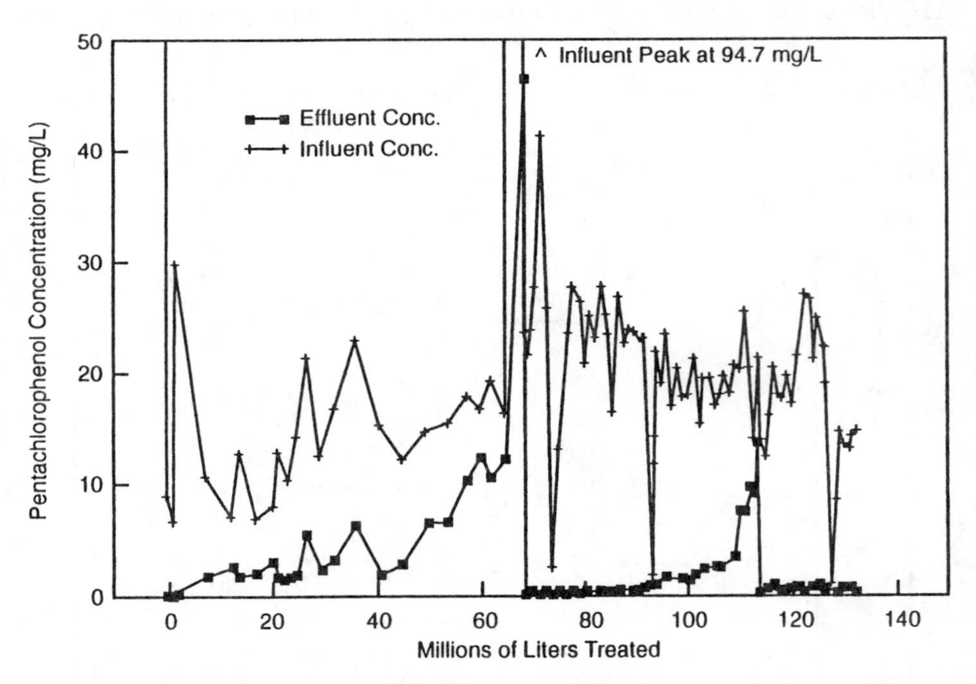

FIGURE 6. Changes in pentachlorophenol concentrations between the influent and effluent groundwater after treatment in the BIFAR column.

CONCLUSIONS

Combined with hydraulic control of the groundwater, a biological treatment train approach has been installed at a former wood-treating site and operated under nitrate-reducing conditions for a total of 4 years. During that time it is estimated that more than 2,800 kg PCP have been biodegraded in the aboveground treatment system, while pumping more than 1.25×10^8 L. Between approximately 23 and 51 kg of PCP are estimated, at a minimum, to have been biodegraded in the soils immediately under the seepage beds. In addition, more than 1,225 kg of PCP have been removed as free product from the surface of the water table. Upon completion of free product removal, full-scale in situ bioremediation is planned for the capillary fringe and saturated zone.

ACKNOWLEDGMENTS

The authors have been assisted during the course of this case study by numerous individuals. Major contributions have been made by Mark Smock, Michael Heyroth, John Marchewka, and Dominick Ciccone who have managed the site, operated the site (MH and JM), and managed the data, respectively. In addition, we thank Suxuan Huang and Dennis Fulmer for conducting the laboratory treatability study. Figures were prepared by Francine Matijak.

REFERENCES

Compeau, G. C., W. D. Mahaffey, and L. Patras. 1991. "Full-Scale Bioremediation of Contaminated Soil and Water." In G. S. Sayler, R. Fox, and J. W. Blackburn (Eds.), *Environmental Biotechnology for Waste Treatment*, pp. 91-109. Plenum Press, New York, NY.

Frick, T. D., R. L. Crawford, M. Martinson, T. Chresand, and G. Bateson. 1988. "Microbiological Cleanup of Groundwater Contaminated by Pentachlorophenol." In G. S. Omen (Ed.), *Environmental Biotechnology. Reducing Risks for Environmental Chemicals through Biotechnology*, pp. 173-191. Plenum Press, New York, NY.

Harmsen, J. 1991. "Possibilities and Limitations of Landfarming for Cleaning Contaminated Soils." In R. E. Hinchee and R. F. Olfenbuttel (Eds.), *On-Site Bioreclamation: Processes for Xenobiotic and Hydrocarbon Treatment*, pp. 255-272. Butterworth-Heinemann, Stoneham, MA.

Hendriksen, H. V., S. Larsen, and B. K. Ahring. 1991. "Anaerobic Degradation of PCP and Phenol in Fixed-Film Reactors: The Influence of an Additional Substrate." *Water Science Technology* 24 (3/4): 431-436.

Mikesell, M. D., and S. A. Boyd. 1988. "Enhancement of Pentachlorophenol Degradation in Soil through Induced Anaerobiosis and Bioaugmentation with Anaerobic Sewage Sludge." *Environmental Science and Technology* 22(12): 1411-1414.

Mueller, J. G., D. P. Middaugh, S. E. Lantz, and P. J. Chapman. 1991. "Biodegradation of Creosote and Pentachlorophenol in Contaminated Groundwater: Chemical and Biological Assessment." *Applied and Environmental Microbiology* 57(5): 1277-1285.

Mueller, J. G., P. J. Chapman, and P. H. Pritchard. 1989. "Creosote-Contaminated Sites." *Environmental Science and Technology* 23(10): 1197-1201.

INVESTIGATIONS ON THE MICROBIAL DEGRADATION OF POLYCYCLIC AROMATIC HYDROCARBONS (PAHs) IN CONTAMINATED SOILS

M. Stieber, P. Werner, and F. H. Frimmel

ABSTRACT

Polycyclic aromatic hydrocarbons (PAHs) are by-products of coal-treatment processes that are found predominantly at abandoned coal gasification and coke oven plants. Due to the high toxicity and mutagenicity of PAHs, there is high interest in remediating those sites. This paper presents results on the biodegradation of PAHs at laboratory scale with respect to practical applications for cleanup. Two different experimental setups are described. One setup allows many experiments simultaneously, providing results on the biodegradation activity in general. The second setup is more sophisticated and allows calculation at a sufficient mass balance and, due to the large volume, the characterization of metabolites and dead-end products and their risk assessment. Some data obtained by those experimental systems are presented, showing different microbial activities during the degradation process.

INTRODUCTION

PAHs are by-products of coal treatment processes and are therefore predominantly found in the soil of abandoned coal gasification and coke oven plants. Depending on the solubility, at least the low-ring system PAHs can be detected in the groundwater. In the presence of BTEX aromatic compounds (benzene, toluene, ethylbenzene, and xylenes) that serve as solubilizers, they also can be found in higher concentrations than the maximum aqueous solubility described in the literature (Werner 1991).

In Germany many coal gasification plants closed down in the late 1960s and early 1970s. These abandoned and highly polluted sites, normally located in the cities, can be reused only after remediation. The predominant contaminants having the highest risk assessment are the PAHs. Concentrations in the soil vary from 10 mg/kg to 10 g/kg dry weight. Residuals also can be found in still existing basements and storage tanks after dismantling, for example. Many research

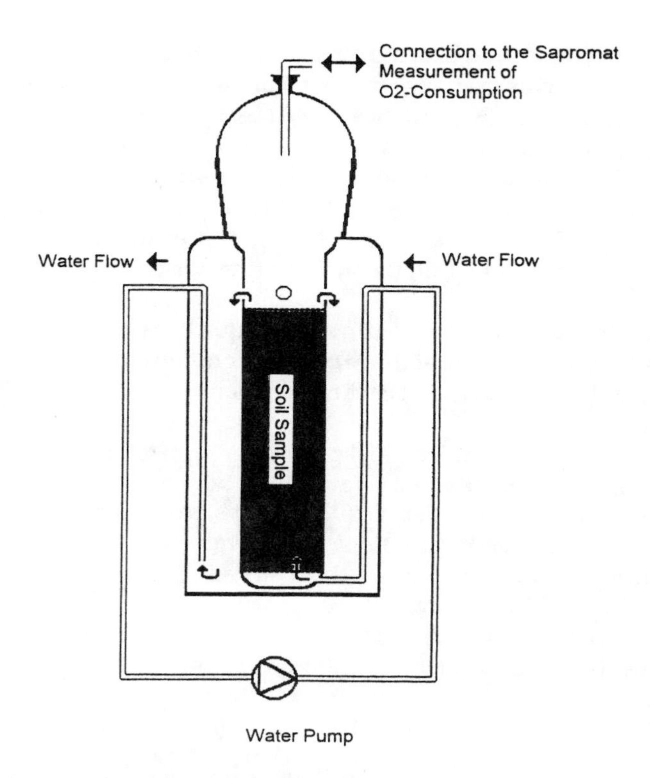

FIGURE 1. Scheme of the Sapromat soil test kit.

programs funded by the Ministry of Research and Technology or by the "Länder" are in progress in Germany concerning the evaluation of the efficiency of soil washing, incineration, and microbial degradation (Stieber et al. 1990).

So far, no known technology works well enough on an economically realistic base (Franzius et al. 1992). The paper points out the microbial processes in the mineralization of PAHs and describes data and experiences obtained in laboratory-scale experiments with respect to practical applicability. It focuses on the necessity of a correct mass balance during the degradation process.

RESULTS AND DISCUSSION

The goal of the investigations presented here is to test the biodegradability of PAHs at laboratory scale under different conditions. The key point is to evaluate the mass balance and to focus on the metabolites and dead-end products and their risk assessment.

To perform the runs, a new test kit based on the Sapromat system was developed. Figure 1 shows the schematic of the test cell which is connected with the genuine Sapromat, in which the oxygen consumed is produced electrolytically with the result of a constant oxygen partial pressure in the system. Figure 2 gives an example of degradation kinetics of PAHs measured using this test kit.

For these experiments, a population of microorganisms from the former coal gasification plant of Karlsruhe served as inoculum. The final concentration of PAH-degraders was in the range of 10^7/mL. It can be seen that naphthalene is the most rapidly degradable PAH. The lag-phases for the other PAHs are up to four times longer. The initial concentrations of the contaminants are given in Figure 2.

This small-scale laboratory setup enables quick test runs with many different soils and allows data on the activity of the microflora in many different soils to be investigated. However, an exact mass balance cannot be expected from the equipment.

Therefore, a more sophisticated percolator system had to be developed and was constructed in our institute (Figure 3). The percolator system is equipped with a computer-controlled data acquisition system, whereby factors such as the rate of aeration, the rate of water flow, the pH values, conductivity, and temperature are automatically controlled and registered. The oxygen consumption and carbon dioxide production are measured with on-line working gas analyzers.

Model experiments were performed using different PAH-mixtures which are degraded by a mixed microflora enriched from a coal gasification plant (see above). The sand filled in the column was artificially contaminated. The results show fairly rapid degradation of the PAHs (Figure 4). The metabolites in the flushing water were determined with gas chromatography (GC) and with high-performance liquid chromatography (HPLC).

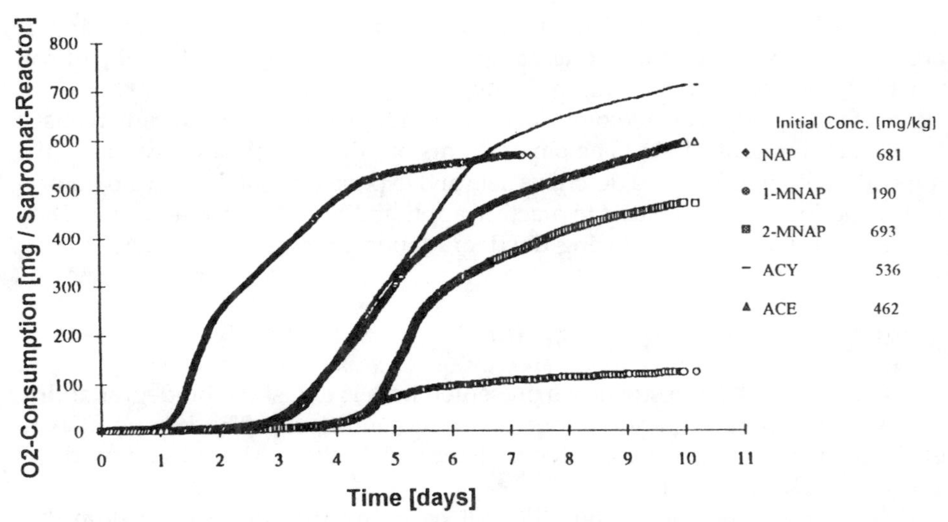

FIGURE 2. Microbial degradation kinetics at 20°C of naphthalene, 1-methylnaphthalene, 2-methylnaphthalene, acenaphthene, and acenaphthylene; each PAH was on sand grains as a sole source of carbon.

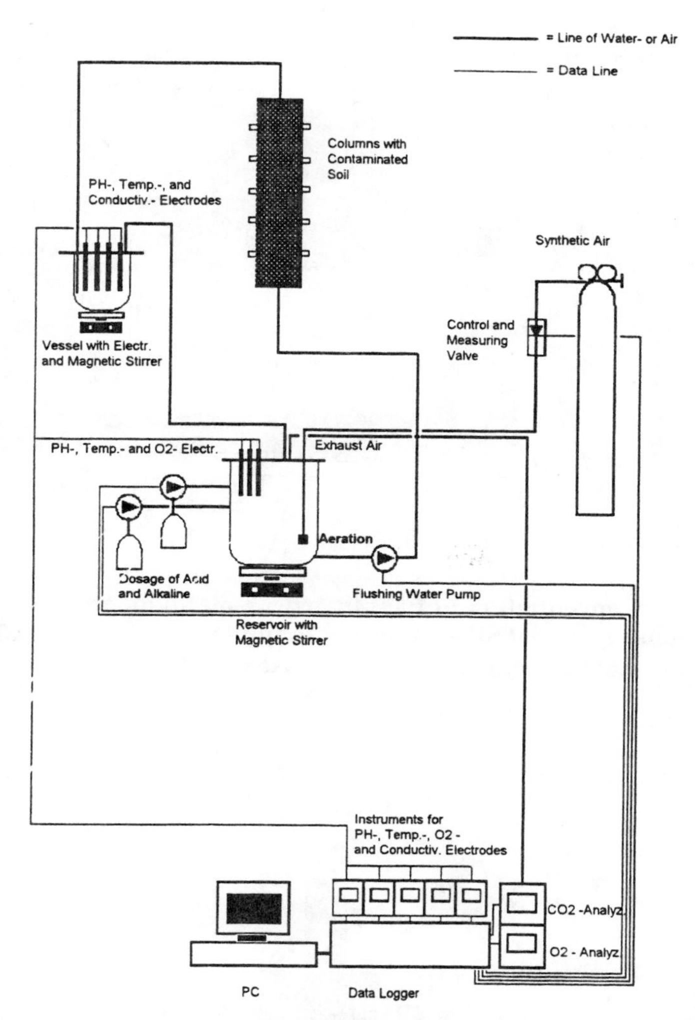

FIGURE 3. Schematic of a laboratory percolator coupled with a data acquisition system.

Figures 5 and 6 show the polaric and nonpolaric metabolites formed during the degradation process. For the polaric ones, no concentration can be determined. The data on oxygen consumption and carbon dioxide production in Figure 7 show the activities of the microflora during the experiment. The activity was spread into three phases:

1. Metabolism of the initial compounds
2. Degradation of the dissolved metabolites, accumulated in the aqueous phase
3. Degradation of the biomass by itself.

Similar experiences are described by Kelley et al. (1990).

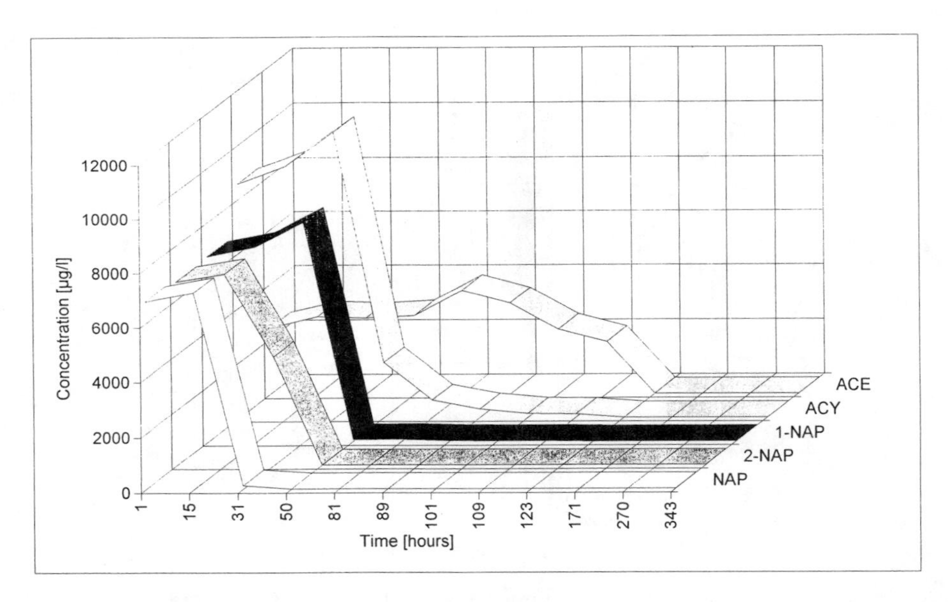

FIGURE 4. Concentration of naphthalene, 1-methylnaphthalene, 2-methylnaphthalene, acenaphthene, and acenaphthylene in the flushing water, extracted with cyclohexane and detected with GC/FID.

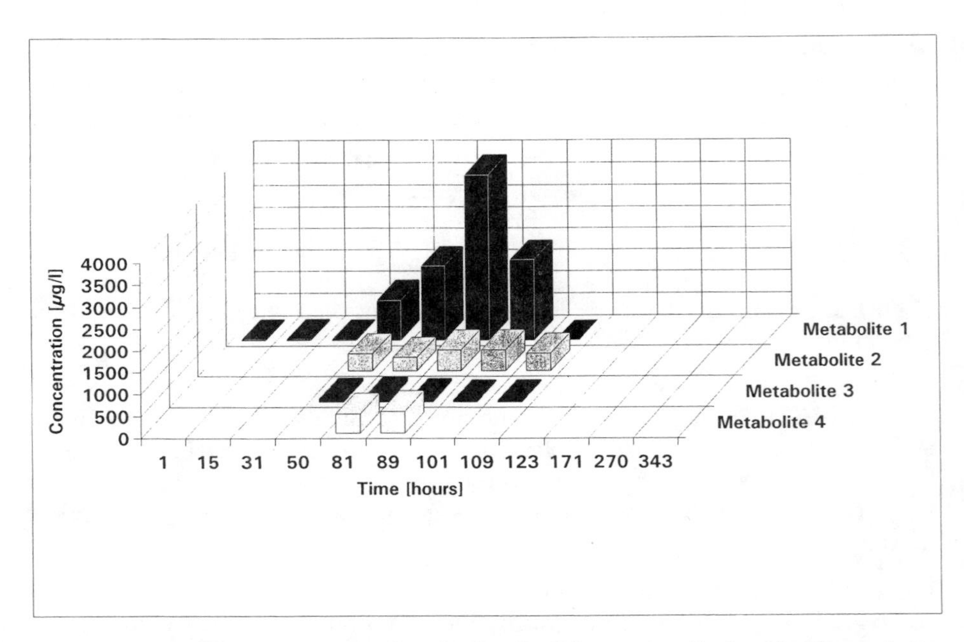

FIGURE 5. Nonpolaric metabolites in the flushing water during the degradation of naphthalene, 1-methylnaphthalene, 2-methylnaphthalene, acenaphthene, and acenaphthylene, detected with HPLC.

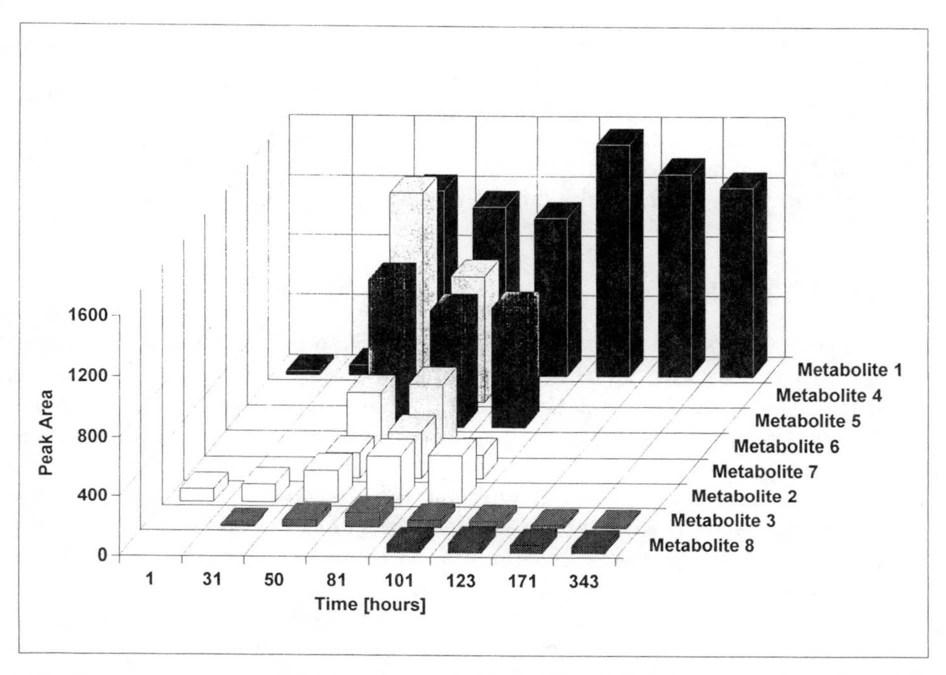

FIGURE 6. Polaric metabolites in the flushing water during the degradation of naphthalene, 1-methylnaphthalene, 2-methylnaphthalene, acenaphthene, and acenaphthylene, detected with HPLC.

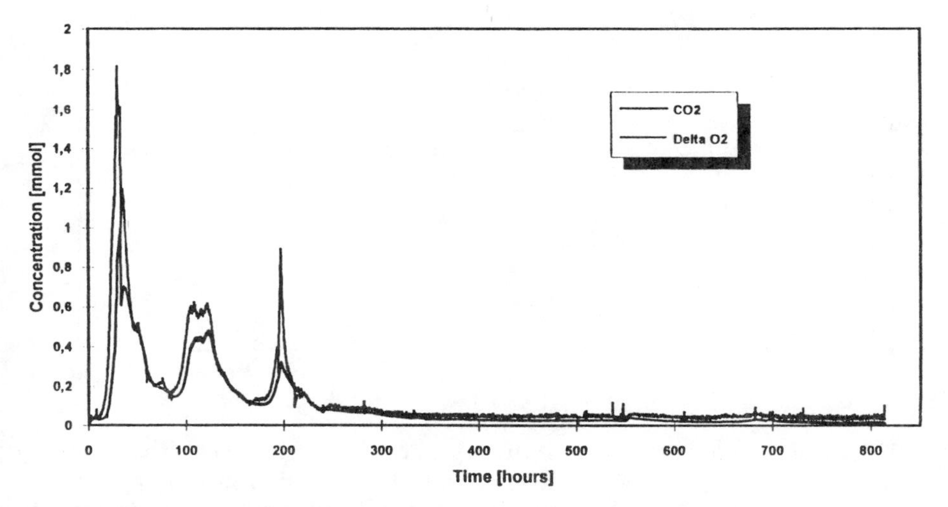

FIGURE 7. Development of consumed O_2 and produced CO_2 during the degradation of naphthalene, 1-methylnaphthalene, 2-methylnaphthalene, acenaphthene, and acenaphthylene.

After a period of accumulation, all of them were degraded, and no peak was evident at the end of the experiment. The dissolved organic carbon (DOC) and the spectral absorption coefficient (SAC) data presented in Figure 8 show an enrichment of organic compounds with a high ultraviolet (UV)-absorption activity, which is typical for aromatic compounds. In contrast to the GC and HPLC analyses the DOC and SAC values at the end of the experiment were still increasing.

CONCLUSIONS

The percolators and the modified Sapromat System described in this paper can be used as a tool for laboratory investigations examining the remediation of contaminated sites. They have been tested intensively with model experiments and now will be applied to characterize the degradation of PAHs in contaminated sites in actual practice. In recent experiments with contaminated soil from a coal gasification plant, it could be shown that the dissolved contaminants were biodegraded in a short time, whereas a large part of the PAHs attached to the soil were recalcitrant. The same observations had been made previously by Stieber et al. (1990) and Park et al. (1990).

In the future, the equipment will be used to measure the influence of ozone and other oxidants, alone or in combination, on the biodegradability, toxicity, and mutagenicity of metabolites and dead-end products.

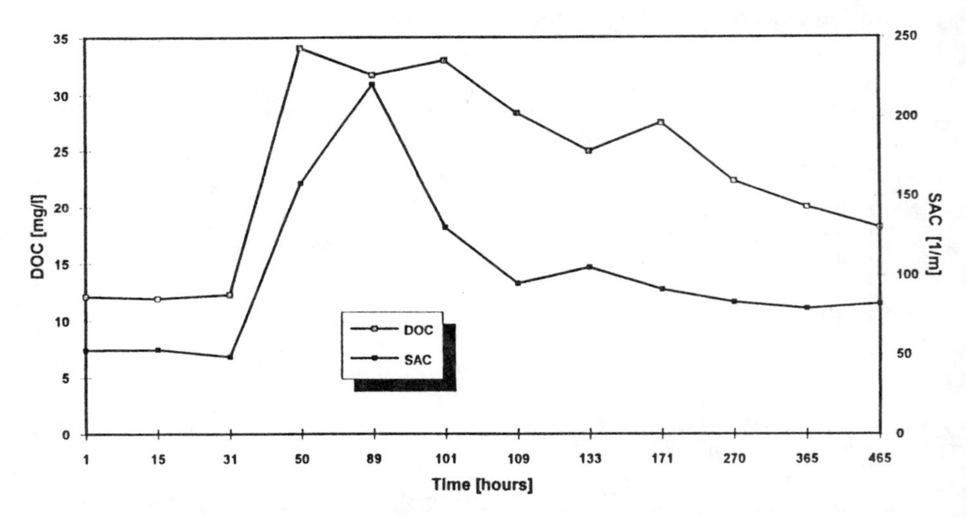

FIGURE 8. Dissolved organic carbon and spectral absorption coefficient (SAC) at 254 nm in the flushing water during the degradation of naphthalene, 1-methylnaphthalene, 2-methylnaphthalene, acenaphthene, and acenaphthylene.

ACKNOWLEDGMENTS

The authors thank the German Ministry for Research and Technology and Project Water Waste and Soil of Baden-Württemberg for the generous financial support.

REFERENCES

Breure, A. M., Sterkenburg A., Volkering F., and J. G. van Alden. 1992. "Bioavailability as a Rate Controlling Step in Soil Decontamination Processes." *Proc. Dechema International Symposium Soil Decontamination.* Karlsruhe, Germany.

Doré, M. 1989. *Chimie des oxydants et traitement des eaux.* Tec. et Doc. 11, éd. Lavoisier, rue Lavoisier, F-75348 Paris Cedex 08, France.

Franzius, V., R. Stegmann, and K. Wolf. 1992. *Handbuch der Altlastensanierung.* R. v. Decker's Verlag, G. Schenck, Heidelberg, Germany.

Kelley, I., J. P. Freeman, and C. E. Cerniglia. 1990. "Identification of Metabolites from Degradation of Naphthalene by a *Mycobacterium* sp." *Biodegradation 1*: 283-290. Kluwer Academic Publishers.

Legube B., S. Guyon, H. Sugimitsu, and M. Doré. 1986. "Ozonation of Naphthalene in Aqueous Solution - I & II." *Wat. Res. 20*: 197-214.

Park, K.S., R. C. Sims, and R. R. Dupont. 1990. "Transformation of PAHs in Soil Systems." *J. Environ. Engin. 116* (3): 632-640.

Stieber, M., K. Böckle, P. Werner, and F. H. Frimmel. 1990. "Biodegradation of Polycyclic Aromatic Hydrocarbons (PAH) in the Subsurface." In F. Arend, M. Hinseveld, and W. J. van den Brink (Eds.), *Contaminated Soil 1990*, pp. 473-479. Kluwer Academic Publishers, Netherlands.

Sturrock, M. G., E. L. Cline, and K. R. Robinson. 1963. "The Ozonation of Phenanthrene with Water as Participating Solvent." *J. Org. Chem. 28*: 2340-2343.

Werner, P. 1991. "German Experiences in the Biodegradation of Creosote and Gaswork-Specific Substances." In R. E. Hinchee and R. F. Olfenbuttel (Eds.), *In Situ Bioreclamation*, pp. 496-517. Butterworth-Heinemann, Stoneham, MA.

BIOREMEDIATION OF WATER AND SOILS CONTAMINATED WITH CREOSOTE: SUSPENSION AND FIXED-FILM BIOREACTORS VS. CONSTRUCTED WETLANDS AND PLOWING VS. SOLID PEROXYGEN TREATMENT

S. C. Tremaine, P. E. McIntire, P. E. Bell, A. K. Siler, N. B. Matolak, T. W. Payne, and N. A. Nimo

ABSTRACT

Contaminated waters and soils at many hazardous waste-contaminated sites can be remediated using biological processes. Bioremediation of polycyclic aromatic hydrocarbon (PAH)-contaminated water was examined to determine the treatment efficiency of suspension and fixed-film bioreactors versus constructed wetlands in a pilot-scale experiment. PAH removal from wastewater was higher for four-ring compounds when a physical support was provided for bacterial growth. Mass balance analysis of PAH removal revealed that fixed-film reactors and wetlands removed a similar amount of contaminants; more than 99% of the PAHs removed in the fixed-film bioreactors were biodegraded, whereas only 1% to 55% of the compounds removed by the wetlands were biodegraded. The volatilized fraction accounted for <5% of PAHs removed in all treatments. Pilot-scale bioremediation of creosote-contaminated soils was studied using a control and three treatments: limed; mechanically oxygenated (plowed, with lime and nutrients); and chemically oxygenated (Permeox™ solid peroxygen, with lime and nutrients). The results indicate that in the first 3 weeks, Permeox™ treatments had the highest degradation rates (122 mg total TPAH [TPAH]/kg•d), whereas rates were lower for plowed and limed treatments (23 and 2.5 mg TPAH/kg•d, respectively). Over the 12-week experiment, losses were highest in the Permeox™ plots.

INTRODUCTION

The chemical composition of creosote is roughly 85% PAHs; 10% phenolic compounds; and 5% N-, S-, and O-heterocyclic compounds (Mueller et al. 1989). Treatment of PAH and creosote-contaminated wastes has included traditional physical and chemical engineering approaches using lagoons, land treatment, filtration, activated carbon adsorption, and incineration (Lynch & Genes 1989). Since the early 1970s, bacterial treatment of creosote wastewaters, soils, and sludges has been applied more extensively and many bench-scale experiments using bacterial treatment have demonstrated significant reductions of PAHs in waters and soils (Lamar & Kirk 1990; Magee et al. 1991; Mueller et al. 1991a, 1991b, 1991c; Qiu & McFarland 1991; Wang & Bartha 1990).

Suspended growth reactors have been commonly used for biological treatment for creosote-contaminated wastewater. For example, Lieberman and Caplan (1988) treated mixed chemical waste (PAH, petroleum hydrocarbons, and grease) in an impoundment lagoon bioreactor that achieved a 10-fold reduction in selected PAHs within 1 week of treatment. Fixed-film bioreactors also have been used; Blais (1988) used 1-inch ring packing material in a bioreactor that removed 70 to 99% of the organic constituents in wastewater contaminated with creosote, PCP, and chromated copper arsenate. Constructed wetlands have been used as an effective treatment option for many kinds of wastewater, including municipal and domestic wastewater, acid and metal mine drainage, landfill leachate, and pulp and paper mill effluent. Wetland substrates act as a physical and biological filter, and wetland plants further sorb contaminants removing as much as 93% of the phenol and 89% of *m*-cresol from simulated wastewater (Wolverton & McDonald 1981).

Land treatment of creosote-contaminated soils has been shown to be a viable treatment option (Ellis et al. 1991, Hildebrandt & Wilson 1991, Smith et al. 1989, Vance 1991). Several studies describe the successful use of on-site landfarming in lined treatment areas. Smith et al. (1989) conducted pilot- and full-scale studies of land treatment of creosote-contaminated soils. Contaminated soil was applied to a lined treatment area and tilled biweekly. During the first year of the pilot-scale study, the concentration of total PAHs decreased by 97%, from 6,600 mg/kg to 176 mg/kg. Ellis et al. (1991) moved creosote-contaminated soil to a concrete treatment bed where it was tilled twice a week. Nutrients and microorganisms were added. During the first 88 days of operation, the total creosote concentration was decreased by 68% to an average concentration of 324 mg/kg.

This paper presents data from pilot-scale studies conducted to evaluate three biological treatment strategies for creosote-contaminated wastewater as well as two treatment strategies for land treatment of creosote-contaminated soils.

MATERIALS AND METHODS

Site and Experimental Design

All studies took place on the site of a former railroad tie preserving facility that had operated for 50 years. The creosote-contaminated wastewater came from a water storage tank located on site, designated Tank 3. Concentrations of contaminants in the wastewater ranged from 100 to 150 mg/L TPAHs. Tank 3 wastewater was first aerated then distributed to either of two wetlands (with or without commercial bacteria) or to three fixed-film treatments (with native bacteria, F1;

with a 50:50 mix of native and commercial bacteria, F2; and commercial bacteria, F3; Figure 1). The commercial bacteria mix (Solmar mix L-104) was added to the water treatment pilot-scale treatments at a rate of 60 g of reconstituted culture per week per treatment (into treatment W1 & F3 and half that amount into F2).

The wetlands were approximately two m^2, filled with soil, and planted with cattails. The fixed-film bioreactors were 690-L troughs filled with 2.5-cm-diameter plastic rings (Flexirings, Aquatic Ecosystems, Florida). Three small secondary treatment units followed bioreactor F3: gravel, wetland, and granular activated carbon (GAC). Details of the treatments have been published elsewhere (Tremaine et al. 1991). Flowrates were adjusted to produce a 5-day residence time, which was maintained for the calibration period and the first 3 weeks of the test. This rate was increased to produce a 2.5-day residence time for the last week of the test. Dissolved oxygen levels in the aeration tank were maintained near 6 mg O_2/L, pH was 9 to 9.5, and nitrogen fertilizer was added to maintain a final concentration of 20 mg/L N.

The area for the land treatment pilot-scale study was in the vicinity of the former drip tracks from the autoclaves. Concentrations of contaminants at the test site ranged between 500 to 14,000 mg/kg TPAHs. The land treatment experiments described here examined biodegradation of creosote-contaminated soils in the summer of 1992. Treatments included only those factors previously shown to provide the greatest stimulation to biodegradation: (1) pH adjustment; (2) nutrient addition; and (3) oxygen addition. Because oxygen availability appeared the most

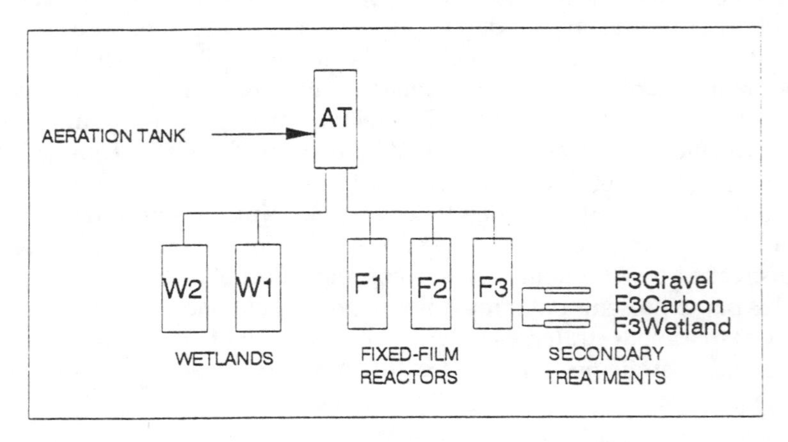

FIGURE 1. Pilot-scale operations area for testing biodegradation methods from creosote wastewater. Wastewater flowed from a storage tank to the aeration tank (AT) and then to the treatments. Treatments included two wetlands (W1 had Solmar bacteria mix L-104 added, W2 had only naturally occurring bacteria), and three fixed-film reactors (F1 had an enrichment culture of native bacteria only, F2 had native enrichment culture plus Solmar bacteria [each at half strength], and F3 had Solmar bacteria only). Three secondary treatments (F3Gravel; F3Wetland; and F3Carbon: granular activated carbon) received effluent from treatment F3.

limiting factor in this system based on the results of a bench-scale study and a winter pilot study, the principal comparison of the experiment discussed in this paper is the relative stimulatory effects of mechanical and chemical oxygenation. The study area was divided into eleven land farming plots that included one control; two minimum-treatment (limed) control plots; four mechanically oxygenated (plowed) plots (with lime and nutrients); and four chemically oxygenated (Permeox™) plots (with lime and nutrients) (Figure 2). Permeox™ Solid Peroxygen (FMC Corporation) was used as the chemical oxygen source; details of the Permeox™ treatment program are proprietary.

Due to large differences in contaminant concentrations, the landfarming site was divided into north and south experimental blocks comprising low- and high-contaminant-concentration regions, respectively. The treatments were randomly assigned within the blocks. Ten sampling points were assigned randomly to each of the eleven plots at the beginning of the experiment; these points were sampled repeatedly throughout the study.

Analytical Methods

Organics. Two different analytical methods were used throughout the course of the wastewater and soils studies. Selected PAH analyses were conducted on an

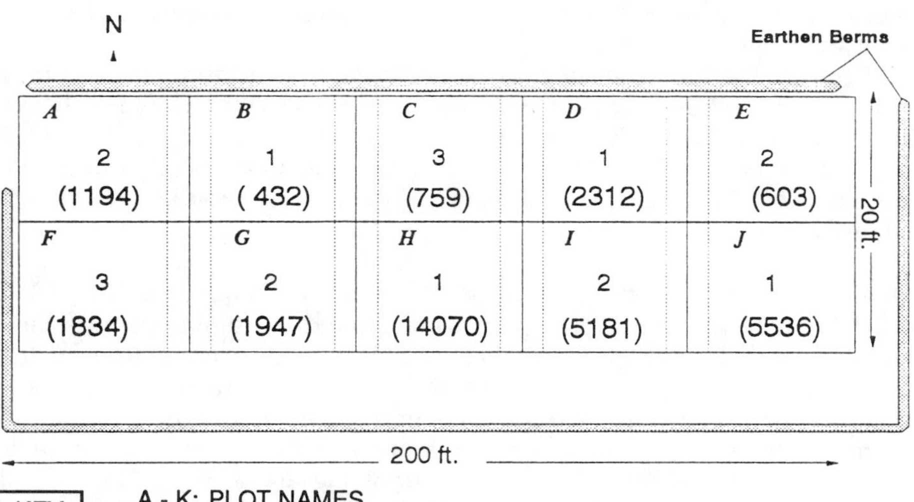

FIGURE 2. Plan of landfarming plots, Phase II. Numbers in parentheses are mean initial TPAH concentrations for the plots at the beginning of the experiment.

HP series 1050 high-performance liquid chromatograph (HPLC) with a programmable fluorescence detector. This method provides concentrations of individual preselected PAHs, in this case naphthalene, acenaphthene, phenanthrene, fluoranthene, and pyrene. The total polynuclear aromatic hydrocarbon (TPNA) method is a screening technique developed in the HYDROSYSTEMS laboratory that provides a rapid and economical monitoring tool (Beach 1991). The TPNA method was used to determine the concentration of PAHs in groups of similar numbers of rings (two to three rings, four rings, five to six rings, and more than six rings) as well as the total PAH concentration.

Mass balance calculations for the loss of contaminants in the wastewater study were performed by measuring volatilization, sorption, and input and output contaminant concentrations; biodegradation was computed by difference. Respirometry was not used to directly determine biodegradation, because without the use of radiolabeled contaminants there would be no way to quantify the CO_2 produced by heterotrophs consuming contaminant compounds vs. other nontarget organics.

PAHs volatilized from the wastewater were measured directly by collecting samples of the air over each of the treatments using glass m (0.8 cm diameter) filled with custom XAD-2 resin (ORBO™ Tubes, Supelco, Inc., Bellefonte, Pennsylvania). The tubes were connected to a Spectrex PAS 3000 air pump that drew air through the tubes at a rate of 1.5 to 2 L/min for 1 hour. Samples were extracted with 2 mL methylene chloride before the solvent was exchanged to acetonitrile and analyzed using the PAH technique.

To determine the amount of PAHs sorbed in the various treatments, substrate samples were collected (as described below), preserved, extracted, and analyzed. Quadruplicate wetland soils samples were collected at two depths in each wetland using a de-tipped 5-mL syringe. Triplicate Flexirings were collected from two sites in each bioreactor and placed in sampling jars with acetonitrile for extraction and analysis.

Nutrients. Dissolved oxygen concentrations were monitored with a YSI model 30 probe and meter. The pH was monitored using pH electrodes and meters, calibrated with two buffers. Percent field capacity was monitored using an E.W. System Soil Tester (Forestry Suppliers). Nutrient concentrations (nitrate, orthophosphate, and ammonia) were monitored using either colorimetric kits (Chemetrics) or an ion-specific electrode (Orions' ammonia electrode). Nutrients and moisture conditions were monitored during the landfarm experiment, and adjustments were made (lime, nutrient, or water additions) to maintain the target treatment conditions. Target nutrient levels in the landfarming experiment were to maintain a C:N:P ratio of 100:10:2.

Bacteria. The abundances of heterotrophic bacteria (HPC) and phenanthrene-degrading bacteria (CSPC) were monitored using standard microbiologic techniques. A ten-fold dilution series (10^{-4} to 10^{-7}) was plated onto replicate half-strength nutrient agar plates. Plates were incubated at room temperature for 6 days. Countable plates were enumerated, and then sprayed with a solution of 10% phenanthrene (w/v) in diethyl ether (Kiyohara et al. 1982). After further incubation, the number of colonies with clearing zones indicative of phenanthrene degradation were counted.

RESULTS AND DISCUSSION

Water Treatment

Concentrations of selected PAHs in the wastewater decreased as the water passed through the treatments. A large mass of sludge developed in the aeration tank, which acted as a suspension reactor. The two- and three-ring compounds (naphthalene, acenaphthene, and phenanthrene) degraded easily; 50% of the losses were in the suspension reactor, while additional loses were seen in the wetlands (20% removal) and fixed-film bioreactors (40% removal) (Figure 3; Table 1). The

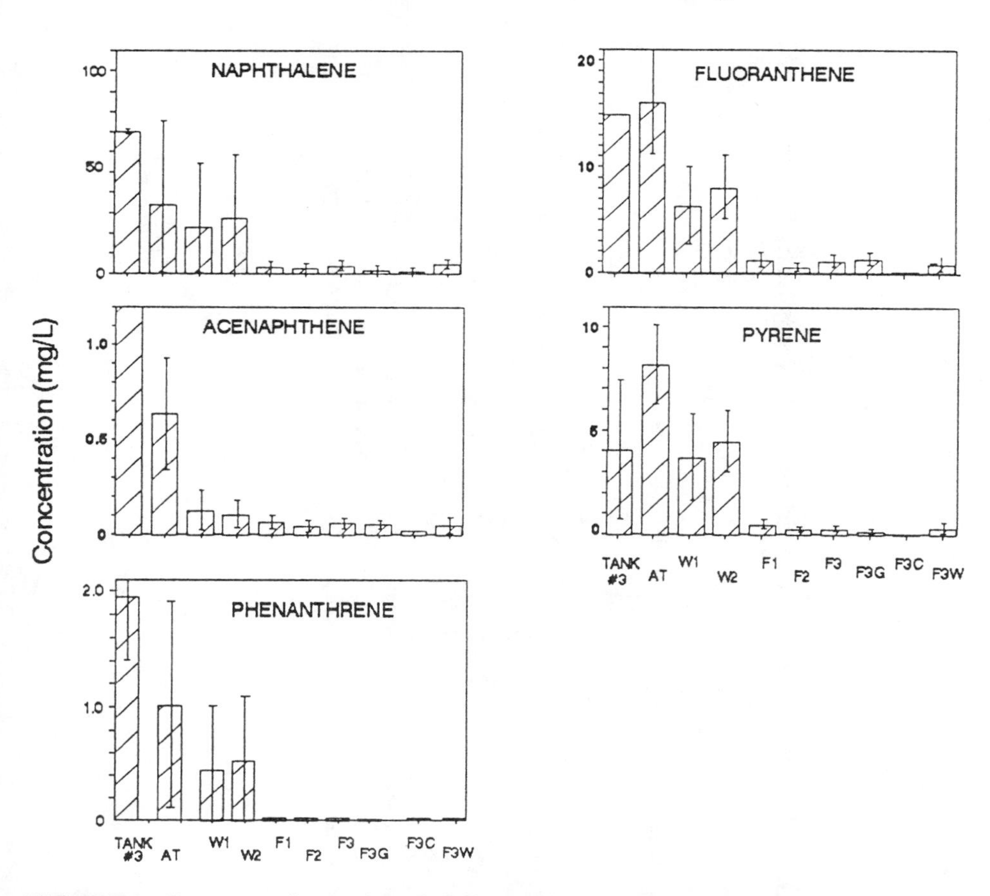

FIGURE 3. Concentrations of naphthalene, acenaphthene, phenanthrene, fluoranthene, and pyrene in samples taken from the pilot-scale wastewater treatment project. Treatment designations are as in Figure 1.

TABLE 1. Percent removal of selected polycyclic aromatic hydrocarbons (PAHs) in pilot-scale wastewater treatments.

Selected PAHs	Pre–treatment Aeration Tank	Primary Treatments					Secondary Treatments		
		Wetland		Fixed–Film			Fixed–Film #3 + (a)		
		W1	W2	F1	F2	F3	Gravel	Carbon	Wetland
Naphthalene	50.67	48.53	20.31	90.10	92.01	89.23	95.65	96.36	86.56
Acenaphthene	47.22	80.00	83.58	89.89	93.05	91.16	91.79	96.84	92.74
Phenanthrene	48.03	55.76	47.70	97.83	97.83	97.43	98.03	98.03	97.43
Fluoranthene	−7.78(b)	60.62	49.77	92.33	96.55	92.95	91.71	99.34	94.62
Pyrene	−101.72(b)	54.88	45.37	94.41	96.95	96.93	98.41	99.50	96.22

(a) These percent removals include removals that occurred in fixed–film reactor #3.

(b) Variations in concentrations of fluoranthene and pyrene in the influent to the aeration tank made calculation of removals of these two compounds inaccurate. These compounds were probably not removed from the wastewater by the aeration tank.

four-ring compounds (fluoranthene and pyrene) were not degraded in the aeration tank and, in fact, apparent concentrations increased in some cases (probably either a sampling artefact or accumulation of metabolic by-products). Concentrations of the selected PAHs decreased in the wetlands (50% removal) and fixed-film bioreactors (95% removal), with a larger reduction in the fixed-film bioreactor treatments. The GAC secondary treatment of treatment F3 (commercial bacteria alone) removed an additional 1 to 7% of contaminant TPAHs.

The source of bacteria made a difference in the wetland, but not in the fixed-film bioreactors. There was better PAH removal by the commercial mix in the wetland (W1) and equal removal in the fixed-film bioreactor. The best percent PAH removal values came in the treatment that mixed both native and commercial bacteria (F2).

The fate of the removed contaminants with regards to volatilization, degradation, and sorption also was compared between the three treatments (suspension bioreactor vs. wetlands vs. fixed-film bioreactor; Table 2). Volatilization was not a significant contributor to PAH removal (<5% TPAHs removed). Although the wetlands removed many of the contaminants, 45 to 100% of the PAHs removed had been sorbed but not degraded. The suspension reactor degraded 96% of the two- and three-ring compounds but none of the four-ring compounds. The fixed-film bioreactors degraded more than 99% of all size classes of PAHs.

Soil Treatment

In the first 3 weeks of the landfarm experiment there were rapid losses of TPAHs in the soils (Table 3). The greatest changes were seen in the Permeox™ plots (D, H, and J) (Figures 4a and 4b). After the third week of the experiment, heavy rains saturated the soils (>80% field capacity) and nutrients became depleted between weeks 3 and 9 (data not shown). Soil moisture conditions improved during the last 3 weeks of the experiment, producing overall reductions in TPAH concentrations (0 to 12 weeks) that were greatest in the Permeox™ plots (33% loss), and nearly equal in the other treatments (plowed, 7%; limed, 5%; and control, 10%) (Table 3). Examination of the TPAH losses from individual plots shows a large variation of responses such that some plowed plots were similar to Permeox™ plots (i.e., Weeks 0-12 North Block = 75%, 34% TPAH losses).

TPAH degradation rates during the first 3 weeks also were highest in the Permeox™ plots (−122 mg TPAH/kg soil•day) and lower for plowed and limed treatments (−23 and −2.5 mg TPAH/kg soil•day, respectively; Table 4). Degradation rates for weeks 6 to 12 show increases in TPAH concentration during the time of soil saturation and then a tendency to remain the same or decrease after the soils dried out (see also Figures 4a & b). The phenomenon of increasing soil contaminant concentrations during soil bioremediation has been widely noted and variously attributed to the effect of biosurfactants and production of daughter products. Data on the fate of individual selected PAHs (Tremaine et al., in preparation) indicate that the hypothesis of daughter products is incorrect, whereas biosurfactants cannot be ruled out.

TABLE 2. Partitioning of selected PAHs removed during biotreatment of creosote-contaminated wastewater. Values for total removal, percent sorbed, and percent volatilized were measured values; percent degraded was computed by difference.

Treatments	Total Removed (mg/L)	Sorbed	Volatilized	Degraded
Naphthalene				
Suspension	35.5	0[a]	1%	96%
Wetlands	9.2	64%	0	36%
Fixed-film	31.2	<1%	<1%	99.8%
Acenaphthene				
Suspension	0.57	0[a]	<1%	95.6%
Wetlands	0.52	100%	0	0
Fixed-film	0.58	<1%	0	99.4%
Phenanthrene				
Suspension	0.94	0[a]	4%	95.6%
Wetlands	0.52	100%	0	0
Fixed-film	0.99	<1%	<1%	99.6%
Fluoranthene				
Suspension	0[b]	0[a]	<1%	0[b]
Wetlands	8.92	45%	<1%	55%
Fixed-film	15.19	4%	<1%	99.7%
Pyrene				
Suspension	0[b]	0[a]	<1%	0[b]
Wetlands	4.11	100%	<1%	0
Fixed-film	7.88	0	<1%	99.6%

(a) Assumed to be zero because of the relatively small surface area for adsorption.
(b) Suspension reactor (aeration tank) pyrene and fluoranthene concentrations were highly variable and sometimes were higher than the source water, making computations highly speculative.

The abundance of heterotrophs and phenanthrene-degrading bacteria increased in all plots in the first week in the experiment and maintained high population levels throughout the experiment, in the range of 10^7 to 10^8 bacteria per gram of soil. Bacteria growth rates were higher in the Permeox™ plots (Tremaine et al., in preparation).

TABLE 3. Mean concentrations of TPAHs with standard deviations and percent change in concentration of TPAHs from the landfarm experiment from weeks 0 to 3 and weeks 0 to 12. Concentrations are in units of mg TPAH/kg soil.

	Permeox™				Plowed				Limed		Control	
	Mean	S.D.	Mean	S.D.	Mean	S.D.	Mean	S.D.	Mean	S.D.	Mean	S.D.
Weeks 0-3, North Block												
Initial conc.	423	27	2,312	563	1,194	751	603	75	759	123		
Wk 3 conc.	270	16	602	48	211	23	281	37	588	118		
% change	-36%		-74%		-82%		-53%		-23%			
Weeks 0-3, South Block												
Initial conc.	14,070	2,784	5,536	1,393	1,947	507	5,181	647	1,834	667		
Wk 3 conc.	8,486	1,391	2,728	536	1,527	366	4,953	764	1,900	656		
% change	-40%		-51%		-22%		-4%		4%			
Weeks 0-12, North Block											Weeks 0-12	
Initial conc.	423	27	2,312	563	1,194	751	603	75	759	123	934	149
Final conc.	449	86	656	51	294	19	395	44	399	32	841	91
% Change	6%		-72%		-75%		-34%		-47%		-10%	
Weeks 0-12, South Block												
Initial conc.	14,070	2,784	5,536	1,393	1,947	507	5,181	647	1,834	667		
Final conc.	8,207	1,999	5,566	1,019	1,487	407	6,170	716	2,083	707		
% Change	-42%		1%		-24%		19%		14%			

Grand Means of Percent Change

	Permeox™	Plowed	Limed	Control
Weeks 0-3	46%	21%	4%	ND*
Weeks 0-12	33%	7%	5%	10%

* ND = not determined.

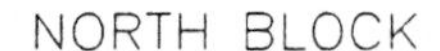

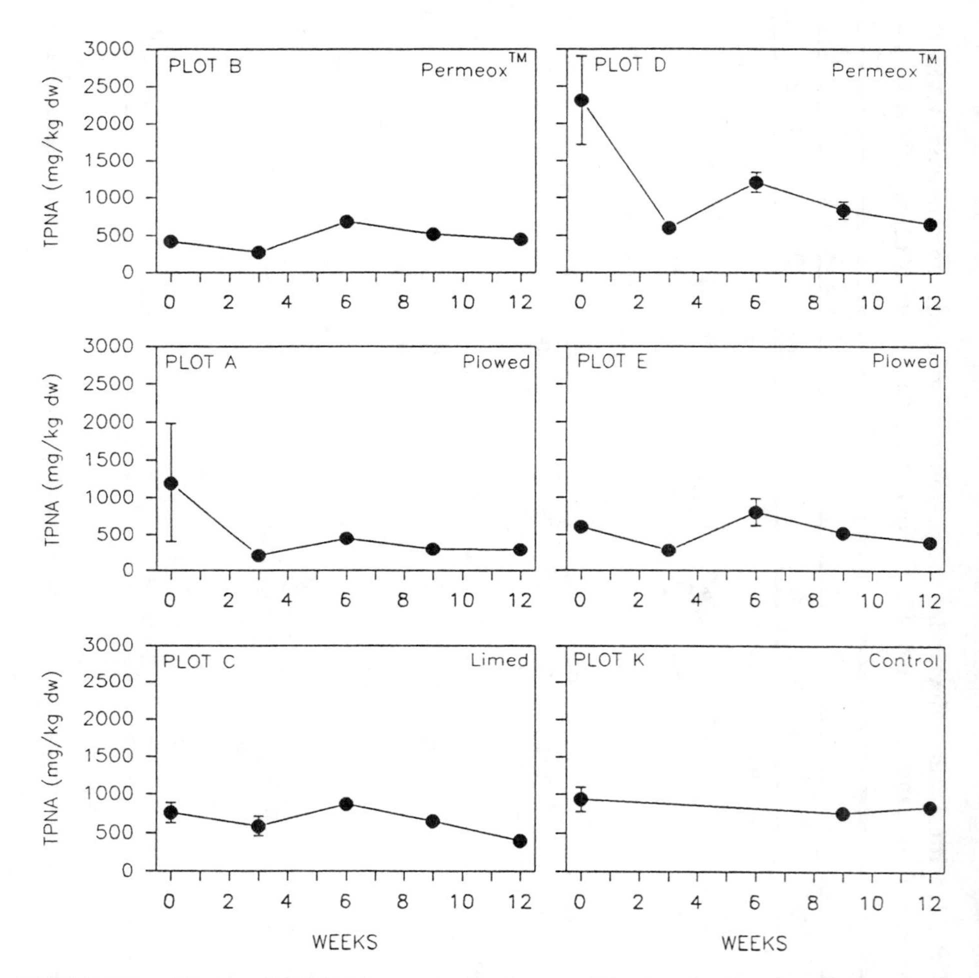

FIGURE 4a. Mean soil TPNA concentrations, with standard errors, measured in the plots in the Northern block during Phase II of the landfarming experiment.

CONCLUSIONS

Conducting treatability optimization studies in the natural environment can be a complex and rewarding task. Alexander and Loehr (1992) wrote in a letter to *Science* concerning bioremediation of the *Exxon Valdez* spill: "Field research in heterogeneous environments exposed to highly variable conditions frequently does not give identical results at different sites or at different times." They further stated that "...the finding that bioremediation worked at two of the sites is considered to be a positive and significant accomplishment." The data presented

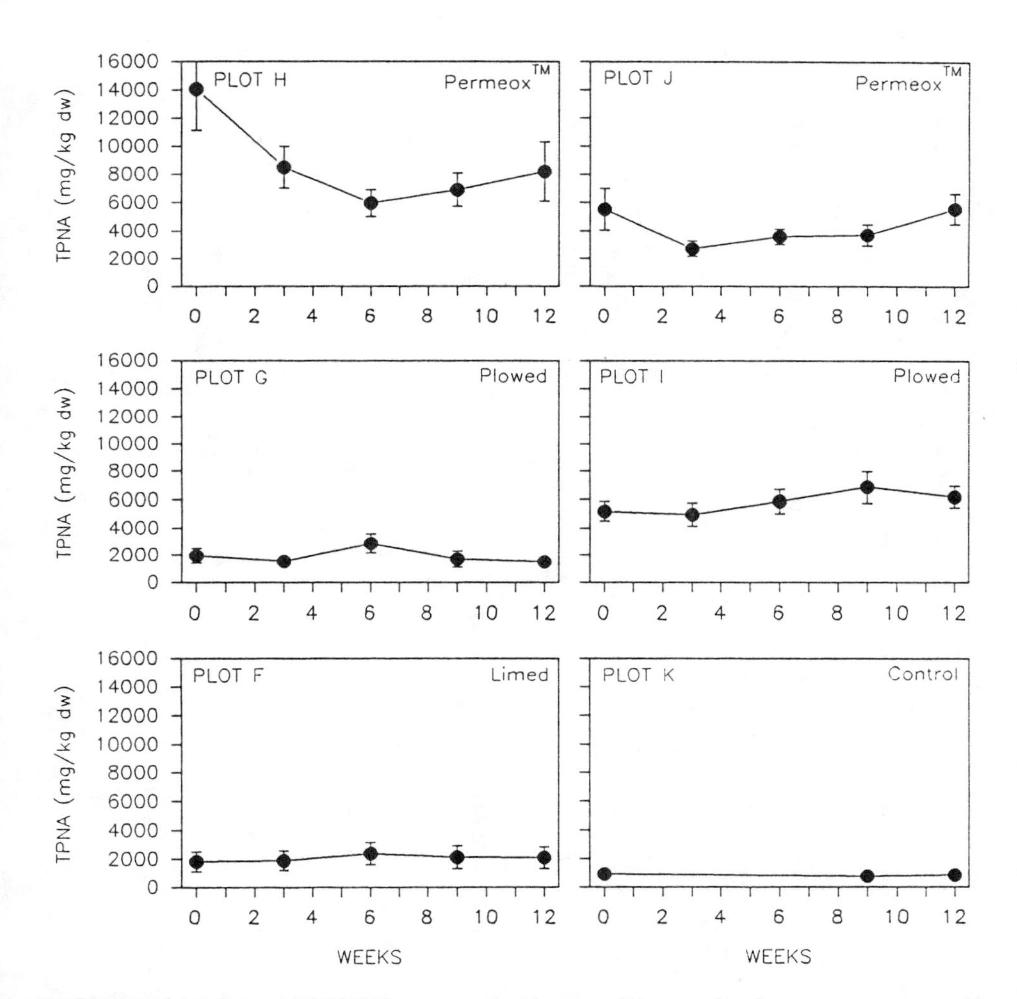

FIGURE 4b. Mean soil TPNA concentrations, with standard errors, measured in the plots in the Southern block during Phase II of the landfarming experiment.

here demonstrate that bioremediation is an effective remediation alternative for creosote compounds in wastewaters and soils. Chemical oxygenation using Permeox™ worked better for soils than the plowed or limed treatments, and fixed-film reactors worked well for wastewater.

Results from these studies emphasized the importance of including a physical support (soil or plastic rings) to enhance PAH removal in wastewater treatment. Fixed-film reactors produced effluent with lower contaminant concentrations than did the wetland treatments. Furthermore, contaminants that were removed

TABLE 4. TPAH degradation rates for weeks 0 to 3 and 6 to 12, the minimum average concentration obtained, the range for that week, and the week in which it occurred for the pilot-scale landfarming experiment.

Treatment	Plot	Block	Weeks 0-3 (ppm/d)		Weeks 6-12 (ppm/d)		Min Avg Conc. (ppm)	Range (ppm)		Week #
			Mean	SEM	Mean	SEM		Min	Max	
Permeox^{TM}	B	N	-7.3	1.85	-5.5	1.47	270	194	371	3
	D	N	-81.4	28.50	-13.2	2.68	602	304	811	3
	H	S	-265.9	135.90	53.8	41.50	5947	1507	10405	6
	J	S	-133.7	45.50	47.1	19.20	2728	857	5951	3
	Mean		-122.1	47.20	20.6	15.06				
Plowed	A	N	-46.8	38.00	-3.7	0.93	211	79	378	3
	E	N	-15.3	3.66	-9.7	4.23	281	111	567	3
	G	S	-20.0	18.41	-32.7	11.30	1488	273	4327	12
	I	S	-10.9	30.80	7.4	8.09	4953	2201	9228	3
	Mean		-23.3	6.99	-9.7	7.32				
Limed	C	N	-8.2	8.92	-11.1	1.47	399	280	589	12
	F	S	-3.1	32.30	-7.2	2.57	1834	377	7630	0
	Mean		-2.5	3.99	-9.2	1.38				

TABLE 5. Comparison of creosote contaminant removal from wastewater and soils as published in the literature.

Wastewater			
Reference	Type of Reactor	Contaminant	Percent Removal
Lieberman & Caplan 1988	Suspended	Select PNAs	-10%
Dust & Thompson [a]	Suspended	Phenols	-99%
Blais 1988	Fixed-Film	Naphthalene	-99%
"	"	Other Organics	-70 to -85%
Wolverton & McDonald 1981	Wetland	Phenol	-93%
"	"	m-cresol	-89%
This Study	Suspended	2-3 ring PNAs	-96%
"	"	4-ring PNAs	0%
"	Wetland	2-4-ring PNAs	-1 to -55%
"	Fixed-Film	2-4-ring PNAs	-99%

SOILS			
REFERENCE	INITIAL CONC. (mg/kg)	CONTAMINANT	DEGRADATION RATE (mg/kg/d)
ECOVA 1988	4,084	Total PNAs	16.5
Smith et al. 1989	6,600	Total PNAs	16
"	17,300	2,3,4-ring TPNAs	173
"	600	5,6-ring PNAs	5
Ellis et al. 1991	7,903	Creosote	43
This Study	500-14,000	TPNAs	
Permeox™	"	"	122
Plowed	"	"	23
Limed	"	"	2

[a] 1973, cited in Mueller et al. 1990

by the fixed-film reactors were degraded, whereas the wetlands primarily removed the larger contaminants by sorption. Volatilization was a relatively unimportant process in removing contaminants from the wastewater in the experiment.

Smaller, less complex compounds are readily degraded, whereas the more complex PAHs are degraded more slowly or require optimal conditions for high bacterial degradation rates. Fixed-film bioreactors generated good conditions for degrading the more recalcitrant four-ring compounds. Their success may be due to the increased surface area available for bacterial growth and the possibility of stratified aerobic/anaerobic bacterial community dynamics.

Pilot-scale land treatment also showed significant reductions in TPAHs in soils, with the chemically oxygenated Permeox™ treatment (with nutrients and pH control) having the highest degradation rates and contaminant losses. Final soil contaminant concentration was dependent on initial concentration, a phenomenon entirely predictable from substrate kinetics. Soil pH buffering alone also stimulated biodegradation but to a lesser extent. Most of the contaminants at this site were in the four-ring group of TPAHs and were degraded. Environmental conditions of poor soil drainage and heavy precipitation as well as nutrient depletion tended to slow down biodegradation. Despite these environmental difficulties, biodegradation of soils contaminated with creosote is a cost-effective alternative to soil incineration. Rates of TPAH losses in the first 3 weeks were similar to degradation rates seen at other sites with similar levels of contamination (Table 5).

In summary, creosote contamination of both wastewater and soils was effectively biodegraded by bacterial communities already present at the site. Addition of a commercial mixture of bacteria did not enhance biodegradation of creosote compounds in wastewater when compared to enriched native populations. Aeration of soils and addition of nutrients increased degradation of creosote compounds in soils, with the addition of chemical oxygen (Permeox™) tending to achieve more degradation than mechanical aeration.

ACKNOWLEDGMENTS

We are grateful for the invaluable technical assistance of Claudio Manissero and Everett Crockett of FMC Corporation, Philadelphia, Pennsylvania, and Arthur J. Bulger of the University of Virginia.

REFERENCES

Alexander, M., and R. C. Loehr. 1992. "Bioremediation Review." Letter to the editor. *Science, 258*:874.

Beach, R. W., K. M. West, L. Silka, M. D. Albertson, and A. Gilchenok. 1991. "A Screening Method for Total Polynuclear Aromatics." *Proceedings of the 14th Annual EPA Conference on Analysis of Pollutants in the Environment*, Norfolk, VA.

Blais, L. 1988. "Pilot-Scale Biological Treatment of Contaminated Groundwater at an Abandoned Wood Treatment Plant." In G. S. Omenn (Ed.), *Environmental Biotechnology Reducing Risks from Environmental Chemicals through Biotechnology*, pp. 445-446. Plenum Press, New York, NY.

ECOVA Corporation. 1988. *Union Pacific Railroad Laramie Tie Plant In Situ Treatment Process Development Program*, Volume 3/4. Milestone II Report.

Ellis, B., P. Harold, and H. Kronberg. 1991. "Bioremediation of a Creosote Contaminated Site." *Environmental Technology* 12:447-459.

Hildebrandt, C. W., and S. B. Wilson. 1991. "On-Site Bioremediation Systems Reduce Crude Oil Contamination." *Journal of Petroleum Technology 43*(1): 18-22.

Kiyohara, H., K. Nagao, and K. Yana. 1982. "Rapid Screen for Bacteria Degrading Water-Insoluble, Solid Hydrocarbons on Agar Plates." *Applied and Environmental Microbiology 43*(2): 454-457.

Lamar, R. T., and T. K. Kirk. 1990. "Use of White-Rot Fungi to Remediate Soils Contaminated with Wood-Preserving Waste." In *Bioremediation of Hazardous Wastes*, pp. 20-22, EPA/600/9-90/041.

Lieberman, M. T., and J. A. Caplan. 1988. "Biological Treatment of Petroleum Hydrocarbons and Polynuclear Aromatic Hydrocarbons in Waste Lagoon." In G. S. Omenn (Ed.), *Environmental Biotechnology: Reducing Risks from Environmental Chemicals through Biotechnology*, p. 453. Plenum Press, New York, NY.

Lynch, J., and B. R. Genes. 1989. "Land Treatment of Hydrocarbon Contaminated Soils." In P. T. Kostecki and E. J. Calabrese (Eds.), *Petroleum Contaminated Soils: Remediation Techniques, Environmental Fate and Risk Assessment*, pp. 163-174. Lewis Publishers, Chelsea, MI.

Magee, B. R., L. W. Lion, and A. T. Lemley. 1991. "Transport of Dissolved Organic Macromolecules and Their Effect on the Transport of Phenanthrene in Porous Media." *Environ. Sci. Technol. 25*: 323-331.

Mueller, J. G., P. J. Chapman, and P. H. Pritchard. 1989. "Creosote-Contaminated Sites." *Environ. Sci. Technol. 23*(10):1197-1201.

Mueller, J. G., P. J. Chapman, and P. H. Pritchard. 1990. "Development of a Sequential Treatment System for Creosote-Contaminated Soil and Water: Bench Studies." In *Bioremediation of Hazardous Wastes*, pp. 42-45. EPA/600/9-90/041, U.S. Environmental Protection Agency.

Mueller, J. G., D. P. Middaugh, S. E. Lantz, and P. J. Chapman. 1991a. "Biodegradation of Creosote and Pentachlorophenol in Contaminated Groundwater: Chemical and Biological Assessment." *Applied and Environmental Microbiology 57*(5): 1277-1285.

Mueller, J. G., S. E. Lantz, B. O. Blattmann, and P. J. Chapman. 1991b. "Bench-Scale Evaluation of Alternative Biological Treatment Processes for the Remediation of Pentachlorophenol- and Creosote-Contaminated Materials: Solid-Phase Bioremediation." *Environ. Sci. Technol. 25*(6): 1045-1055.

Mueller, J. G., S. E. Lantz, B. O. Blattmann, and P. J. Chapman. 1991c. "Bench-Scale Evaluation of Alternative Biological Treatment Processes for the Remediation of Pentachlorophenol- and Creosote-Contaminated Materials: Slurry-Phase Bioremediation." *Environ. Sci. Technol. 25*(6): 1055-1060.

Qiu, X., and J. J. McFarland. 1991. "Bound Residue Formation in PAH Contaminated Soil Composting Using *Phanerochaete chrysosporium*." *Hazardous Waste & Hazardous Materials 8*(2): 115-126.

Smith, J. R., D. V. Nakles, D. F. Sherman, I. F, Neuhauser, R. C. Loehr, and D. Erickson. 1989. "Environmental Fate Mechanisms Influencing Land Treatment of Polynuclear Aromatic Hydrocarbons." *Proc. of the Third International Conference on New Frontiers for Hazardous Waste Management*. Pittsburgh, PA.

Tremaine, S. C., P. E. Bell, N. B. Matolak, and A. Siler. In Preparation. "Biodegradation of Creosote in Contaminated Soil: Pilot-Scale Evaluation of Mechanical vs. Chemical Soil Oxygen."

Tremaine, S. C., P. E. McIntire, and R. B. Beach. 1991. "Comparison of the Best Available Technologies: Creosote Biodegradation." In *Proc. of Hazardous Materials Control/Superfund '91*, pp. 326-332. Hazardous Materials Control Research Institute, Washington, DC.

Vance, D. B. 1991. "Onsite Bioremediation of Oil and Grease Contaminated Soils." *The National Environmental Journal*, pp. 26-30.

Wang, X., X. Yu, and R. Bartha. 1990. "Effect of Bioremediation of Polycyclic Aromatic Hydrocarbon Residues in Soil." *Environmental Science and Technology 24*(7):1086-1089.

Wolverton, B. C., and R. C. McDonald. 1981. "Natural Processes for Treatment of Organic Chemical Waste." *The Environmental Professional 3*:99-104.

RELATIVE RATES OF BIODEGRADATION OF SUBSTITUTED POLYCYCLIC AROMATIC HYDROCARBONS

D. L. Elmendorf, C. E. Haith, G. S. Douglas, and R. C. Prince

ABSTRACT

A consortium of Prince William Sound microorganisms was used to degrade Alaskan North Slope crude oil over a 6-month time period. Microorganisms were inoculated (in triplicate) into flasks containing saline Bushnell-Haas medium with a 1% oil loading. Oil was extracted with methylene chloride approximately every 14 days and incubation continued after addition of fresh Bushnell-Haas medium and consortium. After the second extraction, the oleophilic fertilizer Inipol EAP22™ was added with the oil at a concentration of approximately 10% of the oil. Hydrocarbon degradation was determined by gas chromatographic analyses. The majority (>90%) of the saturates were biodegraded by Day 19, and pristane and phytane were > 90% biodegraded by Day 36. Unsubstituted polycyclic aromatic hydrocarbons (PAHs) also showed extensive degradation, disappearing in the order naphthalenes > fluorenes > dibenzothiophenes > phenanthrenes > chrysenes. Progressive methyl substitution slowed degradation, but even C3-chrysene showed some degradation during the experiment.

INTRODUCTION

The microbiology and chemistry of petroleum degradation in the marine environment has been extensively studied (Atlas 1981a, Fedorak & Westlake 1981, Rambeloarisoa et al. 1984, Walker & Colwell 1976, Zobell 1946). The rates of biodegradation have generally followed the pattern of being highest for saturates followed by the light aromatics (unsubstituted and substituted), then the heavy aromatics (unsubstituted and substituted), and finally the polar compounds (Fusey & Oudot 1984, Jobson et al. 1972, Perry 1977, Walker et al. 1976a).

The biodegradation of petroleum hydrocarbons in the marine environment is affected by a variety of abiotic parameters and biotic processes, including evaporation (Floodgate 1984, Payne et al. 1991), photooxidation (Burwood & Speers 1974), dispersion, emulsification, polymerization (Tagger et al. 1982), temperature

(Atlas & Bartha 1972a, Walker & Colwell 1974), oxygen concentration (Singer & Finnerty 1984), nutrients (Atlas & Bartha 1972b, 1973; Bergstein & Vestal 1978; Boehm & Fiest 1980; Dibble & Bartha 1976; LePetit & N'Guyen 1976; Olivieri et al. 1976), pH, and salinity (Hambrick et al. 1980, Ward & Brock 1978).

Alaskan North Slope crude oil is considered to be a moderately heavy crude oil. Alaskan North Slope crude oil is composed of paraffins, 27.3%; naphthenes, 36.8%; aromatics, 25.3%; and polars and others, 10.6% (Clark & Brown 1977). Payne et al. (1991) estimated that a maximum of 30% of Alaskan North Slope crude oil may disappear by evaporation, and we have used oil that has lost this fraction in the studies reported here. The percentage of polars, asphaltenes, and resins determines whether oils are considered light or heavy. Light oils have low concentrations of these components and are easily biodegraded, whereas heavier oils have higher concentrations and are not as easily degraded (Atlas 1975, Walker et al. 1976a).

Pristane and phytane have been utilized as determinants of the amount biodegradation that occurred because they are considered not to be as readily degraded as the n-alkanes (Atlas 1981b, Fedorak & Westlake 1981, Hughes & McKenzie 1975, Walker et al. 1976b). Hughes and McKenzie (1975) noted that branched alkanes such as pristane were slowly degraded compared to *n*-alkanes. Biodegradation of *Amoco Cadiz* oil in Brittany Coast sediments resulted in only 5 to 15% loss of pristane in 2 weeks at 15°C (Atlas 1981b). Incubation at 30°C for 2 weeks was required before soil microorganisms from a crude oil-contaminated site used pristane and phytane (Jobson et al. 1972). After microbial degradation of South Louisiana crude oil, only traces of *n*-alkanes were detected, but in the case of Kuwait crude oil pristane and phytane remained after biodegradation (Walker et al. 1976b). Water and sediment inocula from Cook Inlet, Alaska, were able to mineralize less than 5% of pristane, in a liquid medium with a 0.25% concentration, after 6 weeks of incubation at 5°C (Roubal & Atlas 1979).

On March 24, 1989, the *Exxon Valdez* spilled approximately 2.85×10^6 L of Alaskan North Slope crude into the subarctic waters of Prince William Sound, Alaska. After evaporation, the only way the majority of spilled oil leaves the marine environment is by physical collection, combustion, or biodegradation. The strategy used in cleaning up the *Exxon Valdez* oil spill was to remove the oil by physically washing the sediment and collecting the oil with skimmers, and subsequently stimulating biodegradation by adding oleophilic [Inipol EAP22™ (CECA S. A., 92062 Paris, La Defense, France)] and slow-release fertilizers [Customblen (Grace-Sierra, Milpitas, California)] (Harrison 1991, Nauman 1991). At the time, bioremediation was a relatively untried technology, and it was necessary to demonstrate its effectiveness in the field (Prince et al. 1990, Pritchard & Costa 1991).

The purpose of this study was to analyze in the laboratory the biodegradation of artificially weathered Alaskan North Slope crude oil by a microbial consortium from Prince William Sound, Alaska, and to assess the biodegradation rates of different components of the oil, to show that pristane and phytane are readily degradable and not reliable markers for determination of biodegradation, and to evaluate the biodegradation of substituted PAHs.

MATERIALS AND METHODS

Media Preparation

Bushnell-Haas (BH) medium was prepared as shown in Table 1, and was sterilized by autoclaving. After cooling, to each liter of BH medium was added 1 mL of Pfenning's vitamin mixture and 5 mL of Wolfe's mineral solution. To a flask containing 1,200 mL of BH medium was added 12 g of Alaskan North Slope (ANS) 521 crude oil (see below) to give a 1% oil loading. Flasks were incubated

TABLE 1. The composition of Bushnell-Haas medium, Pfenning's vitamin solution, and Wolfe's mineral solution.

Ingredient	Amount Added
Bushnell-Haas Medium (pH 7.6)	
KH_2PO_4	1.0 g
K_2HPO_4	1.0 g
$MgSO_4$ $7H_2O$	0.2 g
$CaCl_2$ (anhyd.)	0.02 g
$FeCl_3$ $6H_2O$	0.05 g
NH_4NO_3	1.0 g
$NaCl_3$	0.0 g
Distilled water	1,000.0 mL
Pfenning's Vitamin Mixture	
Biotin	1.0 mg
p-Aminobenz	5.0 mg
Vitamin B-12	5.0 mg
Thiamine	10.0 mg
Distilled water	100.0 mL
Wolfe's Mineral Solution	
Nitrilotriacetic	1.5 g
$MgSO_4$	3.0 g
$MnSO_4$	0.5 g
NaCl	1.0 g
$FeSO_4$ $7H_2O$	0.1 g
$CoSO_4$	0.1 g
$CaCl_2$	0.1 g
$ZnSO_4$	0.1 g
$CuSO_4$ $5H_2O$	0.01 g
$AlK(SO_4)_2$	0.01 g
H_3BO_3	0.01 g
Na_2MoO_4 $2H_2$	0.04 g
Distilled water	1,000.0 mL

at 15°C in a New Brunswick water bath and shaken continuously at 125 rpm for 168 days.

Bushnell-Haas medium was used in the U.S. EPA/Exxon/ADEC/UAF Joint Bioremediation Monitoring Project. This allows comparison of biodegradation in this study with field biodegradation results, which are not discussed because it is not within the scope of this study.

521 Alaskan North Slope Crude Oil

Alaskan North Slope crude oil was distilled in a refluxing apparatus to artificially weather the oil so that the composition would be similar to that of oil spilled in Prince William Sound after natural weathering. The crude oil was heated until the vapor temperature reached 271.6°C (M. Stec, pers. comm.); this oil had lost 30% of its weight. This treatment results in the removal of the volatile components of the oil, which both simulates weathered oil and prevents compositional changes of the oil as a result of the oil extraction procedure.

Inoculum Preparation

Prince William Sound beach material from selected September 1990 A series sampling was used to provide the microbial consortium for the inoculum. The inoculum was prepared by aseptically adding 20 g of each beach sample to a sterile glass bottle containing 80 mL of sterile, saline Bushnell-Haas medium (pH 7.0). Bottles were shaken in a New Brunswick shaker at 160 rpm at room temperature for 20 min, then allowed to settle for 1 min. Supernates were combined in a sterile Erlenmeyer flask to serve as an inoculum; 100 mL of inoculum was added to each experimental flask.

Microbial Enumeration

Estimates of the numbers of oil-degrading and total heterotrophic bacteria were determined using the most-probable-number (MPN) technique described by Brown and Braddock (1990) using marine broth for the enumeration of total heterotrophs. Bushnell-Haas medium containing 10 µL of ANS 521 crude oil was used for the enumeration of oil degraders (Brown & Braddock 1990).

Extractions of Experimental Flasks

Experimental flasks were run in triplicate and extracted every 14 days (approx.) with a minimum of 4 volumes (250 mL/volume) of methylene chloride using a separatory flask. Samples were taken at each extraction for gas chromatographic (GC) analyses. Methylene chloride-oil mixtures were returned to their original flasks, and the methylene chloride was evaporated by flowing nitrogen gas through the flask. Once the methylene chloride was totally evaporated, fresh

Bushnell-Haas medium (pH 7.0) was added and reinoculated. At day 36, 1.2 g of Inipol EAP22™ was added to each flask, to further stimulate biodegradation. After each subsequent extraction, fresh Inipol EAP22™ was added with the amount being reduced to 0.4 g after day 100 and changed to 0.6 g after day 114. Inipol EAP22™ was not added after day 155; this was to simulate the addition of oleophilic fertilizer in PWS, in case cometabolism of oil with more readily degraded compounds played an important role in degradation of the oil.

Oil Analysis

GC/mass spectroscopy (MS) and GC/flame ionization detection (FID) analyses of oil extractions (Table 2) were conducted by Battelle Ocean Sciences, Duxbury,

TABLE 2. Types of aliphatics and PAHs and concentrations (mg/kg) found in ANS 521 crude oil using GC/MS and GC/FID analytical techniques.

Analyte	mg/kg		mg/kg
C13	35.62	Naphthalene	2.65
C14	255.35	C1-naphthalenes	42.23
C15	1276.48	C2-naphthalenes	356.14
C16	2481.92	C3-naphthalenes	937.08
C17	5301.74	C4-naphthalenes	896.72
Pristane	2489.36	Fluorene	82.43
C18	4111.03	C1-fluorenes	339.63
Phytane	2112.22	C2-fluorenes	609.46
C19	4532.12	C3-fluorenes	663.33
C20	3812.93	Phenanthrene	399.33
C21	3338.04	C1-phenanthrenes/anthracenes	1022.73
C22	3254.19	C2-phenanthrenes/anthracenes	1196.16
C23	2809.42	C3-phenanthrenes/anthracenes	857.59
C24	2764.62	C4-phenanthrenes/anthracenes	559.48
C25	2368.11	Dibenzothiophene	364.66
C26	2070.23	C1-dibenzothiophenes	803.43
C27	1481.95	C2-dibenzothiophenes	1103.99
C28	1345.84	C3-dibenzothiophenes	981.23
C29	1198.63	Fluoranthene	3.86
C30	1045.60	Pyrene	15.90
C31	718.43	C1-fluoranthene/pyrene	131.36
C32	558.56	Benzo(a)anthracene	7.74
C33	496.38	Chrysene	72.97
C34	553.03	C1-chrysenes	139.85
		C2-chrysenes	199.10
		C3-chrysenes	153.30
		C4-chrysenes	105.46
		Benzo(b)fluoranthene	8.30
		Benzo(g,h,i)perylene	104.91

Massachusetts (Butler et al. 1991, Douglas et al. 1992). Individual PAHs were identified and quantified by the GC/MS technique, which is sensitive to parts per trillion (ng/L) (Butler et al. 1991, Douglas et al. 1992). Concentrations of *n*-alkanes, pristane and phytane, and total hydrocarbons were determined by GC/FID (Butler et al. 1991, Douglas et al. 1992). Initial ANS 521 crude oil, which served as the reference standard, also was analyzed by GC/MS and GC/FID. The C_{30} 17*a* (H)21*B* (H)hopane concentration was determined for both the reference oil and experimental samples (Butler et al. 1991, Douglas et al. 1992).

RESULTS

Microbial Numbers

Population numbers were relatively low at the beginning of the experiment (3.75×10^6 cells/g & 7.5×10^5 cells/g total heterotrophs/oil-degraders), and increased during the experiment (4.62×10^{11} cells/g and 3.132×10^8 cells/g total heterotrophs/oil-degraders), yet the majority of resolved biodegradation apparently occurred within the first 36 days of the experiment. Total heterotrophic and oil-degrading MPN estimates showed an increase of 2 to 6 orders of magnitude after each new inoculum addition (data not shown).

Biodegradation of Aliphatic and Aromatic Hydrocarbons – Percent Depletion

The majority of the alkanes analyzed for were completely removed by biodegradation within 19 days after inoculation of the experimental flasks. Only the isoprenoid alkanes, pristane and phytane, were not >90% degraded by Day 19; however, pristane was completely degraded and phytane was 91% degraded by Day 36 (Table 3).

Extensive biodegradation of the unsubstituted and substituted PAHs by the Prince William Sound consortium also was observed. Plots of the percent depletion of the substituted aromatics versus unsubstituted aromatics for the phenanthrenes, fluorenes, dibenzothiophenes, and chrysenes are shown in Figures 1 to 5. These plots show that unsubstituted aromatic hydrocarbons, such as naphthalene, fluorene, phenanthrene, and dibenzothiophene are biodegraded more readily than the substituted compounds. The more substituted the compounds, the slower the biodegradation, i.e., C1-phenanthrene/anthracenes > C2-phenanthrene/anthracenes > C3-phenanthrene/anthracenes > C4-phenanthrene/anthracenes (Figure 1). Similar results were observed with substituted fluorenes (Figure 2) and dibenzothiophenes (Figure 3). Biodegradation of the chrysenes was limited and showed a much narrower range between unsubstituted and substituted members (Figure 4). Comparison of the various unsubstituted aromatics shows that the preferential biodegradation is naphthalene > fluorene > dibenzothiophene and phenanthrene > chrysene (Figure 5). An increase in the ring number of a hydrocarbon results in decreased ability of microorganisms to biodegrade

TABLE 3. Percent depleted of aliphatics by biodegradation. Data represent the means of three replicates.

Exp. ID Analyte	Day 19 % depleted	SD	Day 36 % depleted	SD
C13	100	0	100	0
C14	100	0	100	0
C15	100	0	100	0
C16	100	0	100	0
C17	97.67	2.52	100	0
Pristane	85	6	100	0
C18	96.33	0.58	97.67	0.58
Phytane	73.67	5.69	91.33	3.60
C19	91	1	95.33	0.58
C20	96	2.64	99.33	1.15
C21	97.67	0.58	98	0
C22	96.33	0.58	96.33	0.58
C23	100	0	100	0
C24	99	0	99	0
C25	99.33	1.15	100	0
C26	98.33	2.89	100	0
C27	100	0	100	0
C28	100	0	100	0
C29	100	0	100	0
C30	100	0	100	0
C31	100	0	100	0
C32	100	0	100	0
C33	100	0	100	0
C34	100	0	100	0

SD = Standard Deviation.

such materials. Chrysene is one of those hydrocarbons that is generally considered to be nonbiodegradable. These results show that biodegradation of chrysene can occur over time, although it occurs more slowly than naphthalene biodegradation.

DISCUSSION

Microbial Numbers

Microbial numbers increased throughout the course of the experiment, even though the majority of resolved biodegradation occurred within the first 36 days of the experiment. One explanation for the increase in numbers in later incubations might be that the addition of Inipol EAP22™ provided an additional substrate for growth, although it is noteworthy that the population of hydrocarbon degraders increased along with the population of total heterotrophs. Furthermore, in the absence of Inipol EAP22™ the microbial populations remained elevated in the

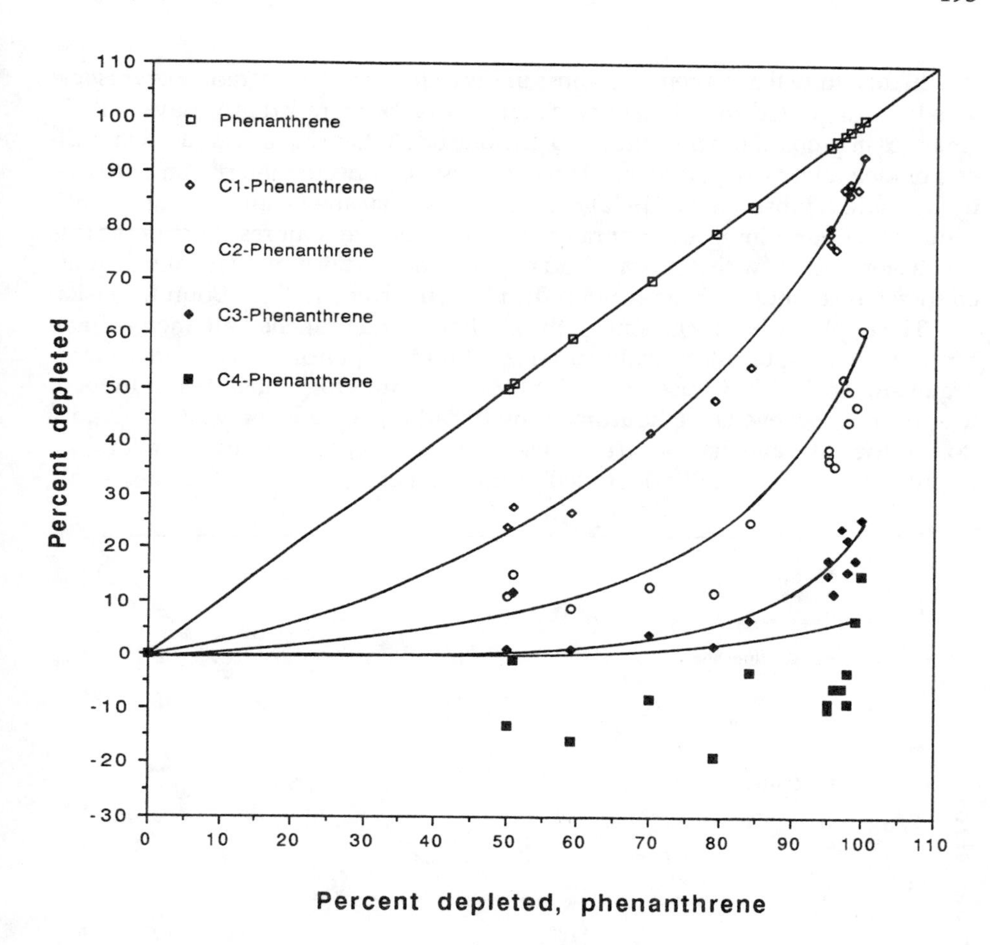

FIGURE 1. Plots of the percent depleted of individual phenanthrenes over the percent depleted of phenanthrene (three replicates).

final incubation (on average, $5.796 \times 10^{14}/2.22 \times 10^{12}$ total heterotrophs/oil-degraders). In any case, no close correlation exists between the absolute numbers of microbes and the amount of biodegradation, which is perhaps not surprising when one considers the complexity of the microbial communities involved in biodegradation and the rather primitive enumeration methods available. Certainly the biodegradation of the complex PAHs proceeded throughout the experiment, as shown in Figures 1 to 5.

Biodegradation of Aliphatic and Aromatic Hydrocarbons

Both the aliphatic and aromatic fractions of the ANS 521 crude oil underwent extensive biodegradation by the Prince William Sound inoculum. The *n*-alkanes C13 to C34 were completely biodegraded (>90%) within 19 days.

In this study the isoprenoid alkanes underwent extensive degradation; pristane was 100% degraded and phytane was more than 90% degraded in 36 days at 15°C. Changes in ratios of other saturates to pristane or phytane have been used to indicate biodegradation of petroleum. However, because pristane and phytane are easily degraded, Jobson et al. (1972) recommended that they should be used with caution for determining levels or rates of biodegradation. Our results support this conclusion and show that pristane and phytane can be much more readily degraded than has sometimes been supposed, at least in Prince William Sound, Alaska.

The aliphatic hydrocarbons — the *n*-alkanes and branched alkanes — have been shown to be the most readily used constituent of petroleums by some microorganisms (Atlas 1975, Fusey & Oudot 1984, Treccani 1964, Zobell 1946). Indeed, it is generally believed that the aromatic hydrocarbons are used only after degradation of the saturates has occurred. However, in several studies the microbial populations preferentially degraded aromatics before saturates (Cooney et al.

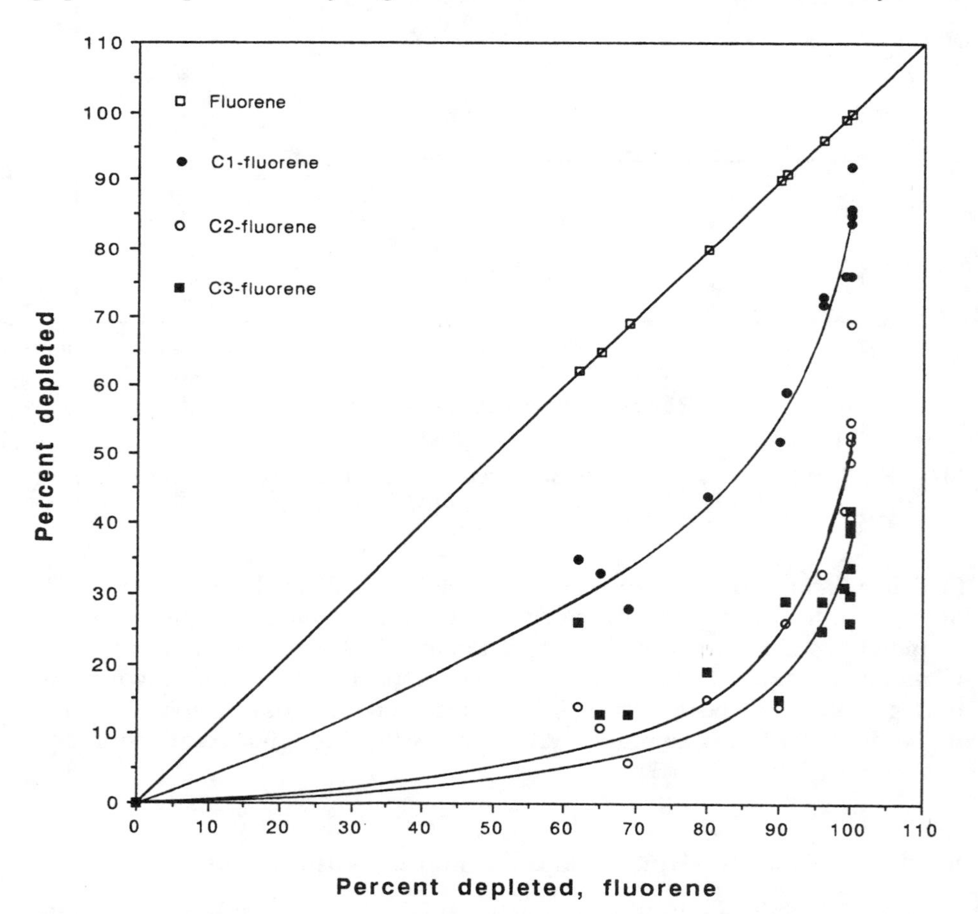

FIGURE 2. Plots of the percent depleted of individual fluorenes over the percent depleted of fluorene (three replicates).

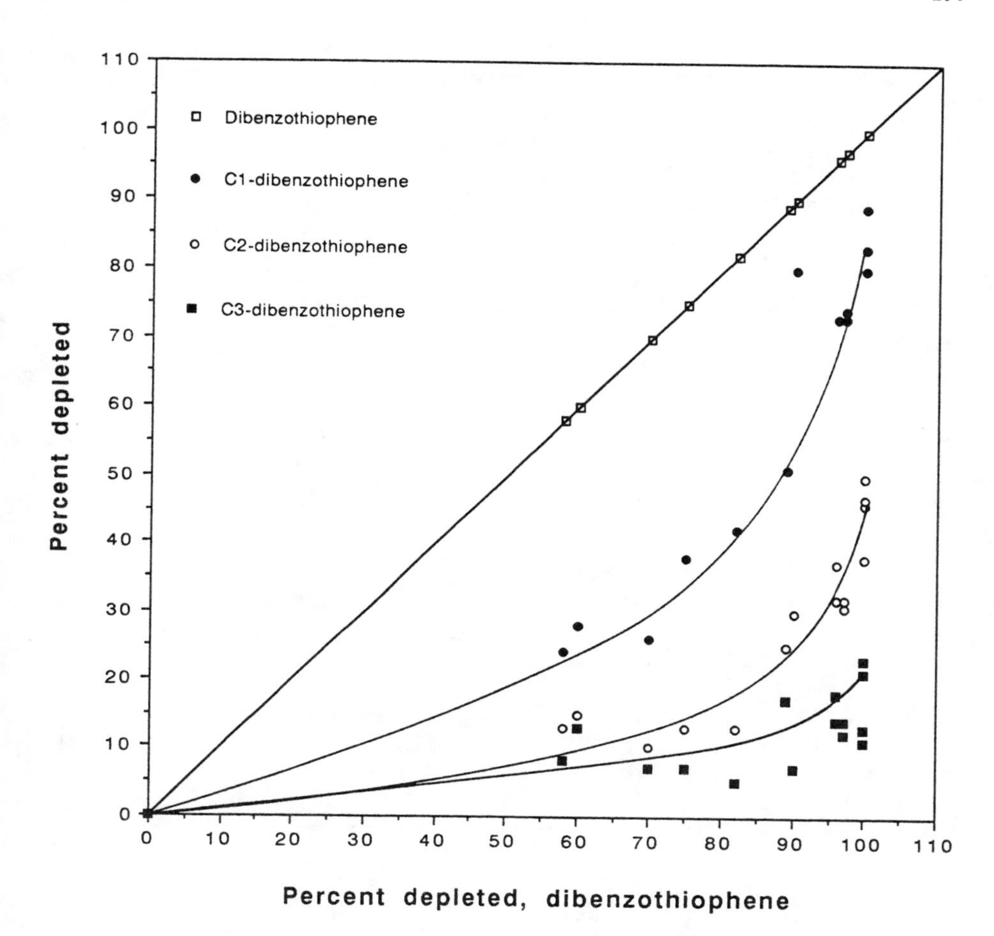

FIGURE 3. Plots of the percent depleted of individual dibenzothiophenes over the percent depleted of dibenzothiophene (three replicates).

1985, Fedorak & Westlake 1981, Jones et al. 1983, Walker et al. 1976b) or degraded them at similar rates (Bertrand et al. 1983, Horowitz & Atlas 1977). A microbial inoculum from Chesapeake Bay showed greater degradation of total aromatics than total saturates in South Louisiana and Kuwait crude oils and a No. 2 fuel oil (Walker et al. 1976b). Prince William Sound inocula readily eliminated the saturates, while at the same time extensive biodegradation of the aromatics occurred (Figures 1 to 5).

The PAHs — naphthalene, fluorene, phenanthrene, and dibenzothiophenes — were heavily degraded (Figures 1 to 5). Pyrene and chrysene were slightly biodegraded. The order of susceptibility to microbial degradation is shown in Table 4. Within families of compounds, the more substituted the compound, the slower the rate at which that constituent was lost by biodegradation (Table 4) and (Figures 1 to 5). Generally, these results are not inconsistent with the widely

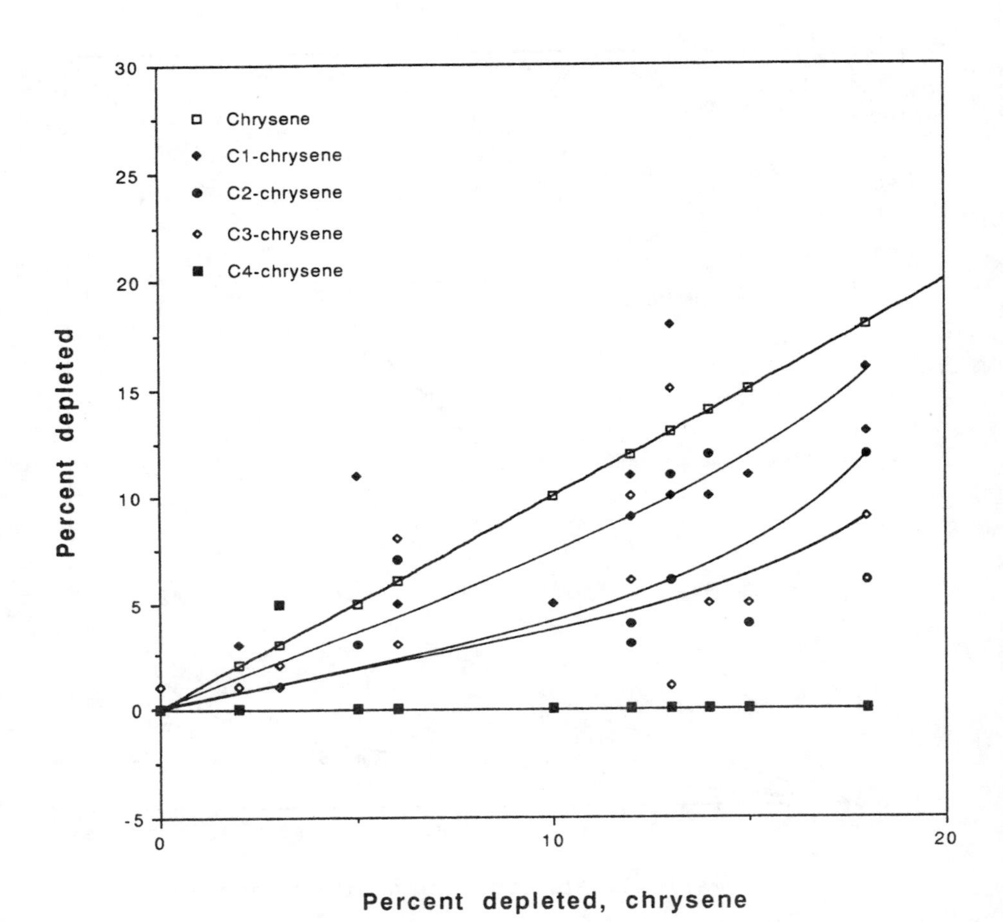

FIGURE 4. Plots of the percent depleted of individual chrysenes over the percent depleted of chrysene (three replicates).

held view that the order of decreasing susceptibility to biodegradation of petroleum constituents has been *n*-alkanes > branched alkanes > low-molecular-weight aromatics > cyclic alkanes > high-molecular-weight aromatics > polar compounds (Fusey & Oudot 1984, Jobson et al. 1972, Perry 1984, Walker et al. 1976a).

In stream sediments contaminated by petroleum (Herbes & Schwall 1978) and coal-coking effluents (Herbes 1981), the turnover rates for four aromatic hydrocarbons were most rapid for naphthalene, followed by anthracene, then benzo(a)anthracene, and then benzo(a)pyrene. The order of biodegradation potentials was determined to be hexadecane ≥ naphthalene > pristane > benzo(a)anthracene (Roubal & Atlas 1979). Low-molecular-weight aromatics such as benzene, toluene, naphthalene, and methyl-naphthalene were degraded by river microorganisms, but the higher-molecular-weight aromatics were recalcitrant (Lee 1977).

Fedorak and Westlake (1981) showed that aromatic degradation preferentially moved from low-molecular-weight, less complex components to larger, complex constituents as the process continued. They determined an approximate series of C2-naphthalenes; phenanthrene and dibenzothiophenes; C3-naphthalenes and methyl-phenanthrenes; and C2-phenanthrenes (Fedorak & Westlake 1981). Atlas (1981b) determined that the C2 and C3-phenanthrenes and C1 through C3-dibenzothiophenes were the most persistent aromatics in crude oil from the *Amoco Cadiz* and *IXTOC I* spills. The order of biodegradation of petroleum hydrocarbons by Prince William Sound microorganisms was hexadecane = naphthalene > pristane > phytane > fluorenes > dibenzothiophene ≈ phenanthrene > chrysene. The Prince William Sound inoculum has the ability to extensively degrade the saturate and aromatic constituents of Alaskan North Slope crude oil, and when provided with sufficient nutrients it degraded a number of substituted aromatic hydrocarbons that are considered resistant to biodegradation. It is important to recognize that

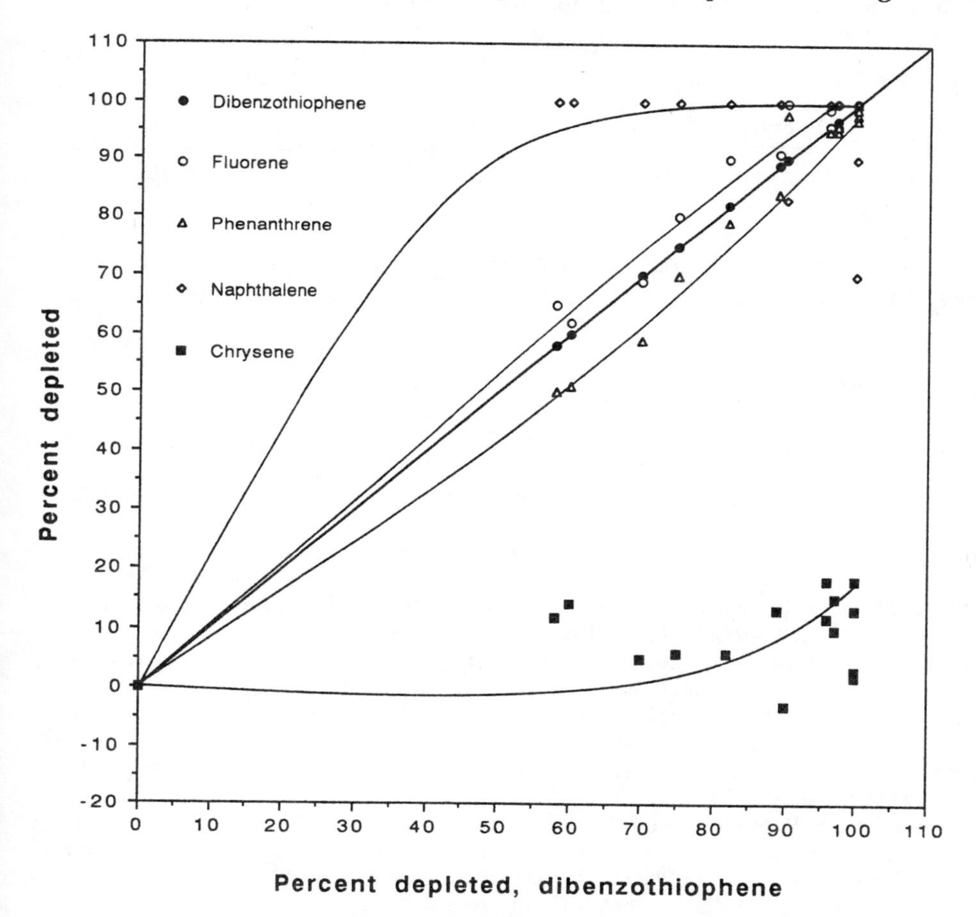

FIGURE 5. Plots of the percent depleted of selected PAHs over the percent depleted of dibenzothiophene (three replicates).

TABLE 4. Order of susceptibility of various PAHs to microbial degradation.

Order of susceptibility to microbial degradation – Highly degradable to least degradable
Naphthalene > Fluorene > Dibenzothiophene > Phenanthrene > Pyrene > Chrysene > Benzo(b)fluoranthene
Naphthalene > C1-naphthalenes > C2-naphthalenes > C3-naphthalenes > C4-naphthalenes
Fluorene > C1-fluorenes > C2-fluorenes > C3-fluorenes
Phenanthrene > C1-phenanthrenes > C2-phenanthrenes > C3-phenanthrenes > C4-phenanthrenes
Dibenzothiophene > C1-diobenzothiophenes > C2-dibenzothiophenes > C3-dibenzothiophenes
Pyrene > C1-fluoranthene/pyrenes
Chrysene > C1-chrysenes > C2-chrysenes > C3-chrysenes > C4-chrysenes

Sources: Fusey and Oudot (1984), Jobson et al. (1972), Perry (1984), Walker et al. (1976a).

saturate and aromatic hydrocarbons were undergoing biodegradation simultaneously; there was no evidence for a stepwise shift from one constituent to another.

REFERENCES

Atlas, R. M. 1975. "Effects of temperature and crude oil composition on petroleum biodegradation." *Appl. Microbiol. 30*: 396-403.

Atlas, R. M. 1981a. "Microbial degradation of petroleum hydrocarbons: An environmental perspective." *Microbiol. Rev. 45*: 180-209.

Atlas, R. M. 1981b. "Fate of oil from two major oil spills: role of microbial degradation in removing oil from the *Amoco Cadiz* and *Ixtoc 1* spills." *Environ. Internat. 5*: 33-38.

Atlas, R. M., and R. Bartha. 1972a. "Biodegradation of petroleum in seawater at low temperatures." *Can. J. Microbiol. 18*: 1851-1855.

Atlas, R. M., and R. Bartha. 1972b. "Degradation and mineralization of petroleum in seawater: Limitation by nitrogen and phosphorus." *Biotechnol Bioeng. 14*: 309-317.

Atlas, R. M., and R. Bartha. 1973. "Stimulated biodegradation of oil slicks using oleophilic fertilizers." *Environ. Sci. Technol. 7*: 538-541.

Bergstein, P. E., and J. R. Vestal. 1978. "Crude oil biodegradation in Arctic tundra ponds." *Arctic 31*: 158-169.

Bertrand, J. C., E. Rambeloarisoa, J. F. Rontani, G. Giusti, and G. Mattei. 1983. "Microbial degradation of crude oil in sea water in continuous culture." *Biotechnol. Lett. 5*: 567-572.

Boehm, P. D., and D. L. Fiest. 1980. "Aspects of the transport of petroleum hydrocarbons to the offshore benthos during the IXTOC-I blowout in the Bay of Campeche." In *Proceedings of the Symposium on Preliminary Results from the September 1979 RESEARCHER/PIERCE IXTOC-I Cruise*, pp. 207-236. National Oceanic and Atmospheric Administration, Boulder, CO.

Brown, E. J., and J. F. Braddock. 1990. "Sheen screen, a miniaturized most-probable-number method for enumeration of oil-degrading microorganisms." *Appl. Environ. Microbiol. 56*: 3895-3896.

Burwood, R., and G. C. Speers. 1974. "Photo-oxidation as a factor in the environmental dispersal of crude oil." *Estuar. Coast. Mar. Sci. 2*: 117-135.

Butler, E. L., R. C. Prince, G. S. Douglas, T. Aczel, C. S. Hsu, M. T. Bronson, J. R. Clark, J. E. Lindstrom, and W. G. Steinhauer. 1991. "Hopane, a new chemical tool for the measurement of oil biodegradation." In R. E. Hinchee and R. F. Olfenbuttel (Eds.), *On-Site Bioreclamation: Processes for Xenobiotic and Hydrocarbon Treatment*, pp. 515-521. Butterworth-Heinemann, Boston, MA.

Clark, R. C., Jr., and D. W. Brown. 1977. "Petroleum: properties and analyses in biotic and abiotic systems." In D. C. Malins (Ed.), *Effects of Petroleum on Arctic and Subarctic Marine Environments and Organisms. Vol. 1*, pp. 1-89. Academic Press, New York, NY.

Cooney,. J. J., S. A. Sliver, and E. A. Beck. 1985. "Factors influencing hydrocarbon degradation in three freshwater lakes." *Microb. Ecol. 11*: 127-137.

Dibble, J. T., and R. Bartha. 1976. "The effect of iron on the biodegradation of petroleum in seawater." *Appl. Environ. Microbiol. 31*: 544-550.

Douglas, G. S., K. P. McCarthy, D. T. Dahlen, J. A. Seavey, W. G. Steinhauer, R. C. Prince, and D. L. Elmendorf. 1992. "Hydrocarbon finger-printing analysis: Fact or fiction?" In P. T. Kostecki and E. J. Calabrese (Eds.), *Contaminated Soils; Diesel Fuel Contamination*, pp. 1-21. Lewis Publishers, Boca Raton, FL.

Fedorak, P. M., and D. W. S. Westlake. 1981. "Microbial degradation of aromatics and saturates in Prudhoe Bay crude oil as determined by glass capillary gas chromatography." *Can. J. Microbiol. 27*: 432-443.

Floodgate, G. D. 1984. "The fate of petroleum in marine ecosystems." In R. M. Atlas (Ed.), *Petroleum Microbiology*, pp. 355-397. Macmillan Publishing Co., New York, NY.

Fusey, P., and J. Oudot. 1984. "Relative influence of physical removal and biodegradation in the depuration of petroleum-contaminated seashore sediments." *Mar. Pollut. Bull. 15*: 136-141.

Hambrick III, G. A., R. D. DeLaune, and W. H. Patrick, Jr. 1980. "Effect of estuarine sediment pH and oxidation-reduction potential on microbial hydrocarbon degradation." *Appl. Environ. Microbiol. 40*: 365-369.

Harrison, O. R. 1991. "An overview of the *Exxon Valdez* oil spill." *Proceedings of the 1991 International Oil Spill Conference*, pp. 313-320. American Petroleum Institute, Washington, DC.

Herbes, S. E. 1981. "Rates of microbial transformation of polycyclic aromatic hydrocarbons in water and sediments in the vicinity of a coal-coking wastewater discharge." *Appl. Environ. Microbiol. 41*: 20-28.

Herbes, S. E., and L. R. Schwall. 1978. "Microbial transformation of polycyclic aromatic hydrocarbons in pristane and petroleum-contaminated sediments." *Appl. Environ. Microbiol. 35*: 306-316.

Horowitz, A., and R. M. Atlas. 1977. "Continuous open flow-through system as a model for oil degradation in the Arctic Ocean." *Appl. Environ. Microbiol. 33*: 647-653.

Hughes, D. E., and P. McKenzie. 1975. "The microbial degradation of oil in the sea." *Proc. R. Soc. Lond. B. 189*: 375-390.

Jobson, A., F. D. Cook, and D. W. S. Westlake. 1972. "Microbial utilization of crude oil." *Appl. Environ. Microbiol. 23:* 1082-1089.

Jones, D M., A. G. Douglas, R. J. Parkes, J. Taylor, W. Giger, and C. Schaffner. 1983. "The recognition of biodegraded petroleum-derived aromatic hydrocarbons in recent marine sediments." *Mar. Pollut. Bull. 14*: 103-108.

Lee, R. F. 1977. "Fate of petroleum components in estuarine waters of the Southeastern United States." *Proceedings of the 1977 Oil Spill Conference*, pp. 611-616. American Petroleum Institute, Washington, D C.

LePetit, J., and M.-H. N'Guyen. 1976. "Besoins en phosphore des bacteries metabolisant les hydrocarbures en mer." *Can. J. Microbiol.* 22: 1364-1373.

Nauman, S. 1991. "Shoreline cleanup: Equipment and operations." *Proceedings of the 1991 International Oil Spill Conference*, pp. 141-147. American Petroleum Institute, Washington, DC.

Olivieri, R., P. Bacchin, A. Robertiello, N. Oddo, L. Degen, and A. Tonolo. 1976. "Microbial degradation of oil spills enhanced by a slow-release fertilizer." *Appl. Environ. Microbiol.* *31*: 629-634.

Payne, J. R., J. C. Clayton, Jr., G. D. McNabb, Jr., and B. E. Kirstein. 1991. "*Exxon Valdez* oil weathering fate and behavior: Model predictions and field observations." *Proceedings of the 1991 International Oil Spill Conference*, pp. 641-654. American Petroleum Institute, Washington, DC.

Perry, J. J. 1977. "Microbial metabolism of cyclic hydrocarbons and related compounds." *Crit. Rev. Microbiol.* *5*: 387-412.

Perry, J. J. 1984. "Microbial metabolism of cyclic alkanes." In R. M. Atlas (Ed.), *Petroleum Microbiology*, pp. 61-98. Macmillan Publishing Co., New York, NY.

Pritchard, P. H., and C. F. Costa. 1991. "EPA's Alaska oil spill bioremediation project." *Environ. Sci. Technol.* *25*: 372-379.

Prince, R. C., J. R. Clark, and J. E. Lindstrom. 1990. *Bioremediation Monitoring Program.* Report to the U.S. Coast Guard and Alaska Dept. of Environmental Protection, Anchorage, AK.

Rambeloarisoa, E., J. F. Rontani, G. Giusti, Z. Duvnjak, and J. C. Bertrand. 1984. "Degradation of crude oil by a mixed population of bacteria isolated from sea-surface foams." *Mar. Biol.* *83*: 69-81.

Roubal, G. E., and R. M. Atlas. 1979. "Hydrocarbon biodegradation in Cook Inlet, Alaska." *Dev. Indust. Microbiol.* *20*: 498-502.

Singer, M. E., and W. R. Finnerty. 1984. "Microbial metabolism of straight-chain and branched alkanes." In R. M. Atlas (Ed.), *Petroleum Microbiology*, pp. 1-60. Macmillan Publishing Co., New York, NY.

Tagger, S., A. Bianchi, M. Julliard, J. LePetit, and B. Roux. 1982. "Traitement biologique de nappes d'hydrocarbures en mer et l'influence de facteurs abiotique." *Deuxième Colloque de Microbiologie Marine, Marseilles*, pp. 143-150. Publications du Centre National pour l'exploitation des Oceans No. 13-1982, Brest, France.

Treccani, V. 1964. "Microbial degradation of hydrocarbons." *Prog. Ind. Microbiol.* *4*: 3-33.

Walker, J. D., and R. R. Colwell. 1974. "Microbial degradation of model petroleum at low temperatures." *Microb. Ecol.* *1*: 63-95.

Walker, J. D., and R. R. Colwell. 1976. "Measuring the potential activity of hydrocarbon-degrading bacteria." *Appl. Environ. Microbiol.* *31*: 189-197.

Walker, J. D., R. R. Colwell, and L. Petrakis. 1976a. "Biodegradation rates of components of petroleum." *Can. J. Microbiol.* *22*: 1209-1213.

Walker, J. D., L. Petrakis, and R. R. Colwell. 1976b. "Comparison of the biodegradability of crude oil and fuel oils." *Can. J. Microbiol.* *22*: 598-602.

Ward, D. M., and T. D. Brock. 1978. "Hydrocarbon biodegradation in hypersaline environments." *Appl. Environ. Microbiol.* *35*: 353-359.

Zobell, C. E. 1946. "Action of microorganisms on hydrocarbons." *Bacteriol. Rev.* *10*: 1-49.

PATHWAYS OF MICROBIAL DEGRADATION OF POLYCYCLIC AROMATIC HYDROCARBONS IN SOIL

B. Mahro, G. Schaefer, and M. Kästner

ABSTRACT

The investigation of the microbial degradation of polycyclic aromatic hydrocarbons (PAHs) in soil is hampered by the fact that it is difficult to distinguish analytically between the PAH disappearance caused by microbial degradation and that one which is caused by adsorption to soil matter. We evaluated the contribution of these processes to PAH depletion in soil more closely by application of different and optimized PAH-extraction protocols and by the use of radiolabeled indicator substances. It is shown that the major fraction of PAH evaded degradation and remained extractable from soil if the soil was supplemented by a specialized PAH-degrading microflora. In contrast, the addition of mature compost led to a complete and accelerated depletion of PAHs. Examinations of the stimulating effect of compost with radiolabeled anthracene showed that compost led to both, an increased mineralization and an increased formation of nonextractable bound residues. We assume that the supplementation of soil with ripe compost may have triggered and stimulated a wide variety of cometabolic or radical oxidative PAH biotransformation processes in soil which in turn led to an enhanced formation of bound residues in the organic soil matrix. The relevance of this conclusion for the assessment of biological soil remediation techniques is discussed.

INTRODUCTION

Polycyclic aromatic hydrocarbons (PAHs) are compounds with a basic structure of two or more condensed aromatic rings. The PAH group has become increasingly important, as the soils of many contaminated sites, especially on the grounds of former gasworks or coking plants, contain high concentrations of PAHs. As some of the PAH compounds have a high mutagenic or carcinogenic potential, any possible hazard to human health must be avoided by decontaminating the affected soil. One way of decontaminating PAH-contaminated soil is to use biotechnical methods, which are offered by a number of companies. These methods are based on findings that certain species of bacteria and fungi are able to metabolize PAH either completely or partially. According to laboratory studies carried out with pure cultures, three types of PAH-degradation can be distinguished (Mahro & Kästner 1993):

1. *Type 1: Complete Mineralization.* The biochemistry of the mineralizing PAH degradation has been studied intensively and is described in various reviews (Cerniglia 1984, Cerniglia & Heitkamp 1989). During PAH degradation by complete mineralization, the oxidation is triggered by the incorporation of an oxygen molecule in the aromatic ring. The reaction is catalyzed by a dioxygenase and leads to *cis*-dihydrodiol intermediates. As a rule, the mineralization of the PAHs proceeds via the a-hydroxy carbonic acids of the (n-1) nuclear compound, which can also accumulate intermediately (Guerin & Jones 1988). Complete mineralization enables the bacteria to grow on PAH as a sole source of carbon and energy. So far, the mineralizing degradation is described only for PAHs with two, three, and four condensed aromatic rings (Heitkamp et al. 1988, Mueller et al. 1990, Walter et al. 1991, Weissenfels et al. 1991)
2. *Type 2: Cometabolic Degradation.* The cometabolic degradation of PAH differs significantly from the mineralizing degradation (Cerniglia & Heitkamp 1989). The organisms do not use PAH for growth and, frequently, the cometabolic metabolism comes to a standstill at a very early stage after initial oxidation. Often the aromatic rings are not even split and phenolic, carboxylic, or chinoic derivatives of the PAHs accumulate as dead-end products. Except the white-rot basidiomycetes, all fungi that have been investigated so far transform PAHs into *trans*-diol intermediates under cometabolic conditions. Unlike most bacteria, the reaction used by these fungi to initially oxidize PAHs is catalyzed by monooxygenases.
3. *Type 3: Unspecific Radical Oxidation.* The unspecific radical oxidation of PAHs was originally described by Bumpus et al. (1985) for white-rot fungi. The most conspicuous characteristic of this type of PAH degradation is that the mechanism is closely related to the ability of the white-rot fungi to degrade lignin (Sanglard et al. 1986). The lignin-degrading enzyme system can attack and oxidize aromatic soil contaminants since ligninases act as peroxidases and are almost substrate-unspecific. The oxidation is initiated by radicals derived from hydrogen peroxide. In addition, these enzymes act extracellularly and can transform non-water-soluble substrates or compounds even if those are adsorbed to solids. The unspecific radical oxidation of PAHs leads to the formation of a wide variety of metabolic products. Apart from phenolic oxidation products, polymerizates and conjugations to other organic molecules also can be produced by these reactions (Bollag & Loll 1983).

However, all this fine work on the biochemistry of PAH-degradation does not answer the question yet, whether these reactions can also take place in soil in a comparable manner. Pure cultures do not occur in soils, and experiments on xenobiotic degradation in soil are complicated by the fact that the microbial activity in soil is significantly affected by parameters such as water content, pore volume, adsorption capacity, ion-exchange capacity, etc. Due to analytical difficulties, only few examinations are published that investigate the microbial PAH-metabolism in soil systems (water contents < maximum water-holding capacity). However, there are some indications that microbial metabolism is in fact altered in soil systems. For example, xenobiotic substances such as pentachlorophenol (Lamar et al. 1990) could be mineralized in liquid systems but were not mineralized in soil systems. It is also known for a long time that pesticides and their oxidation products can be incorporated into the organic soil matrix (Bollag & Loll 1983, Führ et al. 1985). Because the group of PAHs contains some of the most dangerous environmental pollutants, it is necessary to obtain more accurate information on the actual environmental fate of this group of compounds, especially with regard to bioremediation methods.

The scope of the experiments presented in this paper was to evaluate whether the microbial PAH degradation pathways, as they are known from liquid cultures or sediment microcosms, can be considered as a sufficient description of the depletion of PAHs from soils. In that context we were particularly interested to find out to which extent the formation of bound residues may contribute to the depletion of PAHs in soils and how that process is influenced by different soil supplements.

METHODS

Soils

A slightly loamy sand from an Ah-horizon of a Luvisol was used as soil material in the experiments with artificial PAH-contaminations. The material was sieved to a grain size of 2 mm before use. For the soil/compost mixtures, 6-month-old ripe compost, made from organic kitchen refuse and garden waste at the Harburger Compost Plant, Hamburg, Germany, was used. The compost was sieved to a grain size < 4 mm. Contaminated soil was taken from a drilling core of an industrial site in Hamburg, Germany, which was contaminated by tar oil. The soil texture ranged between sandy loam to loamy sand. The material was sieved to a grain size < 4 mm.

Experimental Setup

The homogenization of the soil/bacteria and soil/compost mixtures as well as the addition of water or the contaminants was carried out in a stainless steel pastry-blending machine, type: KM 250 ST (Kenwood, Ltd., Hampshire, UK). The water content of the soil mixtures was adjusted with deionized water to 60% of the maximum water-holding capacity (WHC). Both the soil and the compost showed its maximum respiration activity under these conditions (Stegmann et al. 1991). The WHC was measured as described previously (Mitscherlich 1950).

If necessary, additives such as bacteria, fertilizer, or $HgCl_2$ were added in advance and suspended in deionized water or liquid soil extract.

To contaminate soils artificially, the PAHs were dissolved in dichloromethane and 2 or 4 mL of the solution was trickled (20-mL drops) into the soil material. Dichloromethane was allowed to evaporate while the soil was homogenized (1 h). This procedure allowed recovery of PAHs in soil and soil/compost mixtures with a standard deviation of < 5 %. Except in the experiment presented in Figure 1, the PAH solutions were added after adjusting the water content of the soils.

Batch experiments were carried out at 25°C in closed 1.5-L vessels with 100 g of soil. To provide enough oxygen in the soil cultures, the vessels were opened for 15 min every third day. Sufficient oxygen supply was checked by measuring the respiration activity with carbon dioxide absorption tubes (Dräger, Germany). All quantities stated in the following paragraphs are referred to dry weight (dw) of soil. The experiments were carried out at 25°C.

To evaluate the efficiency of the extraction methods (Figure 1), air-dried soil material of the Luvisol was contaminated with 100 mg of naphthalene/kg of soil or with 25 mg phenanthrene, anthracene, fluoranthene, or pyrene (per kg of soil), respectively. The material was incubated for 25 days. For the degradation experiments presented in Figures 2 and 3, the soil and soil/compost mixtures were contaminated at a concentration of 100 mg of pyrene/kg of soil and incubated over a period of 70 days. Soil and compost were mixed at a ratio of 75%:25%.

The experiments with soil material from the contaminated site (Figure 4) were conducted with five different batch cultures: (a) a sterile control batch, poisoned with $HgCl_2$ (2 g/kg); (b) a nonsupplemented batch culture; (c) a batch

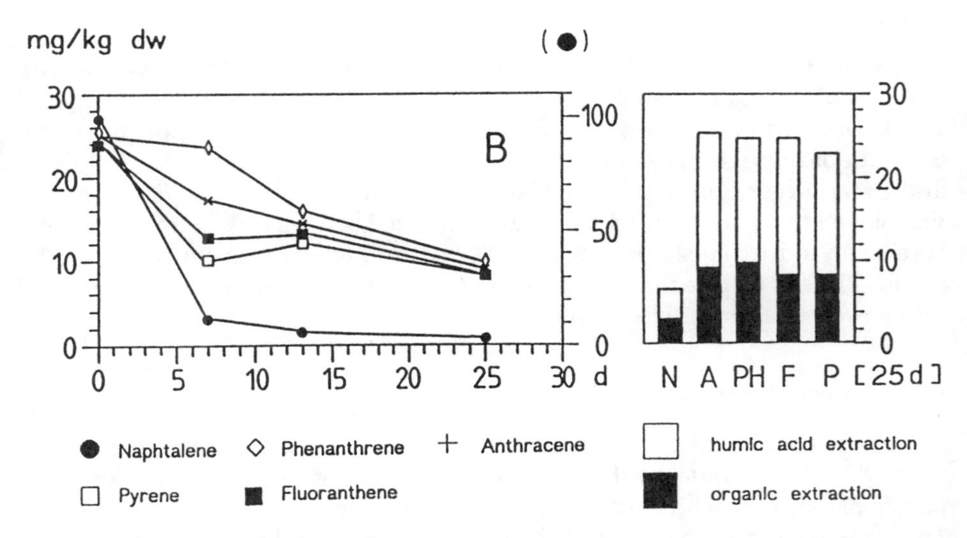

FIGURE 1. Effect of two different extraction methods on the recovery of different PAH in unsterile soil material of an Ah horizon of a parabrown soil (N = naphthalene, Ph = phenanthrene, A = anthracene, F = fluoranthene, P = pyrene).

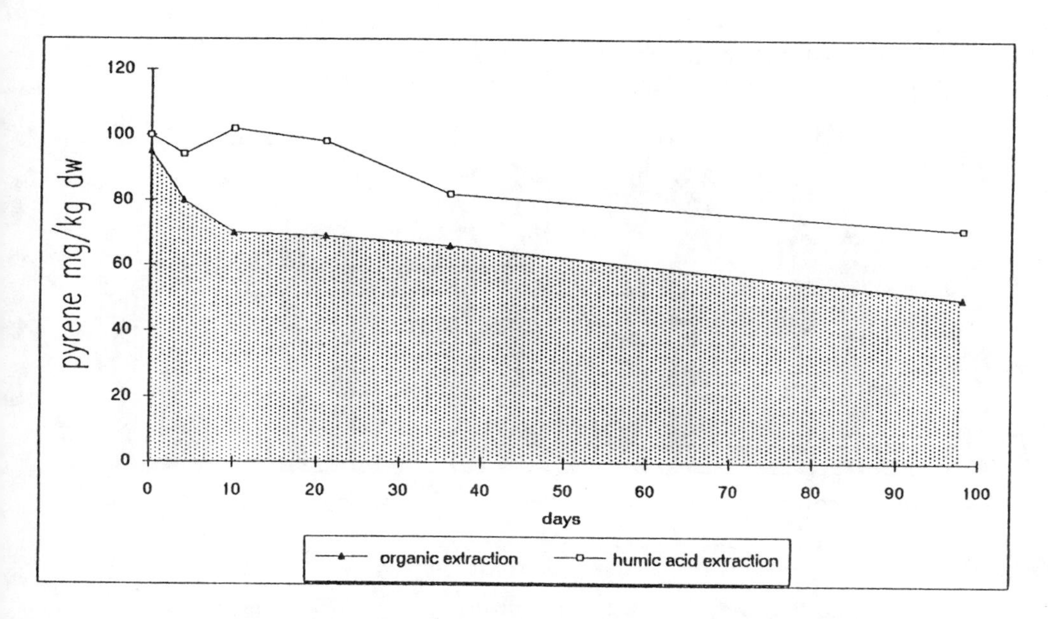

FIGURE 2. Comparison of the extraction-dependent disappearance of pyrene concentrations in unsterile soil material of a Luvisol during a 3-month incubation period (Ah horizon).

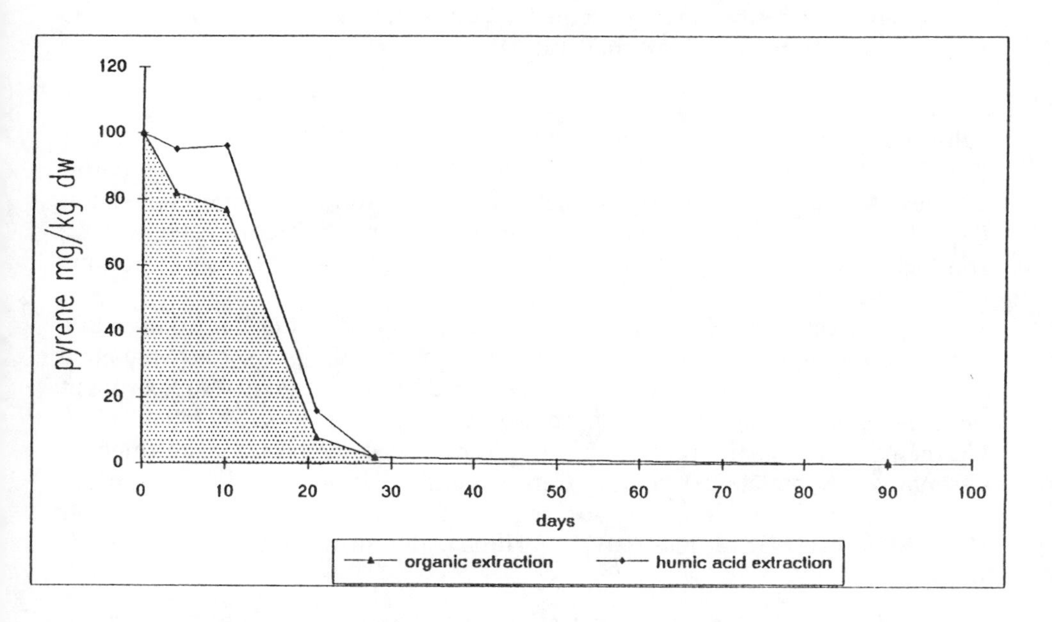

FIGURE 3. Comparison of the extraction-dependent disappearance of pyrene concentrations in a soil/compost mixture during a 3-month incubation period (otherwise same soil as in Figure 2).

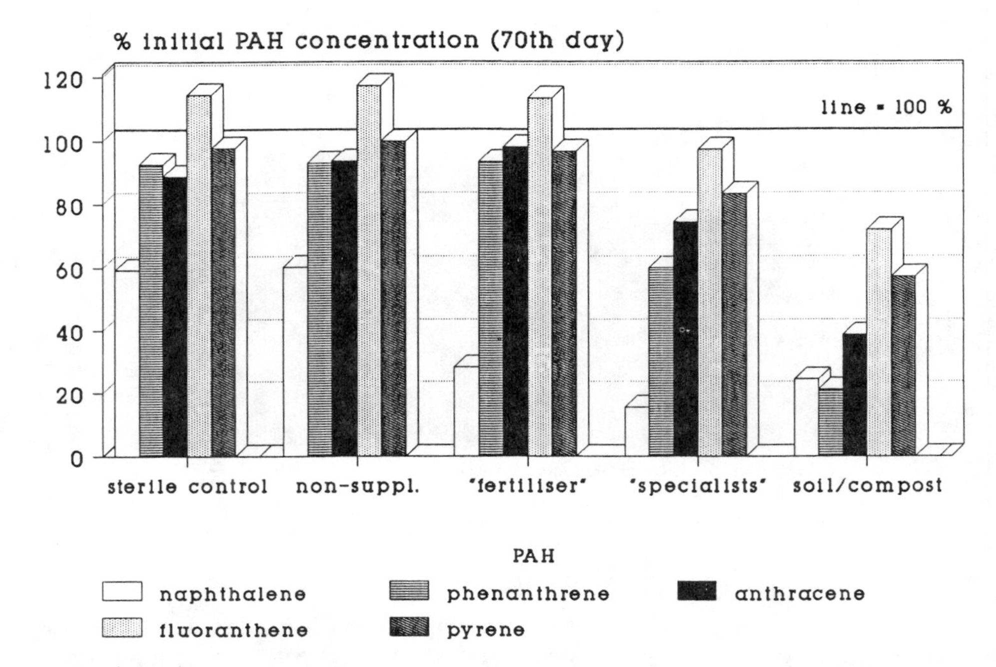

FIGURE 4. Effect of different supplementations on the PAH degradation in soil material of a contaminated site (PAH concentrations are added up from organic extraction and subsequent humic acid extraction; initial PAH concentrations are given in Table 1).

culture supplemented with fertilizer [2 g N/kg, and 0.2 g P/kg; $(NH_4)_2SO_4$ and (K_2HPO_4); C:N:P ratio: 150:10:1]; (d) a batch culture supplemented with a mixture of four PAH-degrading strains of bacteria isolated from the contaminated soil ("specialists"; 5×10^8 cells of each strain/g of soil); (e) a batch culture containing soil and compost mixed at a ratio of 67%:33%. The initial PAH concentrations in each batch culture are given in Table 1.

Experiments with ^{14}C-labeled substances were conducted in 2-L soil bioreactors over a period of 86 days. The basic reactor design and the process scheme have been described by Stegmann et al. (1991). The radiotracers used in the experiment were 1-^{14}C-hexadecane (185 MBq/mmol in toluene; SIGMA Chemie GmbH, Germany) and 9-^{14}C-anthracene (559 MBq/mmol in ethanol; Amersham-Buchler, Germany). Radiolabeled volatile transformation products or ^{14}C-CO_2 in the exhaust gas of the reactors were absorbed in ethylene glycol monomethyl ether (EGME) and in NaOH, respectively. Three different reactors were set up as follows:

1. *Reactor A.* Contained soil material from the Luvisol. The soil was contaminated with 200 mg of anthracene/kg of soil (^{14}C activity =

124 kBq/kg) and supplemented with anthracene-degrading bacteria, strain BA 2 (2.3×10^7 cells/g of soil);
2. *Reactor B.* Contained a soil/compost mixture (80%:20%) and 170 mg anthracene/kg of soil (^{14}C activity = 102 kBq/kg);
3. *Reactor C.* Contained a soil/compost mixture (80%:20%) and had been contaminated with 720 mg hexadecane/kg of soil (^{14}C activity = 67 KBq/kg).

Microorganisms

The PAH-degrading bacteria used in these experiments (strains: BP9, BA2, OR, and FFSPH) are described in detail by Kästner et al. (1991). The strains were isolated from soil of the contaminated site described above and were able to use the respective PAH as sole sources of carbon and energy. The liquid growth medium for the bacteria contained, per liter: 2.1 g Na_2HPO_4, 1.3 g KH_2PO_4, 0.5 g $MgSO_4 * 7H_2O$, 1 g NH_4Cl, 1 mL vitamin and trace element solution, respectively, and 200 mg of either anthracene, phenanthrene, fluoranthene, or pyrene. The pH was adjusted to 6.9. One strain that was able to degrade four PAHs simultaneously, was grown on a mixture of 25 mg of each of the four PAH compounds. The liquid cultures were harvested by centrifugation and the pellet was resuspended in liquid soil extract. This suspension was used as soil inoculum. The inocula were added to the soils before adjusting the water content. The screening for PAH-mineralizing bacteria was carried out with solidified minimal growth medium (15 g of agar/L medium). The surface of the agar medium was coated by a turbid crystalline layer of the respective PAH-compound, which made possible to identify the PAH-degrading colonies due to growth and the formation of clear halos around the colonies.

TABLE 1. Initial PAH concentrations in soil batch cultures set up with soil from a long time contaminated site (see Figure 4).

Initial PAH Concentrations	Soil	Soil/Compost
Naphthalene	1,100 (± 180)	410 (± 80)
Phenanthrene	1,400 (± 160)	950 (± 70)
Anthracene	180 (± 25)	120 (± 15)
Fluoranthene	1,480 (± 130)	1,000 (± 110)
Pyrene	1,250 (± 150)	860 (± 45)

Data in mg/kg of soil (dry weight).

Chemicals

All chemicals except those specified otherwise were purchased from Merck, Darmstadt, Germany.

Analytical Procedures

To extract PAH from soil samples, 3 g of soil and 3 mL ethyl acetate were filled into glass tubes. The tubes were then closed and extracted for 30 minutes in an ultrasonic bath (Super Sonorex, Bendelein, Berlin, Germany). Soil particles were separated from the solvent by centrifugation and the supernatant was analyzed by high-performance liquid chromatography (HPLC) or by β-scintillation counting.

To extract those PAHs still adsorbed to the soil matrix after organic extraction, a subsequent alkaline hydrolysis (humic acid extraction) was conducted. The principle and efficiency of this PAH-extraction method have been described previously with regard to food (Grimmer et al. 1979) and soil samples (Eschenbach et al. 1991). We applied a modified protocol of the technique proposed by Eschenbach et al. (1991). The humic acid extraction was conducted in the same tubes that were used during the organic extraction step. After removal of ethyl acetate, the remaining soil pellets were mixed with 0.2 mL of a NaOH solution (2 mol/L) and 2.8 mL methanol. The pressure-resistant tubes were closed and incubated for 1 hour in a water bath at 95 to 99°C. After cooling, the solvent and the soil were separated and the supernatant was analyzed.

The PAH-compounds of the soil extracts (20 mL injection volume) were separated and quantified by HPLC with UV-detection (pumps D 2248, UV-detector VWM 2141, Pharmacia LKB, Germany) on an RP 18 column (5 mm, Merck, Germany). We applied a linear elution gradient of methanol and water, with the methanol content gradually arising from 70 to 100% after 15 min (flow: 1 mL/min). The PAHs were detected spectrometrically at 254 nm. All PAH determinations were conducted three times and the mean values of these determinations were used for further calculations. The standard deviations of the different PAH-determinations were <5% with artificially contaminated soil and <13% with soil from the contaminated site. The concentrations of ^{14}C radioactivity in the soil extracts, the EGME (volatile ^{14}C- compounds), and the NaOH ($^{14}CO_2$ production) were analyzed by β-scintillation (TRI Carb 1950, Packard Instruments, NL). The determination of ^{14}C-carbon in the soil was conducted by combustion of the material in a combustion unit (Coulomat D 702, Ströhlein Instruments GmbH, Germany). Analyses of samples from the reactors were performed as triplicate determinations of which the mean values are stated in the results section (standard deviations: organic and humic acid extractions <6%; soil combustions <10%). The ^{14}C-content in the biomass was determined by a variation of the fumigation-incubation method of Jenkinson & Powlson (1976). For this purpose, the differences of $^{14}CO_2$ in the fumigated samples compared to the nonfumigated samples were calculated as ^{14}C-biomass.

RESULTS

We had intended to consider the binding and adsorption phenomena of PAH in soils more closely, and our first experiments were carried out to test the reliability of different PAH extraction protocols. We found that, when dry soil matter was artificially contaminated with various PAHs and the water content was adjusted after that, nearly 100% of the compounds could be initially recovered by an organic extraction. However, 25 days later the concentration of the extractable compounds had decreased exponentially by more than 50% (Figure 1), apparently indicating "microbial degradation." Nevertheless, the initial concentrations of all substances, apart from naphthalene, could be recovered when the first organic PAH extraction was followed by a humic acid extraction. This result showed that in artificially contaminated soil samples PAH could obviously "hide" from analysis in the organic soil matrix. The mere disappearance of PAHs in soils should therefore not be equated with microbial degradation.

When the same soil was incubated over a longer period of time (100 days), a slight PAH degradation was observed from day 25 on with both extraction methods, e.g., as for pyrene shown here. But although the portion of PAHs that could be extracted with organic solvents decreased constantly from the very beginning up to the end of the experiment, the absolute amount of PAHs that could be extracted by methanolic saponification remained constant after about 36 days (Figure 2). This result suggests that those PAHs that had strongly adsorbed to the organic soil matrix were not degraded any further in this culture.

When the same soil matter was mixed with ripe compost, the microbial PAH degradation was accelerated significantly and hardly anymore pyrene was detectable after 36 days (the phenanthrene, anthracene, and fluoranthene kinetics were similar; data not shown). But in contrast to the compost-free soil batch, this time we were not able to recover PAHs by humic acid extraction. The adsorbed PAH portion had obviously undergone either true degradation or a more intensified binding process (Figure 3).

The stimulatory effect of compost on the microbial PAH-depletion also was observed with soil material from a long-time PAH-contaminated site. Sterilized and untreated soil material as well as soil material supplemented with fertilizer (nitrogen and phosphorus) showed no significant degradation apart from naphthalene (Figure 4). This result indicates that N and P did not limit degradation in this soil. When a mixture of the bacteria isolated from this site (Kästner et al. 1991) or compost was added to the soil, a significant degradation of PAHs was detected in each of these soil batch cultures. But despite the fact that no microorganisms capable of growing on PAHs could be found in the compost, the soil cultures with added compost showed an even better PAH degradation than those with the added PAH-degrading bacteria.

From the fact, that no PAH-mineralizing bacteria were found in the compost, it must be concluded that the PAH degradation observed in this culture was either due to cometabolic degradation (Type 2) or due to degradation by unspecific radical oxidation (Type 3). Both types of degradation may lead to an accumulation of metabolites in soil. To compare both, the extent of the presumed formation of metabolites in soil and the extent of the complete mineralization pathway, we carried out mass balance experiments with radiolabeled substances. A set of bioreactors was filled with soil or soil/compost mixtures that had been contaminated with anthracene or hexadecane labeled with ^{14}C. After 86 days, almost 100% of the added radioactivity from anthracene was recovered from the soil in reactor A (soil + specialists) by incineration. Likewise, organic or humic acid extraction led to a recovery of 90% (Figure 5). No significant degradation was observed. However, after the same time, only 50% of the added radioactivity was discovered in the soil-compost mixture of reactor B, of which only 10% could be recovered as liquid extract using various extraction methods. The remaining 40% of the activity in the soil/compost mixture was nonextractable and must be considered as bound residues. Similar processes were observed in reactor C (soil/compost) contaminated with the readily degradable ^{14}C-hexadecane. For this substance, the percentage of bound residues was only 16%. Only 3.5% of the radioactivity could be extracted by either of both extraction methods.

The percentage of the different xenobiotic substances that were mineralized into ^{14}C-CO_2 amounted to 0.03% in reactor A (anthracene + soil + specialists), 24% in reactor B (anthracene + soil + compost) and 54% in reactor C (hexadecane + soil + compost).

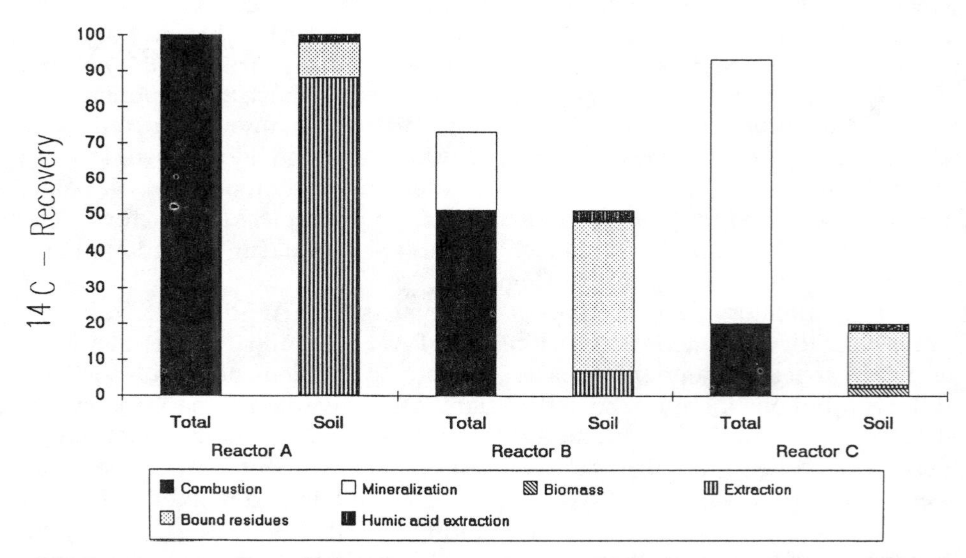

FIGURE 5. Recovery of ^{14}C radioactivity in soil, gas-phase, and liquid extracts of different soil bioreactors. The left bar represents the sum of ^{14}C activities in both, soil and gas stream. The right bar represents ^{14}C activities in different fractions of the soil compartment. *Reactor A*: Soil material contaminated with ^{14}C anthracene (supplemented with anthracene-degrading bacteria). *Reactor B*: Soil/compost-mixture, contaminated with ^{14}C anthracene. *Reactor C*: Soil/compost-mixture, contaminated with ^{14}C hexadecane.

To determine the amount of metabolites produced during degradation of PAHs in reactor B, we compared the amount of the extractable radioactivity with the extractable amount of the parent molecule anthracene in the organic extracts of the soil/compost mixture. The result was that most of the extractable radioactivity (> 90%) was caused by the parent substance anthracene (data not shown). Only toward the end of the experiment, after 86 days, the radioactivity of anthracene comprised a smaller portion (40%) of the total radioactivity in the extracts. However, because in reactor B at this time only 7% of the initial radioactivity still was extractable by organic solvents (Figure 5), it can be concluded that the compost addition did not lead to a significant amount of free anthracene metabolites (< 10%). Instead we found enhanced $^{14}CO_2$-evolution (mineralization) and extensive formation of bound residues.

DISCUSSION

PAHs that cannot be extracted from soil due to their adsorption to the soil matrix have received little attention in the literature so far. We have shown here that parts of the adsorbed portion of PAH can be recovered for analysis from artificially contaminated soils by humic acid extraction (Figure 1 and 2). Bioremediation experiments with soil, whose success has been judged on the mere disappearance of the initial PAH compounds only, should therefore be reconsidered carefully.

One possibility to evaluate this problem more closely could be to investigate the ^{14}C-mass balances of PAH degradation in soil cultures. The results of our mass balance experiments indicate that the mineralization of PAH is only one, and sometimes a very minute, pathway of the microbial PAH degradation in soil (Figure 5). The microbial degradation, which takes place in soils with water contents below the maximum water-holding capacities, seems to be very different from those processes that take place in liquid culture or in soil systems with excess of water such as in sediments. Heitkamp and Cerniglia (1989) showed that, the degree of mineralization of various PAHs could be increased significantly if the sediment microcosms were supplemented with PAH-degrading bacteria. The addition of a pyrene-degrading *Mycobacterium* strain led to a mineralization of 40% of the pyrene compared with almost no CO_2 evolution in the unsupplemented sediment. In addition the authors demonstrated that most of the extractable ^{14}C-activity in the microcosms stemmed from transformed pyrene. The extractable residues were identified as metabolites that also could be found in pure liquid cultures (Heitkamp et al. 1988).

However, the formation of free PAH metabolites was insignificant in soil. The formation of bound residues seems to represent a more relevant depletion pathway for PAHs in soil. In principle, each metabolic transformation product that is formed by one of the three types of microbial PAH degradation (hydroxy carbonic acids, phenolic, or chinoic compounds) can react chemically with the surrounding organic soil matrix and thereby create bound residues. These metabolites could subsequently become included in the synthesis of humic matter. It has been shown with pesticides that, depending on the chemical structure of both, the xenobiotic and the humic molecule, all known types of chemical binding may occur between a xenobiotic and the organic soil matrix (Senesi 1992). The presence of free radicals in the humic matter could even trigger the activation of the parent PAH molecule, for example by the formation of electron acceptor-donator complexes (Ziechmann 1980). The formation of aromatic radical ions also could initiate a covalent binding of PAHs or their metabolites among each other or with humic molecules. Such oxidative coupling processes have been demonstrated with pesticides in humic matter repeatedly (Bollag et al. 1992, Senesi 1992). The highly efficient peroxidase-catalyzed xenobiotic depletion in soil, which has been shown with white-rot fungi, proceeds probably along the same route, as high proportions of soil-bound residues are a characteristic feature in soil systems supplemented with white-rot fungi (Lamar et al. 1990, Qiu & McFarland 1991). Clay minerals and some inorganic salts that occur in soils have also the ability to catalyze oxidative coupling reactions (Bollag & Loll 1983).

We assume that the stimulation of the PAH-degradation by compost supplementation may trigger a similar arsenal of oxidative coupling and binding reactions. The aromatic nuclei seem to be more prone to humic fixation than linear hydrocarbons. We found only very few free PAH metabolites in the extracts of the soil/compost mixture, while the amounts of bound residues derived from anthracene were much higher than the amounts derived from hexadecane.

The initial PAH-degradation in compost must be largely based on a synergistic cometabolic transformation (Type 2) or unspecific radical oxidation (Type 3), because we were not able to find any microorganisms in soil/compost mixture that were able to mineralize PAH as the sole carbon and energy source. In addition, Martens (1982) has shown that even PAH with up to 6 rings can be degraded in ripe compost, although bacteria capable of mineralizing PAHs with more than 4 annealed rings have not been described so far. It is known for a long time, however, that actinomycetes and fungi can be found preferentially in ripe compost (Alexander 1977). These microorganisms are also known to excrete extracellular enzymes such as peroxidases or phenoloxidases and might thereby contribute substantially to the unspecific cooxidation of PAHs in soil/compost mixtures. To explain the relatively high mineralization rates, it must be assumed that parts of the PAH molecules are "synergistically" transformed into CO_2 by unspecific extracellular radical reactions in the soil matrix, by extracellular enzymatic transformation processes, or by cometabolic interaction of different members of the soil microflora.

The nonextractable, bound residues can also be improved substrates for unspecific oxidations and mineralization. Haider and Martin (1988) found with chlorocatechols that *Phanerochaete chrysosporium* was able to free much more $^{14}C-CO_2$ from bound residues of these compounds than from the free soluble parent compounds. If similar phenomena are assumed with the bound residues of PAHs, it is to be expected that this degradation path will lead to a successive transformation into CO_2 too. The turnover rates of this delayed mineralization presumably will be equal to those of the general turnover of organic soil matter of about 2 to 5% per year (Bollag et al. 1992). Further research needs to be done to evaluate the long-term fate of soil-bound PAH residues in more detail.

From the above exposition, the following model of the behavior of PAH degradation in soil can be deduced (Figure 6). PAHs either can be mineralized, which includes the production of biomass, H_2O, and CO_2 (Path 1), or can become incorporated in the humification process, either as original components or as metabolites (Path 2). In soils with low quantities of humic matter, the latter path can also comprise the production of high-molecular structures (polymers) from PAHs and metabolites. The carbon depot in the soil can also serve as an intermediate storage system for PAH. Depending on the binding mechanism, the PAH or PAH-metabolites can undergo either an adsorption-desorption cycle (Path 3) or a delayed mineralization (Path 4).

Bioremediations of PAH-contaminated soils may be based on all four pathways, but the contribution of the single depletion pathways may differ depending on the soil structure, the organic soil matter, the type of contamination, or the prevailing microflora. In the presence of organic humic matter, cometabolic or unspecifically oxidizing microorganisms probably can show a greater efficiency at eliminating aromatic compounds in soil than can mineralizing microorganisms, especially if the contaminants are complex or are adsorbed to the soil matrix. The stimulation of this nonspecialist microflora and the increased input of organic binding sites in the soil matrix might be the reason that additions of organic soil supplements such as compost or shredded bark, used in some soil bioremediation techniques, were so beneficial for efficient PAH depletion. As bound xenobiotic compounds no longer are available for water transport or plant uptake, the biogenic fixation also must have a detoxifying function for soil as an ecosystem. The latter statement remains to be checked carefully, however, by adequate experiments and bioremediation controls.

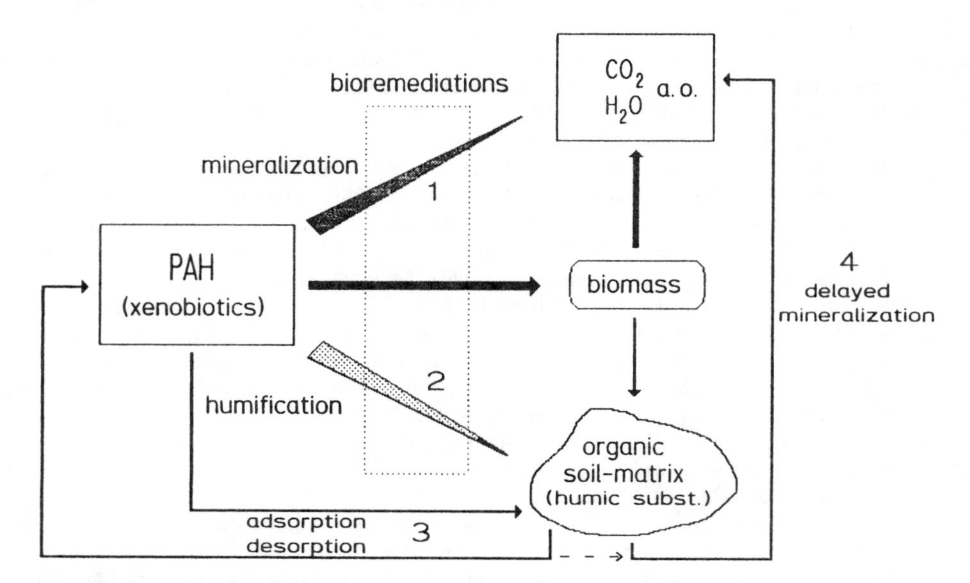

FIGURE 6. Summarizing model on the different microbial and nonmicrobial PAH elimination pathways in soils.

ACKNOWLEDGMENTS

We thank M. Breuer-Jammali, A. Eschenbach, J. Heerenklage, S. Lotter, and Professor Dr. R. Stegmann for experimental support and cooperation. We would also like to thank Professor Dr. V. Kasche, the head of our biotechnology department, for fruitful suggestions on our experimental concept. The research on this subject was financed by the Deutsche ForschungsGemeinschaft (DFG) within the interdisciplinary research project "Sonderforschungsbereich188: Remediation of Contaminated Soils."

REFERENCES

Alexander, M. 1977. *Soil Microbiology*, 2nd ed. John Wiley & Sons, Inc., New York, NY.

Bollag, J.-M., and M. J. Loll. 1983. "Incorporation of xenobiotics into soil humus." *Experientia 39*: 1221-1231.

Bollag, J. M., C. J. Myers, and R. D. Minard. 1992. "Biological and chemical interactions of pesticides with soil organic matter." *The Science of the Total Environment 123/124*: 205-217.

Bumpus, J. A., M. Tien, D. Wright, and S. D. Aust. 1985. "Oxidation of persistant environmental pollutants by a white rot fungus." *Science 228*: 1434-1436.

Cerniglia, C. E. 1984. "Microbial metabolism of polycyclic aromatic hydrocarbons." *Adv. Appl. Microbiol. 30*: 31-71.

Cerniglia, C. E., and M. A. Heitkamp. 1989. "Microbial degradation of polycyclic aromatic hydrocarbons (PAH) in the aquatic environment." In U. Varanasi (Ed.), *Metabolism of polycyclic aromatic hydrocarbons in the aquatic environment*, pp. 41-68. CRC Press, Boca Raton, FL.

Eschenbach, A., P. Gehlen, and R. Bierl. 1991. "Untersuchungen zum Einfluβ von Fluoranthen und Benzo(a)pyren auf Bodenmikroorganismen und zum Abbau dieser Substanzen." *Mitteilungen d. Dt. Bodenkundlichen Gesellschaft 63*: 91-94.

Führ, F., R. Kloskowski, and P.-W. Burauel. 1985. "Bedeutung der gebundenen Rückstände." In Bundesministerium f. Ernährung, Landwirtschaft und Forsten (Hrsg.), *Pflanzenschutzmittel im Boden*, pp. 106-115. Berichte über Landwirtschaft, 198. Sonderheft. Verlag Paul Parey, Hamburg/Berlin.

Grimmer G., A. Hildebrand, and H. Böhnke. 1979. "Profilanalyse der polycyclischen aromatischen Kohlenwasserstoffe in proteinreichen Nahrungsmitteln, Ölen und Fetten (gaschromatographische Bestimmungsmethode)." *Deutsche Lebensmittel-Rundschau 71*, 93-100.

Guerin, W. F., and G. E. Jones. 1988. "Two-stage mineralization of phenanthrene by estuarine enrichment cultures." *Appl. Environ. Microbiol. 54*: 929-936.

Haider, K. M., and J. P. Martin. 1988. "Mineralization of C-labelled humic acids and of humic-acid bound C-xenobiotics by *Phanaerochaete chrysosporium*." *Soil Biol. Biochem. 20*: 425-429.

Heitkamp, M. A., and C. E. Cerniglia. 1989. "Polycyclic aromatic hydrocarbon degradation by a *Mycobacterium* sp. in microcosms containing sediment and water from a pristine ecosystem." *Appl. Environ. Microbiol. 55*: 1968-1973.

Heitkamp, M. A., J. P. Freeman, D. W. Miller, and C. E. Cerniglia. 1988. "Pyrene degradation by a *Mycobacterium* sp.: Identification of ring oxidation and ring fission products." *Appl. Environ. Microbiology 54*: 2556-2565.

Jenkinson, D. S., and D. S. Powlson. 1976. "The effects of biocidal treatments on metabolism in soil — V. A method for measuring soil biomass." *Soil Biol. Biochem. 8*: 209-213.

Kästner, M., M. Breuer, and B. Mahro. 1991. "Bakterien-Isolate aus unterschiedlichen Altlastenstandorten zeigen ein vergleichbares Abbauprofil für PAK und Ölkomponenten." *Gas- und Wasserfach, Wasser - Abwasser 132*: 253-255.

Lamar, R., G. A. Glaser, and T. K. Kirk. 1990. "Fate of pentachlorophenol (PCP) in sterile soils inoculated with the white-rot basidiomycete *Phanaerochaete chrysosporium*: Mineralization, volatilization and depletion of PCP." *Soil Biol. Biochem.* 22: 433-440.

Mahro, B., and M. Kästner. 1993. "Der mikrobielle Abbau polyzyklischer aromatischer Kohlenwasserstoffe (PAK) in Böden und Sedimenten." *Bioengineering 9*: 50-58.

Martens, R. 1982. "Concentrations and microbial mineralization of four to six ring polycyclic aromatic hydrocarbons in composted municipal waste." *Chemosphere 11*: 761-770.

Mitscherlich, E. A. 1950. *Bodenkunde für Landwirte, Forstwirte und Gärtner*. Verlag Paul Parey, Hamburg.

Mueller, J. G., P. J. Chapman, B. O. Blattmann, and P. H. Pritchard. 1990. "Isolation and characterization of a fluoranthene-utilizing strain of *Pseudomonas paucimobilis*." *Appl. Environ. Microbiol. 56*: 1079-1086.

Qiu, X., and M. J. McFarland. 1990. "Bound residue formation in PAH contaminated soil composting using *Phanaerochaete chrysosporium*." *Hazardous Waste & Hazardous Materials 8*: 115-126.

Sanglard, D., M. S. A. Leisola, and A. Fiechter. 1986. "Role of extracellular ligninases in biodegradation of benzo(a)pyrene by *Phanerochaete chrysosporium*." *Enzyme Microb. Technol. 8*: 209-213.

Senesi, N. 1992. "Binding mechanisms of pesticides to soil humic substances." *The Science of the Total Environment 123/124*: 63-76.

Stegmann, R., S. Lotter, and J. Heerenklage. 1991. "Biological treatment of oil-contaminated soils in bioreactors." In R. E. Hinchee and R. F. Olfenbuttel (Eds.), *On Site Bioreclamation: Processes for Xeniobiotic and Hydrocarbon Treatment*, pp. 188-208. Butterworth-Heinemann, Stoneham, MA.

Walter, U., M. Beyer, J. Klein, and H. J. Rehm. 1991. "Degradation of pyrene by *Rhodococcus* sp. UW1." *Appl. Microbiol. Biotechnol. 34*: 671-676.

Weissenfels, W. D., M. Beyer, J. Klein, and H. J. Rehm. 1991. "Microbial metabolism of fluoranthene: Isolation and identification of ring fission products." *Appl. Microbiol. Biotechnol. 34*: 528-535.

Ziechmann, W. 1980. *Huminstoffe: Probleme, Methoden, Ergebnisse.*, Chapter 6. Verlag Chemie, Weinheim , Germany.

STUDIES ON THE MICROBIAL ECOLOGY OF POLYCYCLIC AROMATIC HYDROCARBON BIODEGRADATION

J. G. Mueller, S. E. Lantz, R. Devereux, J. D. Berg, and P. H. Pritchard

ABSTRACT

Soils with known history of exposure to polycyclic aromatic hydrocarbons (PAHs) were collected from Norway, Germany, and the United States and screened for the presence of PAH-degrading bacteria. Purified PAH-degrading isolates were characterized by fatty acid profile analysis (GC-FAME), substrate utilization patterns (Biolog™ assays), 16S rRNA sequence comparisons, and total DNA:DNA hybridizations. Microbial respirometry and chemical analyses also were performed to define the PAH-biodegradation potential of these soils. These studies showed that all soils contaminated with PAHs harbored competent PAH-degrading bacteria that are biochemically similar and phylogenetically related. However, bioremediation strategies relying exclusively on indigenous PAH degraders should be closely evaluated for the ability to achieve site-specific cleanup standards in a timely manner.

INTRODUCTION

The use of specially selected microorganisms to enhance bioremediation efforts has proved effective in a number of applications, especially when combined with bioreactor systems (Mueller et al. 1993, Pritchard 1992). In our studies, the successful use of such isolates to remediate soil and water contaminated with organic wood preservatives (e.g., creosote, and pentachlorophenol [PCP]) has resulted in the opportunity to employ these technologies at similarly contaminated sites throughout the world.

Prior to distribution of these bioremediation strategies, concerns regarding the import of nonindigenous microorganisms needed to be addressed. Toward this end, we embarked on a program to ascertain whether (1) microorganisms similar to those used in our bioremediation strategy could be found in soils far removed from each other geographically; (2) previous exposure of soil microorganisms to PAH mixtures affected their PAH-degrading abilities; and (3) introduction of specially selected inoculant strains would offer any advantages, in terms of operating performance, to bioremediation systems employing indigenous microbiota.

MATERIALS AND METHODS

Acquisition of Soil Samples

Eight samples of soil with a history of exposure to PAH mixtures, such as creosote or diesel fuel, were recovered from a variety of locations. Six of these samples were obtained from creosote-contaminated sites in Norway (long-term exposure), one was sent from a diesel-contaminated site in Germany (recent spill), and one was recovered from an abandoned wood-preserving facility (American Creosote Works [ACW]) in northwest Florida, USA (long-term exposure). Additionally, two soils with no known history of exposure to such chemicals were recovered from agricultural farmland in south-central Illinois, USA.

Soil Analyses

Soil texture, moisture, nutritional status (NH_4-N, NO_3-N, available phosphorus, total phosphorus), water-holding capacity, and pH were determined in accordance with standard methods for soil analyses (Page et al. 1982). Analytical methods for extraction and quantitative determination of pentachlorophenol and 41 creosote constituents by gas chromatography (GC) are described elsewhere (Mueller et al. 1989, 1991).

For microbiological analysis, triplicate 1.0-g samples (wet weight) were placed in 9.0-mL volumes of sterile phosphate buffer (25 mM KH_2PO_4, 25 mM K_2HPO_4, pH 7.1) and shaken vigorously for 15 min (350 rpm). Soil suspensions were allowed to settle for 1 min before serial dilution in the same buffer. Total heterotrophic plate counts were performed with each soil sample using a standard heterotrophic plate count medium (Luria-Bertani agar; Maniatis et al. 1982) and standard microbiological methods (duplicate samples plated in replicate) (Page et al. 1982). Phenanthrene (PHE)- and fluoranthene (FLA)-degrading bacteria were enumerated using an overlay technique (Kiyohara et al. 1981). These values were recorded as the number of colony-forming units (CFUs) on carbon-substrate-free mineral salts agar that cleared the hydrocarbon substrate after 12 to 14 days of incubation at 28°C.

Soil Enrichment with PAHs

Soils were enriched for PAH-degrading microorganisms by adding 5.0 mL of a 10% soil slurry (prepared in sterile mineral salts medium) to 250-mL Erlenmeyer flasks containing 45 mL of sterile mineral salts medium (Mueller et al. 1990). Either PHE or FLA previously had been added to each flask in sterile acetone (evaporated) for a PAH concentration of 500 mg/L. Soils were incubated in the dark with shaking (150 rpm) at 30°C for 14 days. Following 14 days of aerobic incubation and enrichment, cultures were diluted 1:5 (vol/vol) with fresh mineral salts medium including PAHs at 500 mg/L. This transfer procedure was repeated two more times for a total of four enrichments over a 10-week period.

Screening Enrichment Cultures for PAH Degraders

After the second and fourth enrichments, liquid samples were removed from each vessel and screened for the presence of bacteria capable of using PHE and FLA as primary growth substrates. Once individual colonies were single-colony purified, they were transferred to utilizable-carbon-free agar and complex agar, then overlain with PHE or FLA as described by Kiyohara et al. (1981). Plates were incubated for 14 days prior to scoring for individual colonies exhibiting zones of clearing of the PAH substrate.

Colonies demonstrating PAH-clearing abilities were purified and transferred to 125-mL Erlenmeyer flasks containing 25 mL mineral salts broth plus 500 mg/L PHE or FLA as the sole carbon and energy source. Cultures were incubated for 5 to 7 days with shaking (150 rpm) at 30°C. Bacterial growth at the expense of the PAH substrates was measured by visual assessment of turbidity and by monitoring changes in absorbance at 550 nm. As a control, growth in carbon-free mineral salts broth also was monitored.

Microbiological Characterizations

Once the PAH-degrading ability of purified cultures had been validated, cultures were characterized by GC-FAME and the Biolog™ Microplate System™ (Microbe Inotech Laboratories, Inc., St. Louis, Missouri). The taxonomic relationships among these strains was analyzed by evaluating similarity measures from GC-FAME and substrate utilization patterns with principal component analysis (Jacobs 1990).

Phylogenetic relationships were determined by 16S ribosomal RNA sequence comparisons. Universal primers and the polymerase chain reaction were used to amplify 16S rRNA genes (Weisburg et al. 1991) from select PAH-degrading bacterial strains CRE7, CRE11, CRE12, *Pseudomonas paucimobilis* strain EPA505, and the type strain of *Pseudomonas paucimobilis* ATCC 29837. Reverse transcriptase sequencing with the rRNA template as the primer was used to generate 16S rRNA sequence information for select PAH-degrading strains (Lane et al. 1985). Total DNA:DNA hybridizations were performed between select PAH-degraders to define homology (Amann et al. 1992).

Manometric Respirometry Studies

Into 14 flasks (250-mL Erlenmeyer flasks), each containing 10 mL of a 33% soil slurry, was added a predetermined level of inorganic nutrients. Duplicate flasks of each slurry were amended with 500 mg/L naphthalene (NAH), PHE, or FLA; readily utilizable carbon (250 mg/L glucose + 250 mg/L glycerol); or 500 mg/L specification creosote no. 450 (American Wood-Preserver's Association).

Two flasks received no supplemental carbon (to discern the effect of nutrient amendment and aeration), and two more flasks served as killed cell controls (acidified to pH 2.0 with 1 N HCl plus 3.7% formaldehyde) for each soil tested.

Microbial respirometric responses (rate of liberation of CO_2 and the simultaneous consumption of O_2 were determined at 8-hr intervals over an 8-day incubation period (23°C, 100 rpm shaker speed) with a MicroOxymax respirometer (Columbus Instruments, Columbus, Ohio). At the end of each incubation period, slurries from nutrient-amended only, creosote-amended, and killed-cell (control) treatments were extracted and analyzed for the presence of creosote constituents as previously described (Mueller et al. 1991). These values were compared with those determined at time zero for each soil.

RESULTS AND DISCUSSION

Soil Analyses

The results of the physicochemical analyses used to characterize the soils are summarized in Table 1. In general, all measured soil parameters were within a range conducive to biological activity. However, contaminated soils had rather low levels of available nitrogen and, to a certain extent, available phosphorus (Bray P1).

Microbiological analysis of soils prior to enrichment with PAHs showed that all soils harbored culturable heterotrophic bacteria (Table 2). With the exception of the ACW 47 site, all soils with a history of exposure to PAHs also had a discernible number of PHE-degraders, and, with the exception of soils UN 1 and ACW 47, a relatively high number of FLA-degraders. Conversely, the "control soils" (SIU ARC, SIU BRC), with no known history of exposure to PAHs, had no detectable PHE- or FLA-degraders.

The presence of PCP at a relatively high concentration (>250 mg PCP/kg soil dry wt) in soil collected from the abandoned American Creosote Works site (sample ACW 47), in combination with creosote at an average concentration >500 mg creosote PAHs/kg soil dry wt), may represent the reason for the low initial numbers of PHE- and FLA-degraders (i.e., toxicity). In the case of the UN 1 soil sample, the only soil being recently impacted by diesel fuel (accidental highway spill) and not long-term creosote exposure (<1 month exposure at the time of sampling), the high number of PHE-degraders along with a rather low number of FLA-degraders may be the result of low-level exposure to PHE (e.g., diesel vapors) or adaptation to structurally related compounds (Bauer & Capone 1988).

PAH Enrichment Studies

Using PHE as a growth substrate, all soils except SIU ARC and SIU BRC produced turbid cultures within the first week of the first enrichment. All subsequent transfers produced turbid cultures within the first 2 to 3 days of incubation.

TABLE 1. Description and characterization of soils used for the isolation of PAH-degraders.

Origin/Location of Soil	Exposure History	Classification	Nutrient Analysis (mg/kg soil dry wt)				pH	Field Capacity (% wgt)
			NH_4-N	NO_3-N	Bray P1	Bray P2		
Norwegian Soils								
Rade 1	Creosote	Loamy sand	2	10	0	28	6.4	7.1
Rade 2	Creosote	Sand	4	6	6	13	5.9	7.4
Lillestrøm 1	Creosote	Sand	10	6	10	19	6.2	16.0
Lillestrøm 2	Creosote	Loam	3	5	16	28	7.2	37.0
Drammen	Creosote	Sand	1	6	12	25	7.4	8.1
Hommelvik	Creosote	Loamy sand	9	31	3	23	7.2	10.7
German Soil								
UN 1	Mixed PAHs	Sand	1	0	82	87	7.4	23.9
American Soils								
ACW 47	Creosote/PCP	Sand	3	1	20	20	6.7	14.9
SIU ARC	None known	Silt loam	20	13	34	47	6.8	29.2
SIU BRC	None known	Silt loam	0	9	35	48	6.4	20.3

TABLE 2. Enumeration of total aerobic, culturable heterotrophic, phenanthrene-, and fluoranthene-degrading bacteria in soils.

Soil[a]	Total Heterotrophs[b]	Phenanthrene Degraders[c]	Fluoranthene Degraders[c]
	log CFU/mL slurry[d]		
Rade 1	6.47	4.87	5.82
Rade 2	6.00	4.44	5.85
Lillestrøm 1	7.57	3.93	5.26
Lillestrøm 2	7.63	3.65	3.74
Drammen	6.18	4.13	5.16
Hommelvik	6.98	4.04	3.85
UN 1	6.43	5.49	<3.00
ACW 47	6.60	<3.00	<3.00
SIU ARC	7.00	<2.00	<2.00
SIU BRC	7.31	<2.00	<2.00

(a) See Table 1 for description of soils.
(b) Total heterotrophs on Luria-Bertani agar after 5 days incubation at 28°C.
(c) PAH-degraders based on the number of colonies to clear PAH substrates on minimal medium after 14 days incubation at 30°C.
(d) CFU = colony forming units.

Simultaneously, all undissolved PHE crystals were removed, and rapid changes in medium coloration were observed.

Using FLA as an enriching substrate, the observed growth responses were very similar to those recorded in the presence of PHE. Here, fluoranthene biodegradation was apparent within the first 10 days of enrichment for all soils except SIU ARC, SIU BRC, UN 1, and Hommelvik. Growth (as determined by visually apparent increases in turbidity, change in medium coloration, and disappearance of undissolved FLA crystals) with inocula from Hommelvik and UN 1 soil became evident within 21 days of enrichment. Following 40 days of incubation, biodegradation or solubilization of PHE and FLA in liquid medium inoculated with microorganisms recovered from the SIU ARC and SIU BRC soils was not observed.

The enrichment culture conditions used in these studies (e.g., excess inorganic nutrients, elevated temperatures, mixing, and aqueous solutions saturated with PAHs) are substantially different than expected environmental conditions. However, they closely resemble conditions associated with bioreactor operations that are widely used in the bioremediation industry (Berg et al., this volume; Mueller et al. 1993). Thus, while the relatively low number of PAH degraders from unexposed sites may not fully reflect the potential for long-term adaptation, the fact that non-PAH-history soils did not readily yield PHE or FLA degraders suggests that, in the event of recent contamination, bioremediation of PAHs on a short-term basis may require the use of inoculants.

Respirometric Analyses

The respiratory activity of indigenous microflora from all soils with a recorded history of exposure to PAHs was stimulated upon the addition of all organic carbon sources, as well as with the addition of inorganic nutrients alone. Using soil from the Hommelvik site as a typical example, increased respiratory responses were observed upon the addition of individual PAHs (NAH, PHE, or FLA) and creosote (Figure 1, top and middle panels). These responses were shown to be above and beyond that expected from the conversion of resident carbon, which includes creosote PAHs, due to the addition of inorganic nutrients and aeration. Hence, these chemicals represented utilizable carbon sources to the indigenous microflora. Conversely, both soils with no known exposure to PAHs (SIU soils) showed very limited response to the addition of PAHs or inorganic nutrients, but the indigenous soil microflora rapidly mineralized added glucose and glycerol (Figure 1, bottom panel).

In all cases, no activity was observed in the poisoned systems. Compared with analytical chemistry data, increases in respiratory activities generally were associated with accelerated biodegradation of monitored creosote constituents (data not shown).

Microbial Ecology and Bacterial Taxonomy

All soils with a history of PAH exposure, except ACW 47, harbored PHE- and FLA-degrading bacteria of various genera. Of the many isolates recovered, the degradative abilities of 13 PHE-degraders, 14 FLA-degraders, and 1 pentachlorophenol degrader (Resnick & Chapman 1990), isolated from myriad contaminated sites, were positively verified (Table 3, data not shown). These strains were processed for characterization and identification based on GC-FAME and Biolog™ assays (Table 4).

Because the established databases for both GC-FAME and Biolog™ assays are focused predominantly on the identification of pathogenic microorganisms and those of clinical importance, many of the identifications made have low similarity coefficients (Table 4). Hence, most of the identifications were considered to be suggestive rather than conclusive. Principal component analyses using GC-FAME profiles of each of the PAH-degrading bacteria showed that many microorganisms isolated from U.S. soils were closely related to microorganisms recovered from other soils in the United States and Europe (Figure 2).

For example, strain CRE7 (PHE-degrader isolated from the ACW site at Pensacola, Florida, and used commercially in a bioremediation process) was found to be closely related to PHE-degrading strains N2P5 and N2P6 (Rade soil, Norway); strains N3P2 and N3P3 (Lillestrøm soil, Norway); and strains PJC 2288, 2289, and 2295 (PAH degraders also from the Pensacola site). Likewise, FLA-degrading bacteria very similar to strain EPA505 were recovered from Germany (strains G1F1 and G1F2) and geographically separate sites in the United States (strains PJC 2286, 2287). Gram-positive "mycobacterial" strains PJC 2282, PJC 2283, and FDA PYR-1 were found to be very closely related to each other, but, as expected, distinctly different from Gram-negative bacterial isolates.

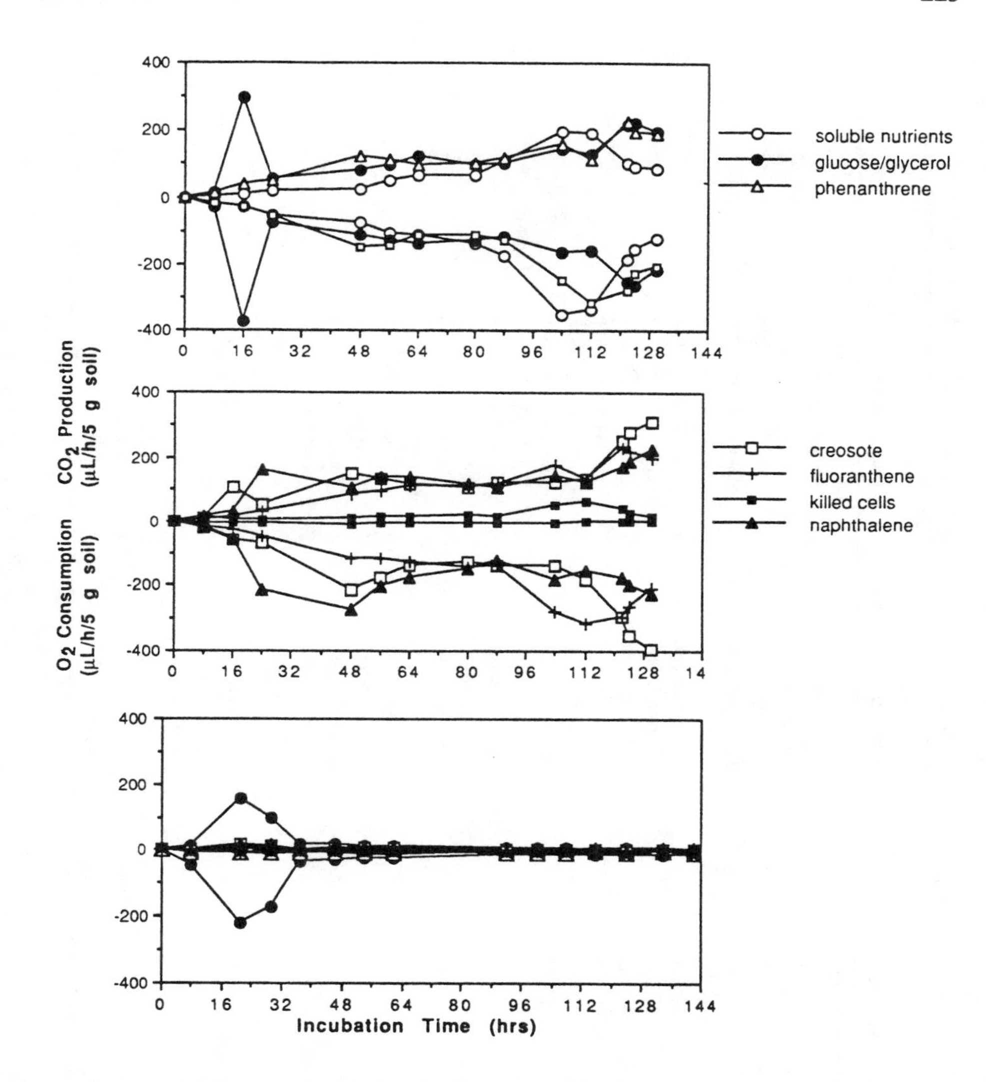

FIGURE 1. Respiratory response of indigenous microorganisms to organic and inorganic amendments. Respiratory activity of soil microorganisms present in Hommelvik soil contaminated with creosote (top and middle panels), and SIU ARC farmland soil with no known history of exposure to PAHs (bottom panel).

Phylogenetic studies using 16S rRNA sequence comparisons showed that two PAH-degrading strains, isolated from creosote-contaminated sites in the USA (strains CRE7 and CRE11), were related to *Pseudomonas aeruginosa.* The 16S rRNA sequences between the PAH-degraders and *P. aeruginosa* were 92 to 93% similar. Correlations developed between 16S rRNA sequence similarity and %DNA relatedness (Amann et al. 1992, Devereux et al. 1990) suggest that, at 92% 16S rRNA

TABLE 3. Bacteria isolated from PAH-contaminated sites for their ability to degrade phenanthrene or fluoranthene.

Isolate Number	Culture Number	Source/Reference	Enrichment Substrate	Soil of Origin
1	EPA505	Mueller et al. 1990	Fluoranthene	Pensacola, Florida
2	PJC 2282	U.S. EPA, GBERL	Fluoranthene/pyrene	Pensacola, Florida
3	PJC 2283	U.S. EPA, GBERL	Fluoranthene/pyrene	Live Oaks, Florida
4	PJC 2285	U.S. EPA, GBERL	Fluoranthene	Live Oaks site, Florida
5	PJC 2286	U.S. EPA, GBERL	Fluoranthene	Live Oaks site, Florida
6	PJC 2287	U.S. EPA, GBERL	Fluoranthene	Pensacola, Florida
7	PJC 2288	U.S. EPA, GBERL	Phenanthrene	Pensacola, Florida
8	PJC 2289	U.S. EPA, GBERL	Phenanthrene	Live Oaks, Florida
9	PJC 2295	U.S. EPA, GBERL	Phenanthrene	Pensacola, Florida
10	CRE7	Mueller et al. 1989	Phenanthrene	Pensacola, Florida
11	CRE11	Mueller et al. 1989	Phenanthrene	Pensacola, Florida
12	CRE12	Mueller et al. 1989	Phenanthrene	Pensacola, Florida
13	AK Phen6	Mueller et al. 1992	Phenanthrene	Prince William Sound, Alaska
14	N1F1	This study	Fluoranthene	Rade, Norway
15	N2P5	This study	Phenanthrene	Rade, Norway
16	N2P6	This study	Phenanthrene	Rade, Norway
17	N3P2	This study	Phenanthrene	Lillestrøm, Norway
18	N3P3	This study	Phenanthrene	Lillestrøm, Norway
19	N3F1	This study	Fluoranthene	Lillestrøm, Norway
20	N3F2	This study	Fluoranthene	Lillestrøm, Norway
21	N4F4	This study	Fluoranthene	Lillestrøm, Norway
22	N5F4	This study	Fluoranthene	Drammen, Norway
23	N6F4	This study	Fluoranthene	Hommelvik, Norway
24	G1F1	This study	Fluoranthene	Germany
25	G1F2	This study	Fluoranthene	Germany
26	G1P1	This study	Phenanthrene	Germany
27	G2P2	This study	Phenanthrene	Germany
28	FDA PYR-1	Heitkamp & Cerniglia 1988	Pyrene	Port Aransas, Texas
29	SR3	Resnick & Champman 1990	Pentachlorophenol	northwest Florida

TABLE 4. Bacterial identifications based on GC-FAME and Biolog™ assays.

Culture Number	GC-FAME Identification (similarity coefficient)	Biolog™ Identification[(a)] (similarity coefficient)
EPA505	*Pseudomonas paucimobilis* (0.13)	No ID [insufficient growth 36 h]
PJC 2282	*Mycobacterium parafortuitum* (0.056)	*Corynebacterium jeikeium* (0.632)
PJC 2283	*Mycobacterium parafortuitum* (0.025)	No ID [*Corynebacterium variabilis* (0.425)]
PJC 2285	No ID [*Pediococcus halophilus* (NA)]	No ID [*Corynebacterium jeikeium* (0.361)]
PJC 2286	*Enterococcus faecium* (0.011)	No ID [*Micrococcus luteus* (0.396)]
PJC 2287	*Pseudomonas saccharophila* (0.441)	No ID [*Moraxella atlantae* (0.358)]
PJC 2288	<u>*Pseudomonas pseudomallei*</u> (0.192)	*Pseudomonas cepacia* (0.557)
PJC 2289	<u>*Pseudomonas pseudomallei*</u> (0.177)	*Pseudomonas cepacia* (0.602)
PJC 2295	*Pseudomonas cepacia* (0.059)	*Pseudomonas cepacia* (0.426)
CRE7	*Pseudomonas cepacia* (0.399)	No ID [*Pseudomonas cepacia* (0.401)]
CRE11	*Pseudomonas aeruginosa* (0.778)	*Pseudomonas aeruginosa* (0.779)
CRE12	<u>*Pseudomonas aeruginosa*</u> (0.366)	*Pseudomonas azelaica* (0.678)
AK PHEN6	No ID [*Pseudomonas saccharophila* (NA)	No ID [*Alteromonas haloplanktis* (0.393)]
N1F1	*Xanthomonas maltophilia* (0.35)	*Pseudomonas corrugata* (0.80)
N2P5	*Pseudomonas cepacia* (0.294)	*Pseudomonas phenazinium* (0.634)
N2P6	*Pseudomonas cepacia* (0.114)	*Pseudomonas gladioli* (0.515)
N3P2	*Pseudomonas cepacia* (0.379)	*Pseudomonas gladioli* (0.404)
N3P3	*Pseudomonas acidovorans* (0.165)	No ID [*Comamonas testosteroni* (0.407)]
N3F1	No ID [*Pseudomonas putida* biotype A]	*Xanthomonas maltophilia* (0.78)
N3F2	*Xanthomonas maltophilia* (0.41)	*Xanthomonas maltophilia* (0.73)
N4F4	*Xanthomonas maltophilia* (0.32)	*Comamonas acidovorans* (0.78)
N5F4	<u>*Pseudomonas putida* biotype B</u> (0.54)	No ID [*Alcaligenes paradoxus*]
N6F4	*Alcaligenes paradoxus* biotype II (0.34)	*Xanthomonas maltophilia* (0.78)
G1F1	No ID [*Pseudomonas delafieldii* (NA)]	*Brucella abortus* biovar 2 (0.79)
G1F2	No ID [*Pseudomonas saccharophila* (NA)]	No ID [*Brucella abortus* biovar 2 (NA)]
G1P1	*Pseudomonas saccharophila* (0.11)	No ID [insufficient growth 36 h]
G1P2	*Acinetobacter calcoaceticus* (0.40)	*Acinetobacter* genospecies 13 (0.65)
FDA PYR-1	*Mycobacterium parafortuitum* (0.059)	*Corynebacterium variabilis* (0.59)
SR3	*Pseudomonas saccharophila* (0.603)	No ID [poor growth]

(a) Closest species identification listed in brackets [ ] when no identification is made. Underscore indicates a match to the clinical database of Microbe Inotech Laboratories, St. Louis, MO.

sequence similarity and ca 50% DNA relatedness, the PAH-degrading strains are related to *P. aeruginosa* at the genus/species level where greater than 20% DNA relatedness indicates a genus-level relationship.

Likewise, results of DNA:DNA hybridization studies showed that strains isolated from Norwegian and U.S. sites also were somewhat related at the genus level. For example, the U.S. isolate strain CRE7 demonstrated 56% and 36% DNA:DNA homology with the Norwegian isolate N2P5 and the German isolate G1P2, respectively, all being isolated for their ability to utilize PHE as a sole carbon source. These relationships follow the GC-FAME principal component analyses presented in Figure 2. Further, the U.S. strain EPA505 demonstrated 24 and 35% DNA:DNA homology with the ATCC type strain of *Pseudomonas paucimobilis* and the Norwegian strain N2P5, respectively.

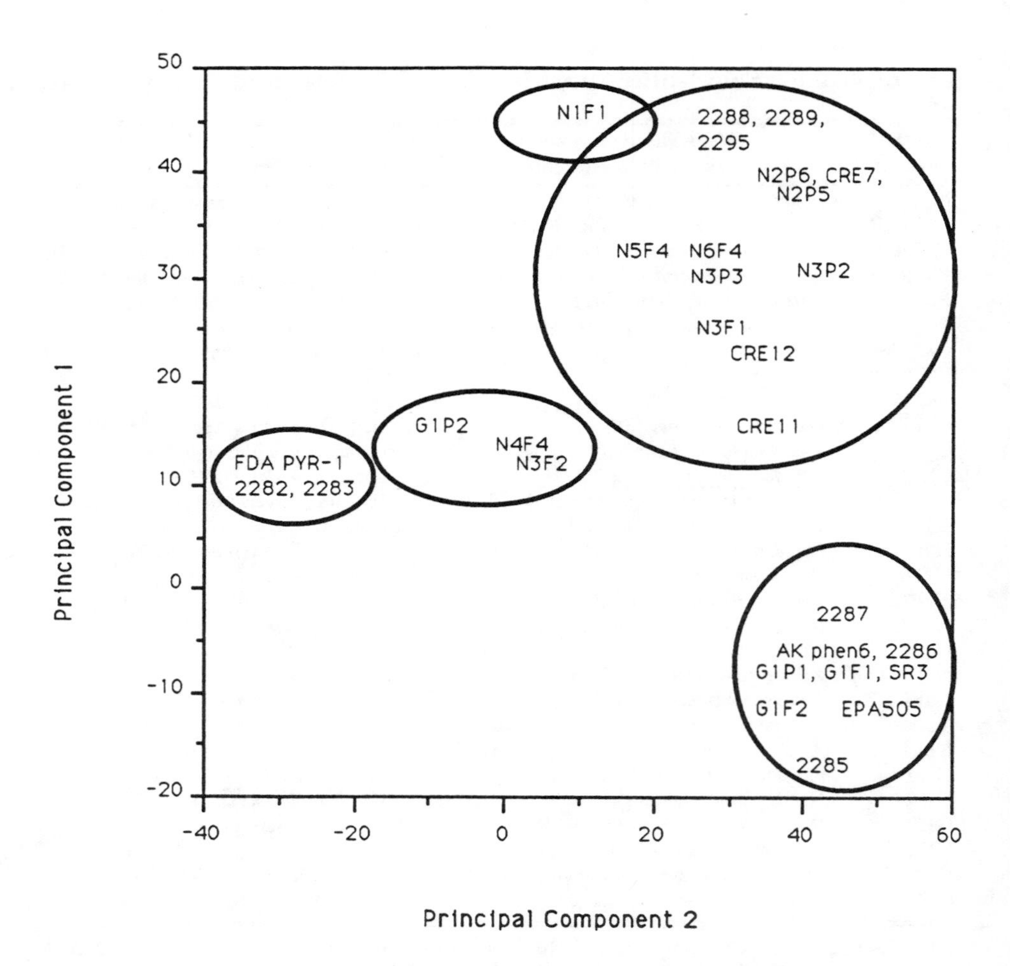

FIGURE 2. Principal component analysis with GC-FAME data from PAH-degrading bacteria isolated from geographically distant PAH-contaminated sites.

CONCLUSIONS

All soils with a history of PAH exposure yielded microbial populations competent for the degradation of the PAHs phenanthrene and fluoranthene. According to GC-FAME, Biolog™, 16S rRNA sequence similarity, and DNA:DNA homology, many of these PAH-degrading microorganisms appeared to be closely related phenotypically and phylogenetically to similar types of organisms isolated from soils at geographically distant sites in the United States. The technique employed to enrich PAH-degrading bacteria thus appeared to select similar types of microorganisms that are indigenous to contaminated soils at each site. Admittedly, this does not necessarily include all types of microorganisms that may play a role in the biodegradation/bioremediation of PAHs (e.g., fungi).

Given that PAH-degraders appeared to be indigenous to geographically diverse PAH-contaminated soils, then if the stimulatory effect of controlled nutriation, mixing, and aeration on the activity of the indigenous microflora results in acceptable rates and extents of biodegradation of targeted chemicals, then, on a site-specific basis, it may be possible to rely solely on the activity of such microorganisms to facilitate site remediation (see Berg et al., this volume). Despite the phylogenetic similarities among these organisms, however, their catabolic abilities would seem to be the most important consideration from a bioremediation perspective. Thus, in the event that indigenous microorganisms do not perform favorably, then utilization of nonindigenous microbes in optimized bioremediation systems could be advantageous for cost-efficient, effective bioremediation. Based on the results of these studies, the export/import of the nonindigenous bacteria used in these studies to augment bioremediation efforts would not seem to represent the introduction of exotic biota, and thus would pose no discernible ecological risk.

ACKNOWLEDGMENTS

We thank Myke'lle Hertsgaard and Barbara Artlet (Technical Resources, Inc., Gulf Breeze, Florida), Sheree Enfinger (U.S. Environmental Protection Agency, Environmental Research Laboratory, Gulf Breeze, Florida), and Stephanie Willis (University of New Hampshire) for technical assistance; Brian Klubek (SIU-Carbondale, Illinois) for soil analyses; Bruce Hemming (Microbe Inotech Laboratories, St. Louis, Missouri) for help in the interpretation of GC-FAME and Biolog™ results; and Peter Chapman (U.S. EPA, ERL, Gulf Breeze, Florida) and Carl Cerniglia (U.S. Food and Drug Administration, National Center for Toxicological Research, Jefferson, Arkansas) for donating PAH-degrading strains for comparative analyses. Soil from Germany was provided by Wolfgang Fabig (Umweltshutz Nord, Germany).

Financial support for these studies was provided by the Norwegian State Railway (NSB) and the U.S. EPA (Gulf Breeze). These studies were performed as part of a Cooperative Research and Development Agreement between the Gulf Breeze Environmental Research Laboratory and SBP Technologies, Inc. (Atlanta, Georgia) as defined under the Federal Technology Transfer Act, 1986 (contract no. FTTA-003).

REFERENCES

Amann, R. I., C. Lin, R. Key, L. Montgomery, and D. A. Stahl. 1992. "Diversity among *Fibrobacter* isolates: Towards a phylogenetic classification." *Syst. Appl. Microbiol.* 15:23-31.

Bauer, J. E., and D. G. Capone. 1988. "Effects of co-occurring aromatic hydrocarbons on the degradation of individual polycyclic aromatic hydrocarbons in marine sediment slurries." *Appl. Environ. Microbiol.* 54:1649-1655.

Berg, J. D., B. Nesgard, R. Gundersen, A. Lorentsen, and T. E. Bennett. 1994. "Washing and slurry phase biotreatment of creosote-contaminated soil." In R. E. Hinchee, A. Leeson,

L. Semprini, and S. K. Ong (Eds.), *Bioremediation of Chlorinated and Polycyclic Aromatic Hydrocarbon Compounds.* Lewis Publishers, Ann Arbor, MI.

Devereux, R., S. H. He, C. L. Doyle, S. Orkland, D. A. Stahl, J. LeGall, and W. B. Whitman. 1990. "Diversity and origin of *Desulfovibrio* species: Phylogenetic definition of a family." *J. Bacteriol. 172*:3609-3619.

Heitkamp, M. A., and C. E. Cerniglia. 1988. "Mineralization of polycyclic aromatic hydrocarbons by a bacterium isolated from sediment below an oil field." *Appl. Environ. Microbiol. 54*:1612-1614.

Jacobs, D. 1990. "SAS/GRAPH software and numerical taxonomy." *In Proceedings of the 15th Annual Users Group Conference*, pp. 1413-1418. SAS Institute, Inc., Cary, NC.

Kiyohara, H., K. Nagao, and K. Yaga. 1981. "Rapid screen for bacteria degrading water-insoluble, solid hydrocarbons on agar plates." *Appl. Environ. Microbiol.* 43:454-457.

Lane, D. J., B. Pace, G. J. Olsen, D. A. Stahl, M. L. Sogin, and N. R. Pace. 1985. "Rapid determination of 16S ribosomal RNA sequences for phylogenetic analyses." *Proc. Natl. Acad. Sci. USA 82*:6955-6959.

Maniatis, T., E. F. Frisch, and J. Sambrook. 1982. *Molecular Cloning: A Laboratory Manual*, p. 68. Cold Spring Harbor Laboratory, Cold Spring Harbor, NY.

Mueller, J. G., P. J. Chapman, B. O. Blattmann, and P. H. Pritchard. 1990. "Isolation and characterization of a fluoranthene-utilizing strain of *Pseudomonas paucimobilis.*" *Appl. Environ. Microbiol. 56*:1079-1086.

Mueller, J. G., P. J. Chapman, and P. H. Pritchard. 1989. "Action of a fluoranthene-utilizing bacterial community on polycyclic aromatic hydrocarbon components of creosote." *Appl. Environ. Microbiol. 55*:3085-3090.

Mueller, J. G., S. E. Lantz, B. O. Blattmann, and P. J. Chapman. 1991. "Bench-scale evaluation of alternative biological treatment processes for the remediation of pentachlorophenol- and creosote-contaminated materials: Solid-phase bioremediation." *Environ. Sci. Technol. 25*:1045-1055.

Mueller, J. G., S. E. Lantz, R. J. Colvin, D. Ross, D. P. Middaugh, and P. H. Pritchard. 1993. "Strategy using bioreactors and specially-selected microorganisms for bioremediation of ground water contaminated with creosote and pentachlorophenol." *Environ. Sci. Technol.* (April issue).

Mueller, J. G., S. M. Resnick, M. E. Shelton, and P. H. Pritchard. 1992. "Effect of inoculation on the biodegradation of weathered Prudhoe Bay crude oil." *J. Indust. Microbiol. 10*:95-105.

Page, A. L., R. H. Miller, and D. R. Keeney. 1982. *Methods of Soils Analysis:* Part 2, *Chemical and Microbiological Properties*, 2nd ed. American Society of Agronomy, Madison, WI, 1159 pp.

Pritchard, P. H. 1992. "Use of inoculation in bioremediation." *Curr. Opinions in Biotechnol. 3*:232-243.

Resnick, S. M., and P. J. Chapman. 1990. "Isolation and characterization of a pentachlorophenol-degrading, Gram-negative bacterium." *Abstr. Ann. Meet. Am. Soc. Microbiol.*, p. 300.

Weisburg, W. G., S. M. Barnes, D. A. Pelleteir, and D. J. Lane. 1991. "16S ribosomal DNA amplification for phylogenetic study." *J. Bacteriol. 173*:697-703.

DEGRADATION OF CHLORINATED PHENOLS AND CHLORINATED DIBENZO-*p*-DIOXINS BY *PHANEROCHAETE CHRYSOSPORIUM*

M. H. Gold, D. K. Joshi, K. Valli, and H. Wariishi

ABSTRACT

Pathways for the degradation of 2,4-dichlorophenol, 2,4,5-trichlorophenol, 2,7-dichlorodibenzo-*p*-dioxin (DCDD), and 2-chlorodibenzo-*p*-dioxin were elucidated by characterization of fungal metabolites and oxidation products generated by purified lignin peroxidase, manganese peroxidase, and intracellular extracts. The results indicate that the degradation of these substrates is initiated by extracellular peroxidase-catalyzed oxidation reactions generating quinone intermediates. This initial step is followed by quinone reduction or by quinone reduction and hydroquinone methylation reactions. The latter reactions regenerate a peroxidase substrate. Several cycles of oxidation and reduction or oxidation, reduction and methylation result in the removal of all the chlorines from the aromatic ring. In subsequent steps, the key intermediates 1,2,4,5-tetrahydroxybenzene and 1,2,4-trihydroxybenzene are subject to ring-opening reactions and further degradation. Furthermore, the results indicate that C-O-C bond cleavage of the chlorinated dioxins is initiated by the lignin peroxidase-catalyzed formation of a substrate aryl cation radical.

INTRODUCTION

Chlorinated phenols and chlorinated dibenzodioxins constitute significant categories of environmental pollutants (Freiter 1979, Keither & Telliard 1979). Chlorinated phenols are major components of paper pulp mill effluents (Valo et al. 1984). They also are used extensively as wood preservatives, pesticides, and herbicide precursors (Freiter 1979, Rappe 1980).

Due to their toxicity, polychlorinated dibenzo-*p*-dioxins (PCDDs) have been recognized as environmental hazards for several decades (Hansen 1991, Rappe 1980). Since PCDDs are chemically stable and fat soluble, they accumulate in the food chain (Hansen 1991, Rappe 1980). PCDDs have been identified in paper pulp mill effluents and as impurities in a variety of industrial chemicals, including polychlorinated biphenyls (PCBs) (Kimbrough & Jensen 1979, Rappe 1980).

The white-rot basidiomycete fungus *Phanerochaete chrysosporium* has potential for application in a variety of bioremediation technologies. This orgainism produces two unique extracellular heme peroxidases, lignin peroxidase (LiP) and manganese peroxidase (MnP), which are believed to constitute the major extracellular components of its lignin and aromatic pollutant degradation system (Gold et al. 1989, Kirk & Farrell 1987). Over the past several years the degradation of a variety of aromatic pollutants by *P. chrysosporium* has been reported (Bumpus & Aust 1987a, Hammel 1989). However, until recently the pathways used by this fungus for chlorinated aromatic pollutant degradation had not been elucidated. Recently, in several papers (Joshi & Gold 1992, Valli & Gold 1991, Valli et al. 1992a, 1992b) and in work described below, we have examined the pathways for the degradation of several chlorinated aromatic pollutants. Pathways for the degradation of anthracene (Hammel et al. 1991) and DDT (Bumpus & Aust 1987b) also have been reported. In this work we demonstrate the involvement of LiP, MnP, quinone reductases, and dioxygenases in the degradation of several chlorinated aromatics by *P. chrysosporium.*

RESULTS AND DISCUSSION

Degradation Pathway for 2,4-Dichlorophenol I

In a recent report we examined the pathway for the degradation of I by *P. chrysosporium* (Valli & Gold 1991). This fungus mineralizes ^{14}C-labeled I only under nutrient nitrogen-limiting conditions, suggesting that the lignin degradative system is responsible (at least in part) for the degradation of this pollutant. Similar results have been reported for pentachlorophenol (Lamar et al. 1990). The degradation pathway for I was elucidated by the identification of fungal metabolites and oxidation products generated by LiP, MnP, and intracellular cell-free extracts. The first step in the proposed pathway is the oxidation of I to 2-chloro- 1,4-benzoquinone II. This reaction is catalyzed by both LiP and MnP. Both 2-chloro-1,4-hydroquinone III and 2-chloro-1,4-dimethoxybenzene IV were identified as metabolites of II. The hydroquinone III and the dimethoxybenzene IV both are metabolized by *P. chrysosporium* to yield 2,5-dimethoxy-1,4-benzoquinone V. In addition, both LiP and MnP oxidize the hydroquinone III to yield a small amount of the nonchlorinated product 2,5-dihydroxy-1,4-benzoquinone VII. LiP also oxidizes 2-chloro-1,4-dimethoxybenzene IV to 2,5-dimethoxy-1,4- benzoquinone V. Both the quinones V and VII are subsequently transformed to the key intermediate tetrahydroxybenzene VIII. The latter is a substrate for intracellular ring-opening dioxygenase(s) that cleave VIII to yield malonic acid. This pathway involves the initial oxidative 4-dechlorination of the substrate by LiP or MnP. Subsequent reduction and/or reduction and methylation results in the regeneration of a substrate for a peroxidase-catalyzed oxidative dechlorination (Valli & Gold 1991). To examine the generality of this oxidation and reduction cycle we examined the degradation of 2,4,5-trichlorophenol.

Degradation of 2,4,5-Trichlorophenol IX

Under ligninolytic conditions, *P. chrysosporium* also mineralizes [U]-^{14}C-labeled IX (Joshi & Gold 1992). The pathway for the degradation of IX was elucidated by the characterization of fungal metabolites and enzyme oxidation products. The first step in the pathway is the peroxidase-catalyzed oxidative 4-dechlorination of IX to generate 2,5-dichloro-1,4-benzoquinone X (Figure 1). Reduction of the quinone X yields 2,5-dichloro-1,4-dihydroxybenzene XI. Methylation of the hydroquinone XI yields 2,5-dichloro-4-methoxyphenol XII. The only metabolic product of XII is the quinone X, suggesting that XII is not part of the primary metabolic pathway. The hydroquinone XI is oxidized by MnP to generate 5-chloro-4-hydroxy-1,2-benzoquinone (XIII). The quinone XIII is reduced to produce 5-chloro-1,2,4-trihydroxybenzene XIV. Finally, the chlorotrihydroxybenzene XIV undergoes another cycle of oxidative dechlorination and reduction to generate tetrahydroxybenzene VIII. As described above, tetrahydroxybenzene VIII is oxidatively cleaved to yield malonic acid, which is subsequently degraded to CO_2.

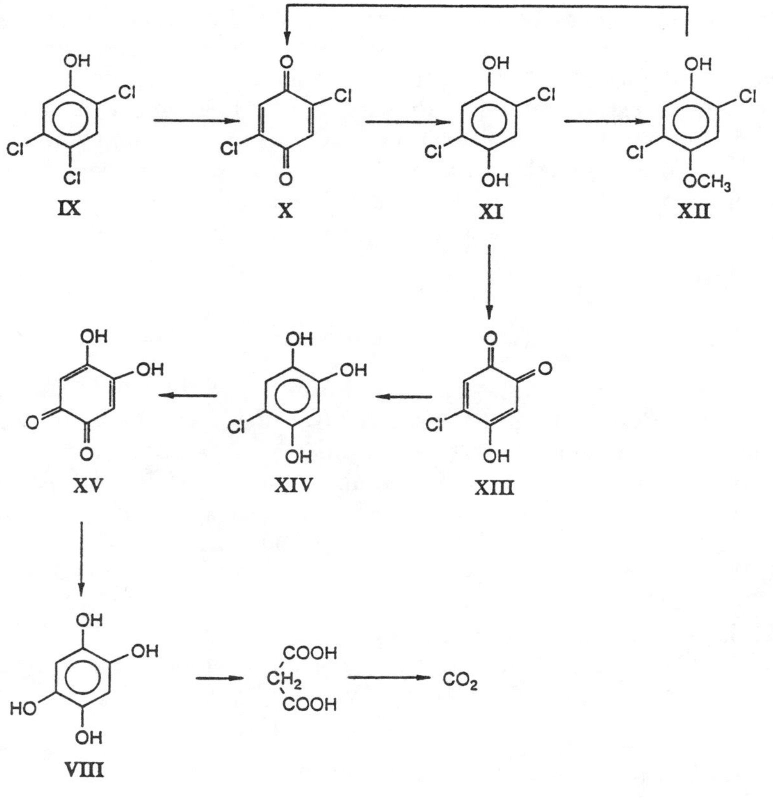

FIGURE 1. Proposed pathway for the degradation of 2,4,5-trichlorophenol IX.

As described above for 2,4-dichlorophenol I, all three chlorines are removed from the aromatic ring of 2,4,5-trichlorophenol X before ring cleavage occurs. The pathway for the degradation of 2,4,5-trichlorophenol also involves several cycles of oxidative dechlorination followed by reactivation of intermediates through quinone reduction.

Degradation of 2,7-Dichlorodibenzo-*p*-dioxin XVI

P. chrysosporium extensively degrades XVI only under ligninolytic conditions (Valli et al. 1992b). The first step in the pathway is the oxidative cleavage of the dioxin ring of XVI by LiP, generating chloride and the monomeric products, 4-chloro-1,2-benzoquinone XVII and 2-hydroxy-1,4-benzoquinone XVIII (Figure 2). In culture, XVIII is reduced to form 1,2,4-trihydroxybenzene XIX. The latter is ring-cleaved to produce, after subsequent reduction, ß-ketoadipic acid. We have shown that intracellular cell free extracts also are able cleave the ring of the trihydroxybenzene XIX. The other primary product of XVI degradation, 4-chloro-1,2-benzoquinone XVII, is reduced to 4-chloro-1,2-dihydroxybenzene XX (Figure 2). The latter also is subject to oxidation to yield XVIII.

Alternatively, XX can be methylated to 4-chloro-1,2-dimethoxybenzene XXI, which is cycled by several oxidation and reduction steps to yield the trihydroxybenzene XIX (Figure 2). Thus the primary quinone products generated by the cleavage of the substrate subsequently are degraded by several cycles of quinone reduction, hydroquinone methylation, and peroxidase-catalyzed oxidation reactions. The key intermediate trihydroxybenzene XIX is ring-cleaved to produce ß-ketoadipic acid. This oxidation, reduction, and methylation cycle apparently is responsible for the degradation of both chlorophenols and PCDDs by *P. chrysosporium*.

Mechanism of C-O-C Bond Cleavage

The nature of the products generated during the oxidation of DCDD XVI by LiP and our understanding of the mechanism of LiP enable us to propose a mechanism for DCDD cleavage (Valli et al. 1992b. C-O-C bond cleavage occurs via the attack of H_2O on cation intermediates (Figure 3). The initial formation of the cation radical A (Figure 3) is followed by the attack of water and the release of one chloride atom. The carbon-centered radical intermediate B is oxidized by LiP/MnP to a cation C which is attacked by water with the cleavage of the first C-O-C bond. Oxidation of the resultant phenol D yields a second carbon-centered radical E', which would be oxidized to a cation F. Finally, F would be attacked by H_2O resulting in the cleavage of the second C-O-C bond and the generation of the monomeric products XVII and XVIII.

LiP Oxidation of 2-Chlorodibenzo-*p*-dioxin XXII

In order to confirm the mechanism of C-O-C bond cleavage of 2,7-dichlorodibenzo-*p*-dioxin XVI we have examined the LiP catalyzed cleavage of

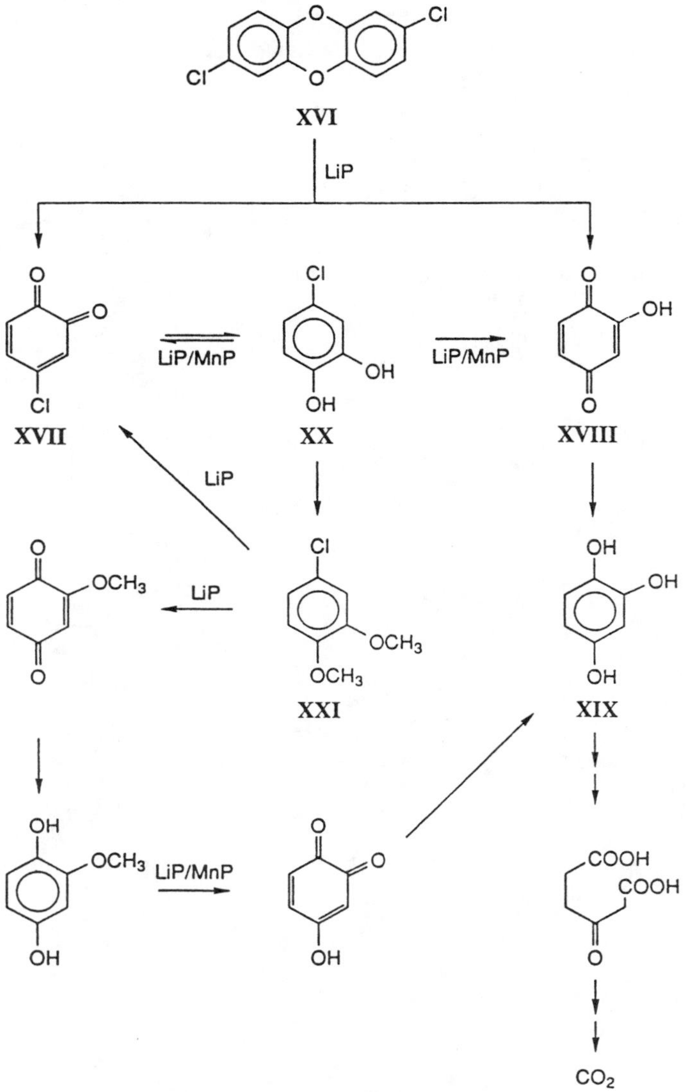

FIGURE 2. Proposed pathway for the degradation of 2,7-dichlorodibenzo-*p*-dioxin XVI.

2-chlorodibenzo-*p*-dioxin XXII. The LiP oxidation of XXII yields four monomeric products that, after reduction, have been identified as catechol XXIII, 4-chlorocatechol XX, trihydroxybenzene XIX, and tetrahydroxybenzene VIII. The nature of these LiP oxidation products as well as the results of our ^{18}O incorporation experiments indicate that LiP is able to catalyze C-O-C bond cleavage of 2-chlorodibenzo-*p*-dioxin XXII in a manner similar to that proposed

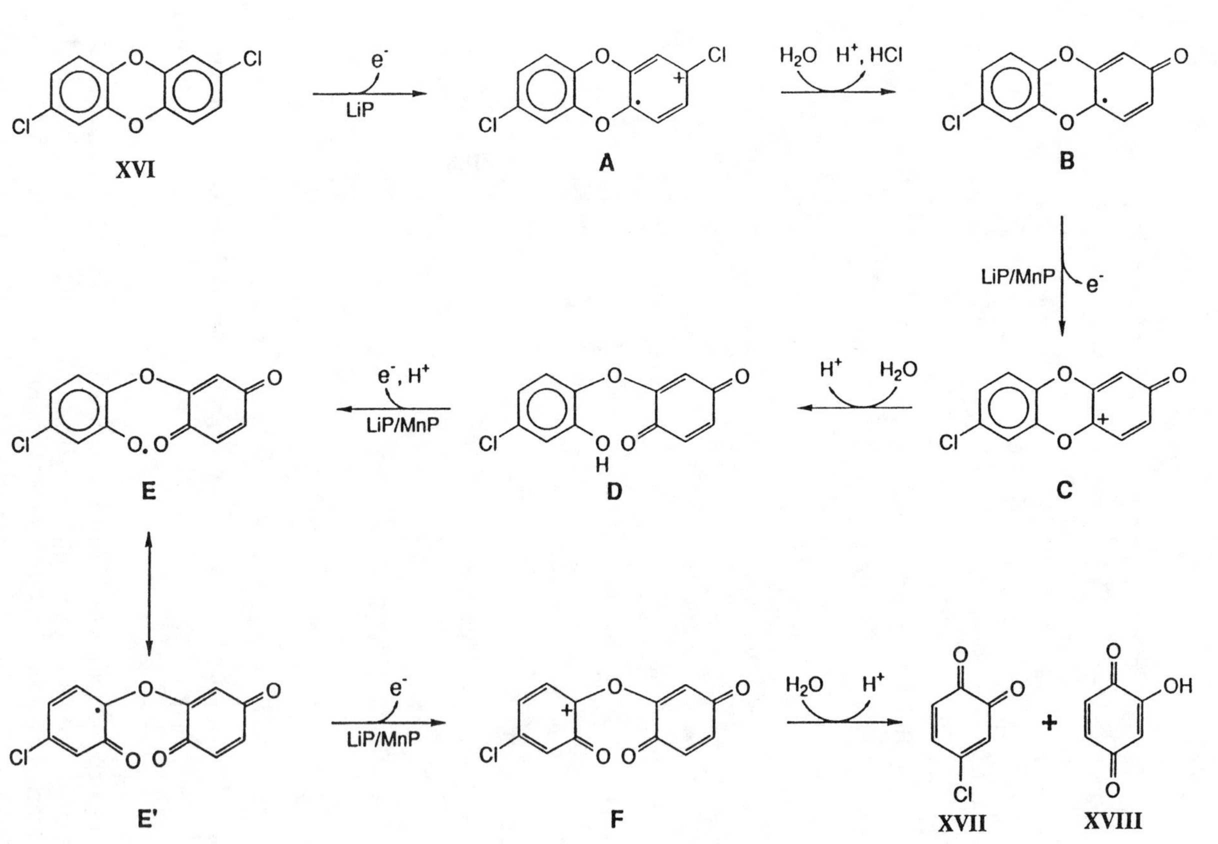

FIGURE 3. **Proposed mechanism for the lignin peroxidase-catalyzed dioxin ring cleavage of 2,7-dichlorodibenzo-*p*-dioxin.**

for the cleavage of XVI (Valli et al. 1992b). The reaction sequence is initiated by the LiP catalyzed oxidation of the substrate XXIIto an aryl cation radical.

We plan to continue to investigate the enzymes involved in the *P. chrysosporium* degradation of chlorinated aromatic pollutants.

ACKNOWLEDGMENTS

This work was supported by the U.S. National Science Foundation, the U.S. Department of Energy (Office of Basic Energy Sciences), the U.S. Department of Agriculture, and Battelle Pacific Northwest Laboratories.

REFERENCES

Bumpus, J. A., and S. D. Aust. 1987a. "Biodegradation of environmental pollutants by the white rot fungus *Phanerochaete chrysosporium*: Involvement of the lignin-degrading system." *Bioessays 6*: 166-170.

Bumpus, J. A., and S. D. Aust. 1987b. "Biodegradation of DDT [1,1,1-trichloro-2,2-bis(4-chlorophenyl)ethane] by the white rot fungus *Phanerochaete chrysosporium*. *Appl. Environ. Microbiol. 53*: 2001-2008.

Freiter, E. R. 1979. "Chlorophenols." In H. F. Mark, D. F. Othmer, C. G. Overberger, and G. T. Seaborg (Eds.), *Encyclopedia of Chemical Technology*, Vol. 5, 3rd ed., pp. 864-872. John Wiley & Sons, Inc., New York, NY.

Gold, M. H., H. Wariishi, and K. Valli. 1989. "Extracellular peroxidases involved in lignin degradation by the white rot basidiomycete *Phanerochaete chrysosporium*." In J. R. Whitaker and P. E. Sonnet (Eds.), *Biocatalysis in Agricultural Biotechnology*, pp. 127-140. ACS Symposium Series No. 389, American Chemical Society, Washington, DC.

Hammel, K. E. 1989. "Organopollutant degradation by lignolytic fungi." *Enzyme Microb. Technol. 11*: 776-777.

Hammel, K. E., B. Green, and W. Z. Gai. 1991. "Ring fission of anthracene by a eukaryote." *Proc. Natl. Acad. Sci. USA 88*: 10605-10608.

Hansen, D. J. 1991. "Dioxin toxicity: New studies prompt debate, regulatory action." *Chem. Eng. News 69*: 7-14.

Joshi, D., and M. H. Gold. 1992. "Degradation of 2,4,5-trichlorophenol by the lignin-degrading basidiomycete *Phanerochaete chrysosporium*." Submitted to *Appl. Environ. Microbiol.*

Keither, L. H., and W. A. Telliard. 1979. "Priority pollutants. I. A perspective view." *Environ. Sci. Technol. 13*: 416-423.

Kimbrough, R. D., and A. A. Jensen (Eds.). 1979. *Halogenated Biphenyls, Terphenyls, Naphthalenes, Dibenzodioxins and Related Products*, 2nd ed. Elsevier, New York, NY.

Kirk, T. K., and R. L. Farrell. 1987. "Enzymatic 'combustion': The microbial degradation of lignin." *Annu. Rev. Microbiol. 41*: 465-505.

Lamar, R. T., M. J. Larsen, and T. K. Kirk. 1990. "Sensitivity to and degradation of pentachlorophenol by *Phanerochaete* sp." *Appl. Environ. Microbiol. 56*: 3519-3526.

Rappe, C. 1980. "Chloroaromatic compounds containing oxygen: phenols, diphenyl ethers, dibenzo-*p*-dioxins, and dibenzofurans." In O. Hutzinger (Ed.), *The Handbook of Environmental Chemistry*, pp. 157-179. Springer-Verlag KG, Berlin.

Valli, K., B. J. Brock, D. Joshi, and M. H. Gold. 1992a. "Degradation of 2,4-dinitrotoluene by the lignin-degrading fungus *Phanerochaete chrysosporium*." *Appl. Environ. Microbiol. 58*: 221-228.

Valli, K., and M. H. Gold. 1991. "Degradation of 2,4-dichlorophenol by the lignin-degrading fungus *Phanerochaete chrysosporium*." *J. Bacteriol. 173*: 345-352.

Valli, K., H. Wariishi, and M. H. Gold. 1992b. "Degradation of 2,7-dichlorodibenzo-*p*-dioxin by the lignin-degrading basidiomycete *Phanerochaete chrysosporium*." *J. Bacteriol. 174*: 2131-2137.

Valo, R., V. Kitunen, M. Salkinoja-Salonen, and S. Räisänen. 1984. "Chlorinated phenols as contaminants of soil and water in the vicinity of two Finnish sawmills." *Chemosphere 13*: 835-844.

FIELD EVALUATIONS OF THE REMEDIATION OF SOILS CONTAMINATED WITH WOOD-PRESERVING CHEMICALS USING LIGNIN-DEGRADING FUNGI

R. T. Lamar and J. A. Glaser

ABSTRACT

The utility of lignin-degrading fungi in remediation of soils contaminated with wood preserving chemicals was evaluated in two independent field investigations. In the first study, the abilities of two lignin-degrading fungi to decrease the pentachlorophenol (PCP) concentration in a strongly alkaline (pH 9.6) sandy gravel soil, contaminated with a commercial wood preservative, were evaluated. Inoculation of soil that contained 250 to 400 $\mu g\ g^{-1}$ PCP with either *Phanerochaete chrysosporium* or *P. sordida* resulted in overall decreases of 88% to 91% of PCP in 6.5 weeks. Fungal inocula consisted of wood chips thoroughly grown through with *P. chrysosporium* or *P. sordida*. The second study was a treatability study conducted at a closed pole treatment facility where PCP and creosote had been used for 40 years. The abilities of the lignin-degrading fungi *P. chrysosporium*, *P. sordida*, and *Trametes hirsuta* to remove PCP and polyaromatic hydrocarbon (PAH) components of creosote from a strongly acidic (pH 3.8) clay soil were evaluated. Inocula consisted of pure cultures of each organism grown on a nutrient-fortified sawdust-grain mixture. The best result was obtained by inoculation of soil that contained 672 $\mu g\ g^{-1}$ PCP and 4017 $\mu g\ g^{-1}$ total PAHs with *P. sordida*. This treatment resulted in an 89% and 75% decreases in PCP and total measured PAHs, respectively. Future directions for further development of a fungal-based soil remediation technology are discussed in light of the results of these studies.

INTRODUCTION

The ability of lignin-degrading fungi to transform a wide variety of hazardous organic compounds to detoxified products has generated interest in using these organisms to treat hazardous materials. We have been developing a fungal technology to remediate soils contaminated with hazardous organics, specifically the wood preservatives pentachlorophenol (PCP) and creosote. The fungal technology works by inoculating contaminated soil with selected species of lignin-degrading fungi that colonize the soil from the inoculum and degrade the contaminants. Results of this work have demonstrated that inoculation of PCP-contaminated soils with such organisms causes a rapid and extensive decrease in PCP concentrations in a variety of soils under both laboratory (Lamar et al. 1990a, Lamar et al. 1990b) and field conditions (Lamar & Dietrich 1990, Lamar et al. unpublished). The results of our work (Davis et al. unpublished) and those of others (Loske et al. 1990) have also demonstrated the ability of these organisms to facilitate the removal of polyaromatic hydrocarbon (PAH) components of creosote.

This paper summarizes the results of two field investigations in which the abilities of several lignin-degrading fungi to deplete PCP and/or creosote from contaminated soils were evaluated. A perspective on the state of development and performance of this technology is offered from experience gained from these studies.

SITE DESCRIPTIONS

The first study was conducted in Oshkosh, Wisconsin, at the site of a former tank farm where a very strongly alkaline (pH 9.6) gravelly sand soil was contaminated with a commercial wood preservative product that contained 84% mineral spirits, 0.8% to 1% paraffin wax, 10% alkyd varnish, and 5% technical grade PCP as the active ingredient. Analyses of soil samples prior to the study showed mineral spirits concentrations ranging from 860 to 6730 $\mu g\ g^{-1}$ and PCP in concentrations ranging from 73 to 227 $\mu g\ g^{-1}$.

The second study was conducted at a site in Brookhaven, Mississippi, where the Escambia Treating Company operated a pole treatment facility from 1946 to 1986. In January 1986, the facility was sold to the employee-owned firm of Brookhaven Wood Preserving Inc. Escambia Treating Company retained ownership of 5.8 ha that included a hazardous waste management unit (HWMU), where K001 sludge accumulated. This material is bottom sediment sludge from the treatment of wastewaters from wood-preserving processes that use creosote or PCP (CFR 1990). The HWMU is an unlined impoundment constructed of a 1- to 2-m-high earthen dike. The impoundment was used to evaporate process wastewater from 1972 to 1984. Escambia Treating Company used two surface impoundments, a condenser pond and a sludge waste pit that were located near the wood-treating process area. Soil and sludge from these impoundments were excavated to a depth exceeding 2.5 m and the material was placed in the southwest corner of the HWMU. This excavated material constitutes the bulk of the waste sludge pile that was the selected soil for this study.

MATERIALS AND METHODS

The Oshkosh study evaluated the abilities of two lignin-degrading fungi, *Phanerochaete chrysosporium* and *P. sordida*, to deplete PCP in the upper 25 cm of the contaminated soil over a 6.5-week period. Preliminary laboratory experiments demonstrated that the high concentration of mineral spirits prevented the growth of *P. chrysosporium*. Therefore, after removal of the gravel pad, the study area was rototilled and the soil was physically mixed to a depth of 30 cm to equalize the concentration of PCP in the soil and to facilitate volatilization of the mineral spirits to a concentration that would allow fungal growth. After soil homogenization, plot borders constructed of galvanized steel and measuring 1 m^2 × 35 cm deep were worked into the soil surface and filled with soil to a depth of 25 cm.

Plots consisting of the two fungal treatments *P. chrysosporium* and *P. sordida* and corresponding controls were established (Table 1). The study was a completely randomized design (CRD), with treatments randomly allocated to experimental plots. The limited availability of test plots allowed the replication of only the two fungal treatments, F1 and F2. Fungal inocula consisted of wood chips thoroughly colonized by either *P. chrysosporium* or *P. sordida*. Inocula (fungal treatments) and sterile chips (control treatments) were applied to the soil at a rate of 3.35% (w/w, dry). Peat, used as a supplemental carbon source, was applied at a rate of 1.93% (w/w, dry). Treatments were applied by tilling the inoculum, sterile chips, and/or peat into the upper 30 cm of soil. The soil water potential in each plot was adjusted to −0.05 MPa with tap water, and the plots were covered with polyethylene to prevent excessive moisture loss. The covers were kept in place except when taking soil samples, when adjusting the soil water potential, and during weekly aeration by manual tilling. Analytical procedures for determination of soil PCP and pentachloroanisole (PCA) concentrations are detailed in Lamar and Dietrich (1990).

At the Brookhaven site, the effects of seven fungal and three control treatments on PCP and PAH concentrations in the soil were assessed in two complementary studies for a 2-month period (Table 2). The first six treatments were assessed in 3-m^2 plots using a CRD. Treatments 7 through 11 were evaluated in 1-m^2 plots in

TABLE 1. Treatment composition for the Oshkosh study.

Treatment	Composition
1	*Phanerochaete chrysosporium* [(a)] + peat[(b)]
2	*Phanerochaete sordida* + peat
3	Wood chips + peat
4	Wood chips[(c)]
5	Peat
6	No amendments

(a) Fungal inocula consisted of aspen wood chips (1.5 by 0.5 by 0.25 cm) thoroughly colonized by *P. chrysosporium* or *P. sordida*. Inoculum was applied to the soil at a rate of 3.35% (w/w,dry).

(b) The peat was sterilized by autoclaving, had a pH of 4.0, and was applied to the soil at a rate of 1.93% (w/w,dry).

(c) Wood chips were sterilized by autoclaving and applied to the soil at a rate of 3.35% (w/w, dry).

TABLE 2. Treatments in the completely random design (CRD) and the balanced incomplete block design (BIB) studies.

Treatment	Inoculum[(a)]	Inoculum Loading Level[(b)] (%)
CRD Treatments		
1	*Phanerochaete chrysosporium*	5
2	*Phanerochaete chrysosporium*	10
3	*Phanerochaete sordida*	10
4	*Phanerochaete chrysosporium*	5
	and *Trametes hirsuta*	5
5	Standard substrate	10
6	No amendment	—
BIB Treatments		
7	*Phanerochaete chrysosporium*	10
8	*Phanerochaete chrysosporium*	13
9	*Phanerochaete chrysosporium*	10 (Day 0)
	Reinoculation with *P. chrysosporium*	3 (Day 14)
10	*Trametes hirsuta*	10
11	Chips only	—

(a) Inocula consisted of pure cultures of each of the three fungi grown on a nutrient-fortified lignocellulosic substrate.
(b) Loading levels are defined as dry weight of inoculum or standard substrate to dry weight of soil on a percentage basis.

a balanced incomplete block (BIB) design, with each treatment replicated four times. Fungal inocula were prepared commercially by aseptic culturing of each of three fungi on a fortified lignocellulosic substrate that is commonly used for commercial mushroom culture. In addition to the inoculum, aspen chips, sterilized with fumigant, were applied to soil at a loading level of 2.5% (w/w, dry) to all plots except the no-treatment control 6. Plots were constructed with a leachate collection system that was overlain with sand. Target soil was excavated from the waste sludge pile, screened through a 1.9-cm mesh screen, mixed to homogenize the contaminants, and applied to the test plots to a depth of 25 cm before treatment application. The day following contaminated soil addition to the plots, sterile aspen (*Populus tremuloides* Michx.) chips were tilled into the soil. One day after chip application, fungal and control treatments were initiated. Tilling equipment was rinsed between application of different fungal strains to avoid cross contamination. Soil water contents were kept above 20% (dry weight basis). Plots were aerated by tilling weekly to a depth of 20 cm. Analytical methods for determination of soil PCP and PCA concentrations are given in Lamar et al. (unpublished). Methods for PAH analyses are given in Davis et al. (unpublished).

RESULTS

Oshkosh

At the beginning of the field study, the soil at the Oshkosh site contained 250 to 400 µg g^{-1} PCP. Inoculation with either *P. chrysosporium* or *P. sordida* resulted in an 82 to 86% decrease of PCP in the soil in 6.5 wk (Figure 1). The results correspond to an 88 to 91% decrease in PCP concentration when corrected for adsorption by wood chips. This decrease was achieved in spite of suboptimal temperatures for the growth and activity of these fungi and without the addition of inorganic nutrients other than those provided by the peat. The lack of a significant difference in the percentage decrease of PCP resulting from inoculation with *P. chrysosporium* compared to inoculation with *P. sordida* indicates the two fungi had similar abilities to deplete PCP in the soil under the conditions encountered during the study. The percentage decrease of PCP in soil receiving the fungal treatments was greater than that in control treatments beginning on day 15. This difference became statistically significant at day 29 and indicates that the dramatic decrease of PCP in soil inoculated with *P. chrysosporium* or *P. sordida* was due to activity of these fungi.

A small percentage of the decrease in the amount of PCP was a result of fungal methylation of PCP to pentachloroanisole (PCA). However, only about 9 to 14% of the PCP was converted to PCA. Thus, methylation was not the major route of PCP transformation and depletion in the contaminated soil.

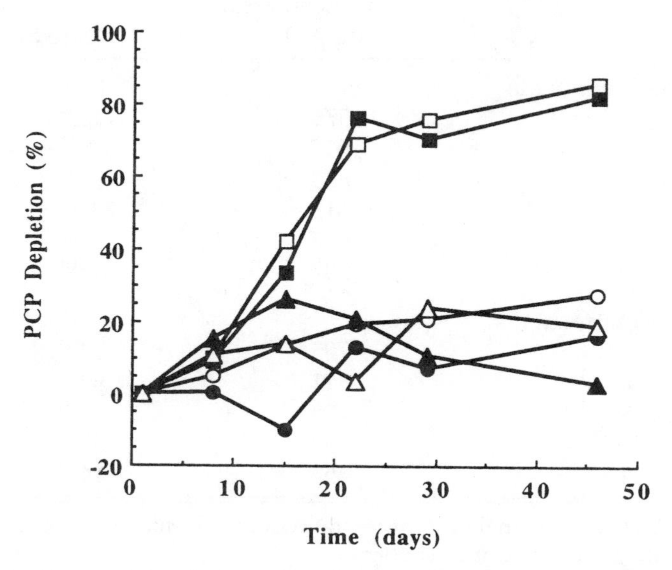

FIGURE 1. The effect of fungal and control treatments on percentage PCP depletion over time in the Oshkosh soil. *Phanerochaete chrysosporium* (□), *P. sordida* (■), chips and peat (○), chips (•), peat (▲), and no treatment (△).

Brookhaven

Initial PCP concentrations and the percentage PCP decrease in fungal and control treated soils after 8 weeks of treatment are shown in Table 3. The greatest depletion (89%) of PCP occurred in soil inoculated with *P. sordida* at an inoculum loading level of 10%. Inoculation with *P. chrysosporium* at 10% also resulted in significant decreases (67% to 72%). Changes in PCP concentrations in soils that received the other fungal treatments were less than in soils receiving the treatments mentioned above. There were no significant changes in the PCP concentrations over the entire 8-week period in soils receiving control treatments (treatments 5, 6, and 11). Thus, the decreases observed in the fungal inoculated soils is attributable to the activity of the fungi.

The effects of treatment with *P. sordida* and the control treatments 5 and 6 on priority pollutant PAH residues of the co-contaminating creosote were also examined (Table 4). Decreases in the concentrations of 3- and 4-ring targeted PAHs were consistently greater after fungal treatment than after control treatments. The 5- and 6-ring PAHs persisted at original concentrations in both fungal and control treatments. However, only 3% of the total weight of the analytes

TABLE 3. Initial PCP concentrations[(a)] and percentage PCP decrease[(b)] after 8 weeks in the Brookhaven study.

Treatment	Initial Concentration ($\mu g\ g^{-1}$)	% Decrease
CRD Treatments		
1	576	15[(c)]
2	1017	67[(a,b)]
3	673	89[(a)]
4	615	23[(b,c)]
5	687	14[(c)]
6	737	15[(c)]
BIB Treatments		
7	705	72[(a)]
8	311	52[(a)]
9	428	55[(a)]
10	399	55[(a)]
11	458	–14[(b)]

(a) Each value for intial PCP concentration and percentage decrease is the mean of 24 observations.

(b) Percentage decrease in the concentration of PCP from 1 day after treatment application. Within experiment, percentage decreases followed by the same letter are not significantly different according to Tukey's multiple comparison procedure = 0.05).

TABLE 4. The effect of 8 weeks of a fungal and control treatment on concentrations of 3- and 4-ring PAHs in the Brookhaven study.

Compound	Initial Concentration[(a)] ($\mu g\ g^{-1}$	% Decrease [(b)] Treatment		
		6	5	3
Acenaphthene	429	47	68	95
Fluorene	225	75	57	95
Phenanthrene	941	69	49	90
Anthracene	684	57	48	85
Fluoranthene	972	23	42	72
Pyrene	572	10	22	52
Benzo(a)anthracene	74	11	13	24
Chrysene	90	6	14	33

(a) Concentration of PAHs in soil samples taken 1 day after treatment application.
(b) Each value for initial concentration and percentage decrease is the mean of 24 observations.

in the waste sludge soil consisted of 5- or 6-ring compounds. Total measured PAHs decreased by 75% after 8 weeks of treatment with *P. sordida*.

Pentachloroanisole was not detected in soil samples taken before treatment application, or at any time in samples taken from control treated soils. However, PCA was observed to accumulate in all fungal treated soils. Accumulation of PCA was coincident with PCP decrease. In both the CRD and BIB studies, the greatest amount of PCA accumulated in soils inoculated with *P. chrysosporium* at 10%. After peak accumulations, there were decreases in the concentration of PCA in the soil in all fungal inoculated plots except those inoculated with *P. chrysosporium* at 5% or *T. hirsuta* at 10%, where PCA accumulations were negligible. The greatest PCA decrease (81%) was observed in soil inoculated with *P. sordida* where the PCA concentration decreased from 84 mg g^{-1} to 16 mg g^{-1} between day 28 and day 56. After 8 weeks, conversion to PCA accounted for only about 3% of the overall 89% PCP decrease in soils inoculated with this fungus. In contrast, 10 to 15% of the PCP decrease in soils inoculated with *P. chrysosporium* at 10% was due to conversion to and accumulation of PCA.

DISCUSSION

The results of these field investigations demonstrate that (1) lignin-degrading fungi can be successfully inoculated into and cultivated in soils contaminated with relatively high concentrations of PCP (up to 1,000 $\mu g\ g^{-1}$) alone or in combination with creosote (4,000 $\mu g\ g^{-1}$ total measured PAHs) and (2) concentrations of PCP and low-molecular-weight PAHs are greatly reduced as a result of the activity of these fungi. The abilities of these organisms to decrease the concentration of

PCP in two soils with very diverse chemical and physical characteristics lends support for the continued development of a fungal-based treatment technology to remediate soils contaminated with organic pollutants. Several important components of the fungal technology require additional development before this technology can become commercially viable. First, fungal inocula that are easy to apply must be able to be produced economically. Next, continued investigation of contaminant fate in fungal inoculated soils and fungal strain selection is required. Finally, the extent of contaminant degradation must be improved.

Currently, the production and delivery of the inoculum is the most expensive component (currently $1.32/kg) of the technology. The development of an inexpensive and effective fungal inoculum is crucial to successful implementation of fungal-based soil remediation. Wood chips and a nutrient-fortified mixture of sawdust and grains have served well as substrates for fungal inocula in these field studies. These materials provide a niche in which the lignin-degrading fungi can outcompete indigenous microbes and survive the toxic environment of the contaminated soil. For large-scale applications, the delivery and application of the necessary quantities of high-quality inocula made with these materials would be difficult to economize in their present form. Novel production and inoculum formulations involving lignocellulosic or other substrates are expected to result in economical inocula that more effectively take advantage of the niche (i.e., the substrate(s)) and growth habits of these organisms.

The effectiveness of remediation treatments is currently assessed by measuring, through direct chemical analysis, the extent of contaminant(s) removal and/or by determining if contaminants are transformed to innocuous products. Very little is known about the fate of contaminants in soil due to the activity of lignin-degrading fungi. We have found that *P. chrysosporium* removes PCP per se from soils primarily by converting it to soil-bound products (PCP-humic material copolymers) (Lamar et al. 1990a). Similarly, *P. chrysosporium* enhanced the polymerization of the creosote component benzo[a]pyrene into soil humic materials (Qiu & McFarland 1991). For certain contaminants the fungal technology works, apparently by transforming pollutants by decreasing their bioavailability through stabilizing them by polymerization with soil organic matter. This type of transformation may not be the fate for all the hazardous compounds that these organisms are able to degrade. The nature and stability of the such pollutant-humic copolymers or soil-bound residues is not known. Work with chemically synthesized humic acid-xenobiotic copolymers suggests that xenobiotics bound to humic materials through enzymatic polymerization reactions are relatively stable (Bollag & Liu 1985, Haider & Martin 1989). The nature, stability, and toxicity of the soil-bound transformation products, under a variety of conditions, require continued evaluation.

More than 1,500 species of lignin-degrading fungi have been estimated to exist in North America alone (Gilbertson 1980). Most of the research on the pollutant-degrading abilities of these organisms has been focused on relatively few species and on *P. chrysosporium* in particular. Fungal strains with superior biotransformation abilities are expected to be identified through simple screening procedures.

Although the most effective fungal treatments in the Oshkosh and Brookhaven studies decreased PCP concentrations by approximately 90%, a significant amount of PCP remained in the soil at the end of each study. Insufficient removal of contaminants is a common drawback of bioremediation treatments in general. In both field studies, the ultimate extent of PCP removal probably was restricted by the extremely low soil temperatures encountered toward the end of the investigations because of the negative effect of these temperatures on fungal growth and activity, and therefore the length of the treatment period. However, the residual PCP simply may have been unavailable to the fungi. In addition to providing optimum environmental conditions for fungal growth and activity, techniques such as inclusion of surfactants as part of the treatment to facilitate the availability of contaminants to fungal or other microbial degradation systems may facilitate more complete contaminant removal.

In summary, the development of a technology that employs lignin-degrading fungi to remediate contaminated soils is promising. Actual implementation of the technology on a commercial scale will require further developments in inoculum production and in techniques to improve the effectiveness of the treatment.

REFERENCES

Bollag, J.-M., and S.-Y. Liu. 1985. "Formation of hybrid oligomers between anthropogenic chemicals and humic acid derivatives." *Organic Geochemistry 8*:131.

CFR. 1990. "Identification and Listing of Hazardous Waste." Code of Federal Regulations, Volume 40, Part 261. Revised July 1, 1990. Office of the Federal Register, National Archives and Records Administration.

Davis, M. W., J. A. Glaser, and R. T. Lamar. 1993. "Ability of lignin-degrading fungi to remove PAH components of creosote in contaminated soil." *Environmental Science and Technology* (Submitted).

Gilbertson, R. L. 1980. "Wood-rotting fungi of North America." *Mycologia 72*:1.

Haider, K. M., and J. P. Martin. 1989. "Mineralization of 14C-labelled humic acids and of humic acid-bound 14C xenobiotics by *Phanerochaete chrysosporium*." *Soil Biology and Biochemistry 20*:425.

Lamar, R. T., J. W. Evans, and J. A. Glaser. 1993. "Use of lignin-degrading fungi in the removal of PCP from contaminated soil: A field treatability study." *Environmental Science and Technology* (Submitted)

Lamar, R. T., and D. M. Dietrich. 1990. "In situ depletion of pentachlorophenol by *Phanerochaete* spp. from contaminated soil." *Applied and Environmental Microbiology 56*:3093.

Lamar, R. T., J. A. Glaser, and T. K. Kirk. 1990a. "Fate of pentachlorophenol (PCP) in sterile soils inoculated with the white-rot basidiomycete *Phanerochaete chrysosporium*: Mineralization, volatilization and depletion of PCP." *Soil Biology and Biochemistry* 22:433.

Lamar, R. T., M. J. Larsen, and T. K. Kirk. 1990b. "Sensitivity to and degradation of pentachlorophenol (PCP) by *Phanerochaete* spp." *Applied and Environmental Microbiology 56*:3519.

Loske, D., A. Huttermann, A. Majcherczky, F. Zadrazil, H. Lorsen, and P. Waldinger. 1990. "Use of white-rot fungi for the clean-up of contaminated sites." In M. P. Coughlan and M. T. Amaral Collaco, (Eds.), *Advances in Biological Treatment of Lignocellulosic Materials.* Elsevier Science, London.

Qiu, X., and M. J. McFarland. 1991. "Bound residue formation in PAH contaminated soil composting using *Phanerochaete chrysosporium*." *Hazardous Waste and Hazardous Materials 8*:115.

A FIELD AND MODELING COMPARISON OF IN SITU TRANSFORMATION OF TRICHLOROETHYLENE BY METHANE UTILIZERS AND PHENOL UTILIZERS

L. Semprini, G. D. Hopkins, and P. L. McCarty

INTRODUCTION

In situ aerobic bioremediation of aquifers contaminated with chlorinated aliphatic hydrocarbons (CAHs) via cometabolism is a promising means of restoring aquifers. In situ cometabolic transformation of CAHs has been studied in a shallow aquifer at a pilot test facility at the Moffett Naval Air Station, California. At that site, in situ biotransformation of trichloroethylene (TCE), *cis*-1,2-dichloroethylene (*c*-DCE), *trans*-1,2-dichloroethylene (*t*-DCE), and vinyl chloride (VC) by a microbial population grown on methane and oxygen (methane-utilizing) was observed (Semprini et al. 1990, 1991). At the same site, the biotransformation of TCE and *c*-DCE by a microbial population grown on phenol and oxygen (phenol-utilizing) was recently observed (Hopkins et al. 1993). Modeling studies of the methane-utilizing biotransformations have been performed (Semprini & McCarty 1991, 1992). This model was used here to evaluate the results from a recent study with phenol-utilizing microorganisms. The biological rate parameters determined from the model simulations for the two different studies are compared.

BIOTRANSFORMATION MODEL

A non-steady-state model (Semprini & McCarty 1991, 1992) was developed to simulate the results from the methane-utilizing field experiments. The model incorporated advection, dispersion, sorption with rate limitation, and the microbial processes of substrate utilization, microbial growth, and CAH cometabolic transformation using competitive inhibition kinetics. The transport was simplified by assuming one-dimensional, uniform flow as a computational compromise to permit more rigorous representation of the biological processes.

Details of the simulations of methane-utilizing biotransformation studies are provided by Semprini and McCarty (1991, 1992). The results indicated that the responses observed in the field could be simulated well by the model, providing that competitive inhibition of the CAH transformation by methane and rate-limited sorption and desorption from the aquifer solids were included. A mechanism for deactivation of the microorganism to the cometabolic transformation also was included in the model formulation.

The same model, with adjusted coefficients as appropriate, was used to simulate the results from the phenol study. The initial population of phenol-utilizing bacteria was allowed to vary as an unconstrained fitting parameter. Both the biological parameters and the rate coefficients for the CAH transformations used were adjusted within a range anticipated from literature values to yield model simulations similar to field observations. The model outputs included the response to biostimulation (phenol and dissolved oxygen [DO] uptake), and the biotransformation of the CAHs as described in the following.

DESCRIPTION OF THE FIELD EXPERIMENTS

The field experiments were performed under induced gradient conditions of injection and extraction. Groundwater was extracted at a rate of 10 L/min; a portion (1.5 L/min) was amended at the surface with the chemicals of interest, including phenol as an electron donor, DO as an electron acceptor, and the target CAH contaminants, and was injected into the test zone. TCE and *c*-DCE, the two CAHs common to both studies, were not present in the groundwater and were continuously added in a controlled manner. The concentration response of phenol, DO, TCE, and *c*-DCE were monitored at the downgradient monitoring wells located 1, 2.2, and 3.8 m from the injection well. Bromide was added as a conservative tracer to determined advective transport velocities and to evaluate the extent of transformation of the TCE and *c*-DCE.

TCE and *c*-DCE were injected until first the test zone was saturated with respect to sorption onto the aquifer solids. Biostimulation was then accomplished through the addition of phenol and DO, while continuing to add the CAHs to the test zone. The phenol and DO were added in alternating pulses to limit biofouling near the injection well and to distribute the microbial growth over the test zone. Phenol was added for 1 hr of an 8-hr pulse cycle, initially at a concentration of 50 mg/L to yield a time-averaged concentration of 6.25 mg/L. After 21 days the phenol injection concentration was increased to 100 mg/L, yielding a time-averaged concentration of 12.5 mg/L.

FIELD AND SIMULATION RESULTS

The measured and simulated phenol and DO response at the S1 well following injection into the test zone are shown in Figure 1. A gradual decrease in DO was observed during the first 20 days of phenol injection. Phenol was infrequently observed at the S1 well, as it was difficult to catch the short high-concentration pulses of only 1 hr in an 8-hr pulse cycle. It was more frequently observed during the first 10 days of injection, with most observations below the detection limit after that, suggesting a phenol-utilizing population had been biostimulated. Model simulations support these observations. The model predicts oscillations in both

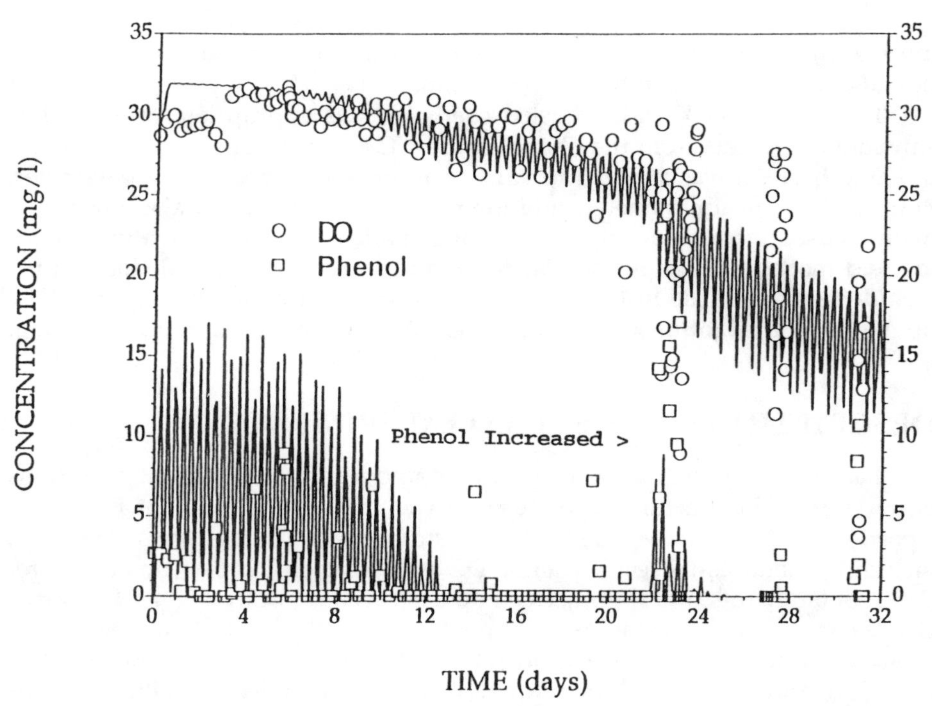

FIGURE 1. Phenol and DO concentration histories at the S1 well and model simulations.

phenol and DO due to the pulsed addition of phenol. After 10 days of injection, the simulated phenol oscillations are greatly reduced due to the biostimulation, with concentrations reductions below the detection limit. The model also simulated a gradual decline in DO, with oscillations in DO increasing during the first 12 days as more phenol was utilized, consistent with the field observations. Upon increasing the phenol injection concentration at 21 days, the DO concentration both decreased and oscillated over a greater concentration range. Here, high concentrations of phenol were observed on days 22 and 23, both because of the higher concentration and the increased monitoring performed at S1. The simulations showed a similar response. Additional increased monitoring around day 28 showed the oscillations in phenol concentration were then more attenuated than immediately after the phenol increase in both model simulations and the field.

The response of *c*-DCE and TCE at the S1 well due to biostimulation is shown in Figures 2 and 3, respectively; *c*-DCE was more rapidly transformed and to a greater extent than TCE. The model simulations showed similar concentration responses. Here CAH cometabolic transformation rates were adjusted somewhat to better fit the observed concentration response. The concentrations of CAHs decreased during the first 20 days of biostimulation, with the most rapid decreases occurring during the first 12 days. Biotransformation was enhanced upon doubling the phenol concentration at 21 days. This enhancement resulted from an increase in biomass. The observed and predicted oscillations in CAH concentration are the result of competitive inhibition between the CAHs and the pulsing phenol.

DISCUSSION

The model developed for the methane-utilizing bacteria provides reasonable simulations to the responses observed in the phenol-utilizing test. The results suggest that similar processes need to be considered with phenol as with methane, including rate-limited sorption, competitive inhibition, and activation and deactivation of the microbial population towards CAH degradation.

The results are similar to those obtained with methane. The model simulations indicate that the cometabolic transformation of *c*-DCE and TCE was associated with the growth of a phenol-utilizing population; *c*-DCE was transformed at a faster rate than TCE. Oscillations in CAH concentration were due to the pulsed addition of phenol, competitive inhibition of the transformation rates, and rate-limited sorption and desorption.

The phenol-utilizing population was more effective at degrading *c*-DCE and TCE than the methane-utilizing population. In the methanotrophic studies, approximately 20 to 30% of the TCE and 45 to 50% of the *c*-DCE were degraded

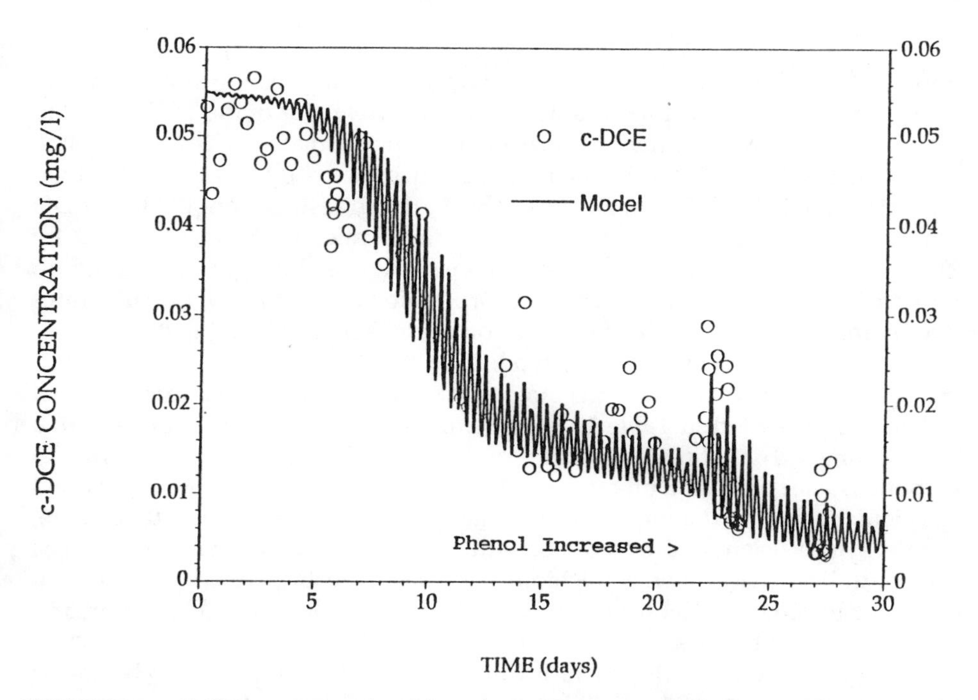

FIGURE 2. *c*-DCE concentration histories at the S1 well and model simulations.

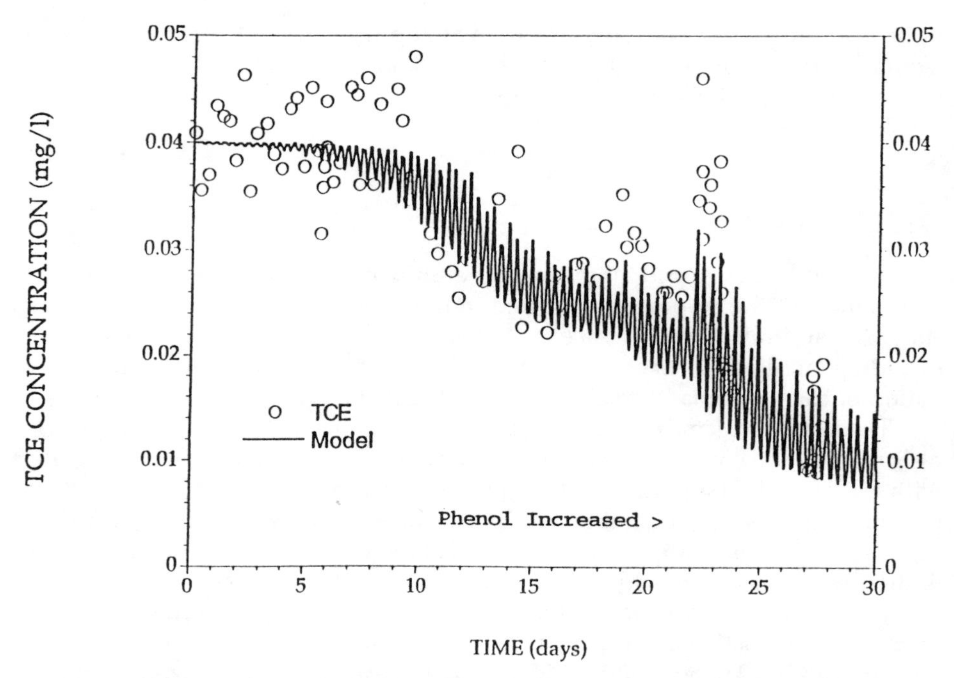

FIGURE 3. TCE concentration histories at the S1 well and model simulations.

upon passage through a 2-m biostimulated zone, while injecting approximately 6 mg/L methane. In the phenol study, approximately 40% of the TCE and 70 to 75% of the *c*-DCE were degraded while injecting approximately 6.25 mg/L phenol.

There was less DO consumption per mass of donor (1.6 g DO/g phenol) than with methane (2.7 g DO/g methane). This partially results from the lower DO requirement for growth and complete substrate oxidation to CO_2 and H_2O of 2.4 g O_2/g phenol, compared to 4.0 g O_2/g methane. The even lower DO consumption observed in the field results from the incorporation of a portion of the substrates into biomass. The lower DO consumption permits more phenol to be added to the aquifer per unit mass of oxygen addition.

A comparison of model-fitted biological parameters for methane and phenol utilizers is presented in Table 1. In general the biological parameters are in a similar range. The estimated initial population of phenol-utilizing bacteria (X_i) is about a factor of 5 greater than that of the methane-utilizing bacteria. The fitted cell yield (Y) was greater for the phenol-utilizers, while the same growth rates (k), cell decay coefficients (b), and donor saturation constants (K_s), were used. The phenol-utilizers had a lower stoichiometric ratio of electron acceptor to electron donor requirement for biomass synthesis (F). The higher Y resulted in a greater phenol-utilizing biomass generated for an equivalent mass of donor injected, compared to methane. The lower F value permitted more phenol to be added for as equivalent amount of DO injected.

TABLE 1. Comparison of biological parameters used in the model simulations.

Parameter	Methane	Phenol Utilizers
X_i (mg/L cells)	0.15	1.0
Y (g cells/g donor)	0.5	0.8
k (g donor/g cells-day)	1.0	1.0
b (day^{-1})	0.10	0.10
K_s (mg donor/L)	1.0	1.0
F (g acceptor/g donor)	2.4	0.8
k_2 (TCE) (d^{-1})	0.01	0.015
K_{s2} TCE (mg/L)	1.0	0.5
k_2 *c*-DCE (day^{-1})	0.05	0.05
K_{s2} *c*-DCE (mg/L)	1.0	0.5
k_2/K_{s2} TCE (L/mg-day)	0.01	0.03
k_2/K_{s2} *c*-DCE (L/mg-day)	0.05	0.10

For the CAH concentration range studied (0.05 to 0.1 mg/L), the biotransformation rate coefficient is given by the ratio k_2/K_{s2}, which was a factor of 3 and was 2 greater for phenol-utilizing bacteria than for the methane-utilizing bacteria for TCE and *c*-DCE, respectively. There is great uncertainty in this estimate, because the fitted parameter depends on many other factors. The comparison, however, suggests that the more effective transformation of TCE and *c*-DCE by phenol-utilizers probably results from the combination of higher rates of transformation and a larger biostimulated microbial population.

ACKNOWLEDGMENTS

This study was supported by the U.S. Environmental Protection Agency (U.S. EPA) Robert S. Kerr Environmental Research Laboratory and the U.S. EPA BioSystems Program, the U.S. Department of Energy, and the U.S. EPA-supported Western Region Hazardous Substance Research Center through grant no. CR-815738. The paper has not been subjected to peer review by these agencies and thus official endorsement of the results and conclusions should not be inferred.

REFERENCES

Hopkins, G. D., L. Semprini, and P. L. McCarty. 1993. "Microcosm and In Situ Field Studies of Enhanced Biotransformation of Trichloroethylene by Phenol-Utilizing Microorganisms." *Applied and Environmental Microbiology* (in press).

Semprini, L., G. D. Hopkins, P. V. Roberts, D. Grbić-Galić, and P. L. McCarty. 1991. "A Field Evaluation of In Situ Biodegradation of Chlorinated Ethenes: Part 3, Studies of Competitive Inhibition." *Ground Water* 29(2):239-250.

Semprini, L., and P. L. McCarty. 1991. "Comparison Between Model Simulations and Field Results for In Situ Biorestoration of Chlorinated Aliphatics: Part 1. Biostimulation of Methanotrophic Bacteria." *Ground Water 29*(3):365-374.

Semprini, L., and P. L. McCarty. 1992. "Comparison Between Model Simulations and Field Results for In Situ Biorestoration of Chlorinated Aliphatics: Part 2. Cometabolic Transformations." *Ground Water 30*(1):37-44.

Semprini, L., P. V. Roberts, G. D. Hopkins, and P. L. McCarty. 1990. "A Field Evaluation of In Situ Biodegradation of Chlorinated Ethenes: Part 2, Results of Biostimulation and Biotransformation Experiments." *Ground Water 28*(5):715-727.

BIOREMEDIATION OF CHLORINATED SOLVENTS IN THE VADOSE ZONE

D. H. Kampbell and B. H. Wilson

INTRODUCTION

Spills of chlorinated solvents to subsurface materials frequently are treated by soil vacuum extraction to remove volatile liquids and vapors in soil air, or by air stripping of contaminated groundwater. These physical treatment processes produce a wastestream of contaminant vapors in air that are discharged to the atmosphere and treated with activated carbon, catalytic combustion, or incineration.

The bioventing technology is widely used to remediate petroleum-derived hydrocarbons. In theory, a corresponding procedure may be used to biologically treat chlorinated solvents that may be cometabolized during microbial growth on another hydrocarbon. Many naturally occurring bacteria in soils and sediments are capable of oxidizing alkanes and aromatics to carbon dioxide. Both groups of compounds have been implicated in the biotransformation of trichloroethylene (TCE), vinyl chloride (VC), the dichloroethylenes, and 1,1,1,-trichloroethane.

TCE, chloroform, 1,1,1,-trichloroethane, 1,2-dichloroethylene, and dichloromethane (methylene chloride) can be biologically cometabolized during microbial growth on a variety of compounds (methane, propane, toluene) that exist as vapors and can be delivered to the subsurface environment through the flow of air (Henson et al. 1988, Wackett & Gibson 1988, Wilson & White 1986). Chlorobenzene, dichloromethane, 1,2-dichloroethane, and VC also can serve as primary substrates for microbial growth (Davis & Carpenter 1990, Janssen et al. 1985, Rittmann & McCarty 1980). Although removal in the presence of a primary substrate may be more rapid, these compounds do not necessarily require a primary substrate for biodegradation. There is no known pathway for aerobic biodegradation of highly oxidized compounds, such as tetrachloroethylene or carbon tetrachloride, due to the number of chlorines in their structure.

The chlorinated solvents, primarily TCE, are toxic to the microorganisms that cometabolize them. The toxic effects of TCE become significant above 6 mg/L water or 2 mg/L air (Table 1, Broholm et al. 1991, Broholm et al. 1990). Air or water in contact with residual-phase TCE frequently exceeds these toxic limits. The toxic effects of VC have not been well documented; concentrations of 1.0 mg/L air generally are tolerated in laboratory microcosms (Table 2). Presented below are preliminary results of microcosm studies that examine the removal of TCE and VC in unsaturated soils using JP-4 or aviation gasoline vapors as the primary substrate for microbial growth.

TABLE 1. Effect of the concentration of TCE on the rate of biodegradation of aviation gasoline[(a)] vapors in soil microcosms.

TCE Concentration		Gasoline Vapor
(mg/kg soil)	(mg/L air, if all TCE volatilized)	Degradation (mg/kg day)
45,600		0.06
4,500	1,000	0.11
13	3	24
4.1	0.9	26
1.2	0.3	31
0.0	0.0	33

(a) The hydrocarbons in gasoline support the cometabolism of TCE.

MICROCOSM STUDIES

The results of laboratory studies using unsaturated soil acclimated to gasoline vapors to remove TCE or VC are presented in Table 2. Soil microcosms were constructed using uncontaminated sand and loam soils. The microcosms were constructed using approximately 35 g (wet weight) per 160 mL serum bottle. The bottles were sealed with a Teflon™-lined silicone septum and crimped with an aluminum cap. The microcosms were acclimated to JP-4 or aviation gasoline vapors with nutrient addition (N,P,K) for a period of at least 4 weeks. Nutrients were added at 0.2 mg each of diammonium phosphate and potassium nitrate into each microcosm. The microcosms were considered acclimated when the gasoline vapors were biologically removed in 2 to 6 hours. After acclimation, TCE vapors were added at a concentration of 17 µg/g soil, and VC vapors were added at a concentration of 4 µg/g soil. No primary substrate (JP-4 or aviation gasoline) was added with the TCE or VC vapors. The control microcosms contained soil that had not been acclimated to the primary substrates.

Removal of the chlorinated solvents was measured over a period of 1.3 to 4 days. Concentrations of JP-4, aviation gasoline, TCE, and VC were measured by headspace chromatography using flame ionization detection. Zero-order rates of removal for the chlorinated solvents were calculated as a difference in initial and final concentrations with respect to time and normalized to the soil mass. The control concentrations were used as the initial values to compensate for sorption to the soil, glass, or septum. Zero-order rates of removal for TCE ranged from 0.53 to 2.0 µg/g soil/day. Zero-order rates of removal for VC ranged from

1.35 to 1.75 μg/g soil/day. These rates of removal are consistent with those calculated from other studies (Speitel & Closmann 1991, Wilson & Kampbell 1993a).

CONCLUSION

Chlorinated solvents can be biologically cometabolized in unsaturated soils using a variety of volatile alkanes and/or aromatics as primary substrates. Uncontaminated soils of various textures and from different locations were able to successfully treat TCE or VC using the volatile constituents of gasoline as primary substrates. These results are in agreement with previous work showing removal of TCE using a mixture of propane and butanes as the primary substrate (Kampbell et al. 1987). The potential for substantial biological treatment of chlorinated solvents in the vadose zone or in prepared soil bed bioreactors does exist. Because acclimation of contaminated soils may be lengthy (Broholm et al. 1991, Speitel & Closmann 1991, Wilson & Kampbell 1993b), this potential treatment may perform more effectively in a prepared soil bed bioreactor starting with

TABLE 2. Removal of vapors of TCE and VC in subsurface material under optimal conditions.

Source	Soil Type	Substrate and Nutrients	Initial Solvent Conc. mg/kg soil	Initial Solvent Conc. μg/L air	Mineral-ization Rate (mg/kg soil/day)
Trichloroethylene					
Tucson, Arizona	Sand	0.6% gasoline vapors, N,P,K	*17*	4,200	1.47
St. Joseph, Michigan	Sand	0.6% gasoline vapors, N,P,K	*17*	4,200	2.0
Racine, Wisconsin	Organic Rich Loam	0.6% gasoline vapors, N,P,K	*17*	4,200	0.53
Vinyl Chloride					
Racine, Wisconsin	Loam	0.6% gasoline vapors, N,P,K	*4.0*	1,000	1.75
Tucson, Arizona	Sand	0.6% gasoline vapors, N,P,K	*4.0*	1,000	1.35

uncontaminated soil or in an uncontaminated portion of the unsaturated subsurface acclimated to treat vapors extracted from the contaminated portion.

REFERENCES

Broholm, K., T. H. Christensen, and B. K. Jensen. 1991. "Laboratory feasibility studies on biological in-situ treatment of a sandy soil contaminated with chlorinated aliphatics." *Environ. Technol. 12*:279-289.

Broholm, K., B. K. Jensen, T. H. Christensen, and L. Olsen. 1990. "Toxicity of 1,1,1-trichloroethane and trichloroethene on a mixed culture of methane-oxidizing bacteria." *Appl. Environ. Microbiol. 56*(8):2488-2493.

Davis, J. W., and C. L. Carpenter. 1990. "Aerobic biodegradation of vinyl chloride in groundwater samples." *Appl. Environ. Microbiol. 56*(12):3868-3880.

Henson, J. M, M. V. Yates, J. W. Cochran, and D. L. Shackleford. 1988. "Microbial removal of halogenated methanes, ethanes, and ethylenes in an aerobic soil exposed to methane." *FEMS Microbiol. Ecol. 52*(3-4):193-201.

Janssen, D. B., A. Scheper, L. Dijkhuizen, and B. Witholt. 1985. "Degradation of halogenated aliphatic compounds by *Xanthobacter autotrophicus* GJ10. *Appl. Environ. Microbiol. 49*(2):673-677.

Kampbell, D. H., J. T. Wilson, H. W. Read, and T. T. Stocksdale. 1987. "Removal of volatile aliphatic hydrocarbons in a soil bioreactor." *J. Air Pollut. Cont. Assoc. 37*(10):1236-1240.

Rittmann, B. E., and P. L. McCarty. 1980. "Utilization of dichloromethane by suspended and fixed-film bacteria." *Appl. Environ. Microbiol. 39*(6):1225-1226.

Speitel, G. E., and F. B. Closmann. 1991. "Chlorinated solvent biodegradation by methanotrophs in unsaturated soils." *J. Environ. Eng. 117*(5):541-558.

Wackett, L. P., and D. T. Gibson. 1988. "Degradation of trichloroethylene by toluene dioxygenase in whole-cell studies with *Pseudomonas putida* F1. *Appl. Environ. Microbiol. 54*(7):1703-1708.

Wilson, B. H., and M. V. White. 1986. "A fixed-film bioreactor to treat trichloroethylene-laden waters from interdiction wells." In *Proceedings of The Sixth National Symposium and Exposition on Aquifer Restoration and Groundwater Monitoring*, pp. 425-435. National Water Well Association. Dublin, OH.

Wilson, J. T., and D. H. Kampbell. 1993a. "Bioventing of chlorinated solvents for ground water cleanup through bioremediation." In *In-situ Bioremediation of Ground Water and Geological Material: A Review of Technologies.* U.S. EPA, Office of Research and Development, R. S. Kerr Lab. Ada, OK.

Wilson, B. H., and D. H. Kampbell. 1993b. Unpublished data.

ADAPTATION OF BACTERIA TO CHLORINATED HYDROCARBON DEGRADATION

F. Pries, J. R. van der Ploeg, A. J. van den Wijngaard, R. Bos, and D. B. Janssen

INTRODUCTION: BIODEGRADATION OF HALOALIPHATICS

Many halogenated hydrocarbons are poorly degraded in the environment and in biological treatment systems as a result of the absence of productive metabolic routes in microorganisms. Examples of biochemical factors that can hinder degradation are (1) inefficient uptake systems, (2) lack of induction of the synthesis of specific catabolic enzymes, (3) absence of enzymes that can carry out the required conversion steps, (4) accumulation of toxic intermediates, and (5) partial conversion without generation of energy for growth.

Of the chlorinated aliphatic hydrocarbons, only a few components can be used as a growth substrate by bacterial cultures. Recalcitrance increases with a higher degree of chlorine substitution, and many attempts to isolate organisms that grow on compounds such as chloroform, 1,1,1-trichloroethane, and 1,1,2-trichloroethane have been unsuccessful. Dehalogenation is a critical step in the conversion of chlorinated aliphatic hydrocarbons. There is a correlation between the activity of dehalogenating enzymes and the ability of the organisms to use compounds as a growth substrate. Compounds not found to be used for growth under aerobic conditions usually are not hydrolyzed by the known dehalogenating enzymes (Janssen et al. 1987a, 1989).

DEGRADATION OF 1,2-DICHLOROETHANE

We have investigated the degradation of 1,2-dichloroethane (DCE), which is not known to occur naturally and is produced by the chemical industry in larger amounts than any other chlorinated compound. It can be converted by bacteria to glycolate in 4 steps (Figure 1). The catabolic pathway in *Xanthobacter autotrophicus* involves two dehalogenating enzymes, encoded by the *dhlA* and *dhlB* genes, which both have been cloned and sequenced (Janssen et al. 1989; van der Ploeg et al. 1991). The haloalkane dehalogenase is, together with an aldehyde dehydrogenase, encoded on a plasmid (Tardif et al. 1991). The same route for TCE degradation was found in different facultative methylotrophic bacteria,

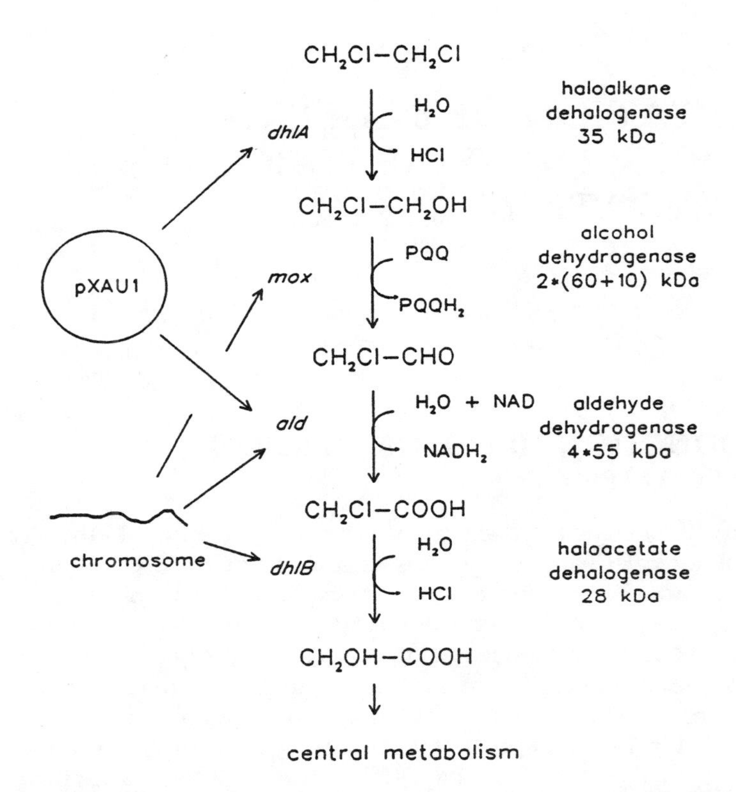

FIGURE 1. The catabolic pathway of 1,2-dichloroethane in *X. autotrophicus* and *A. aquaticus*.

including *Ancylobacter aquaticus* (van den Wijngaard et al. 1992). Strains of *A. aquaticus* selected on 2-chloroethyl vinyl ether possess the same catabolic route for DCE, indicating that the pathways for these compounds are closely related.

The metabolism of DCE involves both general enzymes and enzymes specific to the DCE specific pathway (i.e., present only in strains that are selected on chlorinated compounds), e.g., the chloroacetic acid dehalogenase. The biodegradation of 2-chlorocarboxylic acids is well established and proceeds by dehalogenation by a hydrolytic enzyme, yielding the corresponding hydroxycarboxylic acids. From the DNA sequence of the gene encoding the haloacid dehalogenase, it appeared that the *Xanthobacter* enzyme belongs to the group of L-specific chloropropionate dehalogenases that convert the substrate with inversion of configuration (van der Ploeg et al. 1991). These enzymes show considerable sequence similarity, and can be found in many bacteria that grow on chloroacetate, chloropropionate, and DCE, but do not occur in strains of *Xanthobacter* that do not grow on chloroaliphatics. Thus, one step in the evolution of a catabolic pathway for DCE was the introduction of a dehalogenase gene for chloroacetate.

Adaptation experiments with halocarboxylic acids have been described in Thomas et al. (1992). With *X. autotrophicus* strain GJ10, we found that selection for increased resistance to bromoacetate, which is toxic at a concentration of above 5 mM, yielded mutants that show more than 10-fold increased expression of the dehalogenase due to insertion of a DNA segment closely upstream of the dehalogenase gene (Table 1). This is surprising, because loss or reduced production of haloacid dehalogenase has been found in other cases (Strotmann et al. 1990; Weightman et al. 1985). The physiological cause of this type of adaptation remains

TABLE 1. Adaptation of bacteria that degrade halogenated aliphatic compounds.

Strain	Selection	Result
Loss of Unproductive Routes		
X. autotrophicus GJ10	1,2-dibromoethane resistance	loss of haloalkane dehalogenase
X. autotrophicus GJ10	1,2-dibromoethane resistance	loss of haloalkane dehalogenase[c]
Arthrobacter GJ70	2-bromoethanol resistance	loss of alcohol dehydrogenase[c]
Overexpression		
X. autotrophicus GJ10	bromoacetate resistance	overexpression of haloalkanoic acid[d] dehalogenase
Xanthobacter XD(pPJ66)[a]	bromoacetate resistance	recombination of haloalkanoic acid dehalogenase gene causing over-expression
Pseudomonas GJ1	bromoethanol resistance	aldehyde dehydrogenase overproduction
Enzyme Specificity		
Pseudomonas GJ31(pPJ20)[b]	chlorohexane utilization	modified haloalkane dehalogenase[e] specificity multiple mutations in the *dhlA* gene

(a) Plasmid pPJ66 encodes dhlB.
(b) Plasmid pPJ20 encodes *dhlA* (Figure 1).
(c) See Janssen et al. (1987a) for details.
(d) See van der Ploeg et al. (1991) for details.
(e) Pries et al. 1993.

to be established, but it is most probable that the toxicity of bromoacetic acid is reduced by rapid conversion to glycolate.

ALDEHYDES AS REACTIVE INTERMEDIATES

Haloaldehydes are reactive and toxic intermediates in the biological conversion of several chloroaliphatics, including vinyl chloride (Anders & Pohl 1985). During the conversion of DCE, chloroacetaldehyde is formed as an intermediate. *X. autotrophicus* GJ10 produces two dehydrogenases that are used to convert this compound. One of these enzymes is encoded on the catabolic plasmid (Tardif et al. 1991) and has a high activity towards chloroacetaldehyde. Loss of this enzyme caused extremely high toxicity of 2-chloroethanol (Janssen et al. 1987b).

The lack of an efficient aldehyde dehydrogenase for bromoacetaldehyde probably also explains the high toxicity (5 μM) of 1,2-dibromoethane. The latter compound is readily dehalogenated by different haloalkane dehalogenases, but completely blocks growth of organisms that produce the enzyme. We have found that resistance may occur by loss of the alcohol dehydrogenase that is responsible for the formation of bromoacetaldehyde from 2-bromoethanol. The mutants lacking the alcohol dehydrogenase (Table 1) show improved conversion of 1,2-dibromoethane to the nonhalogenated product ethylene glycol, which, however, did not serve as a growth substrate for the specific strain (Janssen et al. 1987a).

The first step in DCE conversion is the dehalogenation to 2-chloroethanol (Figure 1). The oxidation of 2-chloroethanol is catalyzed by the quinoprotein methanol dehydrogenase that is generally present in methylotrophic bacteria (Anthony 1986). We have found that reconstitution of a 2-chloroethanol catabolic pathway in a *Xanthobacter* strain that does not use this compound for growth can be achieved by introducing both a haloacid dehalogenase gene (*dhlB*) and the plasmid-encoded aldehyde dehydrogenase gene. The latter is required for overcoming the toxicity of 2-chloroethanol.

EVOLUTION AND ADAPTATION OF HALOALKANE DEHALOGENASE SPECIFICITY

The haloalkane dehalogenase is detected only in bacteria that degrade DCE or 2-chloroethyl vinyl ether, and is absent in other *Xanthobacter* isolates. From the three-dimensional structure (Franken et al. 1991), it was proposed that catalysis of chloroalkane hydrolysis proceeds by nucleophilic displacement with an aspartate as the nucleophile. The dehalogenase is not a general hydrolytic protein that fortuitously also converts xenobiotic compounds. Instead, it was evolved to specifically catalyze dehalogenation reactions with xenobiotic haloalkanes.

Various other xenobiotic chlorinated aliphatics can be converted by the enzyme, including ethylchloride, 1,3-dichloropropene, epichlorohydrin, chloroethyl vinyl ether, and chlorobutane. Long-chain chloroalkanes are poorly hydrolyzed. A natural substrate from which the 1,2-dichloroethane degradation ability has evolved is not easy to identify, but it could be methylchloride or dibromomethane, which are known to occur naturally.

The dehalogenase protein appeared to be composed of two domains. The main domain shares structural similarity with other hydrolytic proteins and can be considered as a generally occurring structure in a class of hydrolytic proteins (Ollis et al. 1992). The second domain, the cap domain, forms a separate part of the enzyme that is absent or different in other hydrolytic proteins of the α/β fold type.

We attempted to modify the specificity of haloalkane dehalogenase by selection of mutant enzymes (Table 1). Because 1-chlorohexane is a poor substrate for the wild type enzyme, recombinants of the alcohol-degrading strain *Pseudomonas* GJ31 harboring a plasmid with the *dhlA* gene on it (pPJ20) did not grow on chlorohexane. Selection experiments were carried out in batch cultures containing 1 mM 1-chlorobutane and 2 mM 1-chlorohexane. The appearance of mutants was indicated by growth of some of these cultures to higher cell densities than possible on medium containing only chlorobutane (Figure 2).

Twelve mutants containing a dehalogenase with a very different substrate specificity were characterized. Increased activity with long-chain-length (C5-C9) chloroalkanes was observed. Thus, a typical mutant had V_{max} values with 1-chlorohexane and 1-chloropentane that were 30- to 50-fold higher than with

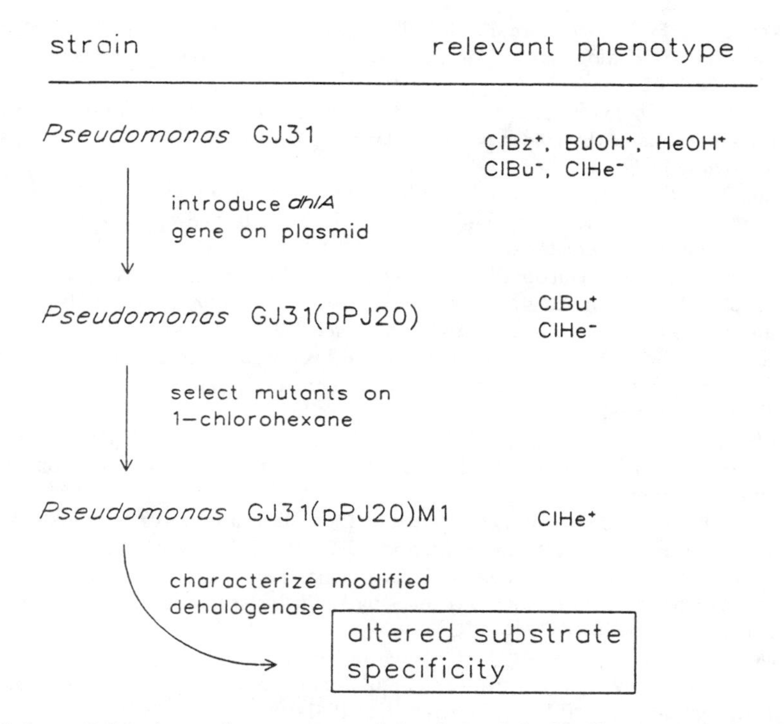

FIGURE 2. Selection of mutants with altered haloalkane dehalogenase specificity.

DCE, whereas the former two compounds were very poor substrates for the wild type enzyme (Pries et al. 1993).

The mutant genes were sequenced after amplification with the polymerase chain reaction and all mutations found in the cap domain of the dehalogenase. The mutants possibly affect the binding of long-chain chloroalkanes as a result of slight differences in the position of the active site residues. The results show that the cap domain plays an important role in modifying the specificity of the haloalkane dehalogenase and suggest that further modification of the activity of the enzyme can be achieved by mutating specific residues of the cap domain.

The effects of various mutations of amino acids forming the active site cavity of the enzyme and the tunnel leading to it are being investigated by site-directed mutagenesis. Modeling mutant proteins and the binding of various substrates will further help to suggest specific mutations that yield dehalogenase activity with substrates that are not hydrolyzed by the present enzyme.

REFERENCES

Anders, M. W., and L. R. Pohl. 1985. "Halogenated alkanes." In M. W. Anders (Ed.), *Bioactivation of Foreign Compounds*, pp. 283-315. Academic Press, New York, NY.

Anthony, C. 1986. "Bacterial oxidation of methane and methanol." *Adv. Microbial. Physiol. 27*: 113-210.

Franken, S. M., H. J. Rozeboom, K. H. Kalk, and B. W. Dijkstra. 1991. "Crystal structure of haloalkane dehalogenase: An enzyme to detoxify halogenated alkanes." *EMBO J. 10*: 1297-1302.

Janssen, D. B., D. Jager, and B. Witholt. 1987a. "Degradation of n-haloalkanes and α,ω-dihaloalkanes by wild-type and mutants of *Acinetobacter* sp. strain GJ70." *Appl. Environ. Microbiol. 53*: 561-566.

Janssen, D. B., S. Keuning, and B. Witholt. 1987b. "Involvement of a quinoprotein alcohol dehydrogenase and an NAD-dependent aldehyde dehydrogenase in 2-chloroethanol metabolism in *Xanthobacter autotrophicus* GJ10." *J. Gen. Microbiol. 133*: 85-92.

Janssen, D. B., F. Pries, J. van der Ploeg, B. Kazemier, P. Terpstra, and B. Witholt. 1989. "Cloning of 1,2-dichloroethane degradation genes of *Xanthobacter autotrophicus* GJ10, and expression and sequencing of the *dhlA* gene." *J. Bacteriol. 171*: 6791-6799.

Ollis, D. L., E. Cheah, M. Cygler, B.W. Dijkstra, F. Frolow, S. Franken, M. Hare, J. Memington, I. Silman, J. Schrag, J. Sussman, and A. Goldman. 1992. "The α/ß hydrolase fold." *Protein Engineering 5*: 197-211.

Pries, F., A. J. van den Wijngaard, R. Bos, and D. B. Janssen. 1993. "The role of spontaneous cap domain mutations in haloalkane dehalogenase specificity and evolution." Manuscript in preparation.

Strotmann, U., M. Pentenga, and D. B. Janssen. 1990. "Degradation of 2-chloroethanol by wild type and mutants of *Pseudomonas putida* US2." *Arch. Microbiol. 514*: 294-300.

Tardif, G., C. W. Greer, D. Labbe, and P. C. K. Lau. 1991. "Involvement of a large plasmid in the degradation of 1,2-dichloroethane by *Xanthobacter autotrophicus* GJ10." *Appl. Environ. Microbiol. 57*: 1853-1857.

Thomas, A. W., A. W. Topping, J. H. Slater, and A. J. Weightman. 1992. "Localization and functional analysis of structural and regulatory dehalogenase genes carried on *DEH* from *Pseudomonas putida* PP3." *J. Bacteriol. 174*: 1941-1947.

van den Wijngaard, A. J., K. van der Kamp, J. van der Ploeg, B. Kazemier, F. Pries, and D. B. Janssen. 1992. "Degradation of 1,2-dichloroethane by facultative methylotrophic bacteria." *Appl. Environ. Microbiol. 58*: 976-983.

van der Ploeg, J., G. van Hall, and D. B. Janssen. 1991. "Characterization of the halocarboxylic acid dehalogenase of *Xanthobacter autotrophicus* GJ10 and sequencing of the *dhlB* gene." *J. Bacteriol. 173*: 7925-7933.

Weightman, A. J., A. L. Weightman, and J. H. Slater. 1985. "Toxic effects of chlorinated alkanoic acids on *Pseudomonas putida* PP3: Selection at high frequencies of mutations in genes encoding dehalogenases." *Appl. Environ. Microbiol. 49*: 1494-1501.

POTENTIAL APPLICATIONS OF PHAGE DISPLAY TO BIOREMEDIATION

V. A. Petrenko and L. Makowski

INTRODUCTION

Phage display refers to the display of foreign peptides on the surface of bacteriophage particles (Scott & Smith 1990). A unique restriction site is engineered into the genome in a region corresponding to the surface of a structural protein. That site is used for the insertion of nucleic acid sequences that result in the formation of fusion proteins with the inserted peptide exposed on the surface of the phage particle. Specific sequences corresponding to peptides or proteins of known activity can be inserted into this site. Alternatively, random, chemically synthesized nucleic acid sequences can be inserted into the sites in a population of phage genomes to produce a library of fusion phages containing very large numbers of peptides (10^9 to at least 10^{12}). Each phage genome has a single random nucleic acid sequence inserted into it, and that particle displays the corresponding encoded peptide sequence on its surface. Screening of libraries for peptides with a particular, desired activity can be followed immediately by the growth of large quantities of the phage displaying the active peptide. Consequently, phage display provides a way to produce very large numbers of potential reactants, and to identify and immediately produce a particular reactant once isolated.

PHAGE DISPLAY VEHICLES

The most commonly used vehicle for phage display is the filamentous *Escherichia coli* phage, M13. The wild-type phage is about 65 Å in diameter and 0.9 μ long, with a total molecular weight of about 16 million. It is composed of 2,800 copies of pVIII (gene 8 protein), 5 copies of pIII, 5 copies each of three other structural proteins, and a single-stranded DNA genome (Makowski 1984, Model & Russel 1988). Figure 1 is a diagram of the structure of the phage particle with the demonstrated sites for peptide display indicated. The similar *Pseudomonas* phage, Pf1, may have special advantages over M13, as discussed below.

Several strategies for the display of peptides have been developed. Peptides may be inserted near the amino terminus of pVIII (Greenwood et al. 1991, Ilyichev et al. 1989). If peptides are inserted into all copies of pVIII, there appears to be a limit of about 6 amino acids for the length of the inserted peptide. The reason for this limit is not known, but may involve the initiation of viral assembly (Makowski

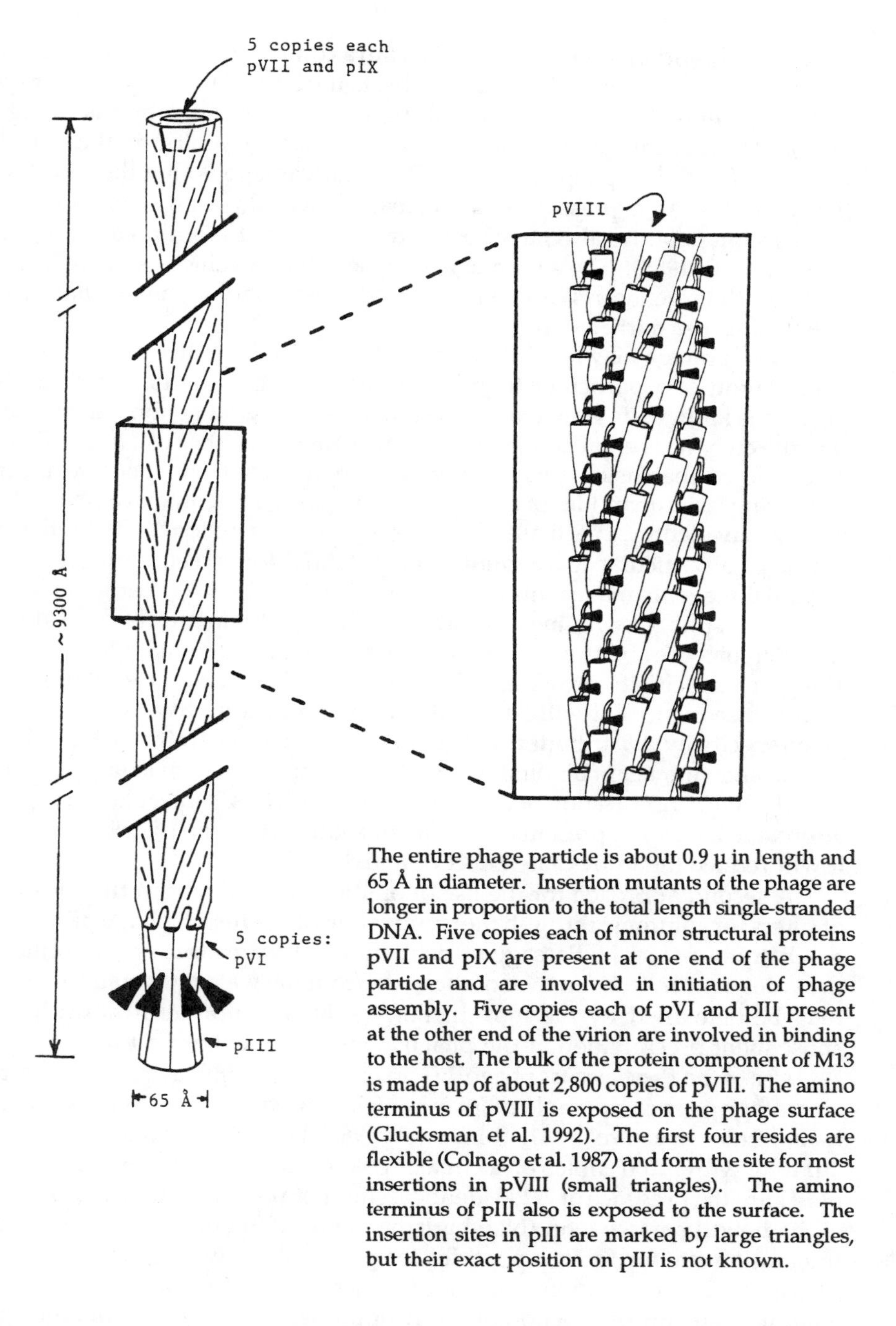

The entire phage particle is about 0.9 μ in length and 65 Å in diameter. Insertion mutants of the phage are longer in proportion to the total length single-stranded DNA. Five copies each of minor structural proteins pVII and pIX are present at one end of the phage particle and are involved in initiation of phage assembly. Five copies each of pVI and pIII present at the other end of the virion are involved in binding to the host. The bulk of the protein component of M13 is made up of about 2,800 copies of pVIII. The amino terminus of pVIII is exposed on the phage surface (Glucksman et al. 1992). The first four resides are flexible (Colnago et al. 1987) and form the site for most insertions in pVIII (small triangles). The amino terminus of pIII also is exposed to the surface. The insertion sites in pIII are marked by large triangles, but their exact position on pIII is not known.

FIGURE 1. Diagram of the structure of filamentous bacteriophage M13 with the sites for incorporation of foreign peptides marked. Figure adapted from Bhattacharjee et al. (1992).

1993). If a hybrid phase is constructed from a combination of native and mutant pVIII (Greenwood et al. 1991), it is possible to introduce large peptides into pVIII, including, for instance, whole variable regions of immunoglobulins (Kang et al. 1991). The advantage of using pVIII as the display vehicle is the very large number of sites per phage particle. The disadvantage is that the insert size is limited unless an expression system creating hybrid phage is used.

Foreign peptides also have been introduced into the amino terminal region of pIII (Smith 1985). The advantage of using pIII as a vehicle is the relative ease with which large inserts can be introduced. The disadvantage is that there are only 5 sites per particle.

Fusion phage may be constructed with activity for either the binding or the degradation of a particular toxin. If an enzyme for a particular degradation process is known, its gene may be inserted into the viral genome in such a way that the enzyme is expressed as a fusion protein with either pIII or pVIII. In this case, use of the fusion phage provides a convenient, inexpensive means of producing large quantities of reagent. Alternatively, screening of a phage display library may identify a particular fusion phage with catalytic activity for the degradation of a particular toxic agent. The potential for screening more than 10^{12} potential reagents makes this a real possibility (Jacobs 1991, Lerner et al. 1991).

Fusion phages with high affinity for a specific toxin are also useful for bioremediation. The ability of the immune system to construct antibodies with almost any given specificity provides strong evidence that, with a repertoire of more than 10^{12} sequences, it will be possible to find a fusion phage with almost any given specificity. Distribution of these phages over a contaminated site will introduce a large number of binding sites that will compete with toxin targets, reducing the effective concentration of the toxins. Although this "molecular sarcophagus" approach does not represent a permanent solution to toxin distribution at a site, it will reduce the immediate effect of a toxin.

A second approach for the use of binding specificity is in the construction of filters (or columns) through the conjugation of the fusion phage particles with a substrate material. Passage of contaminated water through the filter then provides a means for removal of the toxin from the water. Contaminated water may be detoxified by the introduction of the appropriate fusion phage and subsequent precipitation of the phage along with the bound toxin.

Use of the *Pseudomonas* phage, Pf1, in bioremediation may provide substantial advantage over comparable display vehicles based on M13. Pf1 is very similar in molecular structure to M13 (Makowski 1984), being 65 Å in diameter and about 2 μ long with 7,200 copies of the major coat protein. The potential advantages of Pf1 are the presence of 7,200 identical sites per particle, the existence of a non-alpha helical surface loop (Nambudripad et al. 1991) that may be an excellent integration site for foreign peptides (of potentially any size), and the possible distribution in association with *Pseudomonas*, a common soil bacterium, for the continued in situ production of the fusion phage particles. Although Pf1 has not yet been used as a vehicle for phage display, these advantages suggest that its use as a vehicle should be pursued.

Identification of fusion phages exhibiting a particular activity is relatively straightforward. Assay for binding is performed by a slight variation on standard affinity chromatography (Parmley & Smith 1988, Smith 1991). The toxin of interest is linked to, or absorbed onto, a substrate that is then used either as a column material in affinity chromatography, or through a modified procedure referred to as "biopanning" (Parmley & Smith 1988). Identification of phages with particular enzymatic activity requires the development of an assay and may not be substantially different from the isolation of a particular enzymatic activity from a tissue.

A key to identifying peptides with the desired activities is to produce vast libraries of fusion phages for screening. These libraries are created by inserting random oligonucleotide sequences into unique restriction sites in the viral genome. For instance, a library of 3×10^8 recombinants has been generated by cloning randomly synthesized oligonucleotides into a site in pIII (Ćwirla et al. 1990). We have produced a mutant M13, referred to as B1, with a unique Bam1 restriction site between codons 4 and 5 of pVIII (Ilyichev et al. 1989, 1990, 1991). Insertion of random nucleotide sequences into this site provides the possibility of creating a library of up to 20^N peptide sequences where N is the length of the insert. For an insert of 6 amino acids, this is 6.4×10^7 potential peptides. For 10 amino acids, this number rises to 10^{13}. Random oligonucleotides are produced by standard nucleotide synthesis techniques using equimolar mixtures of the nucleotides, biased to minimize the formation of stop codons. Libraries of random peptide sequences can be constructed in this manner and inserted into sites either in pVIII or pIII. An alternative strategy mimics the immune system by inserting random sequences into the hypervariable loops of Fab fragments and then inserting these randomly mutated Fab fragments into pVIII and pIII (Barbas et al. 1991).

CONCLUSION

Phage display offers the prospect of rapid identification of reagents that either bind environmental toxins with high affinity or degrade them. These reagents, located on the surface of viral particles, have the potential for application to a broad range of environmental problems.

ACKNOWLEDGMENTS

We would like to thank Dr. Theodore M. Prociv for suggesting that we apply phage technology to the problems of bioremediation. Funding for the basic research that forms a foundation for these ideas was provided by Russian Academy of Science, National Institutes of Health, and Nation Science Foundation. Collaboration between our two laboratories was made possible by a Fogarty International Research Award from the Public Health Service.

REFERENCES

Barbas III, C. F., A. S. Kang, R. A. Lerner, and S. J. Benkovic. 1991. "Assembly of combinatorial antibody libraries on phage surfaces: The gene III site." *Proc. Nat. Acad. Sci. USA 88*:7978.

Bhattacharjee, S., M. Glucksman, and L. Makowski. 1992. "Structural polymorphism correlated to surface charge in filamentous bacteriophage." *Biop. J. 61*:7625-735.

Colnago, L. A., K. G. Valentine, and S. J. Opella. 1987. "Dynamics of fd coat protein in the bacteriophage." *Biochemistry 26*:847-854.

Cwirla, S.E., E.A. Peters, R. W. Barrett, and W. J. Dower. 1990. Peptides on phage: A vast library of peptides for identifying ligands." *Proc. Natl Acad. Sci. USA 87*:6378.

Glucksman, J. J., S. Bhattacharjee, and L. Makowski. 1992. "Three-dimensional structure of a cloning vector. X-ray diffraction studies of filamentous bacteriophage M13 at 7 Å resolution." *J. Mol. Biol. 226*:455.

Greenwood, J., A. E. Willis, and R. N. Perham. 1991. "Multiple display of foreign peptides on a filamentous bacteriophage." *J. Mol. Biol. 220*:821-827.

Ilyichev, A. A., O. O. Minenkova, S. I. Tatkov, N N. Karpishev, A. M. Eroshkin, V. A. Petrenko, and L. S. Sandakhchiev. 1989. "Construction of M13 viable bacteriophage with the insert of foreign peptides into the major coat protein." *Dokl. Acad. Sci. USSR 307*:481.

Ilyichev A. A., O. O. Minenkova, S. I. Tatkov, N. N. Karpishev, A. M. Eroshkin, V. I. Ofitzerov, Z. A. Akimenko, V. A. Petrenko, and L. S. Sandakhchiev. 1990. "M13 filamentous bacteriophage for protein engineering." *Mol. Biol. (USSR) 24*:530.

Ilyichev, A. A., O. O. Minenkova, G. P. Kishchenko, S. I. Tatkov, N. N. Karpishev, A. M. Eroshkin, V. I. Ofitzerov, Z. A. Akimenko, V. A. Petrenko, and L. S. Sandakhchiev. 1992. "Inserting foreign peptides into the major coat protein of bacteriophage M13." *FEBS Letters 301*:322.

Kang, A. S., C. F. Barbas, K.D. Janda, S. J. Benkovic, and R. A. Lerner. 1991. "Linkage of recognition and replication functions by assembling combinatorial antibody Fab libraries along phage surfaces." *Proc. Nat. Acad. Sci. USA 88*:4363.

Lerner, R. A., S. J. Benkovic, and P. G. Schultz. 1991. "At the crossroads of chemistry and immunology: Catalytic antibodies." *Science 252*:659.

Makowski, L. 1984. "Structural diversity in filamentous bacteriophages." In A. McPherson (Ed.), *Biological Macromolecules and Assemblies*, Vol. 1, *The Viruses*, pp. 203-253. John Wiley and Son, New York, NY.

Makowski, L. 1993. "Structural constraints on the display of foreign peptides on filamentous bacteriophages." *Gene* (in press).

Model, P., and M. Russel. 1988. "Filamentous bacteriophage." In R. Calendar (Ed.), *The Bacteriophages*, Vol. 2, pp. 375-456. Plenum, New York, NY.

Nambudripad, R., W. Stark, and L. Makowski. 1991. "Neutron diffraction studies of the structure of filamentous bacteriophage Pf1. Demonstration that the coat protein consists of a pair of α-helices with an intervening, non-helical surface loop." *J. Mol. Biol. 220*:359.

Parmley, S. F., and G. P. Smith. 1988. "Antibody-selectable filamentous fd phage vectors: Affinity purification of target genes." *Gene 73*:305.

Scott, J. K., and G. P. Smith. 1990. "Searching for peptide ligands with an epitope library." *Science 249*:386.

Smith, G. P. 1985. "Filamentous fusion phage:novel expression vectors that display cloned antigens on the virion surface." *Science 228*:1315.

Smith, G. P. 1991. "Surface presentation of protein epitopes using bacteriophage expression systems." *Curr. Opinion. Biotech.* 2:668-673.

IMPROVING THE BIODEGRADATIVE CAPACITY OF SUBSURFACE BACTERIA

M. F. Romine and F. J. Brockman

INTRODUCTION

The continual release of large volumes of synthetic materials into the environment by agricultural and industrial sources over the last few decades has resulted in pollution of the subsurface environment. Cleanup has been difficult because of the relative inaccessibility of the contaminants caused by their wide dispersal in the deep subsurface, often at low concentrations and in large volumes. As a possible solution for these problems, interest in the introduction of biodegradative bacteria for in situ remediation of these sites has increased greatly in recent years (Timmis et al. 1988).

Selection of biodegradative microbes to apply in such cleanup is limited to strains that can survive among the native bacterial and predator community members at the particular pH, temperature, and moisture status of the site (Alexander 1984). The use of microorganisms isolated from subsurface environments would be advantageous, because these organisms are already adapted to subsurface conditions. The options are further narrowed to strains that are able to degrade the contaminant rapidly, even in the presence of highly recalcitrant anthropogenic waste mixtures, and in conditions that do not require addition of further toxic compounds for the expression of the biodegradative capacity (Sayler et al. 1990).

These obstacles can be overcome by placing the genes of well-characterized biodegradative enzymes under the control of promoters that can be regulated by inexpensive and nontoxic external factors and then moving the new genetic constructs into diverse groups of subsurface microbes. The objective of this research is to test this hypothesis by comparing the activity of two different toluene biodegradative enzymatic pathways from two different regulatable promoters in a variety of subsurface isolates.

CONSTRUCTION OF PLASMIDS

Two plasmids, pMMB66EH (Furste et al. 1986) and pNM185 (Mermod et al. 1986), were chosen as vehicles for transferring genes encoding recruited enzymes into selected subsurface isolates because they can be maintained in a wide variety of microorganisms. Each of these plasmids also encodes both antibiotic resistance markers, which allows us to monitor transfer of the plasmid into the subsurface isolates, and a regulatable promoter to which the recruited genes can be joined. The *Escherichia coli*-derived *ptac* promoter in pMMB66EH is negatively regulated by the product of the plasmid-encoded *lac*I gene. Repression of transcription from this promoter can be relieved by adding the chemical isopropylthiogalactoside (IPTG). The *Pseudomonas putida*-derived *ptol* promoter in pNM185 is positively regulated by the product of the plasmid-encoded *xyl*S gene and the coinducers benzoate and *m*-toluate.

Two enzymes, toluene dioxygenase (*tod*) and toluene-4-monooxygenase (*tmo*), which have the ability to degrade both toluene and trichloroethylene (TCE), were selected as the initial model systems for enzyme recruitment. The DNA encoding *tod* and *tmo* were recruited from *P. putida* F1 and *P. mendocina* KR1, respectively. Four plasmid vehicles were constructed as shown in Figure 1.

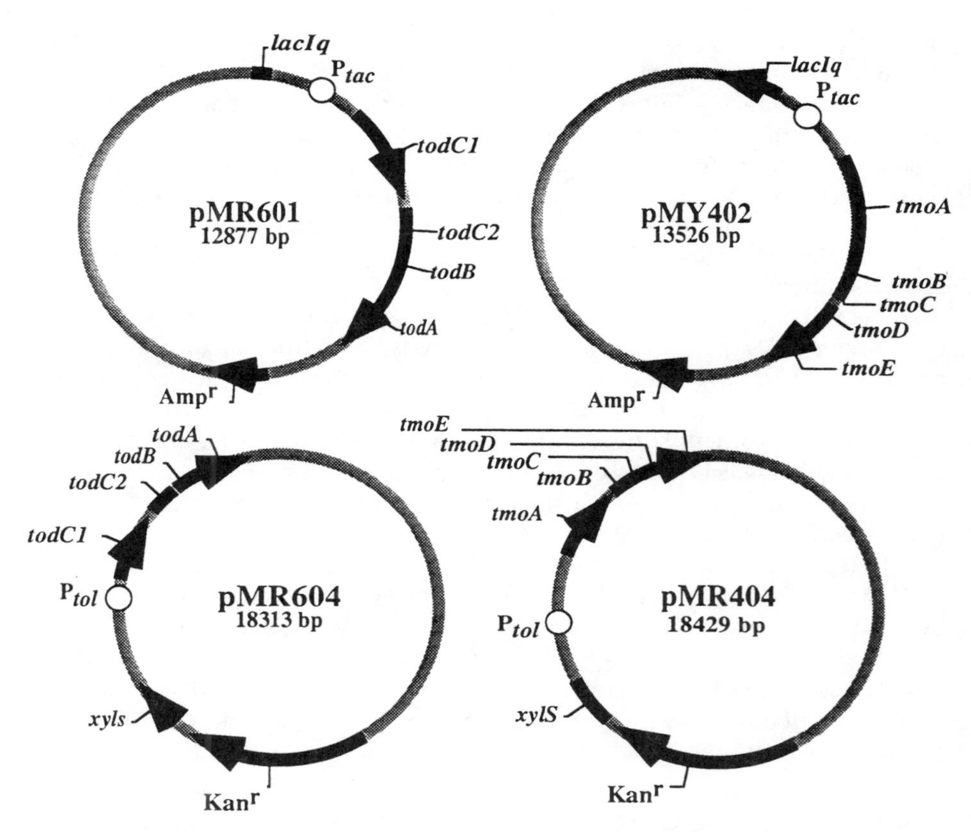

FIGURE 1. Plasmid constructs of recruited genes. The IPTG-regulatable pMMB66EH derivatives are pMR601, encoding the *tod* genes and pMY402, encoding the *tmo* genes. pMY402 was obtained from Burt Ensley at Amgen, Inc. The benzoate- or *m*-toluate-regulatable pNM185 derivatives are pMR604, encoding the *tod* genes and pMR404, encoding the *tmo* genes.

TRANSFER AND EXPRESSION OF RECRUITED ENZYMES IN SUBSURFACE BACTERIA

Fifty-five isolates from the P24 borehole at the Savannah River site in South Carolina were screened for inherent kanamycin and ampicillin resistance. Of the isolates, 29 were unable to grow on the antibiotic-containing media. Based on preliminary typing of these strains by R. Reeves and D. Balkwill at Florida State University, eight antibiotic-sensitive representative strains from different bacterial groups were chosen for enzyme recruitment. These strains are BO615, *Rhodospirillum rubrum*; BO724, *Arthrobacter globiformis*; BO265, *Alcaligenes eutrophus*; BO445, *Acinetobacter calcoaceticus*; BO669 and BO446, *P. testosteroni*; BO259, *P. aeruginosa*; and BO450 (currently not typed).

Cesium-purified plasmid DNA was electroporated into these subsurface isolates. Cells that had received the plasmid were identified by their ability to grow on antibiotic-containing media that they initially were unable to grow on. Then any antibiotic-resistant colonies were tested for presence of the newly acquired plasmid DNA by amplifying the *tod* and *tmo* genes by the polymerase chain reaction. Currently we have all four plasmid constructs in BO445, BO450, BO265, BO669, and BO259. We have successfully transferred pMR604 into BO724 and BO446 at this date. No colonies were formed by BO615 with any of the plasmids after electroporation.

EXPRESSION OF BIODEGRADATIVE CAPACITY

Expression of the *tod* and *tmo* genes was monitored by high-pressure liquid chromatography (HPLC) analysis of culture supernatants. Cultures grown in the presence of toluene and expressing *tod* or *tmo* would be expected to convert toluene to *cis*-dihydrodiol or *p*-cresol, respectively. Detection of toluene breakdown products was much more sensitive and reliable than measurement of toluene loss and therefore results are stated as accumulation *p*-cresol or *cis*-dihydrodiol.

Strains BO445, BO450, BO265, BO669, and BO259 containing each of the four plasmids were tested for their ability to degrade toluene. Culture supernatants were collected from cultures that were grown in minimal media containing 20 mM lactate and 100 ppm toluene. Identical cultures were also set up in the presence of the inducers, IPTG or *m*-toluic acid, at a final concentration of 1 mM.

The results of HPLC of the culture supernatants obtained after 36 hours of growth are depicted in Figures 2 and 3.

None of the parental strains devoid of these 4 plasmids degraded toluene (data not shown) nor did any strain carrying pMY402. Degradation of toluene by *tmo* was exhibited in 4 out of 5 strains and inducibly in 3 of these 4 strains. Degradation of toluene by *tod* was successful in 4 out of 5 strains in the presence of the pMR601 construct and in all five strains carrying the pMR604 construct. Toluene was completely removed from strain BO445 containing pMR601 (induced or noninduced) and by strain BO450 carrying pMR604 (induced or noninduced).

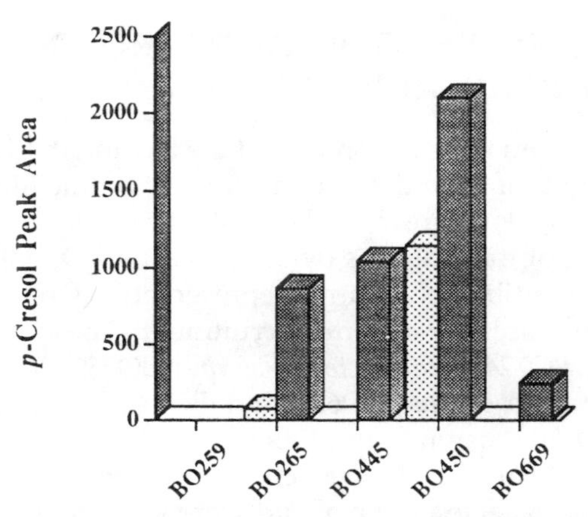

FIGURE 2. ***p*-cresol levels produced by strains BO265, BO259, BO445, BO450, and BO669, with plasmid pMR404 (lightly shaded bar), and pMR404 induced (darkly shaded bar).**

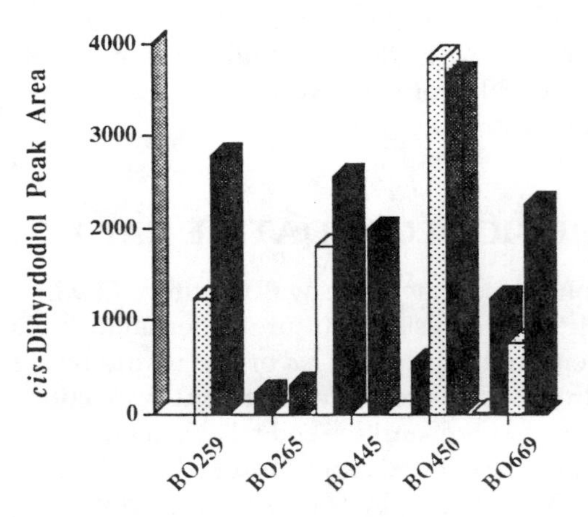

FIGURE 3. ***cis*-dihydrodiol levels produced by strains BO259, BO265, BO445, BO450, and BO669 with plasmids pMR601 (open bar), pMR601 induced (solid bar), pMR604 (lightly shaded bar), and pMR604 induced (darkly shaded bar).**

To examine the kinetics of the *tmo* and *tod* enzymes, a time-course assay was also performed with the parent and derivatives of BO450 that carried plasmid pMR404 (Figure 4) or pMR604 (Figure 5) using benzoic acid as the inducer. The results indicate that degradation of toluene by *tod* (pMR604) is much more rapid than degradation by *tmo* (pMR404) in BO450. After 21 hours only 0.25% of the initial level of toluene remained in the supernatant with pMR604, but more than 90% was still detectable in the supernatant with pMR404. The parental strain BO450 did not produce a detectable level of either compound; thus both plasmids exhibited a considerable level of constitutive activity.

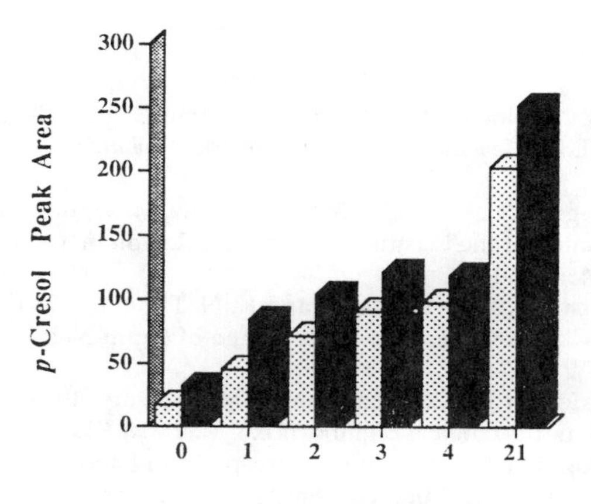

FIGURE 4. Time course of *p*-cresol production by pMR404 in BO450 when induced with benzoic acid (darkly shaded bar) or noninduced (lightly shaded bar).

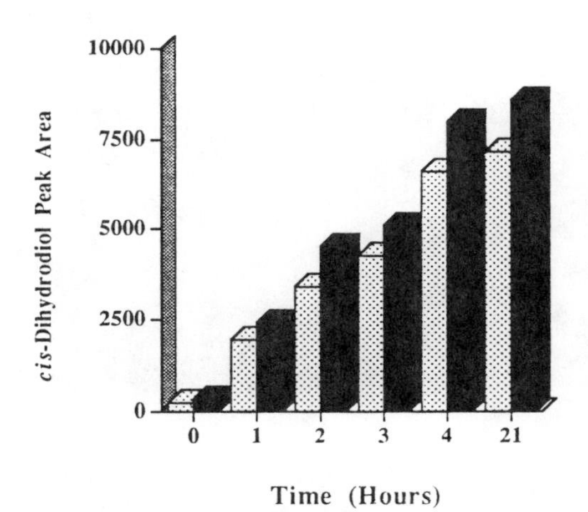

FIGURE 5. Time course of *cis*-dihydrodiol production by pMR604 in BO450 when induced with benzoic acid (darkly shaded bar) or noninduced (lightly shaded bar)

FUTURE RESEARCH

Preliminary results indicate that in some cases recruited *tod* and *tmo* enhanced the biodegradative capacity of subsurface bacteria. Additional experiments will test the effects of cell concentration, growth state, and toluene concentration both on degradation by the engineered strains and on the ability of these strains to degrade TCE. Strains will then be ready for testing in subsurface sediments and groundwater to assess their survival and their ability to express the genes under environmental conditions.

REFERENCES

Alexander, M. 1984. "Ecological Constraints on Genetic Engineering." In G. S. Omenn and A. Hollaender (Eds.), *Genetic Control of Environmental Pollutants*, pp. 151-168. Plenum Press, New York, NY.

Furste, J. P., W. Pansegrau, R. Frank, H. Blöcker, P. Scholz, M. Bagdasarian, and E. Lanka. 1986. "Molecular Cloning of the Plasmid RP4 Primase Region in a Multi-host *tac*P Expression Vector." *Gene 48*: 119-131.

Mermod, N., L. R. Juan, R. P. R. Lehrbach, and K. N. Timmis. 1986. "Vector for Regulated Expression of Cloned Genes in a Wide Range of Gram-Negative Bacteria." *Journal of Bacteriology 167/2*: 447-454.

Sayler, G. S., S. W. Hooper, A. C. Layton, and J. M. H. King. 1990. "Catabolic Plasmids of Environmental and Ecological Significance." *Microbial Ecology 19*: 1-20.

Timmis, K. N., F. Rojo, and J. L. Ramos. 1988. "Prospects for Laboratory Engineering of Bacteria to Degrade Pollutants." In Gilbert S. Omenn (Ed.), *Environmental Biotechnology: Reducing the Risks from Environmental Chemicals through Biotechnology*, pp. 61-79. Plenum Press, New York, NY.

NATURAL ANAEROBIC DEGRADATION OF CHLORINATED SOLVENTS AT A CANADIAN MANUFACTURING PLANT

S. Fiorenza, E. L. Hockman, Jr., S. Szojka, R. M. Woeller, and J. W. Wigger

INTRODUCTION

Reductive dechlorination of tetrachloroethene (PCE) and trichloroethene (TCE) can produce intermediates, such as the dichloroethene isomers and vinyl chloride (VC), with more serious health effects than the starting material, as documented in laboratory studies (Vogel & McCarty 1985) and in field (Molton et al. 1987) studies. A more recent research discovery is that the provision of requisite reducing power, using compounds such as methanol, hydrogen, acetate, formate, and glucose, can yield ethene (Freedman & Gossett 1989), an intermediate of PCE reductive dechlorination. In a later study in which lactate was the electron donor, ethane, the end product of reductive dechlorination of PCE, was formed (de Bruin et al. 1992). The production of ethene also has been observed at field sites contaminated with PCE and TCE (Major et al. 1991, McCarty & Wilson 1992). Trichloroethane (1,1,1-TCA) can be degraded through biological reductive dechlorination to produce 1,1-dichloroethane (1,1-DCA) and chloroethane (CA) or abiotically either by hydrolysis to yield acetic acid, or by elimination to yield 1,1-dichloroethene (1,1-DCE) (Klecka et al. 1990, Vogel et al. 1987). Continued reductive dechlorination of these intermediates could also yield VC, ethane, and ethene.

During a site investigation of a carpet backing manufacturing plant in Ontario, Canada, sampling of groundwater and subsurface materials detected compounds that had never been used at the site, including 1,1-DCE, *cis*- and *trans*-1,2-dichloroethene (*cis*-1,2-DCE and *trans*-1,2-DCE), 1,1-DCA, and VC. An extensive field investigation was conducted to determine site hydrogeology, delineate the contaminant plumes, and identify key organic and inorganic indicators of biological activity. In this paper, we report findings of the ongoing field monitoring program suggesting that 'natural' degradation by indigenous microorganisms at a site contaminated with PCE, TCE, 1,1,1-TCA, dichloromethane (DCM), and naphtha has produced metabolic intermediates and end products indicative of reductive dechlorination.

SITE DESCRIPTION AND HYDROGEOLOGY

The study site is located in the Town of Hawkesbury, Ontario, Canada, approximately 2 km south of the Ottawa River. The facility produces yarns that are either woven into carpet backing or used in other fabrics. In the early 1970s, waste PCE, TCE, DCM, 1,1,1-TCA, naphtha, and latex were placed in on-site lagoons at the east end of the facility and pumped periodically to a surplus yarn storage area at the north-northeast corner of the property. These locations are referred to as the lagoon area and the yarn waste area (see Figure 1). In 1991, most of the source was removed by excavating the lagoons to their manmade base and disposing of the contaminated materials as per regulatory requirements.

The study site surface is a fill consisting of reworked till, sands, and silts (Szojka 1992). Groundwater flow in the fill occurs during recharge events, and its direction is influenced by surface drainage ditches east and southeast of the plant. Beneath the fill is a weathered, glacial till unit (1.2 m to 3.1 m thick) composed

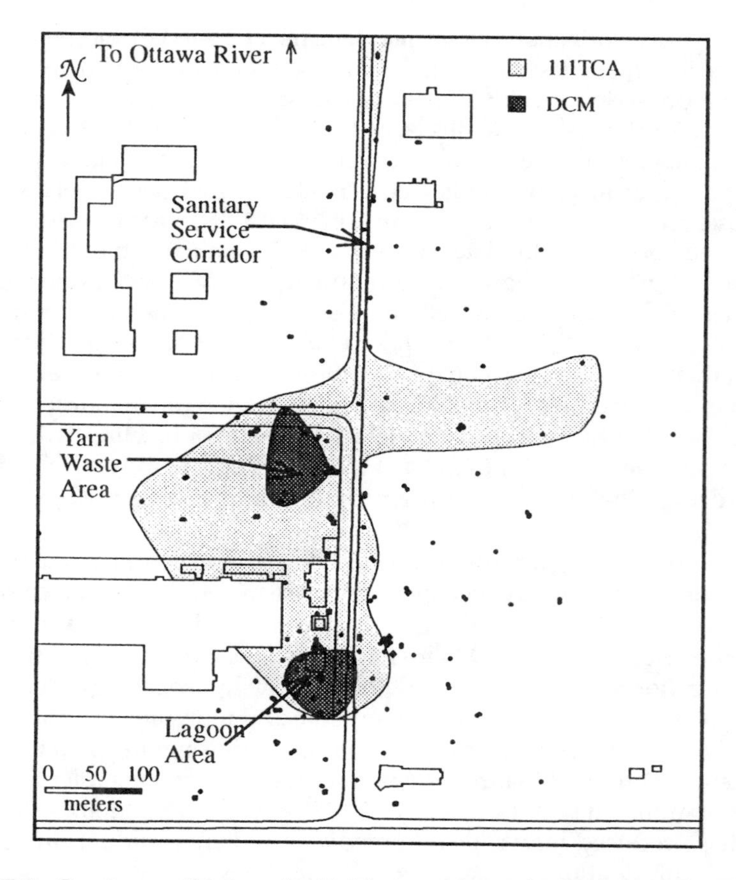

FIGURE 1. Contours of 0.1 mg/L dichloromethane (DCM) and 1,1,1-trichloroethane (11¹TCA) at a fiber manufacturing facility in Ontario, Canada.

of a silty sand, which in the source areas grades to an unweathered, sandy silt (maximum thickness 3.3 m to 4.9 m). The unweathered till above the bedrock at the plant site acts as a confining layer; east and northeast of the plant, the bedrock is overlain only by the weathered till and fine to medium sand, resulting in unconfined hydraulic conditions off site.

The glacial till overlies a fractured shale limestone bedrock. The base of the till and upper, fractured bedrock surface is the permeable zone that is the focus of this study. Regional groundwater flow at the bedrock/till interface is towards the northeast. The horizontal hydraulic gradient averages 0.012. Hydraulic conductivity, determined by bail tests, averages 4.24×10^{-6} m/s in the disposal pit area and 7.14×10^{-6} m/s in the yarn waste area. The organic carbon concentration of the soils is 0.07%; the effective porosity of the bedrock/till interface zone is approximately 0.10.

MONITORING PROGRAM

As a result of the initial site assessment, monitoring wells were installed primarily in the bedrock/till interface zone. However, wells were also completed in the fill/weathered till zone and in the shallow bedrock, with shallow arbitrarily defined as 3 m below the bedrock surface. Selected wells have been sampled biannually since April, 1991; the dissolved volatile organic compound (VOC) concentrations that have been observed during the monitoring program fluctuate with the elevation of the potentiometric surface (data not shown). The results presented are from the most recent sampling event, April, 1992. Using dedicated Waterra™ hand pumps, wells were purged of three well volumes, or until dry, prior to sample collection. It is believed that dissolved oxygen (DO) was introduced during sampling, so no DO data are given.

Groundwater samples for VOC analysis were collected without headspace in 40-mL glass vials provided by the analytical laboratory (Environmental Health Laboratories, South Bend, Indiana) and preserved with HCl. Samples for pH, alkalinity, conductivity, chloride, and sulfate were collected in 1-L polyethylene bottles; for calcium, magnesium, sodium, total iron, and total manganese, samples were collected in 250-mL bottles that were rinsed six times with groundwater and preserved with HNO_3; samples for dissolved iron and manganese were filtered in the field (0.45 μm filter membrane) and collected in 250-mL bottles that were rinsed six times with filtered groundwater and preserved with HNO_3. Groundwater samples for methanol, ethanol, and dissolved gases were collected without headspace in duplicate 100-mL amber bottles provided by the analytical laboratory (Mann Testing Laboratories Ltd., Mississauga, Ontario). The quality assurance/quality control (QA/QC) protocol called for the submittal of five matrix spikes, five replicate samples (by sample splitting), five blind duplicate samples, four field blanks, and four travel blanks to the analytical laboratories (Szojka 1992).

Groundwater was analyzed for VOCs with U.S. Environmental Protection Agency (U.S. EPA) Method 8260 (U.S. EPA 1990). Inorganic parameters and volatile fatty acids (VFAs) were analyzed according to the methods contained in the American Public Health Association Standard Methods (APHA 1989) by

Accutest Laboratories, Ltd., Nepean, Ontario. Methanol and ethanol and the dissolved gases, ethane and ethene, were analyzed with gas chromatography/flame ionization detection (GC/FID); methane was determined by a modified combustible gas method (APHA 1989) by Mann Testing Laboratories Ltd., Mississauga, Ontario. Methanol and ethanol were analyzed by direct injection of the aqueous sample (Hewlett Packard Model 5880 GC, DB 624 megabore, 30-m column from J & W Scientific, initial time = 4 min, temperature = 30 to 100°C, 10°C/min ramp); ethene and ethane were measured by headspace analysis (HP 5890, Porapak packing type T, mesh 50-80, Waters Associates, Inc., Milford, Massachusetts, 70°C isothermal and GSQ megabore column, Chromatographic Specialties, Brookville, Ontario, 40°C isothermal for resolution of ethane and ethene). The gas concentrations were converted to dissolved concentrations using Henry's constants of 1.34, 0.442, and 24.65 atm $\cdot m^3 \cdot mol^{-1}$ for methane, ethene, and ethane, respectively. Method detection limits for the monitored parameters are given in Table 1. Solubility (S) data from the *Handbook on Environmental Chemical Data* (Verschuren 1977) were used in the estimation of Koc (Lyman et al. 1990) and calculation of retardation factors (R).

MONITORING RESULTS

The most soluble primary contaminant at this site is DCM (S = 20,000 mg/L, R = 1.21). A contour plot of the dissolved 0.1-mg/L DCM plume during April, 1992, is shown in Figure 1. DCM was confined to the property, despite its high solubility and low retardation factor. The DCM concentrations observed during the monitoring program were an order of magnitude greater in the lagoon area wells than in the yarn waste area (data not shown).

The 0.1-mg/L plume of 1,1,1-TCA (S = 4,400 mg/L, R = 1.64) that is displayed in Figure 1 extended off site to the east of the lagoon area and north-northeast of the yarn waste area. The 1,1,1-TCA concentrations observed during the monitoring program were 1 order of magnitude greater in the yarn waste area than in the lagoon area. Although the elimination product, 1,1-DCE, and the reductive dechlorination product, 1,1-DCA, were both observed in monitoring samples, 1,1,1-TCA plume size was not restrained by degradation to the extent of the DCM plume. Reductive dechlorination appears to be the major pathway for 1,1,1-TCA degradation, as more 1,1-DCA than 1,1-DCE was detected in monitoring samples (see Table 1). The 1,1,1-TCA plume to the east and west of the yarn waste area seemed to flow along fracture lines.

The dissolved plumes of the primary contaminants, PCE and TCE, and their reductive dechlorination products, *cis*-1,2-DCE and VC, are shown in Figure 2 (April, 1992 data, 0.1-mg/L contours). Although other DCE isomers, *trans*-1,2-DCE and 1,1-DCE, were detected in the monitoring program, their concentrations were 2 orders of magnitude less than that of *cis*-1,2-DCE (see Table 1). At this site, the *cis*-1,2 isomer of DCE was the dominant degradation product.

The 0.1-mg/L plumes of the primary contaminants, PCE (S = 150 mg/L, R = 5.75) and TCE (S = 1,100 mg/L, R = 2.38), were smaller in magnitude than

TABLE 1. Inorganic and organic parameters analyzed in groundwater samples collected at a fiber manufacturing facility in Ontario, Canada. Concentrations are in mg/L, unless noted.

Location		Up-gradient	Lagoon Area	Down-gradient[(a)]
Well		46-2B	18-2A	25-2A
		mg/L		
Constituent	MDL			
Sodium	1	209	69	26
Chloride	1	759	338	81
Chloride[(b)]		443	238	47
Calcium	1	254	318	71
Magnesium	1	78	75	11
Sulfate	3	18	ND	15
Fe, total/dissolved	0.01	2.10/0.24	19.5/16.8	1.75/6/76
Mn, total/dissolved	0.01	1.02/0.24	10.11/9.93	2.54/8.49
Alkalinity as $CaCO_3$	1	320	650	146
VFAs as acetic acid	1	ND	492/453[(c)]	ND
Methane[(d)]	10 ppmv	BMDL	0.060/0.065[(c)]	0.010
Ethene	1 ppmv	ND	0.076/0.076[(c)]	0.023
Ethane	1 ppmv	ND	8.5E-6/7.5E-6[(c)]	ND
Methanol	1.0	ND	4.20/3.80	ND
Ethanol	1.0	ND	ND	ND
pH		7.08	6.57	7.18
Conductivity, μmhos/cm	10	2806	2214	569
PCE/TCE	0.001/0.0005	ND/ND	0.016/1.50	ND/ND
cis-1,2-DCE	0.001	ND	56.0	4.5
trans-1,2-DCE	0.001	ND	0.57	0.067
VC	0.0005	ND	4.20	5.2
1,1-DCE	0.0005	ND	0.40	0.023
1,1,1-TCA	0.0005	ND	5.50	0.33
1,1-DCA	0.001	ND	7.20	2.1
CA	0.002	ND	0.190	0.043
DCM	0.001	ND	37.0	0.022

(a) Downgradient 50 m from Well 18-2A.
(b) Corrected for road salt by converting to molarity and subtracting the chloride equivalent to the moles of sodium detected above background.
(c) Blind duplicate sample.
(d) MDL of headspace analysis in ppmv; table values are dissolved concentrations in mg/L.

those of their transformation products. In fact, dissolved PCE was no longer detected in the lagoon area at or above the 0.1-mg/L concentration level, but was greater than 1.0 mg/L in the yarn waste area (data not shown). TCE was detected at concentrations greater than 1.0 mg/L in the lagoon area, but was less than 1.0 mg/L in the yarn area. The 0.1-mg/L *cis*-1,2-DCE (S = 800 mg/L, R = 2.39)

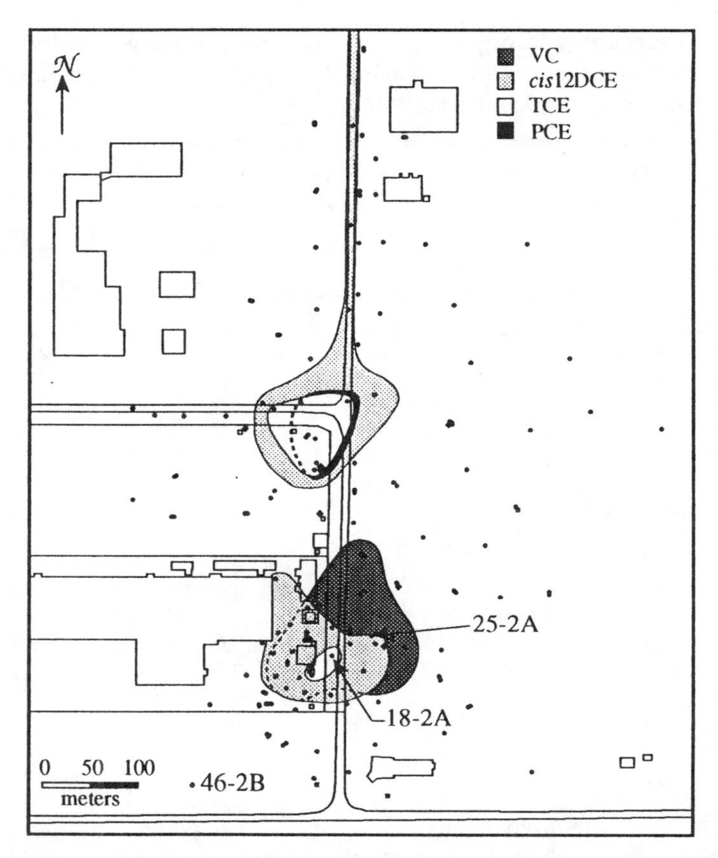

FIGURE 2. Contours of 0.1 mg/L tetrachloroethene (PCE), trichloroethene (TCE), *cis*-1,2-dichloroethene (*cis*-1,2-DCE), and vinyl chloride (VC) at a fiber manufacturing facility in Ontario, Canada.

plume extended north of each source area. No VC (S = 2,700 mg/L, R = 1.92) contamination was detected at or above the 0.1-mg/L concentration level in the yarn waste area, but the 0.1-mg/L VC plume extended off site to the northeast of the lagoons. From these data, it appears that the conversion of PCE and TCE to *cis*-1,2-DCE and VC was rapid in the lagoon area, but that VC production in the yarn waste area was limited.

Further analysis has shown that the lagoon and yarn waste areas also were sites of methane fermentation (selected data are given in Table 1). Higher concentrations of methane and VFAs (1 to 2 orders of magnitude) were found near the lagoons than the yarn waste area. When ethane and ethene were resolved with the GSQ column, it was apparent that the detection of ethene at a monitoring point coincided with the detection of methane; however, ethane was not always associated with methane. Methane and ethene were observed in groundwater from both source areas and immediately downgradient of the lagoon area. Ethene concentrations were highest in the lagoon area. Low concentrations of ethane were detected within and immediately downgradient of both source areas.

Groundwater from Well 18-2A in the lagoon area (see Table 1) had reduced pH, no detectable sulfate, and elevated alkalinity, VFAs, chloride, and total and dissolved Fe and Mn, in addition to methane, ethane, and ethene. The downgradient well, 25-2A, had increased pH and sulfate and decreased concentrations of the inorganic compounds, VOCs, methane, and ethene. The elevated chloride and sodium in the upgradient well, 46-2B, are the result of road salting on the adjacent, major highway. Between Well 46-2B and the highway is a drainage ditch whose invert is seasonally above the groundwater elevation, leading to flow from the ditch to the monitoring well.

DISCUSSION

Dichloromethane is known to degrade under both aerobic (Kohler-Staub et al. 1986) and anaerobic conditions (Freedman & Gossett 1991). The initial degradation of DCM may have been aerobic, causing a depletion in DO and leading to reduced conditions. Aerobic degradation of the naphtha components also could have consumed DO. One potential source of methane in the source areas is anaerobic DCM degradation, which produces methane and CO_2 (Freedman & Gossett 1991).

The transformation products identified at the manufacturing facility are those associated with anaerobic, reductive dechlorination, although abiotic mechanisms could cause some of the observed intermediates (Klecka et al. 1990, Vogel et al. 1987). For example, the methanol detected in the lagoon area could have occurred by hydrolysis of DCM. The observed 1,1-DCE could be the result of abiotic decomposition of 1,1,1-TCA. In the case of 1,1,1-TCA, high concentrations of 1,1-DCA compared to 1,1-DCE and the detection of chloroethane (CA) suggest that reductive dechlorination is the major degradation pathway. The major DCE isomer produced was *cis*-1,2-DCE. During reductive dechlorination, other investigators also have observed that *cis*-1,2-DCE is the dominant isomer formed (Barrio-Lage et al. 1987, de Bruin et al. 1992); however, predominance of *trans*-1,2-DCE has been noted (Freedman & Gossett 1989).

The seasonal fluctuation in VOC concentration, coupled with the relatively short time period of the monitoring program (2 years), hinder the analysis of source concentration change over time. Differences in the sizes of primary contaminant plumes suggest that degradation has limited the extent of certain plumes but not others, especially when individual solubility and retardation factors are considered. The DCM, PCE, and TCE plumes are restricted to the source areas, but 1,1,1-TCA extends off site. If just dispersion and retardation had affected plume size, the DCM plume should be the most extensive, but it is restricted in size, suggesting that degradation is limiting the extent of DCM contamination. The intermediate solubilities and retardation factors of TCE and 1,1,1-TCA and detection of reductive dechlorination intermediates imply that degradation has affected their plume sizes, but the relative importance of degradation and sorption remains to be determined.

The occurrence of methane and VFAs in groundwater from the source areas indicated reducing conditions and anaerobic activity, as did the lowered pH, increased dissolved and total metals, and increased alkalinity. These areas are likely to be the location of active degradation, with the downgradient occurrence of the metabolic intermediates likely the result of transport.

The detection of ethene, an intermediate of PCE and TCE dechlorination, coincided with the presence of methane in groundwater samples. The primary contaminant, PCE, can undergo sustained reductive dechlorination under methanogenic conditions when an electron donor, such as methanol (Freedman & Gossett 1989) or lactate (de Bruin et al. 1992), is supplied. However, reductive dechlorination of 55 mg/L PCE to ethene has occurred without the production of methane when excess methanol was added (DiStefano et al. 1991).

At this site, it is possible that the organic contaminants, such as naphtha components and perhaps the VFAs, have served as electron donors for reductive dechlorination. In one study, addition of toluene to methanogenic aquifer slurries may have provided an initial source of reducing power (Sewell & Gibson 1991). In a different study, the addition of lactate, propionate, crotonate, butyrate, and ethanol stimulated reductive dechlorination in methanogenic aquifer slurries, whereas acetate and methanol addition did not (Gibson & Sewell 1992). A potential source of acetic acid and reducing equivalents at this site could be the anaerobic degradation of DCM. DCM degradation by a methanogenic consortium produced CO_2 by oxidation and acetic acid by fermentation (Freedman & Gossett 1991); in this model, the reducing equivalents from CO_2 production and the acetic acid were both used for methanogenesis. Another possible source of acetic acid is hydrolysis of 1,1,1-TCA.

It has been observed that the ease of reductive dechlorination decreases with decreasing number of chlorine substituents (Fathepure et al. 1987), and that the rate-limiting step is the conversion of VC to ethene (Freedman & Gossett 1989). Because exogenous electron donors are required, the decreased concentrations of VC and ethene detected in the yarn waste area may be caused by lower initial concentrations of DCM, and thus, supplies of electron donors. Alternatively, the minimal concentrations of VC and ethene observed in the yarn waste area could be due to lower initial concentrations of PCE and TCE there than found in the lagoons.

Because the highest concentrations of ethene, methane, and VFAs were restricted to the lagoon area, electron donor availability is probably limiting the continued reduction of the less-chlorinated metabolic intermediates. It is possible that the downgradient edges of the plumes are aerobic. The relative ease of aerobic degradation increases with decreasing number of chlorine substituents; thus, VC might be undergoing aerobic degradation off site. Aerobic degradation of VC has been documented in laboratory studies (Davis & Carpenter 1990, Phelps et al. 1991). To further elucidate the degradation mechanisms at this site, additional research will be required.

ACKNOWLEDGMENTS

The authors would like to thank A. Stabenau, D. Litchfield, J. MacDonald, and M. Priddle for their aid in the preparation of this manuscript.

REFERENCES

American Public Health Association. 1989. *Standard Methods for the Examination of Water and Wastewater*, 17th ed., Washington, DC.

Barrio-Lage, G. A., F. Z. Parsons, R. S. Nassar, and P. A. Lorenzo. 1987. "Biotransformation of trichloroethene in a variety of subsurface materials." *Environ. Toxicol. Chem. 6*: 571-578.

deBruin, W. P. D., M. J. J. Kotterman, M. A. Posthumus, G. Schraa, and A. J. B. Zehnder. 1992. "Complete biological reductive transformation of tetrachloroethene to ethane." *Appl. Environ. Microbiol. 58*(6): 1996-2000.

Davis, J. W., and C. L. Carpenter. 1990. "Aerobic biodegradation of vinyl chloride in groundwater samples." *Appl. Environ. Microbiol. 56*(12): 3878-3880.

DiStefano, T. D., J. M. Gossett, and S. H. Zinder. 1991. "Reductive dechlorination of high concentrations of tetrachloroethene to ethene by an anaerobic enrichment culture in the absence of methanogenesis." *Appl. Environ. Microbiol. 57*(8): 2287-2292.

Fathepure, B. Z., J. P. Nengu, and S. A. Boyd. 1987. "Anaerobic bacteria that dechlorinate perchloroethene." *Appl. Environ. Microbiol. 53*(11): 2671-2674.

Freedman, D. L., and J. M. Gossett. 1989. "Biological reductive dechlorination of tetrachloroethylene and trichloroethylene to ethylene under methanogenic conditions." *Appl. Environ. Microbiol. 55*(9): 2144-2151.

Freedman, D. L., and J. M. Gossett. 1991. "Biodegradation of dichloromethane and its utilization as a growth substrate under methanogenic conditions." *Appl. Environ. Microbiol. 57*(10): 2847-2857.

Gibson, S. A., and G. W. Sewell. 1992. "Stimulation of reductive dechlorination of tetrachloroethene in anaerobic aquifer microcosms by addition of short-chain organic acids or alcohols." *Appl. Environ. Microbiol. 59*(4): 1392-1393.

Klecka, G. M., S. J. Gonsior, and D. A. Markham. 1990. "Biological transformations of 1,1,1-trichloroethane in subsurface soils and groundwater." *Environ. Toxicol. Chem. 9*: 1437-1451.

Kohler-Staub, D., S. Hartmans, R. Gälli, F. Suter, and T. Leisinger. 1986. "Evidence for identical dichloromethane dehalogenases in different methylotrophic bacteria." *J. Gen. Microbiol. 132*: 2837-2843.

Lyman, W. J., W. F. Reehl, and D. H. Rosenblatt. 1990. *Handbook of Chemical Property Estimation Methods.* American Chemical Society, Washington, DC.

Major, D. W., E. W. Hodgins, and B. J. Butler. 1991. "Field and laboratory evidence of in situ biotransformation of tetrachloroethene to ethene and ethane at a chemical transfer facility in North Toronto." *On-Site Bioreclamation: Processes for Xenobiotic and Hydrocarbon Treatment.*, pp. 147-171. Butterworth-Heinemann, Boston, MA.

McCarty, P. L., and J. T. Wilson. 1992. "Natural anaerobic treatment of a TCE plume. St. Joseph, Michigan, NPL site." *Symposium on Bioremediation of Hazardous Wastes: U.S. EPA's Biosystems Technology Development Program*, pp. 68-72. May 5-6, Chicago, IL.

Molton, P. M., R. T. Hallen, and J. W. Pyne. 1987. *Study of Vinyl Chloride Formation at Landfill Sites in California.* BNWL-2311206978, Battelle Pacific Northwest Laboratories, Richland, WA.

Phelps, T. J., K. Malachowsky, R. M. Schram, and D. C. White. 1991. "Aerobic mineralization of vinyl chloride by a bacterium of the order *Actinomycetales*." *Appl. Environ. Microbiol. 57*(4): 1252-1254.

Sewell, G. W., and S. A. Gibson. 1991. "Stimulation of the reductive dechlorination of tetrachloroethene in anaerobic aquifer microcosms by the addition of toluene." *Environ. Sci. Technol. 25*(5): 982-984.

Szojka, S. 1992. *Hydrogeological Summary Report.* Water and Earth Science Associates Ltd. for Amoco Fabrics and Fibers Ltd., Hawkesbury, Ontario, Canada.

U.S. Environmental Protection Agency. 1990. *Test Methods for Evaluating Solid Waste.* Volume 1B: *Laboratory Manual*, Physical/Chemical Methods, SW-846, 3rd ed., Rev 1. Office of Solid Waste and Emergency Response.

Verschuren, K. 1977. *Handbook of Environmental Data on Organic Chemicals.* Van Nostrand Reinhold Co., New York, NY.

Vogel, T. M., C. S. Criddle, and P. L. McCarty. 1987. "Transformations of halogenated aliphatic compounds." *Environ. Sci. Technol.* 21: 722-736.

Vogel, T. M., and P. L. McCarty. 1985. "Biotransformation of tetrachloroethylene to trichloroethylene, dichloroethylene, vinyl chloride, and carbon dioxide under methanogenic conditions." *Appl. Environ. Microbiol.* 49(5): 1080-1083.

ENVIRONMENTAL RESTORATION USING PLANT-MICROBE BIOAUGMENTATION

M. T. Kingsley, J. K. Fredrickson, F. B. Metting, and R. J. Seidler

INTRODUCTION

Land farming, for the purpose of bioremediation, refers traditionally to the spreading of contaminated soil, sediments, or other material over land; mechanically mixing it; incorporating various amendments, such as fertilizer or mulch; and sometimes inoculating with degradative microorganisms. In general, living plants are not involved in the process. Populations of bacteria added to soils often decline rapidly and become metabolically inactive. To efficiently degrade contaminants, microorganisms must be metabolically active. Thus, a significant obstacle to the successful use of microorganisms for environmental applications is their long-term survival and the expression of their degradative genes in situ.

In bulk soil, carbon is often a limiting resource and its absence can lead to a failure to bioremediate (Boething & Alexander 1979, Bolton et al. 1992, Goldstein et al. 1985, Schmidt & Alexander 1985). Rhizosphere microorganisms are known to be more metabolically active than those in bulk soil, because they obtain carbon and energy from root exudates and decaying root matter (Bolton et al. 1992). In addition to being more active, rhizosphere populations are more abundant, often containing 10^8 or more culturable bacteria per gram of soil, and bacterial populations on the rhizoplane can exceed 10^9/g root (Fredrickson & Elliott 1985, Bolton et al. 1992).

Many of the critical parameters that influence the competitive ability of rhizosphere bacteria have not been identified, but microorganisms have frequently been introduced into soil (bioaugmentation) as part of routine or novel agronomic practices. These include the application of *Rhizobium* sp. to improve nitrogen fixation in leguminous plants, biocontrol of root pathogens via inoculation with a variety of bacteria, and inoculation to stimulate root and plant growth (Bolton et al. 1992; De Freitas & Germida 1991; Macdonald 1989; Sivan & Chet 1992; Zablotowicz et al. 1991, 1992). However, the use of rhizosphere bacteria and their in situ stimulation by plant roots for degrading organic contaminants has received little attention. The few studies that have been published demonstrate that the use of plants enhances the rate of loss of organic contaminants (Aprill & Sims 1990, Hsu & Bartha 1979, Knaebel 1992, Reddy & Sethunathan 1983, Walton & Anderson 1990).

Published studies have demonstrated the feasibility of using rhizobacteria (*Pseudomonas putida*) for the rapid removal of chlorinated pesticides from contaminated soil (Short et al. 1990), and to promote germination of radish seeds in the presence of otherwise phytotoxic levels of the herbicide 2,4-dichlorophenoxyacetic acid (2,4-D) (Short et al. 1990), and phenoxyacetic acid (PAA) (Short et al. 1992). The present investigation was undertaken to determine if these strains (*Pseudomonas putida* PPO301/pRO101 and PPO301/pRO103) could be used to bioremediate 2,4-D-amended soil via plant-microbe bioaugmentation.

Methods. Microcosms were prepared in 50 mL sterile, disposable, polypropylene centrifuge tubes containing 50 g (wet weight) of Burbank sandy loam (see Cataldo et al. 1990 for soil characteristics) at 30% moisture. The soil treatments were prepared in bulk, in 1-quart canning jars and uniformly mixed on a roller mill, prior to dispensing into tubes. For the initial 30 day greenhouse trial, the following treatments were employed: 0 (control) or 500 µg/g 2,4-D; uninoculated or uniformly inoculated with an equal mixture of *Pseudomonas putida* strains PPO301/pRO101 and PPO301/pRO103{both are Nal^r Tc^r} (Harker et al. 1989, Short et al. 1990) to 10^9 colony-forming units (cfu)/g; unplanted or planted with three seeds each of *Triticum aestivum* L. cv. Edwall) inserted three cm into the soil.

A second microcosm-based experiment utilized a soil-overlay inoculation method to examine the colonization ability of the *Pseudomonas putida* strains. Burbank sandy loam soil (non-sterile) was amended to either 0 or 500 µg/g 2,4-D and brought to 30% moisture with sterile distilled water, and then 40 g was added to microcosm tubes. This layer of soil was overlain with 10 g of moist sterile soil (= uninoculated) or 10 g of inoculum. The inoculum consisted of sterilized Burbank sandy loam (autoclaved 1 h on each of 3 separate days), brought to 30% moisture and inoculated with an equal mixture of *P. putida* PPO301/pRO101 and PPO301/pRO103, incubated 48 hours at 30°C (population of 10^9 cfu/g determined by viable counts) prior to dispensing into microcosm tubes. The microcosms either remained unplanted or received three wheat seeds inserted into the upper (overlay) soil layer.

Microcosms were analyzed for plant top and root dry weight, soil and rhizosphere counts of inoculum strains and populations of total soil heterotrophic bacteria. Visible estimates of root length were made. Analysis of 2,4-D in soil was performed via high-performance liquid chromatography (HPLC) according to Short et al. (1990).

RESULTS

In uninoculated, 2,4-D-amended microcosms wheat seeds failed to germinate. Microbial inoculation protected the wheat seeds in the 500 µg/g 2,4-D-amended treatments. However, the seeds germinated later and grew more slowly and produced less shoot, but more root dry matter than plants from the unamended control microcosms as measured after 30 days (see Figure 1). The roots near the crowns of these plants were visibly deformed and distorted; the roots were also shorter than the roots of control plants. The roots of the control plants extended to the bottom of the microcosm tubes (ca. 60 mm) within the first 2 weeks and

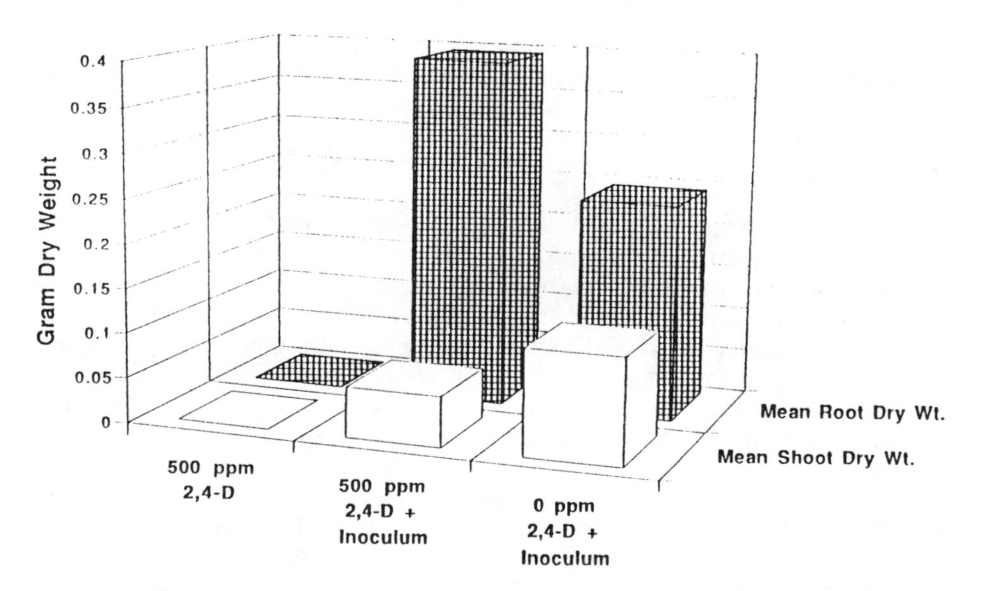

FIGURE 1. Mean wheat plant shoot and root dry weights following a 30-day greenhouse trial; seeds failed to germinate in uninoculated microcosms.

continued elongating, forming compacted root masses at the bottoms of the microcosm tubes. The longest roots in the experimental treatment (inoculated, 500 µg/g 2,4-D) never reached the bottom of the microcosms (i.e. < 60 mm). In inoculated treatments, 2,4-D was reduced 99% in 30 days, i.e., from 500 µg/g to 5 µg/g. There was not a measurable difference in the level of 2,4-D remaining in the inoculated microcosms that contained wheat plants or those without plants (i.e., ca. 5 µg/g 2,4-D remained). In the uninoculated treatments, the indigenous soil microflora decreased the initial level of 2,4-D from 500 µg/g to only 250 µg/g (50%) after 30 days.

In the overlay inoculation trial wheat seeds germinated readily in all treatments; however, in uninoculated, 2,4-D-amended treatments the seedlings were killed within ca. 2 weeks of germination as a result of 2,4-D toxicity. Viable counts of root and soil samples from inoculated control treatments (i.e., 0 µg/g 2,4-D) and inoculated experimental (500 µg/g 2,4-D) (see Table 1) indicated that the inoculum readily colonized the wheat roots. Unlike the previous experiment, there was not a significant difference in shoot dry weights between control microcosms (0 µg/g 2,4-D, either inoculated or uninoculated) and experimental microcosms (500 µg/g 2,4-D, inoculated). However, there was a significant difference in root growth between the two treatments. In control microcosms, the roots reached the bottom of the tubes (as observed in the previous experiment); in experimental treatments (2,4-D, inoculated), despite being well colonized the roots were confined to the upper inoculum layer and to approximately the first cm of the 2,4-D-containing layer. The root systems of these plants were grossly deformed. The level of 2,4-D remaining in these microcosms after 30 days was ca. 300 μg/g (whether inoculated or uninoculated).

DISCUSSION

The use of rhizosphere-competitive bacteria for bioremediation offers several advantages. Many organic contaminants cannot serve as sole carbon and energy sources (or are not present in high enough concentrations to do so) and are degraded only when additional electron donors are supplied. However, rhizosphere bacteria derive most substrates and carbon from the plant root and remain metabolically active for extended periods, and thus they can continue to degrade the contaminants for an extended period of time. As plant roots grow through the soil, microorganisms continuously colonize the growing root tips. Depending on the plant species, root growth can be extensive both horizontally and vertically. It is not uncommon for roots of legumes and plants from arid regions to penetrate to 10-15 m below the soil surface. Grass species typically have dense rooting patterns and can spread by rhizomes. For example, it has been estimated that a single wheat plant (*Triticum aestivum*) can produce a total root length approaching 71,000 m, which constitutes a large surface area dispersed throughout the soil (Pavlychenko 1937). The few studies that have been published demonstrate that the use of plants enhances the rate of loss of organic contaminants (Aprill & Sims 1990, Hsu & Bartha 1979, Knaebel 1992, Reddy & Sethunathan 1983, Walton

TABLE 1. Viable counts of inoculum strains and heterotrophic bacteria from upper and lower roots and soil of overlay inoculated microcosms.

	Inoculum[a]		Total Heterotrophs[b]	
Inoculated Control Plants	Root counts[c] cfu/g	Soil Counts[d] cfu/g	Root counts cfu/g	Soil Counts cfu/g
upper	1×10^8	8×10^6	NT[e]	8×10^7
lower	4×10^6	1×10^3	NT	3×10^7
Inoculated/500 μg/g 2,4-D				
upper	7×10^7	1×10^7	NT	3×10^8
lower	NA[f]	$< 10^1$	NA	6×10^7

(a) Inoculum plated on media containing Nal (100 μg/ml) and Tc (20 μg/ml)
(b) Plated on TSA
(c,d) Root and soil counts on dry weight basis
(e) NT = not tested
(f) NA = not applicable

& Anderson 1990). These studies relied upon the rhizosphere stimulation of the indigenous microflora and dealt with relatively low concentrations of compounds.

Uniform soil inoculation to 10^9 cfu/g, as in the first microcosm trial described here or in a previous studies (Short et al. 1990, 1992), is readily achievable in the laboratory, but it is unrealistic for the field. For field-scale applications of rhizosphere bioaugmentation some form of seed/plant inoculation, such as by seed coating or in furrow inoculation will be required. Therefore aggressive, rhizosphere-competent, root-colonizing organisms will be necessary. In both trials inoculation was required for wheat seed germination and growth in the presence of 500 µg/g 2,4-D. In the absence to 2,4-D the *P. putida* inoculum colonized roots in the control microcosms at ca. 10^6 cfu/g dry root in the presence of ca. 10^8 cfu/g (soil) of competing heterotrophic bacteria (Table 1). However, in the experimental treatments an inoculum population of 10^7 cfu/g root was apparently insufficient to allow wheat roots to penetrate more than a few mm into the 2,4-D-containing soil layer. The distorted and deformed root systems were confined to the inoculum layer and the upper few mm of the 2,4-D layer. The overall level of 2,4-D to which the plant roots were exposed as they attempted to penetrate downward was high (ca. 300 µg/g), resulting in the distortion of the root system and lack of vertical root growth. The differences observed between root growth in the two microcosm trials can be attributed to the mode of inoculation. In the uniformly inoculated trial the 2,4-D was degraded globally throughout the microcosm; the lowered levels of 2,4-D allowed for more extensive vertical root growth. By contrast, in the overlay trial the inoculum/roots were exposed to a steep vertical gradient of 2,4-D.

CONCLUSIONS

The current experiments demonstrate that bioaugmentation with rhizosphere-competent bacteria, capable of degrading the herbicide 2,4-D can protect seeds, allowing germination and plant growth to proceed in the presence of an otherwise phytotoxic level of herbicide. Because 2,4-D is a synthetic auxin, even modest amounts can affect root growth and morphology. Plant-microbe bioaugmentation should work more effectively with other organic pollutants, especially those that are not phytotoxic and do not have plant growth regulatory effects.

ACKNOWLEDGMENT

Pacific Northwest Laboratory is operated for the U.S. Department of Energy by Battelle under contract DE-AC06-76RLO 1830.

REFERENCES

Aprill, W., and R. C. Sims. 1990. "Evaluation of the use of prairie grasses for stimulating polycyclic aromatic hydrocarbon treatment in soil." *Chemosphere 20*: 253-265.

Boething, R. S., and M. Alexander. 1979. "Effect of concentration of organic chemicals on their biodegradation by natural microbial communities." *Appl. Environ. Microbiol. 37*: 1211-1216.

Bolton, H., J. K. Fredrickson, and L. F. Elliott. 1992. "Microbial ecology of the rhizosphere." In F. B. Metting (Ed.), *Soil Microbial Ecology. Applications in Agriculture, Forestry, and Environmental Management*, pp. 27-63. Marcel Dekker, Inc., New York, NY.

Cataldo, D. A., S. D. Harvey, and R. J. Fellow. 1990. *An Evaluation of the Environmental Fate and Behavior of Munitions Material (TNT, RDX) in Soil and Plant Systems: Environmental Fate and Behavior of RDX*. PNL-7529, Pacific Northwest Laboratory, Richland, WA.

De Freitas, J. R., and J. J. Germida. 1991. "*Pseudomonas cepacia* and *Pseudomonas putida* as winter wheat inoculants for biocontrol of *Rhizoctonia solani*." *Can. J. Microbiol. 37*(10): 780-784.

Fredrickson, J. K., and L. F. Elliott. 1985. "Colonizing of winter wheat roots by inhibitory rhizobacteria." *Soil Sci. Soc. Am. J. 49*: 1172-1177.

Goldstein, R. M., L. M. Mallory, and M. Alexander. 1985. "Reasons for possible failure of inoculation to enhance biodegradation." *Appl. Environ. Microbiol. 50*:977-983.

Harker, A. R., R. H. Olsen, and R. J. Seidler. 1989. "Phenoxyacetic acid degradation by the 2,4-dichlorophenoxyacetic acid (TFD) pathway of plasmid pJP4: Mapping and characterization of the regulatory gene, *tfdR*." *J. Bacteriol. 171*: 314-320.

Hsu, T.-S., and R. Bartha. 1979. "Accelerated mineralization of two organophosphate insecticides in the rhizosphere." *Appl. Environ. Microbiol. 37*: 36-41.

Knaebel, D. B. 1992. "The effects of intact rhizosphere microbial communities on the mineralization of surfactants in surface soils." Abstr. p. 397. Am. Soc. Microbiol. Annual Meeting.

Macdonald, R. M. 1989. "An overview of crop inoculation." In R. Campbell and R. M. Macdonald (Eds.), *Microbial Inoculation of Crop Plants*, pp. 1-10. IRL Press, Oxford, UK.

Pavlychenko, T. K. 1937. "Quantitative study of the entire root system of weed and crop plants under field conditions." *Ecology 13*: 391-396.

Reddy, B. R., and N. Sethunathan. 1983. "Mineralization of parathion in the rice rhizosphere." *Appl. Environ. Microbiol. 45*: 826-829.

Schmidt, S. K., and M. Alexander. 1985. "Effects of dissolved organic carbon and second substrates on the biodegradation of organic compounds at low concentrations." *Appl. Environ. Microbiol. 49*: 822-827.

Short, K. A., R. J. Seidler, and R. H. Olsen. 1990. "Survival and degradative capacity of *Pseudomonas putida* induced or constitutively expressing plasmid-mediated degradation of 2,4-D in soil." *Can. J. Microbiol. 36*: 821-826.

Short, K. A., R. J. King, R. J. Seidler, and R. H. Olsen. 1992. Biodegradation of phenoxyacetic acid in soil by *Pseudomonas putida* PPO301(pRO103), a constitutive degrader of 2,4-dichlorophenoxyacetate. *Mol. Ecol. 1*: 89-94.

Sivan, A., and I. Chet. 1992. "Microbial control of plant diseases." In R. Mitchell (Ed.), *Environmental Microbiology*, pp. 335-354. Wiley-Liss, Inc., New York, NY.

Walton, B. T., and T. A. Anderson. 1990. "Microbial degradation of trichloroethylene in the rhizosphere: Potential application to biological remediation of waste sites." *Appl. Environ. Microbiol. 56*: 1012-1016.

Zablotowicz, R. M., E. M. Tipping, F. M. Scher, M. Ijzerman, and J. W. Kloepper. 1991. "In-furrow spray as a delivery system for plant growth-promoting rhizobacteria and other rhizosphere-competent bacteria." *Can. J. Microbiol. 37*(8): 632-636.

Zablotowicz, R. M., C. M. Press, N. Lyng, G. L. Brown, and J. W. Kloepper. 1992. "Compatibility of plant growth promoting rhizobacterial strains with agrichemicals applied to seed." *Can. J. Microbiol. 38*(1): 45-50.

TRANSFORMATION OF TETRACHLOROMETHANE UNDER DENITRIFYING CONDITIONS BY A SUBSURFACE BACTERIAL CONSORTIUM AND ITS ISOLATES

E. J. Hansen, D. L. Johnstone,
J. K. Fredrickson, and T. M. Brouns

INTRODUCTION

In groundwater contaminated with both nitrate and tetrachloromethane (CCl_4), bacteria that can transform tetrachloromethane under denitrifying conditions could remove both contaminants simultaneously. A pure culture has been reported to transform tetrachloromethane under denitrifying conditions (Criddle et al. 1990).

In this study, it was discovered that the denitrifiers of a subsurface consortium were responsible for the transformation of tetrachloromethane. A consortium isolated from groundwater and individual isolates of denitrifying and nondenitrifying bacteria from the consortium were used for this study. To determine if the denitrifying isolates from the consortium possess the ability to transform tetrachloromethane or if the intact consortium is required, a screening of pure cultures of the isolates and their ability to transform CCl_4 was conducted. Also, to determine if an auxiliary electron donor was necessary for the transformation of tetrachloromethane, degradation in the absence of an electron donor was determined.

All experiments were conducted in batch bottles using a sterile simulated groundwater media (SGM) which was purged with filtered nitrogen gas. The SGM was formulated to approximate the major ion concentrations of the groundwater at the U.S. Department of Energy's Hanford site, including nitrate (400.0 mg/L). Acetate (381.8 mg/L) was used for a growth substrate. Inocula of 1% stationary phase cultures of the individual isolates or the consortium were used which were then incubated anaerobically (Bouwer & McCarty 1983) at 24°C. Killed cell and no cell controls were performed on all experiments. A concentration of 1.0 mg/L CCl_4 was used to approximate the median concentration of the contaminated groundwater at the Hanford site.

RESULTS

The consortium was composed of eight individual isolates designated as DN-1 through DN-8. Two isolates from the consortium, DN-2 and DN-5, were able to denitrify. Both isolates were identified tentatively as *Pseudomonas stutzeri* using the API® system (Amy 1992). Isolate DN-1 was also identified as *P. stutzeri* using the API system, but this isolate did not denitrify. Isolate DN-3 was identified as *P. paucimobilis*, and DN-6 was identified to the genus *Acinetobacter* and confirmed by fatty acid analysis. The isolates DN-4, DN-7, and DN-8 were not identified using the API (NFT® and 20E®) system. Transmission electron microscopy (TEM) indicates that DN-2 and DN-5 (0.7 µm × 1.7 µm and 0.7 µm × 1.6 µm, respectively) have similar cell morphology, including a polar flagellum and a capsular exopolymer. Conventional negative staining of the isolates DN-2 and DN-5 with India ink confirmed the presence of cell capsules.

DN-2, DN-5, and the consortium concomitantly reduced nitrate during the transformation of CCl_4. After 3 days, nitrate was reduced to less than 1.0 mg/L by the denitrifiers. The production of nitrous oxide in the presence of acetylene, along with the reduction of nitrate, confirmed that DN-2 and DN-5 were dissimilatory nitrate reducers which reduced nitrate at approximately the same rate as each other and the consortium.

Experiments using $[^{14}C]CCl_4$ were conducted to determine if CCl_4 was mineralized by the individual groundwater isolates under denitrifying conditions. The results showed that only isolates DN-2 and DN-5 were capable of transforming CCl_4 to CO_2. The total recovery of ^{14}C ranged from 85% to 96% after 4 days.

The maximum rate of CCl_4 transformation occurred during the active phase of cell growth, Day 0 to Day 1 (Figure 1), and dropped to near zero after the rate-limiting concentrations of acetate and nitrate were depleted, Day 3 to Day 4 (Figure 1). The addition of acetate and nitrate stimulated further CCl_4 transformation with an increase of 130% in mineralization (Figure 2). Duplicate serum

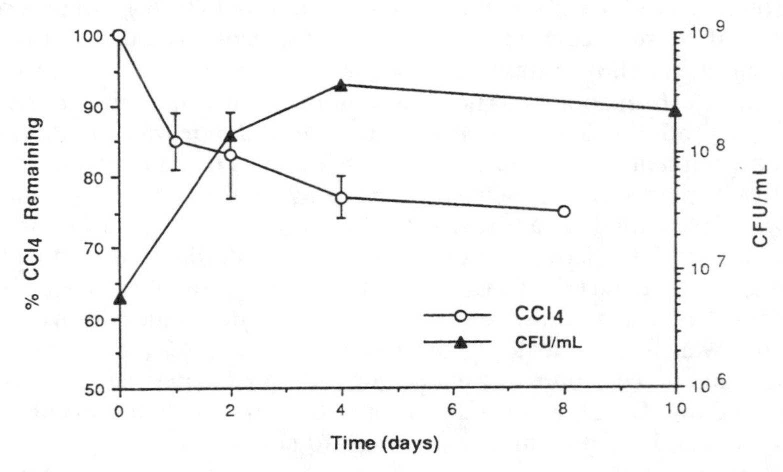

FIGURE 1. Average percent of tetrachloromethane remaining in acetate/nitrate-limited batch cultures of the consortium (data are expressed as percentages relative to corresponding average values in sterile controls). Average growth response of isolates DN-2, DN-5, and the consortium in 5.0 mg/L CCl_4 and simulated groundwater medium under denitrifying conditions.

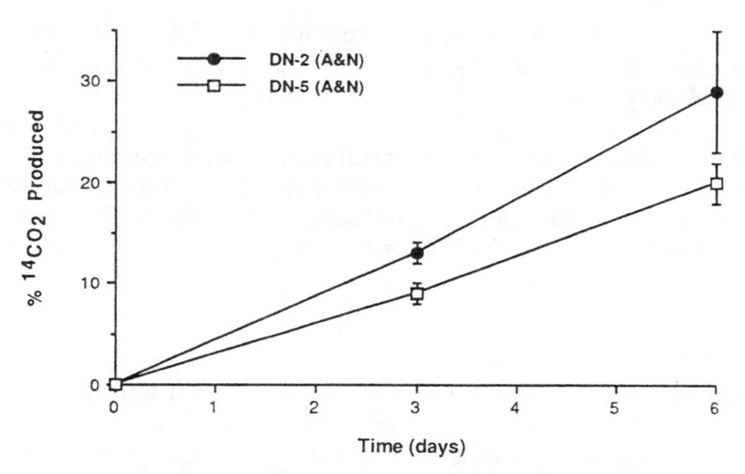

FIGURE 2. Percent of $^{14}CO_2$ produced from $^{14}CCl_4$ by isolates DN-2 and DN-5 after the addition at Day 3 of nitrate (N) and acetate (A).

bottles that received additional acetate (381.8 mg/L) and nitrate (400.0 mg/L) on Day 3 transformed an average of 14% of the remaining CCl_4 after an additional 3 days (Day 6) compared to an average of 11% transformed in 3 days ($p = 0.09$).

Controls using [^{14}C]CCl_4 were performed to determine if the individual isolates as well as the consortium could mineralize CCl_4 without an auxiliary carbon source (no acetate, but with nitrate). Serum bottles were sampled at 4 days and 8 days. There was no significant mineralization (0% to 1.5% of initial CCl_4 concentration) after either 4 days or 8 days with any of the individual isolates or the consortium.

To determine if acetate or nitrate was limiting the mineralization of CCl_4 by the denitrifiers, DN-2 and DN-5 experiments using [^{14}C]CCl_4 were conducted. The results demonstrated that there was little or no significant difference in CCl_4 mineralization between 3 days and after 6 days when either acetate or nitrate was added individually at Day 3. The addition of both acetate and nitrate at Day 3 caused a significant increase (130%) in the mineralization of CCl_4, indicating that both are required for mineralization (Figure 2). Also, the isolates DN-2 and DN-5 differed in the extent to which they mineralized CCl_4. The isolate DN-2 mineralized 20 to 50% more CCl_4 than did DN-5 after Day 6 ($p = 0.07$).

During the transformation of tetrachloromethane by DN-2, DN-5 and the groundwater consortium, trichloromethane ($CHCl_3$), were not detected until Day 4. At 4 days, an average of 4.0%, 3.5%, and 3.7% of the initial soluble CCl_4 concentration in the batch cultures of DN-2, DN-5, and the groundwater consortium, respectively, was transformed to $CHCl_3$. Tetrachloromethane was not transformed to $CHCl_3$ in killed cell controls. After 8 days, an average of 5.8%, 4.9%, and 5.1% was present as $CHCl_3$. Trichloromethane-loss controls (at $CHCl_3$ concentrations of those detected after 3 days in test batch bottles) showed that less than 1% was lost in sterile SGM after four days (at Day 8).

The highest observed level of transformation of CCl_4 by the isolates and the consortium was 35% (29% to CO_2 and 6% to $CHCl_3$) of the initial CCl_4 concentration after 6 days of incubation with acetate as the electron donor and nitrate as the electron acceptor (Figure 1). Mineralization ranged from 9% in 3 days to 29% after 6 days. This range of mineralization was a function of the initial CCl_4 concentration and the concentrations of the electron donor (acetate) and acceptor (nitrate). Trichloromethane production ranged from 3.5% of initial CCl_4 concentration after 4 days to 5.8% after 8 days.

DISCUSSION

Two strains of *Pseudomonas stutzeri* obtained from a tetrachloromethane-degrading groundwater consortium were capable of CCl_4 transformation under denitrifying conditions. One other pure culture has been reported to transform CCl_4 under denitrifying conditions (Criddle et al. 1990). This denitrifier was also isolated from an aquifer and also characterized as *Pseudomonas stutzeri*. However, the aquifer from which this isolate was obtained had no known previous exposure to CCl_4, unlike DN-2 and DN-5 which were isolated from tetrachloromethane- and nitrate-contaminated groundwater. They may have developed the capacity to transform tetrachloromethane to their advantage, possibly for detoxification.

Although the transformation of CCl_4 has been demonstrated with denitrifying mixed cultures (Bouwer & McCarty 1983), it was generally considered to be the nondenitrifying "secondary" bacteria that were responsible for the transformation (Egli et al. 1988). However, in this study it was observed that only two out of eight isolates of a groundwater consortium were capable of transforming CCl_4 and both were denitrifiers, although an auxiliary electron donor was necessary for the transformation.

The transformation of CCl_4 by DN-2 and DN-5, although not as effective as the stain used by Criddle et al. (1990), there were similar products. In this study, the only products discovered trichloromethane and carbon dioxide (which was the major product). A relatively small percentage of $CHCl_3$ was produced during the biotransformation of CCl_4 suggesting that the metabolism may proceed by two possible pathways: either CCl_4 to $CHCl_3$ then to carbon dioxide, or a rapid two-electron reduction of CCl_4 to a dichlorocarbene radical followed by spontaneous hydrolysis to formate (Egli et al. 1988, Criddle et al. 1990). Formate could then be oxidized to carbon dioxide. Criddle et al. (1990) suggested that the transformation of CCl_4 may be dependent on such mechanisms as those for the scavenging of trace nutrients, since their findings indicated a dependence on iron availability.

Bioremediation of tetrachloromethane- and nitrate-contaminated groundwater by bacteria such as DN-2 and DN-5 under denitrifying conditions could be achieved by in situ and pump-and-treat approaches. Experiments at Battelle, Pacific Northwest Laboratory are currently being conducted to establish the efficacy of such approaches.

REFERENCES

Amy, P. S., D. L. Haldeman, D. Ringelberg, D. H. Hall, and C. Russell. 1992. "Comparison of identification systems for characterization of bacteria isolated from water and endolithic habitat within the deep subsurface." *Appl. Environ. Microbiol. 58*:3367-3373.

Bouwer, E. J., and P. L. McCarty. 1983. "Transformations of halogenated organic compounds under denitrification conditions." *Appl. Environ. Microbiol. 45*:1295-1299.

Criddle, C. S., J. T. DeWitt, D. Grbić-Galić, and P. L. McCarty. 1990. "Transformation of carbon tetrachloride by *Pseudomonas* sp. strain KC under denitrification conditions." *Appl. Environ. Microbiol. 56*:3240-3246.

Egli, C., T. Tschan, R. Schlotz, A. M. Cook, and T. Leisinger. 1988. "Transformation of tetrachloromethane to dichloromethane and carbon dioxide by *Acetobacterium woodii*." *Appl. Environ. Microbiol. 54*:2819-2823.

CHANGE IN TRICHLOROETHYLENE DECOMPOSITION ACTIVITY OF *METHYLOCYSTIS* SP. M DURING BATCH CULTURE

T. Shimomura, F. Okada, K. Mishima, H. Uchiyama, and O. Yagi

INTRODUCTION

It is difficult for common bacteria in soil to degrade trichloroethylene (TCE) to carbon dioxide. However, some methanotrophic bacteria can degrade TCE aerobically. Degradation of TCE by methanotrophs results from a cometabolic reaction catalyzed by methane monooxygenase (MMO). MMO is an enzyme produced in the cell to oxidize methane to methanol. TCE degradation has been achieved in serum bottles (Little 1988) and in fixed film bioreactors containing a methanotrophic consortium (Strandberg 1989). To apply the TCE biodegradation process by methanotrophs for practical use, it is necessary to maintain their degradation activity at a high level. Further study of the fermentational characteristics of isolated methanotrophic bacteria may lead to the development of effective methods for maintaining and reactivating their activity.

After screening tests of methanotrophic bacteria, *Methylocystis* sp. M (strain M), which can degrade TCE aerobically, was isolated from soil in Japan (Uchiyama 1989). This paper reports on the changes observed in the TCE degradation activity by strain M during the culture period and describes an improved method of culturing methanotrophic bacteria that can degrade TCE rapidly.

MATERIALS AND METHODS

Methylocystis sp. M (strain M) was used in all experiments. Strain M was subcultured in 20 mL of mineral salt medium at 30°C in a 125-mL serum bottle under an air/methane (8:2 in volume) atmosphere, as described in Uchiyama (1989). The fermenter scale cultivations were performed at 30°C in a 2.5-L jar fermenter (model M-100 Tokyo Rika Co., Tokyo) at a stirring speed of 600 rpm. The subculture (20 mL) was inoculated into 1.75 L of fresh, sterilized medium, and the pH was maintained between 6.5 and 7.0 through automatic addition of 1 N NaOH.

Two methods of cultivation were tested. In one cultivation, a continuous supply of air/methane mixture (8:2 volume ratio) was added at the rate of 250 mL/min. In the other, methanol (with initial concentration of 1,000 mg/L) was added to the medium and culture was done with continuous supply of air. In methanol cultivation, methanol was added to the medium periodically to maintain the concentration of methanol so that it did not limit the growth rate. After a period of cultivation with methanol, the air/methane mixture (8:2 volume ratio) was supplied continuously to activate the TCE degradation activity of strain M.

During the culture period, samples were removed periodically from the fermenter to investigate TCE degradation activity. Strain M was harvested by centrifugation at 5,000 xg, and suspended in 20 mL of fresh medium in a 120-mL serum bottle at a cell density of 1,000 mg/L. The bottles were sealed with Teflon™-coated butyl stoppers, which were then crimped with aluminum caps. TCE (0.1 mL) from a 600-mg/L aqueous stock solution was added by means of a syringe through the septa of the bottles at a liquid concentration of 1 mg/L. The bottles were then incubated at 28°C with shaking at 120 rpm. TCE degradation was monitored by measuring the TCE concentration in the bottle's headspace gas by gas chromatography (GC), as described in Uchiyama (1989). The control experiment was done in the same way with no cell. All the liquid TCE concentrations were calculated using Henry's law.

A GC (model G-3000; Hitachi Co., Tokyo) equipped with a flame ionization detector (FID) was used for quantitative determination of TCE. The capillary column (inner diameter 0.53 mm, length 15 m) was coated with TC-WAX (GL Science Co., Tokyo). The injector, oven, and detector temperatures were kept at 200, 50, and 200°C, respectively. The carrier gas was helium (18 mL/min). The volume of gas sample injected into the gas chromatograph was 0.2 mL.

TCE degradation activity was estimated as a first-order reaction expressed as

$$\Delta C/\Delta t = k_1 \, C \, Xt$$

where $\Delta C/\Delta t$ = degradation rate of TCE (mg TCE/g cell - h)
k_1 = first-order reaction constant for TCE degradation (L/g cell – h)
C = concentration of TCE (mg/L)
X = cell density (g/L)
t = reaction time (h)

Cell densities were determined by measuring the absorbance at 580 nm (A580) using a spectrophotometer (model U-100; Hitachi Co., Tokyo). Harvested cells were washed twice and dried at 110°C for 2 h, and the dry cell weight was measured.

RESULTS AND DISCUSSION

Strain M proliferated to a cell density of about 5 g/L in 170 h when it was cultured in the jar fermenter with methane as described above. Cell density was limited to 5 g/L, even though methane and air were continuously supplied (Figure 1).

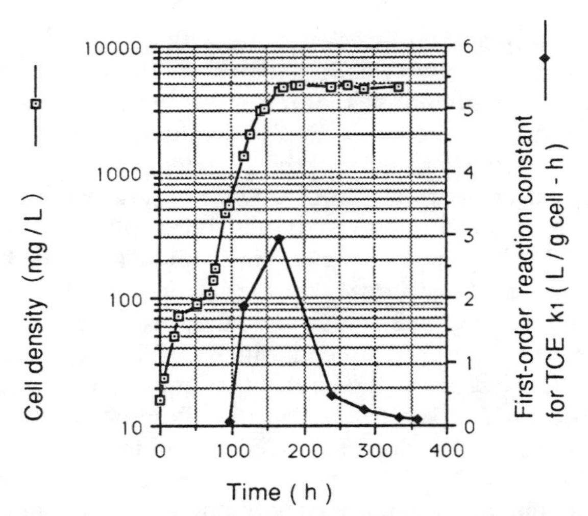

FIGURE 1. Change in TCE degradation activity of strain M during batch culture with a continuous supply of methane.

When incubated in the fermenter with a continuous supply of methane, the TCE degradation activity of strain M increased rapidly in the later half of the exponential growth period, and reached its maximum value at a cell density of 4 g/L just after the end of the exponential period (Figure 1).

The maximum first-order reaction constant for TCE degradation, k_1, was 3 (L/g cell – h). When strain M having this activity was incubated in a Teflon™–sealed serum bottle at a cell density of 1 g/L as described above, it degraded TCE within 2 h from 1 mg/L to 0.0025 mg/L (Figure 2). The linearity of the concentration of the residual TCE (mg/L) on semilogarithmic graph (Figure 2) indicated that TCE degradation was a first-order reaction. It has also been reported that the transformation of chlorinated compounds is a first-order reaction (Arvin 1991).

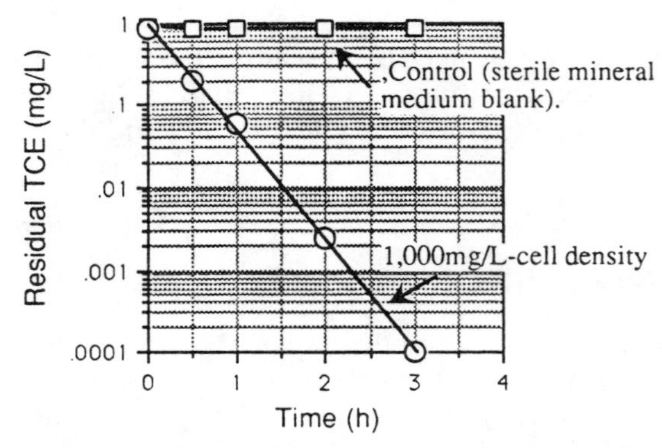

FIGURE 2. Time course of TCE degradation by strain M at the end of its exponential growth period.

When the proliferation of strain M stopped, the TCE degradation activity decreased even though methane and air were still supplied continuously. This decrease of TCE degradation activity most likely resulted from deactivation of MMO by the culture. Strain M incubated in ordinary water, which was periodically replaced by freshwater, was able to maintain its TCE degradation activity for more than 400 h at 28°C under an air/methane (8:2 in volume) atmosphere (data not shown).

Some methanotrophic bacteria can proliferate using methane or methanol as the only carbon source. In large-scale batch culturing, methanol is a better substrate than methane because it is safer to handle. However, a methanotrophic consortium containing strain M was unable to degrade TCE when cultivated with methanol (Uchiyama 1989). It is likely that MMO is not produced in the cell because it is an enzyme produced to oxidize methane, not methanol. When strain M was cultured in a jar fermenter by supplying methanol, as described above, it proliferated to a cell density of about 2.5 g/L after 143 h of cultivation (Figure 3). However, little TCE degradation activity was observed until a continuous supply of methane was started. A continuous supply of air/methane mixture to this strain M resulted in recovery of TCE degradation activity, k_1, up to about 3 (L/g cell - h) in 77 h (Figure 3). The maximum activity value was as high as that when strain M was initially cultured with methane (Figure 1). These results suggest that methanotrophic bacteria cultured with methanol can produce MMO in the cell and that its TCE degradation activity can be activated by stimulation with methane.

Strain M at the end of its exponential growth period was found to be able to degrade TCE very rapidly. For practical application of the TCE biodegradation

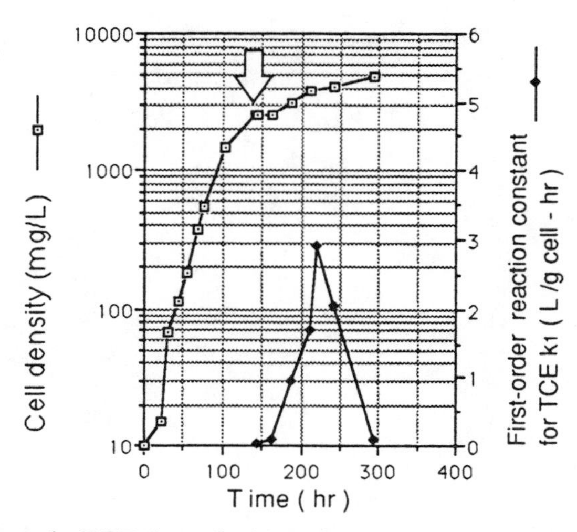

FIGURE 3. Change in TCE degradation activity of strain M during batch culture with methanol, and during activation supplying methane. The broad arrow represents the time when continuous supply of methane was started.

process by strain M, it will be necessary to maintain the degradation activity at a high level even after this particular period. Our investigations are now directed toward finding ways to maintain and reactivate the activity.

REFERENCES

Arvin, E. 1991. "Bio Degradation Kinetics of Chlorinated Aliphatic Hydrocarbons with Methane Oxidizing Bacteria in an Aerobic Fixed Biofilm Reactor." *Wat. Res. 25*: 873.

Little, C. D., A. V. Palumbo, S. E. Herbes, M. E. Lidstrom, R. L. Tyndall, and P. J. Gilmer. 1988. "Trichloroethylene Biodegradation by a Methane-Oxidizing Bacterium." *Appl. Environ. Microbiol. 54*: 951.

Strandberg, G. W., T. L. Donaldson, and L. L. Farr. 1989. "Degradation of Trichloroethylene and *trans*-1,2-Dichloroethylene by a Methanotrophic Consortium in a Fixed Film, Packed-Bed Bioreactor." *Environ. Sci. Technol. 23*: 1422.

Uchiyama, H., T. Nakajima, O. Yagi, and T. Tabuchi. 1989. "Aerobic Degradation of Trichloroethylene by a New Type II Methane-Utilizing Bacterium, Strain M." *Agric. Biol. Chem. 53*: 2903.

FACTORS AFFECTING TRANSFORMATION OF CHLORINATED ALIPHATIC HYDROCARBONS BY METHANOTROPHS

M. E. Dolan and P. L. McCarty

INTRODUCTION

Groundwater contamination by chlorinated aliphatic hydrocarbons (CAHs) has resulted from their widespread use and disposal in the environment. Aerobic methanotrophic oxidation of CAHs is possible (Fogel et al. 1986; Wilson & Wilson 1985) and has been the subject of many studies. One factor found to affect the methanotrophic transformation of CAHs is transformation product toxicity (TPT). The activity of resting cells has been found to decrease by TPT in proportion to the amount of contaminant transformed (Alvarez-Cohen & McCarty 1991a; Oldenhuis et al. 1991), resulting in a finite transformation capacity (T_c), expressed as the mass of contaminant transformed per mass of cells inactivated. Competitive inhibition between CAHs also has been noticed (Alvarez-Cohen & McCarty 1991b) and is believed to be the result of more than one contaminant competing for the active site of methane monooxygenase (MMO), which is responsible for the initial step in the methanotrophic transformation of CAHs. When assessing the applicability of a treatment scheme involving methanotrophic oxidation of CAHs, it is beneficial to know which, if any, of these factors may influence the outcome. Toward this end, results are presented from studies on methanotrophic transformation of chloroethenes, including trichloroethylene (TCE); 1,2-*trans*-dichloroethylene (*t*-DCE); 1,2-*cis*-dichloroethylene (*c*-DCE); 1,1-dichloroethylene (1,1-DCE); and vinyl chloride (VC), comparing results obtained from aqueous batch systems with those obtained in soil column tests.

MATERIALS AND METHODS

Transformation capacity experiments were conducted in a 20°C environmental chamber using 62-mL glass bottles sealed with Teflon™-lined Mininert caps containing 25 mL cell and mineral media solution with the balance as headspace similar to that described by Alvarez-Cohen & McCarty (1991a). Cells from a methanotrophic mixed culture were obtained from a 10-L, continuous-gas-feed chemostat (10% methane in air at 280 mL/min) operated with a 9-day detention time (Alvarez-Cohen & McCarty 1991a). The bottles were shaken by circular action at 300 RPM to ensure sufficient mass transfer rates between headspace and solution. Headspace samples were analyzed for CAHs by gas chromatography (GC) using either a photoionization detector or an electrolytic conductivity detector. Aqueous CAH concentrations and total CAH mass were calculated using Henry's constants. Controls with no cells were treated identical to other samples. Inhibition experiments were conducted similarly, except 250-mL bottles with 105 mL cell and mineral media solution were used.

Soil column experiments were conducted using aquifer material obtained aseptically from a chloroethene-contaminated site in St. Joseph, Michigan (McCarty et al. 1991). The fine sand was aseptically packed in 15-mL glass test tubes (13 mm I.D.) and operated in sequential batch exchanges. The feed consisted of a mix of oxygen- and methane-saturated, filter-sterilized, uncontaminated groundwater from the site spiked with solutions containing VC and/or TCE. Pore water samples were analyzed for CAH content by GC. Methane was determined by analyzing headspace from a vial containing the pore water sample on a GC with a flame ionization detector. The columns were incubated at room temperature (~23°C) in the dark.

RESULTS

Aqueous batch transformation tests at high CAH concentrations (from 15 mg/L for 1,1-DCE to 65 mg/L for *t*-DCE), in the presence of excess formate as an external source of reducing power but without methane, resulted in the maximum transformation capacity of the cells being expressed within 5 to 10 hr. The results are presented in Table 1. At the end of the 24-hr transformation test, cells exposed to high CAH concentrations were found to have little or no methane-oxidizing capability remaining, whereas cells not exposed to CAHs retained most of their original methane-oxidizing capability. The loss of methane-oxidizing ability following CAH transformation was believed to result from TPT. This was

TABLE 1. Transformation capacities for a mixed methanotrophic culture.

CAH	Transformation Capacity (μmoles/mg TSS[a])	Transformation Yield (mg CAH/mg CH_4)
t-DCE	4.8	0.16
c-DCE	3.6	0.12
VC	2.3	0.05
TCE	0.85	0.04
1,1-DCE	0.13	0.004

(a) TSS is total suspended solids.

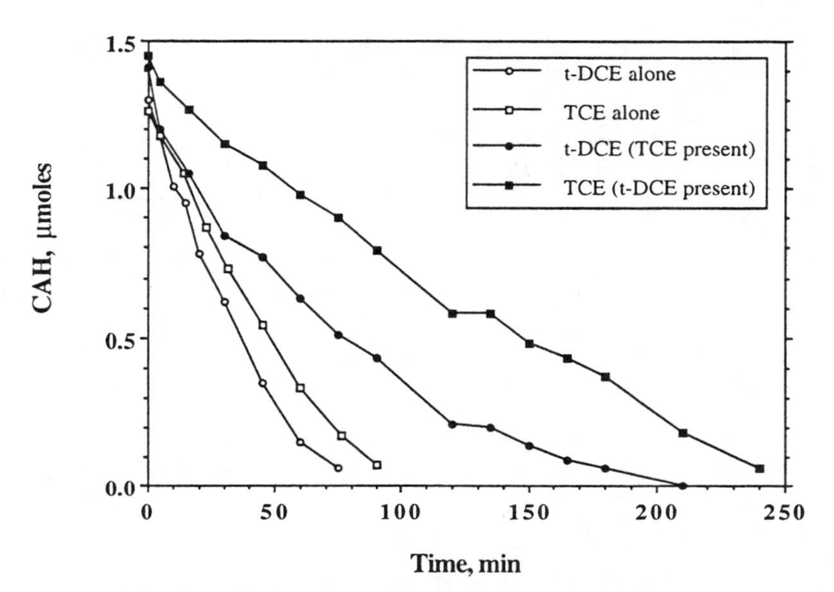

FIGURE 1. CAH transformation by a mixed methanotrophic culture.

confirmed using the approach by Alvarez-Cohen & McCarty (1991b) in which the formate using ability of cells decreased following CAH transformation, while cells exposed to acetylene to inhibit MMO showed no CAH transformation, but could oxidize formate at rates similar to controls.

When TCE and *t*-DCE were transformed together, a decrease in the transformation rate for each compound was exhibited, with the presence of *t*-DCE having a greater impact on TCE transformation than the reverse (Figure 1). Additional tests were conducted to determine how VC affected the transformation rates of the other chloroethenes. Typical results, shown for TCE in Figure 2, indicate a decrease in TCE transformation rates as the VC concentration was increased. The effects of VC concentration on transformation rates of the other CAHs studied, determined by least squares linear regression, are shown in Figure 3. Here, 1,1-DCE was transformed slower than other CAHs and exhibited significant TPT (data not shown), even at these lower concentrations (450 μg/L).

The soil columns were enriched for methanotrophic activity with repeated batch feeding of filter-sterilized groundwater containing dissolved oxygen and methane. An initial feeding of approximately 3.2 mg/L methane and 1.0 mg/L TCE and/or VC was followed by 60 days of incubation to stimulate the indigenous methanotrophic population. The columns were fed again and incubated for approximately 3 weeks. Although little TCE degradation was observed, approximately 70% of the VC was transformed in both the presence and absence of TCE when compared to control columns that were fed only dissolved oxygen (Table 2). Transformation yields of 0.8 mg VC transformed per mg methane used were determined for this culture.

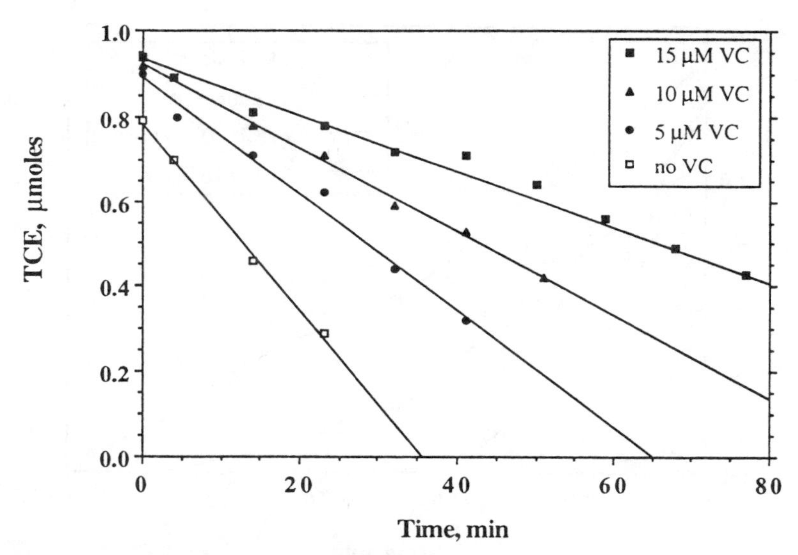

FIGURE 2. TCE transformation when mixed with VC.

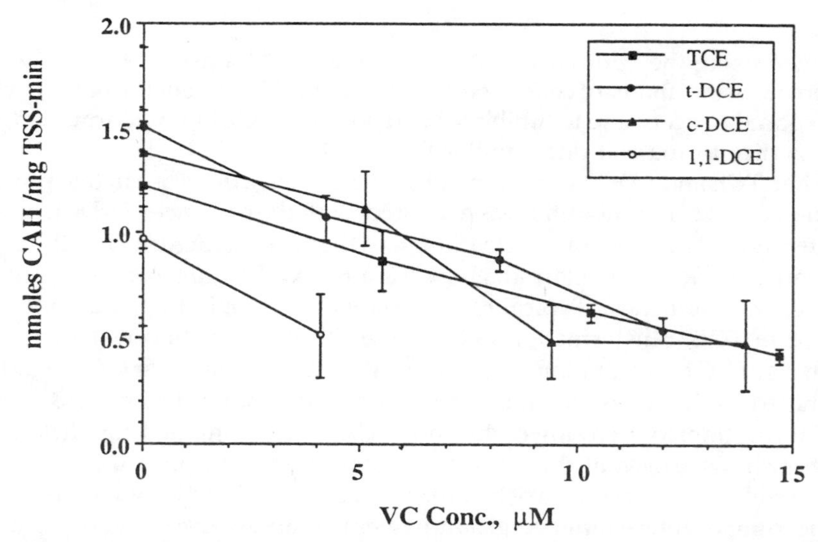

FIGURE 3. Initial CAH transformation rates as a function of initial VC concentration with 95% confidence intervals.

DISCUSSION

In batch transformation tests with a mixed methanotrophic culture, both transformation product toxicity and competitive inhibition were found to affect the extent and/or rate of transformation. Although TPT was evident for all of the chloroethenes studied, the most toxic products were produced from the compound

TABLE 2. Results of small column incubation with VC and TCE.

	VC			TCE		
Infl CH_4[a] mg/L	Influent mg/L	Effluent mg/L	Decrease %	Influent mg/L	Effluent mg/L	Decrease %
n/a	0.93	0.82	0.12	n/a	n/a	n/a
n/a	0.97	0.78	0.20	n/a	n/a	n/a
3.16	0.97	0.15	0.85	n/a	n/a	n/a
3.80	1.04	0.16	0.85	n/a	n/a	n/a
n/a	n/a	n/a	n/a	0.90	0.67	0.26
n/a	n/a	n/a	n/a	0.94	0.62	0.34
3.09	n/a	n/a	n/a	0.95	0.60	0.37
3.06	n/a	n/a	n/a	0.93	0.59	0.37
n/a	0.80	0.68	0.15	0.97	0.65	0.33
n/a	0.80	0.68	0.15	0.99	0.68	0.31
3.15	0.83	0.10	0.88	1.03	0.54	0.47
3.09	0.82	0.09	0.89	1.01	0.61	0.40
n/a	7.35	6.75	0.08	n/a	n/a	n/a
n/a	7.26	6.12	0.16	n/a	n/a	n/a
3.21	8.06	4.19	0.48	n/a	n/a	n/a
3.02	7.48	4.14	0.45	n/a	n/a	n/a

(a) Effluent CH_4 in all cases was 0.00 mg/L.
n/a – compound not included in feed solution.

with the most asymmetrical chlorine distribution, 1,1-DCE. When chloroethenes were mixed together, competitive inhibition between the two was exhibited. A decrease in the transformation rate of one CAH with increased initial concentrations of another CAH (VC) was observed.

Soil column transformation studies indicated significant VC transformation occurred with or without TCE present, while little TCE transformation occurred. TPT did not seem to be a major factor in soil column operations since the methane-oxidizing capability of the organisms remained high. It is not known whether the presence of TCE slowed the rate of VC transformation in the soil columns, but it did not affect the extent of VC transformation. While transformation yields of 0.04 mg TCE/mg CH_4 from batch studies were in general agreement with the results from the column tests, batch test transformation yields for VC (0.05 mg VC/mg CH_4) were significantly lower than those obtained from the soil column tests (0.8 mg VC/mg CH_4). Thus, batch transformation studies may greatly underestimate the potential for in situ transformation of VC by methanotrophs.

REFERENCES

Alvarez-Cohen, L. M., and P. L. McCarty. 1991a. "Effects of toxicity, aeration, and reductant supply on trichloroethylene transformation by a mixed methanotrophic culture." *Appl. Environ. Microbiol. 57* (1): 228-235.

Alvarez-Cohen, L. M., and P. L. McCarty. 1991b. "Product toxicity and cometabolic competitive inhibition modeling of chloroform and trichloroethylene transformation by methanotrophic resting cells." *Appl. Environ. Microbiol.* 57(4): 1031-1037.

Fogel, M. M., A. R. Taddeo, and S. Fogel. 1986. "Biodegradation of chlorinated ethenes by a methane-utilizing mixed culture." *Appl. Environ. Microbiol.* 51(4): 720-724.

McCarty, P. L., L. Semprini, M. E. Dolan, T. C. Harmon, C. Tiedeman, and S. M. Gorelick. 1991. "*In situ* methanotrophic bioremediation for contaminated groundwater at St. Joseph, Michigan." In R. E. Hinchee and R. F. Olfenbuttel (Eds.), *On-Site Bioreclamation: Processes for Xenobiotic and Hydrocarbon Treatment*, pp. 16-40. Butterworth-Heinemann, Stoneham, MA.

Oldenhuis, R., J. Y. Oedzes, J. J. van der Waarde, and D. B. Janssen. 1991. "Kinetics of chlorinated hydrocarbon degradation by *Methylosinus trichosporium* OB3b and toxicity of trichloroethylene." *Appl. Environ. Microbiol.* 57(1): 7-14.

Wilson, J. T., and B. H. Wilson. 1985. "Biotransformation of trichloroethylene in soil." *Appl. Environ. Microbiol.* 49(1): 242-243.

TRANSFORMATION OF TETRACHLOROETHENE TO ETHENE IN MIXED METHANOGENIC CULTURES: EFFECT OF ELECTRON DONOR, BIOMASS LEVELS, AND INHIBITORS

G. Rasmussen, S. J. Komisar, and J. F. Ferguson

INTRODUCTION

Tetrachloroethene (PCE), a common industrial solvent now widespread in the environment, resists aerobic degradation. However, several studies have shown that PCE is transformed via sequential reductive dechlorination under anaerobic conditions both in laboratory experiments (Bouwer & McCarty 1983, de Bruin et al. 1992, DiStefano et al. 1991, Freedman & Gossett 1989) and in natural environments (Parsons et al. 1984). This degradation is believed to be biological, but dehalogenation catalyzed by reduced transition-metal-bearing corrins (Vit B-12) and porphyrins (F430) in the absence of biota also has been observed (Gantzer & Wackett 1991). The degradation of PCE usually was incomplete, resulting in accumulations of trichloroethene (TCE), dichloroethene (DCE) isomers, and/or vinyl chloride (VC). A few studies have shown complete dechlorination to ethene (ETH) (DiStefano et al. 1991, Freedman & Gossett 1989), whereas de Bruin et al. (1992) found that ethene was further reduced to ethane.

Little is known about the dehalogenation reactions in anaerobic consortia. Evidence about whether the dehalogenating organisms are methanogens, acetogens, sulfidogens, or facultative anaerobes is inconclusive. Nor are the mechanisms of the reactions explicated.

This research employed various electron donors and inhibitors in an attempt to understand dehalogenation reactions. Complex substrates that can be used by fermenters and acetogens were compared to simple electron donors that can be used only by methanogens and by a few acetogens. Inhibitors were added to mixed methanogenic cultures in an attempt to determine which bacteria group(s) are responsible for or involved in PCE degradation. The PCE degradation efficiency was also tested at different bacterial biomass levels to determine changes in degradation rates and patterns.

METHODS

All experiments were conducted in 160-mL serum bottles containing anaerobic biomass (originally from a sludge digestor), electron donor, bicarbonate, resazurin, reducing agent, and nutrient solution, a total of 100 mL, generally following the procedure of Shelton and Tiedje (1984). The bottles were incubated at 36°C. Live bottles were compared to blanks and killed controls. The bottles were analyzed for gas production, gas composition, volatile fatty acids (VFA), and chlorinated compounds every 3 to 6 days. Ethene was analyzed for on a few occasions in the end of the experiments. The chloroethenes were determined with EPA Method 601. Ethane was measured by gas chromatography with flame ionization detection (GC-FID), but not as often as the chloroethenes. Usually the experiment lasted until the bacteria used all of the chemical oxygen demand (COD) (gas production stopped). The bottles were then refed and respiked for a new experiment.

RESULTS AND DISCUSSION

Use of Different Electron Donors. Different parts of the consortia were fed by providing different electron donors: fructose, phenol/benzoate, hydrogen, and acetate. Three refeedings were conducted with increasing PCE concentrations (0.5 to 5 mole/bottle). Use of different electron donors initially showed only small differences in degradation rate, but the degradation rates improved for subsequent refeedings in the bottles fed more complex compounds (fructose, benzoate) compared to the bottles with simple compounds (acetate, hydrogen).

The transformation patterns in general were the same in all bottles, independent of electron donor. Figure 1 shows PCE degradation in one of the fructose-fed bottles and represents the general biotransformation pathway of PCE (theoretical PCE concentration added is shown with an arrow). The degradation proceeded all the way to ETH. TCE and *cis*-1,2-DCE were produced in trace amounts, whereas VC was produced after a few days and accumulated; ETH also was seen (analyzed on 3 occasions only). The transformation of VC to ETH apparently was rate limiting.

Several tests, including autoclaved, cyanide killed, salt control, unfed, inhibited by both 2-bromoethanesulfonic acid (BESA), and vancomycin, were conducted to determine if abiotic reactions took place. No significant degradation was observed in any of the bottles, although trace amounts of TCE were seen in some tests. It is suggested that no significant abiotic or biologically mediated abiotic reactions, as seen in the study of Gantzer and Wackett (1991), took place.

Degradation of PCE at Different Biomass Levels. Biomass (originally 80 mg/L volatile suspended solids [VSS]) was diluted into ½, ¼, ⅛, and 1/16 fractions; placed in new serum bottles; respiked with PCE; and refed phenol and benzoate.

Figure 2a shows that with ½ the original biomass, PCE degraded to significant amounts of TCE and traces of *cis*-DCE before the appearance of VC. In the second

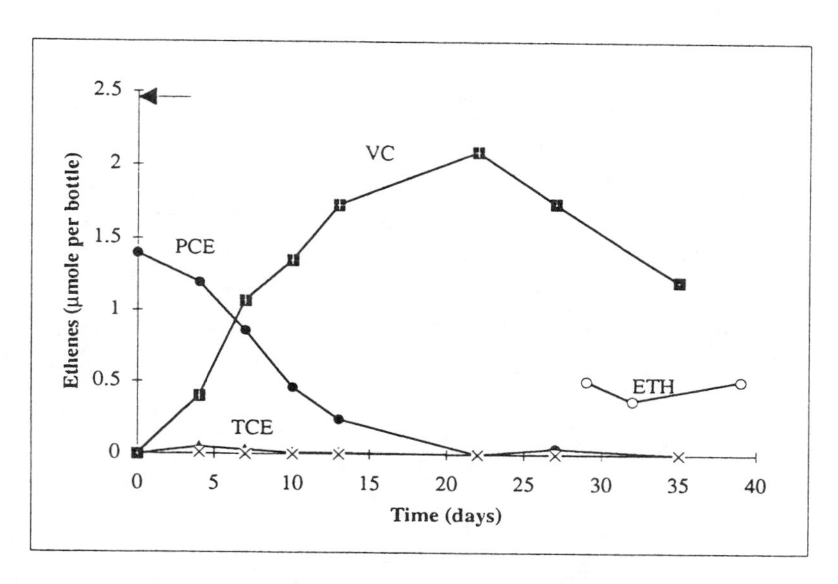

FIGURE 1. Reductive dechlorination of PCE in a mixed culture fed benzoate and phenol as the electron donors. Amount theoretically added is shown with an arrow.

dilution (¼), TCE accumulated to a larger extent and persisted to the end of the experiment; *cis*-DCE also persisted. VC was produced in trace amounts but accumulated only late in the experiment. Only a trace of ETH was present (analyzed on one occasion only). With $^1/_8$ of the original biomass (Figure 2b), PCE persisted throughout the experiment, and TCE was just beginning to accumulate. The degradation with $^1/_{16}$ of the biomass was virtually identical to that shown in Figure 2b.

Although the extent of dehalogenation was markedly affected, the PCE degradation rates in the ½ and ¼ dilutions were the same, and the rates in the $^1/_8$ and $^1/_{16}$ dilutions were reduced by only about $^1/_3$. Thus, the specific rate of PCE degradation increased as the biomass was reduced.

Then 35 days after the last analysis, the bottles were refed with benzoate, respiked with PCE, and analyzed after 23 days of incubation. All the PCE was converted to VC and ETH in all bottles, indicating acclimation or growth of bacteria active in dehalogenation.

Effect of BESA and Vancomycin. BESA, an inhibitor of methane formation, stopped methane formation and severely retarded but did not stop PCE transformation. This was observed in bottles fed fructose, benzoate/phenol, acetate, and hydrogen. Irrespective of electron donor, BESA reduced dehalogenation rates by more than 50%, and TCE was the only significant metabolite. The methanogens were completely inhibited and presumably could not be responsible for the degradation of PCE. Accumulation of acetate showed that the acetogens were still active and may have been responsible for the degradation.

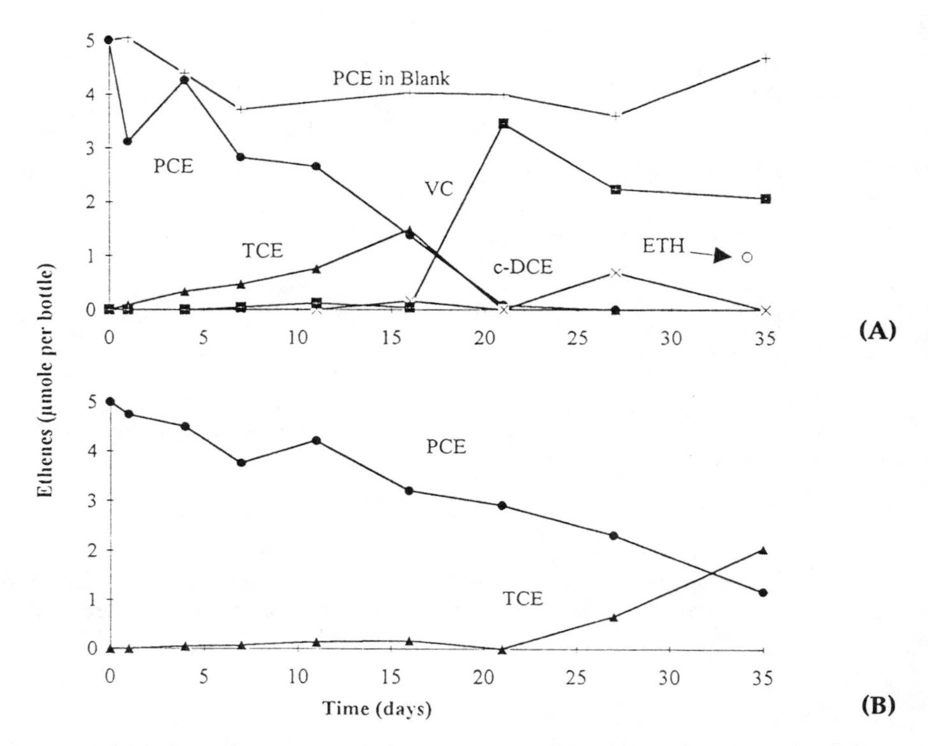

FIGURE 2. Reductive dechlorination of PCE in a mixed culture fed fructose as the electron donors, containing (A) ½ of the original bottle biomass (40 mg VSS/L), and (B) ⅛ of the original biomass (10 mg VSS/L).

In some BESA-inhibited bottles (0.5 to 1 mmol), small amounts of methane were produced after 2 to 4 weeks of incubation. Large amounts of ETH were produced at the same time, indicating breakdown of BESA (Belay & Daniels 1987).

Vancomycin (100 mg/L), an inhibitor of eubacteria, was added to benzoate-fed cultures. PCE degradation slowed, but complete conversion to VC and ETH occurred within 34 days. Vancomycin did not completely stop production of VFAs, indicating that the eubacteria were not completely inhibited. These inhibition results were inconclusive. Overall results do show that the degradation was most effective when neither inhibitor was present, which may indicate a cooperation between the bacterial groups.

CONCLUSIONS

PCE was dechlorinated to ethene under anaerobic conditions within 2 to 4 weeks, but the degradation of VC to ETH was rate limiting. Only trace amounts of TCE and DCE usually were seen. This conversion was demonstrated with acetate, hydrog n, fructose, and benzoate as electron donors.

of TCE and DCE usually were seen. This conversion was demonstrated with acetate, hydrogen, fructose, and benzoate as electron donors.

The degradation became more effective in the bottles successively fed the more complex compounds (fructose, benzoate).

Serial dilutions led to slower PCE transformation and buildup of TCE and *cis*-DCE. On refeeding and respiking, the differences in degradation largely disappeared, indicating acclimation or growth of the PCE degradation activity.

It cannot be concluded that either the methanogens or eubacteria were solely responsible for the PCE dechlorination. Both BESA and vancomycin retarded PCE dehalogenation. However, degradation proceeded slowly with each inhibitor; and there was reason to believe that neither inhibitor was completely effective against the targeted groups of bacteria.

ACKNOWLEDGMENTS

This research was supported by the Valle Exchange Program, The Chlorine Institute, and the National Institute of Environmental Health Sciences (Grant 2P42 ES 04696).

REFERENCES

Belay, N., and L. Daniels. 1987. "Production of Ethane, Ethylene, and Acetylene From Halogenated Hydrocarbons by Methanogenic Bacteria." *Applied Environ. Microbiol. 53*: 1604-1610.

Bouwer, E. J., and P. L. McCarty. 1983. "Transformation of 1- and 2-Carbon Halogenated Aliphatic Organic Compounds under Methanogenic Conditions." *Applied Environ. Microbiol. 45*: 1286-1294.

de Bruin, W. P., M. J. J. Kotterman, M. A. Posthumus, G. Schraa, and J. B. Zehnder. 1992. "Complete "Biological Reductive Transformation of Tetrachloroethene to Ethane." *Applied Environ. Microbiol. 58*: 1996-2000.

DiStefano, T. D., J. M. Gossett, and S. H. Zinder. 1991. "Reductive Dechlorination of High Concentrations of Tetrachloroethene to Ethene by an Anaerobic Enrichment Culture in the Absence of Methanogenesis." *Applied Environ. Microbiol. 57*: 2287-2292.

Freedman, D. L., and J. M. Gossett. 1989. "Biological Reductive Dechlorination of Tetrachloroethylene and Trichloroethylene to Ethylene under Methanogenic Conditions." *Applied Environ. Microbiol. 55*: 2144-2151.

Gantzer, C. J., and L. P. Wackett. 1991. "Reductive Dechlorination Catalyzed by Bacterial Transition-Metal Coenzymes." *Environ. Sci. Technol. 25*: 715-722.

Parsons, F., P. R. Wood, and J. DeMarco. 1984. "Transformation of Tetrachloroethene and Trichloroethene in Microcosms and Groundwater." *Journal of American Water Works Assoc.* February, pp. 56-59.

Shelton, D. R., and J. M. Tiedje. 1984. "General Method for Determining Anaerobic Biodegradation Potential." *Applied Environ. Microbiol. 47*: 850.

EVALUATING IN SITU TRICHLOROETHENE BIOTRANSFORMATION USING IN SITU MICROCOSMS

E. E. Cox, D. W. Acton, and D. W. Major

INTRODUCTION

A recent investigation by Major et al. (in submission) at a study site in Merced, California, demonstrated that the microbial biomass in trichloroethene (TCE) contaminated aquifer sediments could mineralize TCE to CO_2 under both aerobic and anaerobic conditions. Anaerobic mineralization of TCE was observed with and without the production of dechlorination intermediates (*c*DCE, *t*DCE, and VC). Based on these results, six in situ microcosms (ISMs) were installed at a depth of 21.3 m below ground surface to verify that the TCE biotransformation observed in these laboratory experiments can occur under field conditions.

MATERIALS AND METHODS

ISM Design and Installation. The ISM units used in this study are similar to those described by Gillham et al. (1990). Each ISM was modified to include a stainless steel, in-line, nitrogen-drive pump that was connected to the sample spike and the sample lines (see Figure 1). Teflon™ lines (0.3175 cm O.D.) were used to sample groundwater from the test chamber, and nylon lines (0.48 cm O.D.) were used to operate the nitrogen-drive pumps.

A boring was drilled to the top of the confined aquifer (approximately 20.7 m below ground surface) using hollow-stem augers (10.8 cm I.D.). An ISM was attached to BW drill casing and was pushed into undisturbed aquifer material in advance of the lead augers using drill rig hydraulics. A reverse thread breakaway coupling allowed detachment of the drill casing from the ISM head. Sample lines were run to the surface through Schedule 40 PVC (5 cm O.D.) protective riser attached to the head of the ISM. A sand pack (#1 Monterey sand) was placed in the boring annulus above the ISM head, and the boring was grouted to the surface.

Experimental Treatments. Three treatments were investigated in duplicate to assess the biotransformation potential of TCE by the indigenous microbial

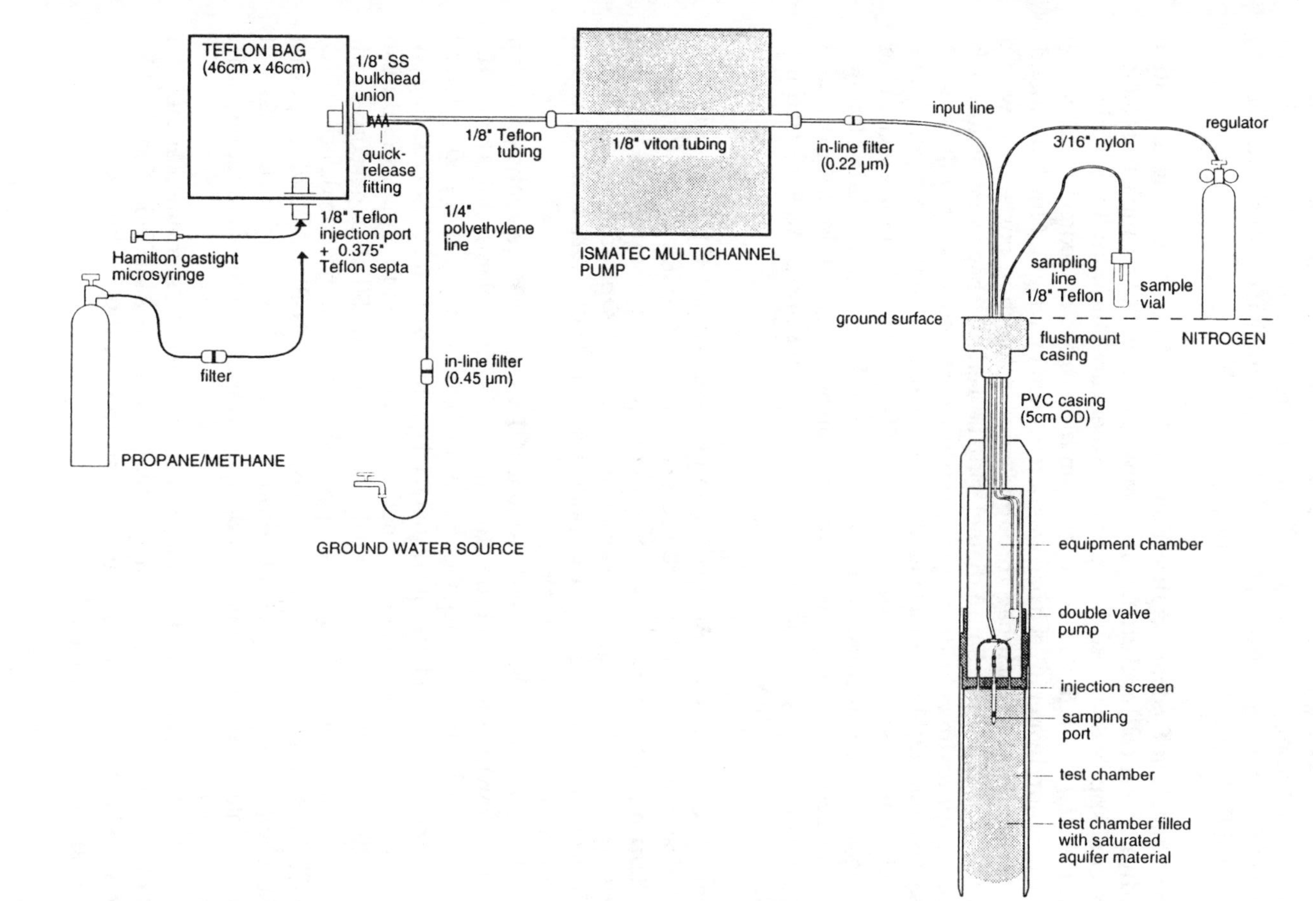

FIGURE 1. Schematic of the in situ microcosm (ISM) injection system.

community in the confined aquifer: (1) an unamended control supplemented with 1.0 mg/L TCE, and 100 mg/L NaBr; (2) an aerobic treatment amended with 1.0 mg/L TCE, 2.5 mg/L propane, 2.5 mg/L methane, 10 mg/L $NaNO_3$, 5 mg/L K_2HPO_4, and 100 mg/L NaBr; and (3) an anaerobic treatment amended with 1.0 mg/L TCE, 50 mg/L $NaCH_3CO_2^-$, 500 mg/L methanol, 10 mg/L $NaNO_3$, 5 mg/L K_2HPO_4, and 100 mg/L NaBr.

ISM Injection System. Figure 1 shows the ISM injection system. Native groundwater was collected directly into preweighed Teflon™ sample bags (46 cm by 46 cm, approximate 10-L capacity, American Durafilm Inc., Holliston, Massachusetts) through an in-line 0.45 µm high capacity Geotech™ filter; the volume of groundwater in each bag was determined by weight. Nutrient amendments were added to each bag through Teflon™ injection ports lined with a Teflon™-faced septum (9.5 mm diameter). The Teflon™ bags were vigorously shaken for approximately 10 minutes to ensure complete mixing of the amendments. An Ismatec™ 12-channel peristaltic pump (Cole Palmer, Chicago, Illinois) was used to inject the solutions from the Teflon™ bags into the ISM test chamber. Each injection line was fitted with a sterile in-line 0.2 µm filter (Lida Manufacturing, Chicago, Illinois) to prevent introduction of microorganisms into the ISM. One pore volume (approximately 1.0 L) of injection solution was injected into each ISM, as determined by bag weight.

Sample Collection and Analysis. Samples were collected periodically from each ISM as the experiment progressed to determine if TCE biotransformation was occurring. Groundwater samples were collected without headspace into 20-mL glass vials. Each vial was sealed with a Teflon™ septum-lined screw cap. Samples were shipped on ice to the laboratory and generally were analyzed within 48 hours from the time of sampling.

Methane, ethane, ethene, and propane (C1-C4) gases were analyzed with a Hewlett Packard 5840A gas chromatograph (GC) with a flame ionization (FID) detector, and a 30-m GS-Q megabore column. The GC was calibrated (external standard or ESTD method) using an analyzed gas mixture, and dissolved concentrations were determined using the method described in Moraghan and Buresh (1977). Chlorinated ethenes were analyzed by injecting 50 µL of gaseous headspace onto a Photovac™ 10S70 GC equipped with a photoionization detector (PID) and a CP-Si15 capillary column. The Photovac™ was calibrated with external standards for each sampling event. Bromide was analyzed at the University of Waterloo, Ontario, using a Dionex ion chromatography unit with a Millipore AS4A ion exchange column.

Dissolved oxygen concentrations in each ISM were determined at the start and the end of the experiment using a Hach™ OX-DT test kit (modified Winkler method) and a digital titrator. The detection limit for this method was approximately 0.01 mg/L of dissolved oxygen.

RESULTS AND DISCUSSION

This paper presents preliminary data obtained from six sample events over a 101-day period. The TCE and bromide concentrations shown are an average of duplicate ISMs for each treatment investigated.

Unamended In Situ Microcosms. Figure 2 shows that TCE is not being biotransformed in ISMs that have not received nutrient amendments. Trace concentrations of methane were detected at days 11, 38, and 52 (6.8, 14.5, and 3.4 µg/L, respectively). The presence of methane in the control ISMs may be the result of anaerobic (methanogenic) activity that develops in these microcosms as dissolved oxygen concentrations become depleted.

Anaerobic In Situ Microcosms. Figure 3 demonstrates that TCE is being rapidly biotransformed in ISMs amended with acetate and methanol. Trace concentrations of *cis*-1,2-DCE and *trans*-1,2-DCE ranging from 1.0 to 11.5 µg/L were detected at days 24, 38, 52, and 101. Vinyl chloride was identified at trace concentrations at days 38, 52, and 101 (1.1 to 7.9 µg/L). Methane was detected at day 38 (61.7 µg/L) and day 52 (114.9 µg/L). Ethene was not detected in groundwater samples from the anaerobic microcosms.

These results are similar to those of Cox et al. (1993) who found that radiolabeled TCE could be mineralized under anaerobic conditions. As with our results, Cox et al. (1993) detected only trace concentrations of *cis*-1,2-DCE, *trans*-1,2-DCE, and VC produced during radiolabeled TCE mineralization to $^{14}CO_2$. Concentrations of CO_2 were not quantified in the anaerobic ISMs.

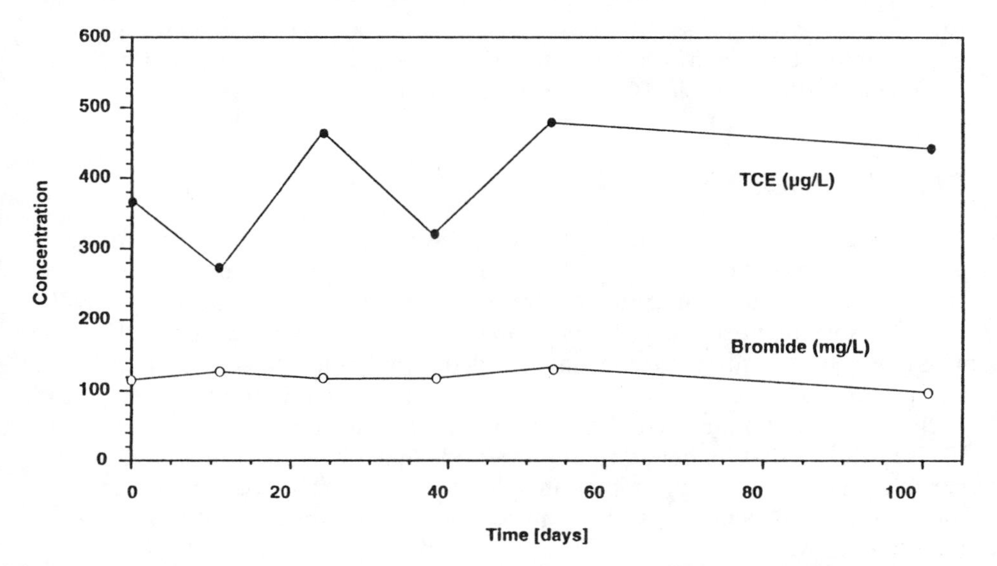

FIGURE 2. In situ TCE biotransformation in the unamended ISMs.

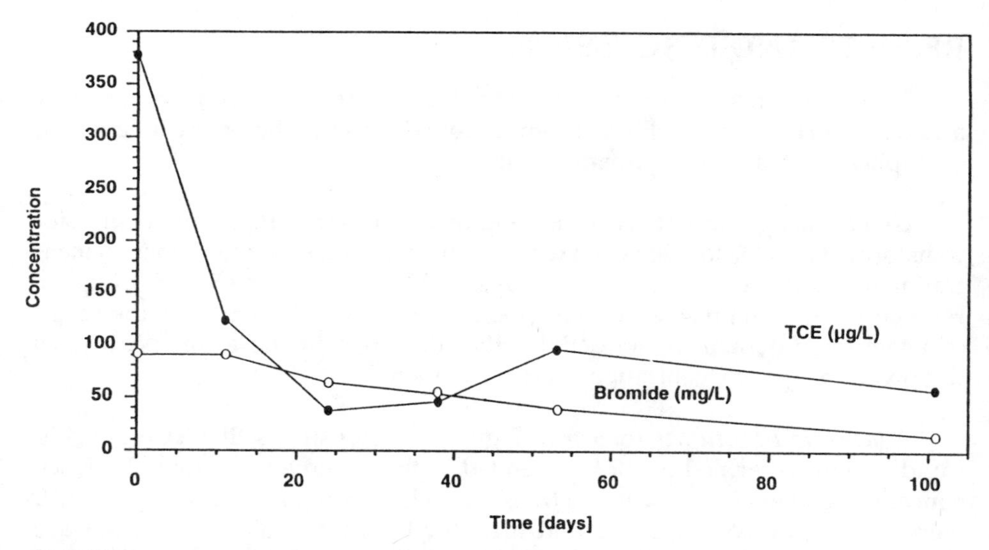

FIGURE 3. In situ TCE biotransformation in the anaerobic ISMs, in the presence of methanol and acetate.

Aerobic In Situ Microcosms. The addition of methane and propane to the aerobic ISMs was unsuccessful in enhancing TCE biotransformation in these microcosms. The concentrations of methane and propane did not decrease over the 101-day period. The lack of TCE biotransformation in the aerobic microcosms may be attributable to insufficient dissolved oxygen concentrations. It is possible that the background dissolved oxygen concentrations (~7.0 mg/L) used in this study rapidly became limiting due to either biological or abiotic reactions, thus inhibiting further aerobic TCE biotransformation.

CONCLUSIONS

The data of the preliminary screening study suggest that the ISMs are good test cells for assessing the in situ biotransformation rates of chemicals in the subsurface. The screening demonstration showed that TCE can be relatively quickly biotransformed in situ under anaerobic conditions induced at this site. The detection of only trace concentrations of vinyl chloride suggest that VC is being biotransformed; however, ethene was not detected in the anaerobic ISMs and CO_2 was not quantified. The lack of TCE biotransformation in the aerobic ISMs indicates that longer adaptation periods may be required to stimulate the aerobic populations (e.g., methanotrophs) to biotransform TCE in situ and/or more frequent injections may be required to maintain the aerobic conditions in the microcosms.

The in situ TCE biotransformation study is continuing, focusing primarily on anaerobic TCE biotransformation in order to obtain a carbon mass balance and to ensure that TCE is being completely mineralized to innocuous products at this site.

ACKNOWLEDGMENTS

We thank the General Electric Company for funding this research, Dr. Deborah Hankins for her support, and Phil Walsack and Brian Jackson for their assistance in collecting samples.

REFERENCES

Cox, E. E., D. W. Major, D. W. Acton, T. J. Phelps, and D. C. White. 1993. "Evaluating trichloroethene biodegradation by measuring the in situ status and activities of microbial populations." *Proceedings of the Second International Symposium on In Situ and On Site Bioreclamation.* (paper E-19, Vol. 1).

Gillham, R. W., M. J. L. Robin, and C. J. Ptacek. 1990. "A device for *in situ* determination of geochemical transport parameters: 1. Retardation." *Groundwater 28*(5):666-672.

Gillham, R. W, R. C. Starr, and D. J. Miller. 1990. "A device for *in situ* determination of geochemical transport parameters: 2. Biochemical reactions." *Groundwater 28*(6):858-862.

Moraghan, J. T., and R. Buresh. 1977. "Correction for dissolved nitrous oxide in nitrogen studies." *Soil Sci. Soc. Am. J. 41*(6):1201-1202.

IN SITU REMEDIATION USING ANAEROBIC BIOTRANSFORMATION OF GROUNDWATER CONTAMINATED WITH CHLORINATED SOLVENTS[1]

G. J. Smith and G. A. Ferguson

INTRODUCTION

When closing down a manufacturing facility in the midwestern United States, chlorinated solvents in the form of trichloroethene (TCE) and 1,1,1-trichloroethane (1,1,1-TCA) were discovered in the soils and groundwater beneath the plant. The solvents were observed to be undergoing a natural transformation, resulting in the detection of dichlorinated and monochlorinated aliphatics in the soils and groundwater. It was suspected that the transformation was biologically mediated.

During the investigation and testing activities, it was discovered that in a significant portion of the area, the solvents were present in an undissolved state, or as a dense, nonaqueous phase liquid (DNAPL). The DNAPL was found to be toxic to the microorganisms responsible for biotransformation. The remedial approach combines a thermal approach for the removal of the DNAPL (not discussed herein), coupled with nutrient treatment of the aqueous phase to promote biotransformation of TCE and 1,1,1-TCA to daughter compounds that are more easily removed from the subsurface.

BIOTRANSFORMATION TO ACCELERATE GROUNDWATER CLEANUP

Compounds such as TCE and 1,1,1-TCA are readily biodegradable under anaerobic conditions (Vogel & McCarty 1985), producing dichlorinated and monochlorinated alkenes and alkanes. The compounds have different chemical characteristics than their parent compounds (such as aqueous solubilities, vapor pressures, and octanol-water and octanol-carbon partition coefficients), and hence have different fate and transport characteristics. This has been noted by Tsentas and Supkow (1985), who observed that the biodegradation daughter products predominate at the front of a plume of chlorinated aliphatics in groundwater.

---

[1] Patent applied for, for process and apparatus.

Jackson and Patterson (1989) observed this phenomenon and likened it to a chromatographic dispersion in an aquifer. These authors developed a model (modified herein for chlorinated alkenes and alkanes) where the number of pore volumes to obtain 90% decontamination of an aquifer is presented against the octanol-water partition coefficient (Figure 1). According to this model, 30.5 pore volumes would be required to remove 90% of TCE in an aquifer, versus 4.4 and 5.3 pore volumes for *trans*- and *cis*-1,2-dichloroethene, respectively. Our approach is to enhance the natural biotransformation of the compounds producing the dechlorinated daughter compounds to reduce the number of pore volume flushes, shorten the remedial time frame, and therefore, reduce the cost of remediation.

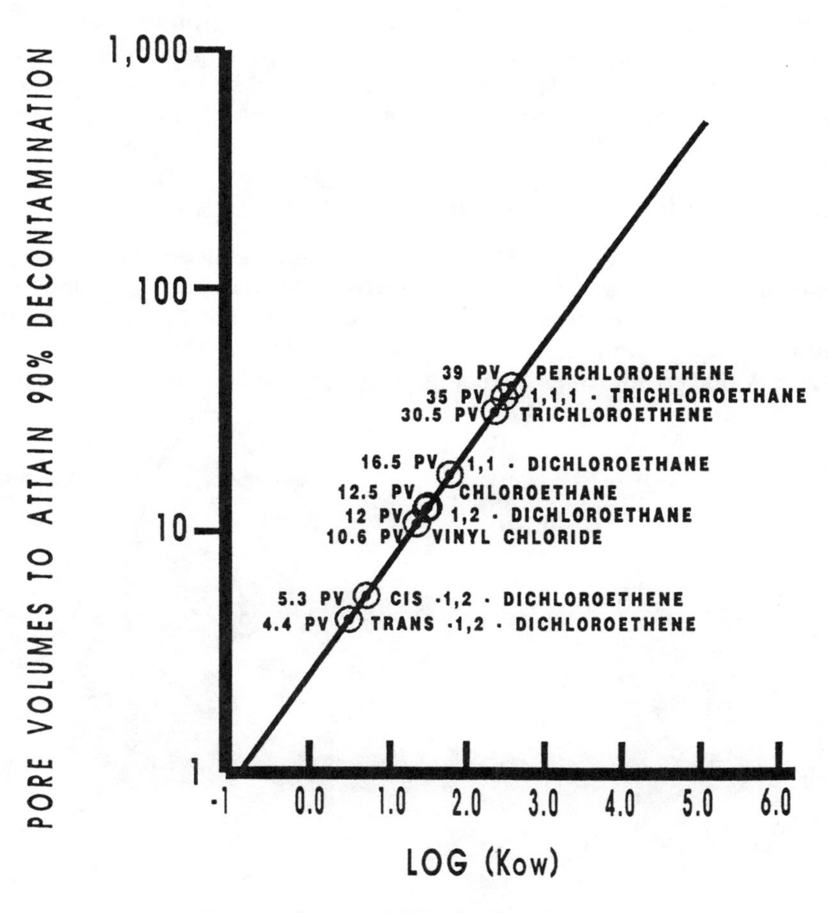

FIGURE 1. Estimation of number of pore volumes to attain 90% aquifer decontamination. Modified from Jackson and Patterson (1989).

PROJECT BACKGROUND

The manufacturing facility is located on a lacustrine sequence of sediments consisting of fine sands and silts and clay interbeds, and underlain by dense glacial clay till. The lacustrine sequence extends to a depth of approximately 6 m, with the water table found at a depth of approximately 2 m below the floor of the facility. Underlying the lacustrine sequence is dense glacial clay. The lacustrine sediments have measured hydraulic conductivities in the range of 10^{-5} to 10^{-4} cm/sec. With the available drawdown in the lacustrine sequence, sustained yields for a given groundwater extraction well were estimated to be very low, at less than 0.026 L/min.

Our approach for the groundwater remediation is to heat the subsurface and simultaneously provide nutrients while performing groundwater and vapor extraction. This has been done to modify subsurface conditions to produce compounds more amenable to vapor and groundwater extraction. A conceptual depiction of the equipment process is presented in Figure 2.

OBSERVATIONS ON REMEDIAL PROGRESS

To cost effectively monitor the progress of the remediation, we analyzed samples from the individual extraction wells using a headspace extraction technique, with the chemical analysis performed using a Hewlett Packard HP-5980 gas chromatograph equipped with an electron capture detector. Standards for TCE, 1,1,1-TCA, and *cis*- and *trans*-dichloroethene were used to calibrate the instrument.

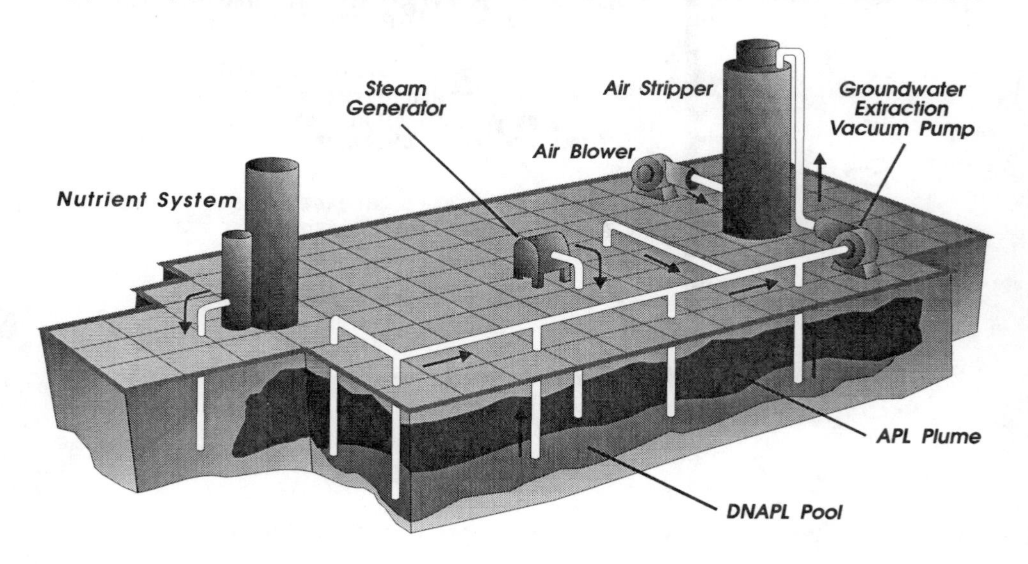

FIGURE 2. Conceptual view of process and equipment.

We sampled and analyzed groundwater from the extraction wells in December 1991 and April, July, and October 1992. From these samplings, it is interesting to note that *cis*-1,2-dichloroethene was detected in 86% of the extraction wells in December 1991, but dropped to 37% in April 1992, 20% in July 1992, and 7.5% in October 1992. The corresponding percentage of extraction wells showing the detection of *trans*-1,2-dichloroethene was 89% in December 1991, 91% in April 1992, 82% in July 1992, and 82% in October 1992. Wood et al. (1985) indicates that *cis*-1,2-dichloroethene is produced in preference to other daughter compounds from the biotransformation of TCE. Therefore, we believe that the reduction in concentration and quantity of *cis*-1,2-dichloroethene may be a result of its transport and fate characteristics. The *cis*-isomer has a higher Henry's law constant than the *trans*-isomer. The most significant reduction in *cis*-1,2-dichloroethene occurred from December 1991 to April 1992. During this period, there was significant removal of hydrocarbon as a vapor (see Figure 3).

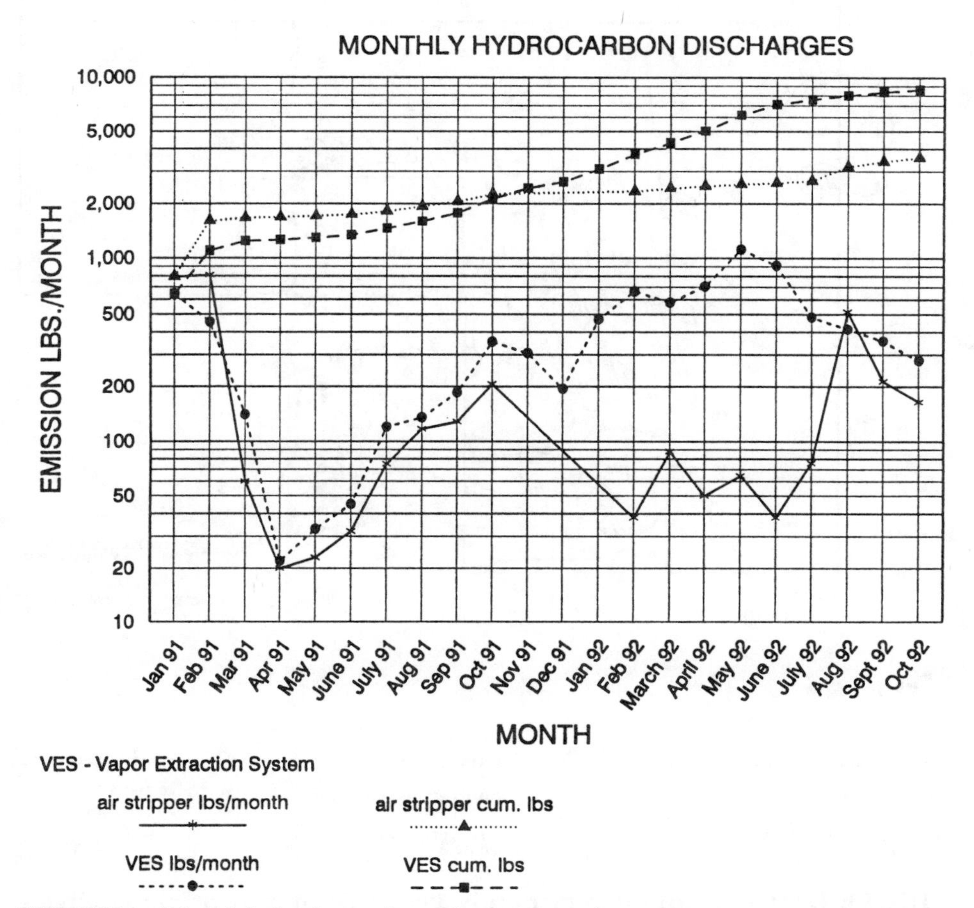

FIGURE 3. Monthly hydrocarbon discharges. (To convert lb into kg, divide by 2.21.)

The analytical results from the extraction well sampling (described above) were used to construct concentration distribution maps for TCE and 1,1,1-TCA (see Figures 4 and 5). The areas above the cleanup criteria have been dramatically reduced in size from the original plume. Of the 94 wells sampled in July 1992, 86.2% were below the cleanup criteria for 1,1,1-TCA, versus 75% in April 1992 and 64% in December 1991. Further, 58.8% of the wells sampled in July 1992 were below the cleanup criteria for TCE, versus 48% in April 1992 and 24% in

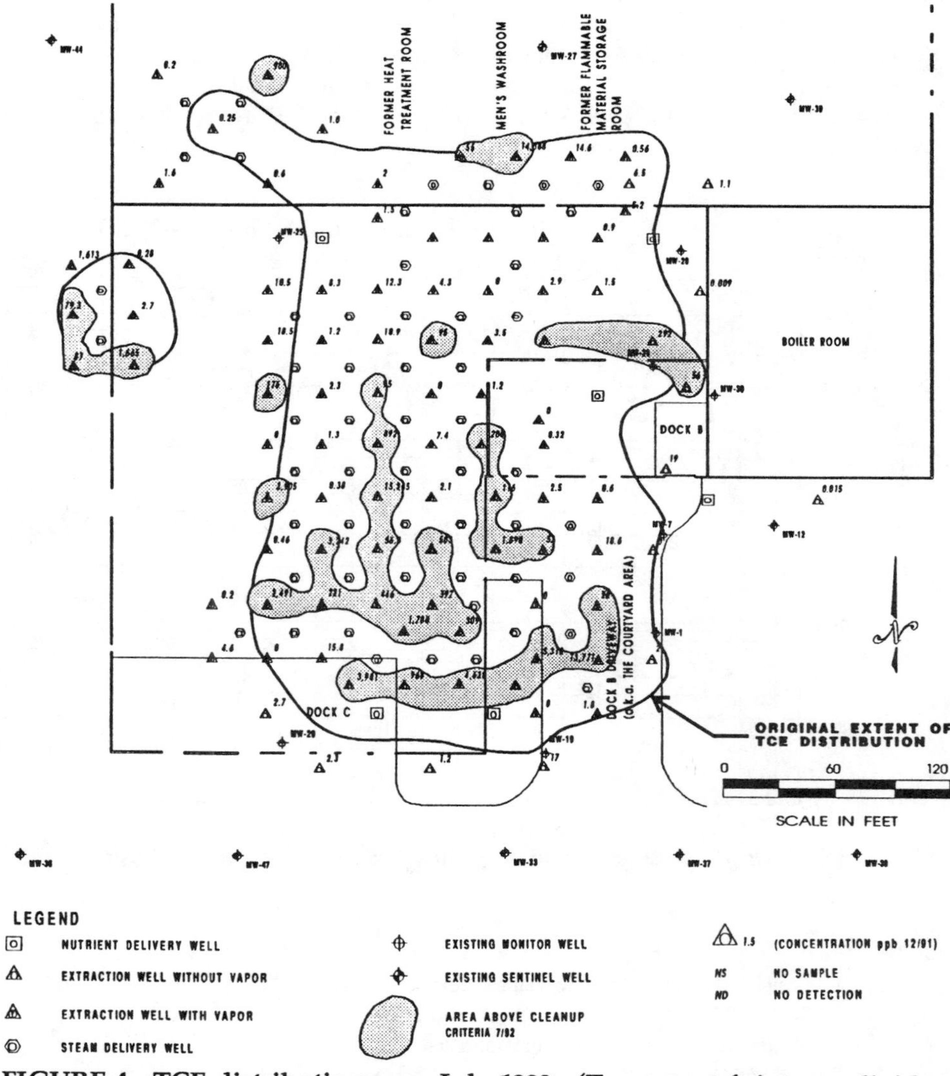

FIGURE 4. TCE distribution map, July 1992. (To convert ft into m, divide by 3.28.)

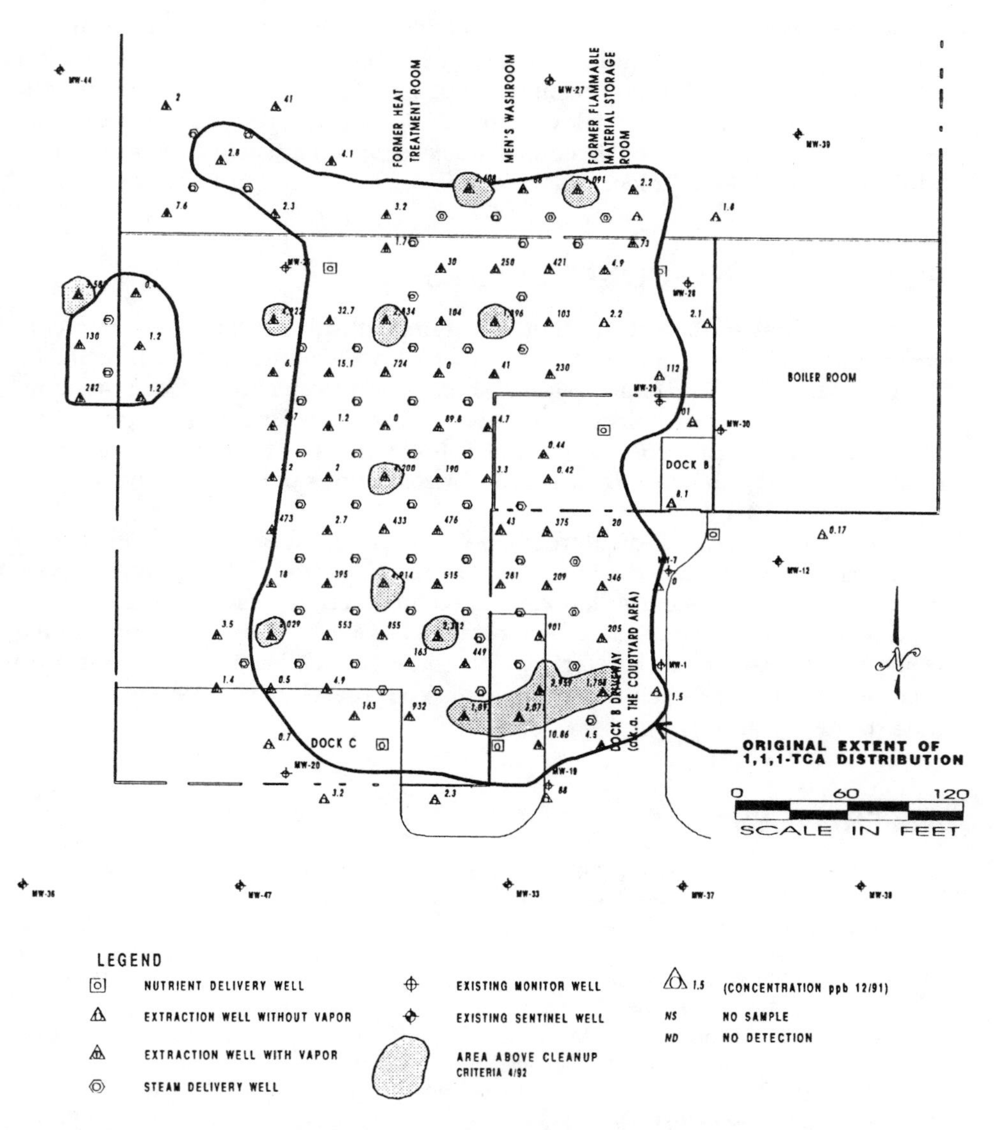

FIGURE 5. 1,1,1-TCA distribution map, July 1992. (To convert ft into m, divide by 3.28.)

December 1991. Maximum concentrations of 1,1,1-TCA have been reduced by an order of magnitude in each of these samplings.

The progress of the remedial effort was further evaluated by comparing the ratios of the concentrations of the daughter to parent compounds. At many locations it was observed that the ratios of the dichlorinated ethenes to TCE were decreasing as the remediation progressed. It was observed that these areas contained concentrations of 1,1,1-TCA at concentrations greater than 5 mg/L. Vargas and Ahlert (1987) indicate that toxic inhibition of the biotransformation of TCE can occur with the concentration of 1,1,1-TCA as low as 2 mg/L with bioactivity returning to normal levels after 20 days. Observations from test areas where nutrients have resided in the subsurface for approximately 2 years indicated inhibition is still taking place.

SUMMARY AND CONCLUSIONS

We have taken the approach of modifying subsurface conditions to create compounds that are more easily extracted to enhance remedial efforts. Contaminant removal is taking place via both groundwater and vapor extraction.

Dramatic reductions in plume size have been observed in the application of this remedial approach. As the area of the plume has decreased, there have been significant reductions in chemical concentrations in the groundwater (see Figures 4 and 5).

We found that *cis*-1,2-dichloroethene has been removed from the subsurface at a much faster rate than the *trans*-isomer. The *trans*-isomer has a higher vapor pressure, but a lower Henry's law constant than the *cis*-isomer. Because a greater percentage of the solvents have been removed as a vapor, the differences in the Henry's law constant may be a factor. If conditions can be manipulated to produce predominantly *cis*-1,2-dichloroethene, it may be possible to shorten the remedial time frame.

REFERENCES

Jackson, R. E., and R. J. Patterson. 1989. "A Remedial Investigation of an Organically Polluted Outwash Aquifer." *Groundwater Monitoring Review, 3*:(9).

Tsentas, C., and D. J. Supkow. 1985. "Migration and Apparent Biodegradation of Organic Compounds in a Fractured Bedrock Aquifer." *Proceedings of the NWWA/API Conference on Petroleum Hydrocarbons and Organic Chemicals in Groundwater — Prevention, Detection, Restoration*, pp. 77-89, Houston, TX.

Vargas, C., and R. C. Ahlert. 1987. "Anaerobic Degradation of Chlorinated Solvents." *Journal of the Water Pollution Control Federation 59*(11): 964-968.

Vogel, T. M., and P. L. McCarty. 1985. "Biotransformation of Tetrachloroethylene to Trichloroethylene, Dichloroethylene, Vinyl Chloride, and Carbon Dioxide Under Methanogenic Conditions." *Applied Environ. Micro. 49*(5): 1208-1213.

Wood, P. R., R. F. Land, and I. L. Payan. 1985. "Anaerobic Transformation, Transport, and Removal of Volatile Chlorinated Organics in Groundwater." In: C. H. Ward, W. Giger, and P. L. McCarty (Eds.), *Groundwater Quality*, pp. 493-511. John Wiley & Sons, New York, NY.

SOLUBLE METHANE MONOOXYGENASE ACTIVITY IN *METHYLOMONAS METHANICA* 68-1 ISOLATED FROM A TRICHLOROETHYLENE-CONTAMINATED AQUIFER

S.-C. Koh, J. P. Bowman, and G. S. Sayler

It has been postulated that certain habitats, in particular groundwater, lacking readily available copper but with abundant methane and oxygen, may provide a suitable niche for methanotrophs producing soluble methane monooxygenase (Dalton 1992). Soluble methane monooxygenase (sMMO), which occurs only in a select variety of methanotrophs, has an extremely broad substrate specificity. Most recent research has focused on sMMO degradation of chlorinated aliphatic compounds, in particular trichloroethylene. In situ bioremediation efforts involving methane/air injection into subsurface sediments and groundwaters (Hazen 1992) are focused on stimulating the growth of methanotrophs that produce this enzyme.

It has been thought that sMMO-producing methanotrophs belong only to two of the three methanotroph taxonomic groups or types, including types II and X. As part of a larger investigation on methanotroph population dynamics, this study reports on the first type I methanotroph able to produce sMMO. Little et al. (1988) isolated a number of type I methanotrophs not suspected of producing sMMO from an aquifer contaminated with TCE and a wide variety of other contaminants. The isolate primarily investigated, 46-1, degraded TCE only slowly at a level comparable to that of particulate methane monooxygenase (pMMO) (DiSpirito et al. 1992). Another strain was isolated by Little et al. (1988), but was not further investigated. This strain, designated 68-1, was a pink-pigmented, rod-shaped bacterium identified as a *Methylomonas methanica* (Koh et al. 1993). It was capable of rapidly oxidizing naphthalene to naphthol and degrading TCE when grown in copper-free or copper-limiting conditions (Figure 1; Koh et al. 1993). When grown with 1 µM $CuSO_4$, naphthalene oxidation was completely absent. As it is known that copper represses the synthesis of sMMO, it was believed 68-1 harbored both a pMMO and an sMMO, which are both regulated by copper availability as seen in *Methylosinus trichosporium* OB3b and *Methylococcus capsulatus* Bath. Naphthalene oxidation has been used to indicate the presence of sMMO activity (Brusseau et al. 1990). Naphthalene is specifically oxidized to 1- or 2-naphthol. Naphthol can be detected colorimetrically after reacting it with Fast B Blue salt. The reaction with Fast B Blue salt yields a high extinction

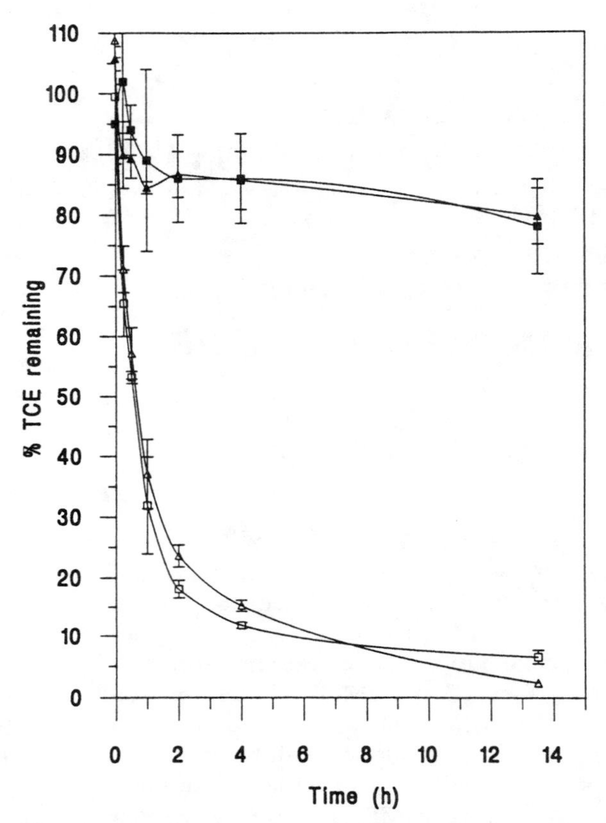

FIGURE 1. TCE degradation by viable cell suspensions of *Methylomonas methanica* 68-1 (□) and *Methylosinus trichosporium* OB3b (Δ) in early stationary growth phase was incubated with 20 mg/L TCE. Heat-killed controls of 68-1 (■) and OB3b (▴) also are shown for comparison (data from Koh et al., 1993).

coefficient diazo compound (Brusseau et al. 1990). Using a modified version of the naphthalene oxidation assay (Brusseau et al. 1990), it was possible to accurately quantify sMMO activity.

Various comparisons were made between *Methylomonas methanica* 68-1 and the much more heavily studied sMMO-producing, type II methanotroph *Methylosinus trichosporium* OB3b. These strains have been compared here with regard to growth kinetics, naphthalene oxidation and trichloroethylene (TCE) degradation kinetics, response to TCE toxicity and TCE degradation capacity, and chlorinated aliphatic compound degradation rates.

Growth kinetic experiments were performed in nitrate mineral salts medium (Koh et al. 1993) in which the strains were cultivated under a 1:4 methane:air atmosphere at 25°C. Methane consumption was determined using gas chromatography equipped with a flame ionization detector. Growth kinetic data suggested that conditions required for maximal sMMO production were slightly suboptimal for both the strains (Table 1). Interestingly, higher specific growth rates (μ_{max}) and growth yields (Y_s) were seen with 68-1 when compared to OB3b in both copper-rich and copper-limiting conditions (Table 1).

Naphthalene oxidation and TCE degradation kinetic experiments were performed using the procedure of Brusseau et al. (1990) using resting cell cultures supplemented with 20 mM sodium formate. TCE degradation was monitored using a gas chromatogram equipped with an electron capture detector. Both K_m and V_{max} values were calculated from equations derived from rectangular hyperbolic curve fits performed by the computer program DeltaGraph (DeltaPoint, Inc., Monterey, California). The substrate affinity (K_m) of the sMMO of 68-1 for both naphthalene and TCE was somewhat lower than the K_m of the sMMO of OB3b; however, the maximal specific activity (V_{max}) of the sMMO in 68-1 was significantly higher than that of *Methylosinus trichosporium* OB3b (Table 1). A possible reason for this could be that 68-1 has a higher proportion of its total cell protein consisting of sMMO compared to OB3b's proportion. Naphthalene

TABLE 1. Growth and naphthalene and TCE degradation kinetic parameters of *Methylomonas methanica* 68-1 and *Methylosinus trichosporium* OB3b.

Parameters	*Methylomonas methanica* 68-1	*Methylosinus trichosporium* OB3b
NMS[(a)] + 1 µM $CuSO_4$:		
μ_{max} (h^{-1})	0.074	0.060
Y_s (g cells/g CH_4)	0.80	0.67
NMS:		
μ_{max} (h^{-1})	0.060	0.053
Y_s (g cells/g CH_4)	0.73	0.63
Naphthalene oxidation[(b)]:		
K_m (µM)	70 ± 4	40 ± 3
V_{max} ($nmol^{-1}$ h^{-1} mg $protein^{-1}$)	551 ± 27	321 ± 16
TCE degradation[(b)]:		
K_m (µM)	225 ± 13	126 ± 8
V_{max} ($nmol^{-1}$ min^{-1} mg $protein^{-1}$)	2325 ± 160	995 ± 160

(a) NMS - nitrate mineral salts medium (Koh et al. 1993).

(b) V_{max} values were based on 1-hour incubations and are not initial rates. Protein levels in 68-1 and OB3b were similar with the growth conditions used. Data are from Koh et al. (1993).

oxidation was found to occur linearly over at least a 1 hour period when its concentration exceeded the K_m value. The naphthalene oxidation procedure could be thus used as an accurate means of ascertaining sMMO activity of cultures. It would be an ideal procedure for monitoring and optimizing sMMO levels in bioreactor systems.

Genetic homology studies between the sMMO genes of 68-1 and OB3b found the level of similarity to be rather low. A 2.2-kb sMMO probe, based on a part of the *Methylosinus trichosporium* OB3b sMMO gene cluster (Tsien & Hanson 1992) was used for comparisons under both high and low stringency conditions (Tsien & Hanson 1992). This gene probe has been found to hybridize to a wide range of sMMO-producing methanotrophs because genes of the sMMO cluster are relatively conserved. In our study, the high and low stringency conditions allowed for 30% and 40% nucleotide mismatch, respectively. In any case the sMMO probe did not hybridize with 68-1 chromosomal DNA in either conditions. The overall genetic and amino acid homology could be low simply due to the rather different mol% guanine + cytosine values of 68-1 and OB3b (approx. 52% and 62%, respectively). On the other hand, it is possible that sMMO diversity may be higher than previously thought.

TCE has been shown to have a toxic effect on the activity of sMMO (Alvarez-Cohen & McCarty, 1991). To determine the resilience of 68-1 to TCE toxicity the testing procedure devised by Alvarez-Cohen & McCarty (1991) was used, with *Methylosinus trichosporium* OB3b used as the comparison strain. The relative TCE resilience and the TCE transformation capacity of 68-1 was found to be similar to that of OB3b. In our experiments, 68-1 was able to degrade 0.095 mg TCE/mg of cells, whereas OB3b was able to degrade 0.081 mg TCE/mg of cells.

The ability of 68-1 to degrade several chlorinated aliphatic compounds also was assessed. The concentration of all the compounds tested in this study (Table 2) was 100 μM. The degradation of these compounds was monitored using gas chromatography. The degradation rates were calculated from 1-hour incubation periods and are not initial rates. As mentioned above 68-1 proved to have a higher specific rates of degradation for several of these compounds (Table 2) when compared to OB3b. On the other hand, 68-1 did not seem to be as effective as OB3b for the degradation of some of the more heavily chlorinated aliphatic compounds, including 1,1,1- and 1,1,2-trichloroethane, and 1,1,1,2- and 1,1,2,2-tetrachloroethane (Table 2). This difference may relate to variations in the active site structure of the sMMOs of 68-1 and OB3b.

This study indicates that strain selection is one possible path for optimizing sMMO-based pump-and-treat bioreactor systems. With the apparent diversity of sMMO-producing methanotrophs, the isolation of superior TCE-degrading strains, such as *Methylomonas methanica* 68-1, is possible. However, it is important to note the performance of these strains in natural and rigorous settings still requires study. Future work with 68-1 will focus on how to optimize its sMMO productivity and competitiveness (to other methanotrophs) in simulated groundwater systems.

TABLE 2. Chlorinated aliphatic compound degradation by *Methylomonas methanica* 68-1 and *Methylosinus trichosporium* OB3b.

Compound[(a)]	*Methylomonas methanica* 68-1	*Methylosinus trichosporium* OB3b
	Specific degradation rate ($nmol^{-1}$ h^{-1} mg $protein^{-1}$)[(b)]	
Dichloromethane	1450 ± 89	899 ± 35
Chloroform	432 ± 35	227 ± 21
1,1-Dichloroethane	174 ± 18	110 ± 14
1,2-Dichloroethane	127 ± 28	89 ± 10
1,1-Dichloroethylene	4 ± 2	11 ± 1
1,2-Dichloroethylene (*cis*:*trans* 1:1)	14 ± 3	30 ± 2
1,1,1-Trichloroethane	2 ± 1	15 ± 2
1,1,2-Trichloroethane	48 ± 6	62 ± 20
1,1,1,2-Tetrachloroethane	0	8 ± 1
1,1,2,2-Tetrachloroethane	3 ± 1	15 ± 1

(a) Compounds were tested at an initial concentration of 100 μM. No detectable degradation of the compounds carbon tetrachloride or tetrachloroethylene was detected.
(b) Specific activity is based on 1-hour incubation periods, not initial rates.

ACKNOWLEDGMENTS

This work was supported by the Waste Management Research and Education Institute, the University of Tennessee, Knoxville, and by the Savannah River Laboratory of the U.S. Department of Energy and U.S. Air Force subcontracts DE-AC05-840R21400 and F49620-92-J-0147 through the Oak Ridge National Laboratory, respectively.

REFERENCES

Alvarez-Cohen, L., and P. L. McCarty. 1991. "Effects of Toxicity, Aeration, and Reductant Supply on Trichloroethylene Transformation by a Mixed Methanotrophic Culture." *Appl. Environ. Microbiol.* 57:228-235.

Brusseau, G. A., H. C. Tsien, R. S. Hanson, and L. P. Wackett. 1990. "Optimization of Trichloroethylene Oxidation by Methanotrophs and the Use of a Colorimetric Assay to Detect Soluble Methane Monooxygenase Activity." *Biodegradation* 1:19-29.

Dalton, H. 1992. "Methane-oxidation by Methanotrophs: Physiological and Mechanistic Implications." In J. C. Murrell and H. Dalton (Eds.), *Methane and Methanol Utilizers*, pp. 85-114. Plenum Press, New York, NY.

DiSpirito, A. A., J. Gulledge, J. C. Murrell, A. K. Shiemke, M. E. Lidstrom, and C. L. Krema. 1992. "Trichloroethylene Oxidation by the Membrane Associated Methane Monooxygenase in Type I, Type II, and Type X Methanotrophs." *Biodegradation* 2:151-164.

Hazen, T. 1992. *Test Plan for In-situ Bioremediation Demonstration of the Savannah River Integrated Demonstration.* Project DOE/OTD TTP No. SR 0566-01 (U). Westinghouse Savannah River Co., Savannah River Site, Aiken, SC.

Koh, S.-C., J. P. Bowman, and G. S. Sayler. 1993. "Soluble Methane Monooxygenase Production and Rapid Trichloroethylene Degradation by a Type I Methanotroph, *Methylomonas methanica* 68-1." *Appl. Environ. Microbiol.* In press.

Little, C. D., A. V. Palumbo, S. E. Herbes, M. E. Lidstrom, R. L. Tyndall, and P. J. Gilmer. 1988. "Trichloroethylene Biodegradation by a Methane-Oxidizing Bacterium." *Appl. Environ. Microbiol. 54*:951-956.

Tsien, H. C., and R. S. Hanson. 1992. "Soluble Methane Monooxygenase Component B Gene Probe for Identification of Methanotrophs that Rapidly Degrade Trichloroethylene." *Appl. Environ. Microbiol. 58*:953-960.

TRICHLOROETHYLENE COMETABOLISM BY PHENOL-DEGRADING BACTERIA IN SEQUENCING BIOFILM REACTORS

G. E. Speitel, Jr., R. L. Segar, Jr., and S. L. De Wys

INTRODUCTION

Biofilm reactors for the treatment of aqueous streams contaminated with chlorinated solvents, such as trichloroethylene (TCE), are a potentially attractive alternative to conventional treatment methods. The ability of microorganisms to cometabolize TCE has been studied extensively, yet development of engineered reactors to exploit this capability has proceeded slowly. Issues such as slow degradation kinetics, enzyme competition, and by-product toxicity have hampered reactor development.

Staged reactor configurations have been favored recently as a means of increasing reactor efficiency (Speitel & Leonard 1992). In this research, "staging" was accomplished within biofilm reactors by sequencing between a growth stage and a cometabolic degradation stage. Phenol was used as the growth substrate to induce oxygenases that cometabolize chlorinated solvents, particularly TCE. Degradation took place in the absence of phenol to avoid enzyme competition. A desirable feature of sequencing is that contamination by phenol of the treated water is avoided.

Because the microorganisms are effectively starving during TCE degradation, sequencing reactors have to be periodically rejuvenated with a short growth stage to maintain activity. The design of this non-steady-state process is not straightforward. Sequencing bioreactor design parameters probably include frequency and duration of rejuvenation, phenol and oxygen concentration, flow rate and detention time, and chlorinated solvent concentration and detention time. Feeding strategies that varied the first three parameters were investigated in this research to identify combinations that may be attractive for larger scale testing.

RESULTS

Preliminary studies were conducted to identify promising cultures for reactor development. A mixed culture of phenol-utilizing organisms was selected to inoculate the bioreactors. Greater than 95% removal and 80% mineralization of TCE were obtained in ^{14}C-TCE studies. The culture also degraded 1,1-dichloroethylene and *trans*-dichloroethylene at a rate similar to TCE, whereas *cis*-dichloroethylene was degraded more rapidly. Little or no degradation of tetrachloroethylene, chlorinated ethanes, or chlorinated methanes was observed in batch studies.

The biofilm reactors were 1.5-cm in diameter by 40-cm in length glass chromatography columns filled with 3-mm glass beads. The sequencing experiments began with an initial growth period of 3 days. Nutrient water containing phenol and inoculum was pumped upflow through the reactors at a hydraulic residence time (HRT) of about 0.5 minute to establish a uniform biofilm over the length of the reactor. Reactor effluent was reaerated and recirculated after phenol and nutrient amendment. Once the biofilm was established, application of phenol was halted, and water containing 100 µg/L TCE was treated in a single pass, upflow manner at a flowrate yielding an HRT of 14 minutes. The low flowrate during the TCE degradation stage was necessary to compensate for the slower kinetics of cometabolism in comparison to growth substrate metabolism. Phenol was withheld during the TCE degradation stage to avoid enzyme competition. Degradation periods of 12 to 48 hours were followed by 1- to 4-hour periods of phenol rejuvenation. The reactors cycled between TCE degradation and phenol rejuvenation for 6 to 30 days in each experiment.

Reactor performance was characterized in several ways. The percent TCE removals after and before rejuvenation were averaged to give an average TCE removal for each cycle. As a measurement of the change in TCE removal during a cycle, a quantity Δr was defined as the difference between the percent removals after and before rejuvenation for each cycle. The use of two sample points to represent a degradation cycle was validated in preliminary studies (Segar et al. 1992). Finally, the pseudo first-order TCE degradation rate constants (k_1) within the biofilm reactor were estimated using a biofilm reactor model (Rittmann 1982). The rate constant was obtained by using it as a fitting parameter to match the predicted TCE removal with the measured TCE removal.

The time-averaged TCE removals ranged from 60 to 90% at a detention time of 14 minutes. Testing of several cycle frequencies and durations showed that 2 to 3 hours of rejuvenation per day sustained TCE removal. Examples of reactor performance are provided in Figures 1 and 2. Three reactors were operated in parallel, each receiving a different phenol concentration (5, 25, or 100 mg/L) during rejuvenation. Figure 1 shows the average removal over time when rejuvenated 3 hours every 21 hours of degradation (termed a 3/24 feeding strategy). The TCE removal was stable for all phenol feed concentrations; however, the column fed 25 mg/L performed better than the others with an average removal of 69%. When the rejuvenation frequency was increased to twice a day with a 1.5/12 strategy, the average removal was greater than 90% during the first 7 days of operation for the reactor fed 25 mg/L. Several of the measurements after rejuvenation indicated greater than 99% TCE removal, which translated into rate constants greater than 160 L/gVSS (volatile suspended solids)/day, indicating a high level of cometabolic activity. However, after this initial period the removals dropped precipitously, then leveled out at about 70%. Concurrently, gas bubble formation in the biofilm was observed, possibly indicating anoxic conditions within the biofilm although the bulk flow remained aerobic. A conductivity tracer test indicated that the reactor detention time had been reduced by more than 50%

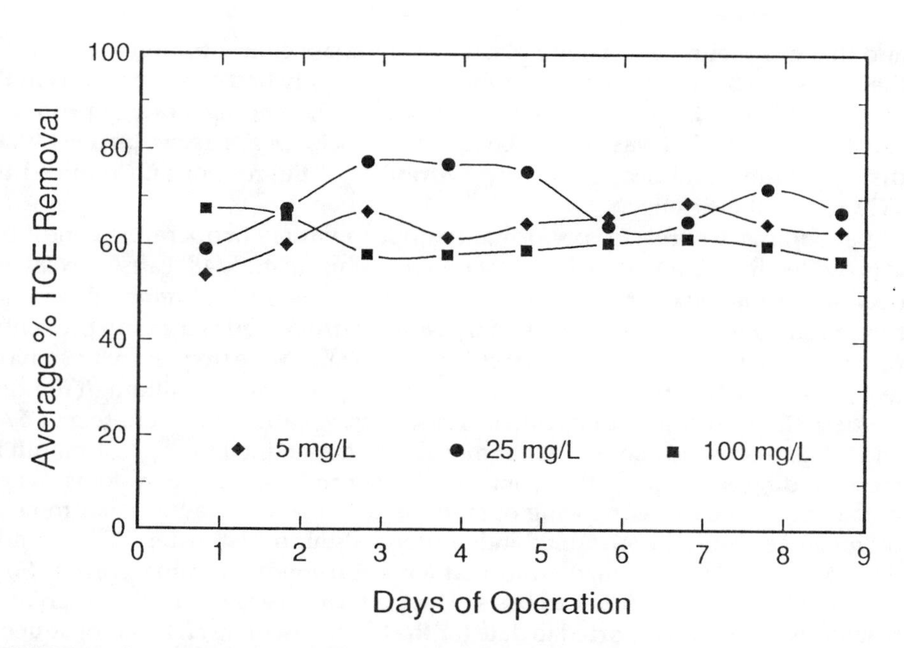

FIGURE 1. Average TCE removal for 3/24 feeding strategy.

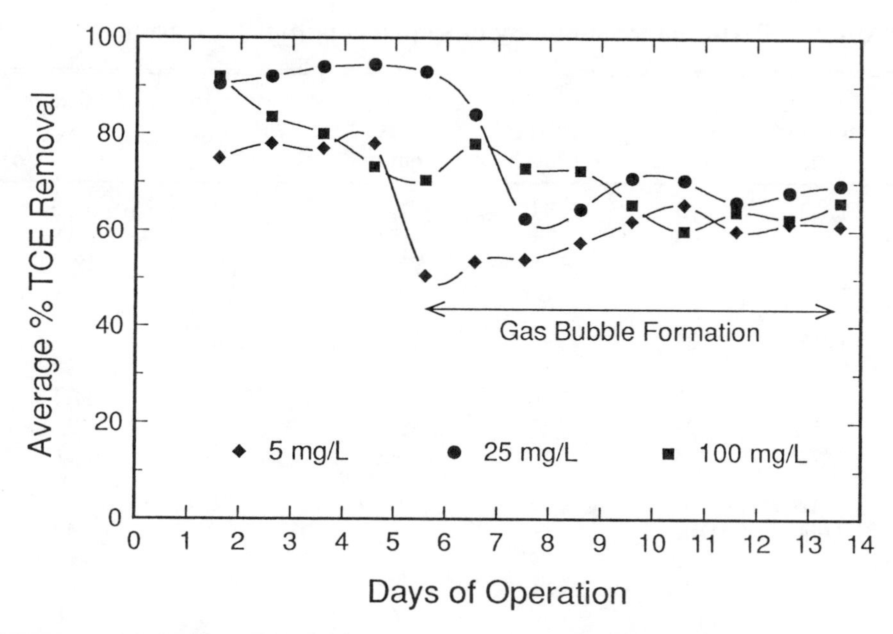

FIGURE 2. Average TCE removal for 1.5/12 feeding strategy.

and that channeling was taking place. The biomass concentration at the end of the run was 3.8 mg/mL, whereas other runs typically had concentrations in the range of 1.5 to 2.2 mg/mL. Apparently, the 1.5/12 feeding strategy promoted excessive growth that was initially beneficial to reactor performance but eventually disrupted the usual near-plug flow hydraulics of the reactor and reduced the TCE removal efficiency.

A summary of reactor performance under a variety of operating conditions is provided in Table 1. Less frequent rejuvenations (e.g., 4/48) caused excessive decline in removal over the course of a cycle. Initial levels of removal were not recovered by rejuvenation presumably because prolonged starvation prevented recovery of oxygenase and endogenous energy levels. More frequent rejuvenations (e.g., 1/12 and 2/24) provided reasonably good performance, although removal was less stable over time than with increased rejuvenation durations (e.g., 1.5/12 and 3/24) discussed above. TCE removal was maintained for up to 1 month by these feeding strategies. It is conceivable that in larger scale systems, where greater control over sequencing operations could be maintained, that removal performance could be sustained indefinitely. Pseudo first-order TCE degradation rate constants obtained from mathematical modeling ranged from 50 to 110 L/gVSS/day on a sustained basis. This performance is among the best biofilm reactor technologies reported to date for the treatment of μg/L levels of aqueous chlorinated ethene contamination.

TABLE 1. Performance summary of sequencing biofilm reactors.

Feeding Strategy (hr/hr)	Phenol Conc. (mg/L)	Duration of Exprt. (days)	Average % TCE Removal	Average Δr Value (%)	Average k_1 (L/gVSS/d)
1/12	5	6, 11[(a)]	71, 66	16, 9	52, 30
2/24	5	29	52	21	34
1.5/12	5	5, 9[(b)]	77, 59	10, 2	58, 31
3/24	5	8	63	19	48
1/12	25	6, 11[(a)]	70, 66	16, 6	54, 47
2/24	25	29	69	19	61
4/48	25	5, 7[(a)]	66, 73	25, 38	44, 58
1.5/12	25	7, 7[(b)]	91, 67	15, 9	112, 40
3/24	25	8	69	21	57
1/12	100	6, 11[(a)]	71, 55	7, 12	43
2/24	100	29	61	15	49
4/48	100	5, 7[(a)]	73, 77	25, 50	68, 82
1.5/12	100	14	72	2	63
3/24	100	8	61	10	38

(a) Repeat of same experiment.
(b) Same run, first and second half.

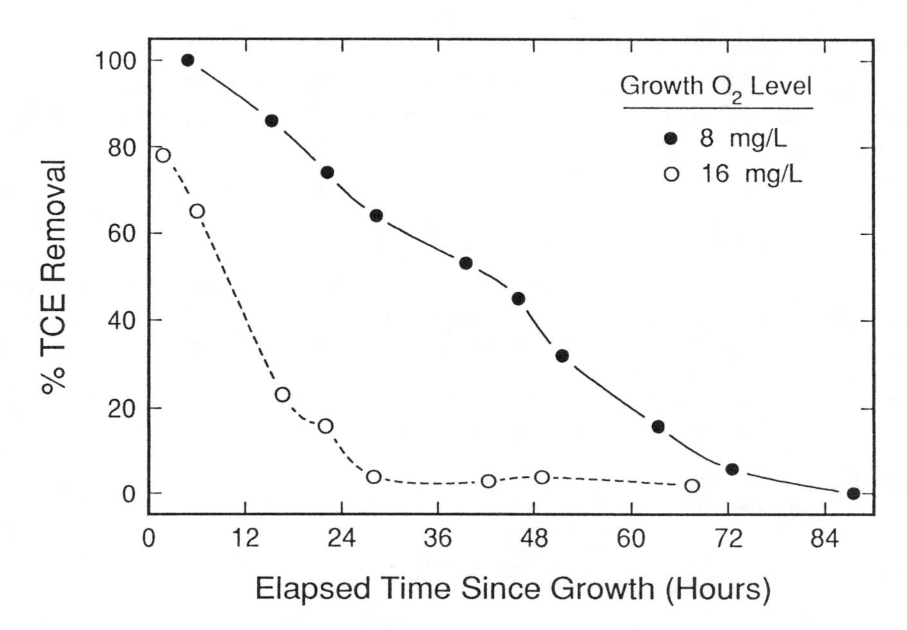

FIGURE 3. Degradation stage TCE removal vs. growth oxygen level.

Ongoing studies are focusing on the impact of environmental conditions during the initial growth stage by monitoring TCE removal in the following degradation stage. An example is provided in Figure 3 for two bioreactors grown in parallel at different oxygen concentrations and then switched to TCE removal under identical conditions. TCE removal and capacity were lower after growth at 16 mg/L oxygen than at 8 mg/L oxygen. During the starvation conditions of the degradation stage, organisms probably rely on various storage materials such as complex carbohydrates or lipids to provide an endogenous source of reducing power needed for cometabolism. Also, substantial decay of oxygenase enzymes needed for cometabolism may occur during starvation. The observed performance difference may be related to the effect of oxygen level on the regulation of oxygenase expression and storage product formation (Dawes & Senior 1973).

CONCLUSIONS

Bacteria were grown on phenol in bench-scale biofilm reactors, followed by cycling between treatment of aqueous TCE at 100 µg/L and rejuvenation with a phenol solution. TCE degradation rate constants of 50 to 110 L/gVSS/day were sustained for 10 to 30 days. Two to three hours of rejuvenation per day of TCE degradation were required. The best feeding strategies yielded time-averaged TCE removals of 70 to 90% at an HRT of 14 minutes.

ACKNOWLEDGMENTS

This research was funded by the Gulf Coast Hazardous Substance Research Center and by the Abel Wolman Doctoral Fellowship for R.L.S. from the American Water Works Association.

REFERENCES

Dawes, E. A., and P. J. Senior. 1973. "The Role and Regulation of Energy Reserve Polymers in Microorganisms." *Adv. Microb. Physiol. 10*: 135.

Rittmann, B. E. 1982. "Comparative Performance of Biofilm Reactor Types." *Biotechnol. Bioeng. 24*: 1341.

Segar, R. L. Jr., S. L. De Wys, and G. E. Speitel Jr. 1992. "Feeding Strategies That Sustain TCE Cometabolism in Sequencing Biofilm Reactors." *Proceed. Ann. Conf. Water Environ. Fed.* New Orleans, LA.

Speitel, G. E. Jr., and J. M. Leonard. 1990. "A Sequencing Biofilm Reactor for the Treatment of Chlorinated Solvents Using Methanotrophs." *Water Environ. Res. 64*: 712.

PHENOL-INDUCED TCE DEGRADATION BY PURE AND MIXED CULTURES IN BATCH STUDIES AND CONTINUOUS-FLOW REACTORS

C. G. Coyle

INTRODUCTION

Widespread contamination of groundwater by trichloroethylene (TCE) has given rise to research into methods for removing TCE from contaminated water. Under proper conditions, some *Pseudomonas* species produce toluene oxygenase enzymes. Some toluene oxygenase enzymes have been shown to fortuitously oxidize TCE (Nelson et al. 1987). Phenol has been shown to induce some toluene oxygenase enzymes that degrade TCE (Nelson et al. 1988, Wackett & Gibson et al. 1988).

Because of anticipated difficulties in maintaining pure cultures in full-scale treatment systems, mixed cultures were included in TCE biodegradation experiments. Aerobic, suspended growth reactors were operated in a completely mixed, continuous-flow mode with 5-day hydraulic and solids retention times. Pure culture (*P. putida* F1) and mixed culture reactors were maintained. The mixed culture was seeded with samples from municipal wastewater plants and landfill leachate. Details of the mass balance experiments were presented previously (Coyle et al. 1993), and are not shown here. TCE degradation also was observed in mixed culture batch experiments. Experimental results were compared to results, reported by Folsom and co-workers, of pure culture experiments using *P. cepacia* G4.

RESULTS

Mass balance experiments were performed to determine the extent to which TCE was being degraded in the reactors (Coyle 1990, Coyle et al. 1993). When acetate (530 mg/L) and glucose (470 mg/L) were the primary substrates and phenol was provided at 20 mg/L, significant levels of TCE degradation were observed after more than 4 weeks in either the pure or mixed culture reactors.

After switching to phenol as the primary substrate, microscopic observations indicated that the pure culture remained pure for more than 4 weeks. Three days after inoculation, the color of the pure culture changed from yellowish-tan to dark brown. This may have been due to buildup of a catechol derivative. Catechol is a suspected intermediate of phenol degradation by *P. putida* F1 (Spain et al. 1989).

When phenol was the sole primary substrate and under essentially identical conditions, the mixed culture reactor effected substantially greater reductions of phenol and TCE than the pure culture reactor (see Table 1). Also the estimated percent of TCE biodegraded by the mixed culture reactor was much greater than that of the pure culture reactor. (Subsequent experiments indicated that the mixed culture reactor was biodegrading 91.4 ± 3.9% of the incoming TCE [Shurtliff 1992].) Evidence of phenol inhibition, catechol accumulation, or toxicity due to TCE degradation was not observed in the mixed culture reactor. No attempts were made to optimize degradation rates or to minimize inhibition between phenol and TCE in the reactors.

TCE degradation also was observed in mixed culture batch studies (Coyle et al. 1993). Mixed culture suspensions were incubated on phenol for 1 hour and resuspended in buffer, prior to initiation of the batch experiments. During incubation, the initial phenol concentration was 80 mg/L. The yield (Y) for growth of the mixed culture on phenol was determined to be 0.61 g volatile suspended solids (VSS) per g of phenol (Shurtliff 1992). Before incubation, the VSS level was 243 mg/L. Assuming complete phenol utilization during incubation, the post-incubation VSS level was calculated to be 292 mg/L. The rate of TCE degradation in batch experiments was calculated assuming a VSS level of 292 mg/L (see Table 2). Measurements showed that ca. one-half of the phenol was removed after 1 hour. Thus the actual VSS level was probably less than the calculated volatile suspended solids (VSS) level.

Because it has been shown that transfer of TCE from air to water was faster than biodegradation of TCE by *P. cepacia* G4 (Folsom et al. 1990), it was assumed that phase transfer would not limit TCE biodegradation by mixed cultures. Batch experiments were performed at 25°C in vials with a total volume of 122 mL (see Figure 1). The volume of the suspension was 25 mL. Assuming a Henry's constant of 0.4 (Folsom et al. 1990), the initial aqueous TCE concentration was calculated to be 0.94 mg/L. Known volumes of TCE, in methanol standards, were added prior to sealing the vials; phenol was not provided. Control vials contained TCE and buffer solution sans culture.

Others have shown that adsorption to biomass is not a significant removal mechanism for TCE (Folsom et al. 1990, Shurtliff 1992). Except for an initial lag period, the mixed culture TCE degradation rate was linear for most of the first 3 hrs (see Figure 1 and Table 2). The batch degradation experiments were performed with one initial TCE level, and no attempts were made to determine the maximum rate of TCE degradation (V_{max}) by varying initial TCE levels.

The mixed culture degraded TCE much more rapidly in batch experiments than in the reactor (see Table 1 and 2). Clearly operation of the mixed culture reactor did not take full advantage of the mixed culture's capacity for degrading TCE.

DISCUSSION

P. cepacia G4 uses toluene monooxygenase enzymes to degrade phenol and TCE (Shields et al. 1989), unlike *P. putida* F1 which uses toluene dioxygenase

TABLE 1. Summary of results from mass balance experiments.

Culture	VSS (mg/L)	Phenol Levels in Influent & Reactor, Respectively (mg/L)	TCE Levels in Influent & Reactor, Respectively (mg/L)	Estimated % of TCE Biodegraded	TCE Degradation Rates in Reactors	
					(mg/d/g of VSS)	(mg/g of phenol)
P. putida F1	80	420, 25	16.0, 0.8	47	19	21
Mixed Culture	150	420, <0.5	18.8, 0.1	85	20	37

TABLE 2. TCE degradation rates.

Culture	Degradation Rates in Batch Studies		Degradation Rates in Reactors	
	(g/d/g of VSS)	(μmol/min/g of protein)	(μmol/min/g of protein)	(mg/g of phenol)
P. cepacia G4	—	8 (V_{max})[a]	6.3[b]	not reported
Mixed Culture	39.4	347[c]	0.18[c]	37

(a) Folsom et al. (1990).
(b) Folsom & Chapman (1991).
(c) These rates were calculated assuming 0.6 g protein per g of VSS.

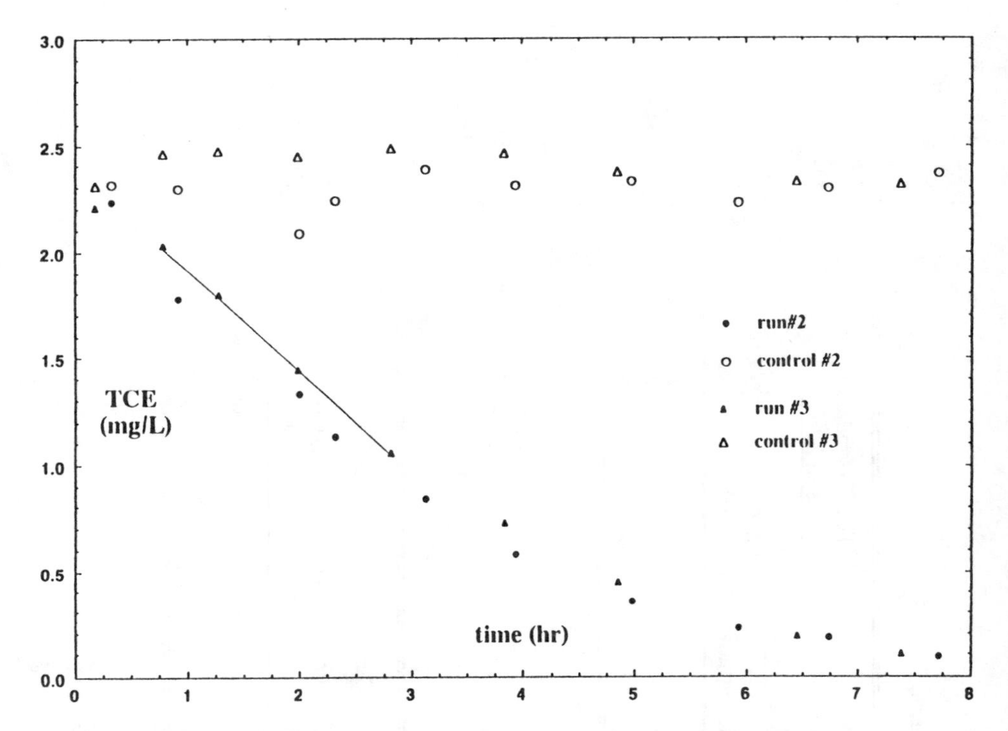

FIGURE 1. TCE degradation during mixed culture batch experiments. Concentrations were reported as if all TCE were present in the aqueous phase.

enzymes (Zylstra et al. 1989). Phenol is an effective inducer of the toluene monooxygenase enzyme system of *P. cepacia* G4 (Nelson et al. 1987). Researchers have shown that *P. putida* F1 exhibits poor growth on phenol relative to toluene (Spain et al. 1989). Thus *P. cepacia* G4 appears to be better suited to growth on phenol than does *P. putida* F1.

Folsom and co-workers reported kinetic coefficients (V_{max} and K_s) for TCE degradation by *P. cepacia* (Folsom et al. 1990). The reported maximum rate of TCE degradation, V_{max}, for *P. cepacia* G4 (see Table 2) was similar to, or greater than, those reported for other pure cultures (Folsom et al. 1990). Folsom and Chapman used phenol-induced *P. cepacia* G4 cultures to study TCE degradation in a two-chamber reactor system, designed to minimize inhibition between phenol and TCE (Folsom & Chapman 1991). The reported TCE degradation rates for *P. cepacia* G4, in the reactor system, were similar to the reported V_{max} value (see Table 2), indicating that the reactor provided nearly optimal conditions for *P. cepacia* G4 to degrade TCE (in terms of mass of TCE degraded per mass of cells).

Comparison of degradation rates from batch experiments indicates that mixed cultures have a greater capacity than *P. cepacia* G4, for degrading TCE. However, because the experiments were performed independently, the cultures should be compared under identical conditions to confirm the comparison of degradation rates.

It should be noted that maximizing the rate of TCE degradation, in terms of mass of TCE degraded per mass of cells, does not necessarily maximize the efficiency of a TCE-degrading reactor system. An economically efficient reactor system, for biodegradation of TCE, should be designed to concomitantly minimize generation of excess biomass, minimize phenol consumption, minimize aeration requirements, and maximize TCE degradation rates.

ACKNOWLEDGMENTS

This work was funded by the U.S. Environmental Protection Agency through the Region 7 & 9 Hazardous Substance Research Center at Kansas State University and in part from grant AFOSR-88-6225 from the U.S. Air Force Office of Scientific Research.

REFERENCES

Brock, T. D., and M. T. Madigan. 1988. *Biology of Microorganisms.* Prentice Hall, Englewood Cliffs, NJ.

Coyle, C. G. 1990. M.S. Thesis, University of Iowa, Iowa City, IA.

Coyle, C. G., G. F. Parkin, and D. T. Gibson. 1993. "Aerobic Phenol-Induced TCE Degradation in Completely Mixed, Continuous-Culture Reactors." *Biodegradation* Vol. 4, Issue no. 1.

Folsom, B. R., and P. J. Chapman. 1991. "Performance Characterization of a Model Bioreactor for the Biodegradation of Trichloroethylene by *Pseudomonas cepacia* G4." *Applied Environ. Microbiol. 57*:1602.

Folsom, B. R., P. J. Chapman, and P. H. Pritchard. 1990. "Phenol and Trichloroethylene Degradation by *Pseudomonas cepacia* G4: Kinetics and Interactions Between Substrates." *Applied Environ. Microbiol. 56*:1279.

Nelson, M. J. K., S. O. Montgomery, W. R. Mahaffey, and P. H. Pritchard. 1987. "Biodegradation of Trichloroethylene and Involvement of an Aromatic Pathway." *Applied Environ. Microbiol. 53*:949.

Nelson, M. J. K., S. O. Montgomery, and P. H. Pritchard. 1988. "Trichloroethylene Metabolism by Microorganisms that Degrade Aromatic Compounds." *Applied Environ. Microbiol. 54*:604.

Shields, M. S., S. O. Montgomery, P. J. Chapman, S. M. Cuskey, and P. H. Pritchard. 1989. "Novel Pathway of Toluene Catabolism in the Trichloroethylene-Degrading Bacterium G4." *Applied Environ. Microbiol. 55*:1624.

Shurtliff, M. 1992. M.S. Thesis, University of Iowa, Iowa City, IA.

Spain, J. C., G. J. Zylstra, C. K. Blake, and D. T. Gibson. 1989. "Monohydroxylation of Phenol and 2,5-Dichlorophenol by Toluene Dioxygenase in *Pseudomonas putida* F1." *Applied Environ. Microbiol. 55*:2648.

Wackett, L. P. and D. T. Gibson. 1988. "Degradation of Trichloroethylene by Toluene Dioxygenase in Whole-Cell Studies with *Pseudomonas putida* F1." *Applied Environ. Microbiol. 54*:1703.

Zylstra, G. J., L. P. Wackett, and D. T. Gibson. 1989. "Trichloroethylene Degradation by *Escherichia coli* Containing the Cloned *Pseudomonas putida* F1 Toluene Dioxygenase Genes." *Applied Environ. Microbiol. 55*:3162.

COMPARISON OF METHANOTROPHIC AND ANAEROBIC BIOREMEDIATION OF CHLORINATED ETHENES IN GROUNDWATER

R. Legrand

INTRODUCTION

Aerobic methanotrophic bacteria cometabolize trichloroethene (trichloroethylene, TCE) in the presence of methane and oxygen. Methanotrophic treatment technology (MTT) is based on this process and the Gas Research Institute (GRI) sponsors its development. Various reactor designs have been developed at bench or pilot scale, including systems in which methane and oxygen are supplied in the gas phase, and submerged reactors where these gases are dissolved in the feedwater. An example of a submerged reactor is the fluidized-bed reactor (FBR) developed at Cornell University and Michigan Biotechnology Institute (MBI). Granular activated carbon (GAC) was found to be a better support medium than diatomaceous earth for a methanotrophic FBR. On GAC, faster microbial growth, greater extent of TCE degradation, and lower methane demand were observed (Wu et al. 1992). The apparent Michaelis-Menten half-velocity constant (K_s) for TCE (290 µg/L) was one order of magnitude less than previously reported.

The economics of two physical technologies (air stripping with vapor-phase GAC, and liquid-phase GAC or LGAC) were compared to those of MTT (Legrand & Jackson 1992). As can be seen from Figure 1, the projected cost of MTT is comparable to that of air stripping and is clearly lower than the cost of LGAC. Additionally, MTT destroys the contaminant on site rather than transferring it to GAC. Recent FBR work indicates that coarser media lead to substantially improved methane demand and biological kinetics, which likely will decrease costs further.

The need to transfer large volumes of sparingly soluble oxygen and methane into the feedwater is a significant challenge of FBR technology. It can be obviated if chlorinated ethenes are instead reductively dechlorinated in an anaerobic environment and in the presence of an electron donor. This approach eliminates mass transfer problems because an easily soluble electron donor such as sucrose or ethanol can be used and no oxygen needs to be dissolved. Potential drawbacks include the persistence of an organic residual and possibly vinyl chloride (VC) in the effluent, as well as competition from the reduction of sulfate, a common constituent of groundwater. The frequent presence of PCEs along with TCE in groundwater is an additional incentive for investigating anaerobic biodegradation

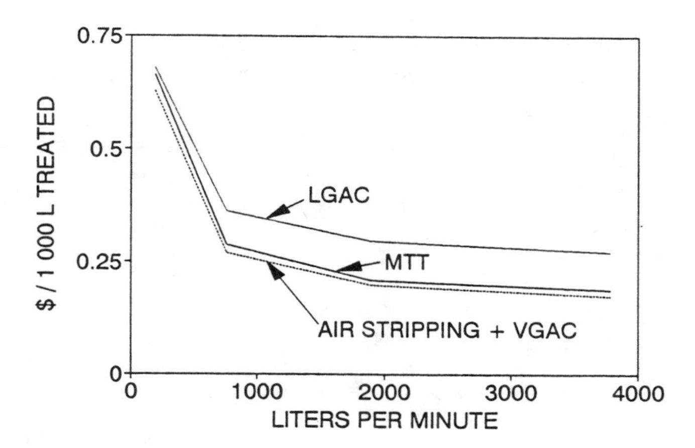

FIGURE 1. Projected operational cost for three remediation technologies; influent = 1 mg TCE/L, effluent < 4.5 µg TCE/L; adapted from Legrand and Jackson (1992).

(PCE is rapidly dechlorinated anaerobically but is refractory to aerobic degradation).

Jewell et al. (1991) documented a rapid anaerobic dechlorination of PCE (from 10,000 to under 50 µg/L in less than 2 h) and generally observed continuing dechlorination to VC, which degraded much more slowly. VC is the most toxic chlorinated ethene and it doesn't adsorb well to activated carbon. It can be rapidly oxidated to µg/L levels in a methanotrophic second stage, but this oxidation is subject to strong competitive inhibition from methane.

De Bruin et al.(1992) developed an anaerobic culture from fluvial sediment and crushed reactor sludge granules that reduces 9 µM (1.5 mg/L) of PCE to ethene and ethane. They documented the disappearance of VC in 1.2 h at 10°C with 1 mM (90 mg/L) lactate as the electron donor. It is unclear whether operation at this low temperature is a requirement. If it is, low growth rates may inhibit the full-scale implementation of this process.

COMPARISON OF BIOREMEDIATION APPROACHES

Three approaches to TCE bioremediation were compared: MTT, hybrid anaerobic-aerobic, and anaerobic. An existing mathematical model of MTT was used (Legrand & Jackson 1992); it was customized for the different biochemical processes considered, and published experimental data were used. Each process was defined at full scale and costs were determined. The results are summarized in Table 1. Due to the early stage of development of these technologies, these comparisons are preliminary. The MTT data (Wu et al. 1992) have the most credibility because they were obtained at a full-size bed depth of 3 m.

TABLE 1. Preliminary economic evaluation of 3 biochemical routes to TCE degradation; groundwater flow = 1,893 L/min (500 gpm), influent = 1 mg TCE/L, effluent < 4.5 µg TCE/L.

Process	MTT[a]	Hybrid Anaerobic and Aerobic[b]	Anaerobic[c]
Reactor Depth (m), Stage 1	4.6	8.9	8.0
Stage 2	4.6	10.7	–
Diameter (m), Stage 1	3.0	5.4	5.8
Stage 2	3.0	8.2	–
Influent µg TCE/L assumed	1000	1000	1000
µg PCE/L assumed	0	100	100
Capital Cost ($1000 U.S.)	411	1267	448
Operational Cost[d] ($/1000 L)	0.21	0.30	0.16
% of Operational Cost for			
Methane	6%	1%	0%
Electron donor (ethanol)	–	4%	19%
LGAC for effluent polishing	31%	–	–

(a) 2 methanotrophic FBR stages in series, based on Wu et al. (1992).
(b) Anaerobic stage followed by methanotrophic (aerobic) stage, based on Jewell et al. (1991).
(c) Single anaerobic stage, based on de Bruin et al. (1992).
(d) Including capital recovery at i = 10% for 10 years.

DISCUSSION

Methanotrophic Oxidation of TCE (MTT). The configuration assumed is two FBRs in series, because the model predicts that this is more cost-effective than just one stage, at least using the kinetics derived from the MBI experiments.[1] The first MTT stage brings TCE concentration down from 1,000 to 300 µg/L, the second stage to 40 µg/L, and LGAC polishes the effluent to 4 µg/L (below the NAS recommended drinking water limit of 4.5 µg/TCE/L). From Table 1 it is clear that the cost of methane is negligible. The capital cost of LGAC is moderate, however the cost of replacing LGAC is high; indeed LGAC is responsible for 31% of the treatment cost. Research should focus on further lowering the treatment endpoint.

---

[1] Extensive conversion requires high recirculation, which dilutes the influent concentration. Dilute TCE concentrations are associated with higher methane demands and lower reaction rates, according to Michaelis-Menten kinetics. These two problems compound each other and result in exponentially increasing reactor sizes and recirculation as lower effluent concentrations are pursued. This limits the extent of biodegradation that can cost effectively be achieved in one stage.

Hybrid Anaerobic-Aerobic Process. No recirculation is used in the posited anaerobic first stage. The influent will be undiluted, resulting in high Michaelis-Menten reaction rates; biomass concentration in the reactor will be maximized by using very fine media. The dechlorination is modeled as a series of first-order reactions using coefficients estimated from the results in Jewell et al. (1991). The second stage is modeled as three concurrent oxidations (of TCE, DCE, and VC residuals) following Michaelis-Menten kinetics. The limiting factor is the need to degrade VC below 2 μg/L. As mentioned before, this cannot be achieved practically with LGAC. The support medium in both reactors is diatomaceous earth, because it is the only medium with documented performance in a hybrid system.

As can be seen from Table 1, the capital cost is projected to be three times higher than that of the methanotrophic facility; two-thirds of the hybrid facility's capital cost is in the second, aerobic stage. The cost of the second stage is so high because VC oxidation down to the 2 μg VC/L drinking water standard must be achieved through biological treatment alone. A high recirculation rate is assumed, to both minimize methane concentration and avoid methane inhibition. As in the pure methanotrophic process, methane cost is negligible. The cost of adding an organic electron donor is also minor; note that it is assumed that 31 mg COD (chemical oxygen demand)/L is added to the first stage influent. The second stage accounts for 60% of the total treatment cost. Switching from diatomaceous earth to GAC may significantly enhance performance and reduce costs.

Anaerobic Dechlorination. If an anaerobic culture could be induced to dechlorinate VC rapidly, there would be no need for a costly VC-oxidating second stage. Such rapid VC dechlorination was achieved by de Bruin et al. (1992), and their data were extrapolated into a hypothetical reactor that would reductively dechlorinate TCE and PCE to ethene and ethane with a VC residual under 2 μg/L. The kinetics are estimated from the same source, specifically from a reactor operating at 10°C and a retention time of 6 hours that they describe. First-order kinetics are assumed for the successive dechlorinations. I assumed that the performance described by de Bruin et al. (1992) can be preserved in a full-scale reactor with superficial velocities two orders of magnitude higher.

The electron donor addition before the anaerobic stage amounts to 76 mg COD/L but can probably be reduced, which would have a minor beneficial impact on cost. The cost estimate should include a brief aerobic posttreatment, even if only to reduce the organic residual, here estimated at 20 mg COD/L. The capital cost is similar to the estimate for a methanotrophic facility. The operating cost is much lower, even if the minor cost of aerobic posttreatment is added.

REFERENCES

de Bruin, W. P., M. J. J. Kotterman, M. A. Posthumus, G. Schraa, and A. J. B. Zehnder. 1992. "Complete Biological Reductive Transformation of Tetrachloroethene to Ethane." *Appl. and Envir. Microbio. 58(6)*: 1996-2000.

Jewell, W. J., S. R. Carter, K. H. Chu, D. E. Fennell, Y. M. Nelson, T. E. White, and M.S. Wilson. 1991. *Methanotrophs for Biological Pollution Control: TCE Removal and Nutrient Removal with the Expanded Bed*. Contract No. 5089-260-1798. Ann. Rep. for 1990. Cornell Univ. for GRI.

Legrand, R., and D. R. Jackson. 1992. "Methanotrophic Biodegradation of Chlorinated Ethenes: Design Constraints Affecting the Expanded Bed Process; Cost Competitiveness and Sensitivities." In G. L. Smith (Ed.), *Implementation of Biotechnology in Industrial Waste Treatment and Bioremediation*, Lewis Publishers, Ann Arbor, MI (in press).

Wu, W. M., D. Wagner, J. Krzewinski, and R. Hickey. 1992. *Pilot Test of the Granular Activated Carbon Fluidized Bed Reactor (GAC-FBR) and Pelletized Diatomaceous Earth Fluidized Bed Reactor (PDE-FBR) for Methanotrophic TCE Degradation*. Quar. Rep. thru 10-92, MBI; for Radian Corp.

LABORATORY EVALUATION OF AEROBIC TRANSFORMATION OF TRICHLOROETHYLENE IN GROUNDWATER USING SOIL COLUMN STUDIES

L. T. LaPat-Polasko, G. A. Fisher, and V. H. Bess

To evaluate the potential for bioremediation of trichloroethylene (TCE)-contaminated groundwater, a 14-week bench-scale study was implemented using site soil and groundwater. Several studies have shown that aerobic microorganisms are capable of degrading TCE and other chlorinated organics by cometabolizing toluene or phenol (Folsom et al. 1990, Nelson et al. 1986, Nelson et al. 1987, Wackett & Gibson 1988). Soil column microcosm studies using phenol or toluene supplements demonstrated greater than 65% reduction in TCE-contaminated groundwater due to cometabolic degradation (Hopkins et al. 1992). Further in situ studies showed that the field results were qualitatively consistent with the microcosm study (Hopkins et al. 1992). The objectives of this study were to stimulate an indigenous aerobic microbial population capable of degrading TCE in groundwater and to evaluate potential toluene oxygenase inducers.

MATERIALS AND METHODS

Six soil columns were prepared using vadose zone soil from the contaminated site. Low-pressure chromatography columns, 2.5 cm diameter by 50 cm height (Bio-Rad), were autoclaved prior to soil packing. The glass columns, Teflon™ tubing, and fittings were sterilized to prevent the introduction of microorganisms from the column equipment. The columns initially were packed by aseptically filling them with sieved soil to a height of 45 cm (280 g of soil at 15% moisture). Groundwater was pumped in an upflow direction at 8 mL/min until the columns received 6 pore volumes of groundwater. The columns were covered with aluminum foil to prevent the growth of photosynthetic organisms, and kept fully water-saturated at room temperature (22°C) for the duration of the study. Column exchange was accomplished using a peristaltic pump and silicone tubing fitted with Teflon™ connectors. Exchange of the column fluid contents was performed weekly at 5 mL/min. After initial equilibration, the groundwater was aerated for at least 1 hour immediately prior to exchange. TCE was injected at a final concentration ranging from 250 μg/L to 1,250 μg/L.

Table 1 lists the amendments for each of the six soil columns. The amendments for the columns were chosen based on a stepwise progression, beginning with the most critical parameter for aerobic processes, oxygen. The major nutrients, nitrogen and phosphorus, were the second amendment to the bench-scale system, followed by three potential inducers of the toluene oxygenase enzyme: phenol (Nelson et al. 1987), salicylic acid (personal communication, Dr. Ogunseitan, University of California, Irvine), and tyrosine (United States Patent No. 4,925,802). The purpose of the sterile column was to monitor the level of chemical and physical removal of TCE in the groundwater.

RESULTS

Figure 1 shows the percent TCE removed from the groundwater effluent from Column 1 – sterile soil and groundwater, Column 2 – hydrogen peroxide

TABLE 1. Column identification key — weeks 1 to 8.[(a)]

Column	Amendment	Description
Column 1	Sterile Control	Contains sterile soil and receives a weekly exchange of sterile groundwater amended with TCE and 70 mg/L hydrogen peroxide.
Column 2	Nonsterile Control with Hydrogen Peroxide	Receives a weekly exchange of groundwater amended with TCE and 70 mg/L hydrogen peroxide.
Column 3	Nutrient Amended (Nonsterile)	Receives a weekly exchange of groundwater amended with TCE, 70 mg/L hydrogen peroxide, and the following nutrients, 3.4 mg/L ammonium phosphate and 4.0 mg/L ammonium chloride.
Column 4	Phenol Amended (Nonsterile)	Receives a weekly exchange of groundwater amended with TCE, 70 mg/L hydrogen peroxide, nutrients, and 100 μM phenol.
Column 5	Salicylic Acid Amended (Nonsterile)	Receives a weekly exchange of groundwater amended with TCE, 70 mg/L hydrogen peroxide, nutrients, and 100 μM salicylic acid.
Column 6	Tyrosine Amended (Nonsterile)	Receives a weekly exchange of groundwater amended with TCE, 70 mg/L hydrogen peroxide, nutrients, and 100 μM tyrosine.

(a) All groundwater mentioned in the key was aerated for at least 1 hour immediately prior to the column exchange to remove native TCE.

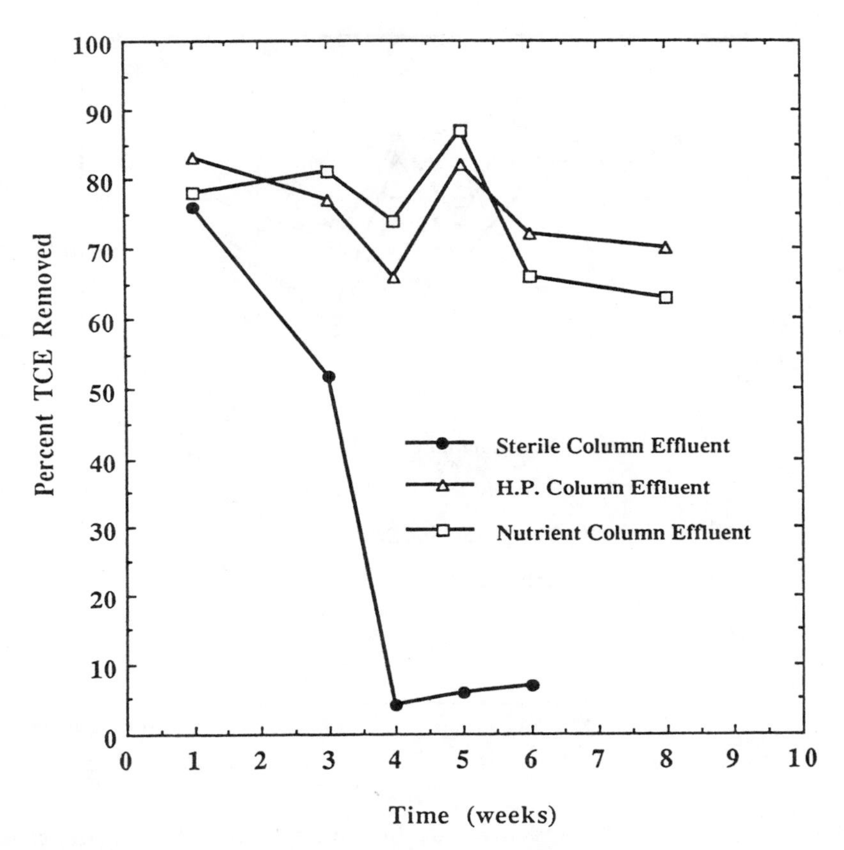

FIGURE 1. Percent TCE removed versus time in Column 1 – sterile column effluent; Column 2 – hydrogen peroxide (H.P.) column effluent; and Column 3 – nutrient column effluent.

amended, and Column 3 – hydrogen peroxide and nutrient amended. This figure indicates that there were significant losses (greater than 50%) of TCE during the first 3 weeks of the bench-scale study due to physical and chemical mechanisms. Although there were major losses of TCE in the groundwater effluent from the sterile column during the initial 3-week period, there was less than an 8% difference between the influent and effluent concentration of the sterile column for the next 4 weeks. The percent TCE removed from Columns 2 and 3 ranged from approximately 65% to 85%. The addition of nutrients to the groundwater did not significantly increase the percent removal of TCE from the groundwater.

Figure 2 shows the percent TCE removed from the groundwater effluent from Column 2 – H.P.; Column 4 – H.P., nutrients and phenol amended; Column 5 – H.P., nutrients and salicylic acid amended; and Column 6 – H.P., nutrients and tyrosine amended for the 8-week bench-scale study. Figure 2 indicates that the three potential oxygenase inducers (phenol, salicylic acid, and

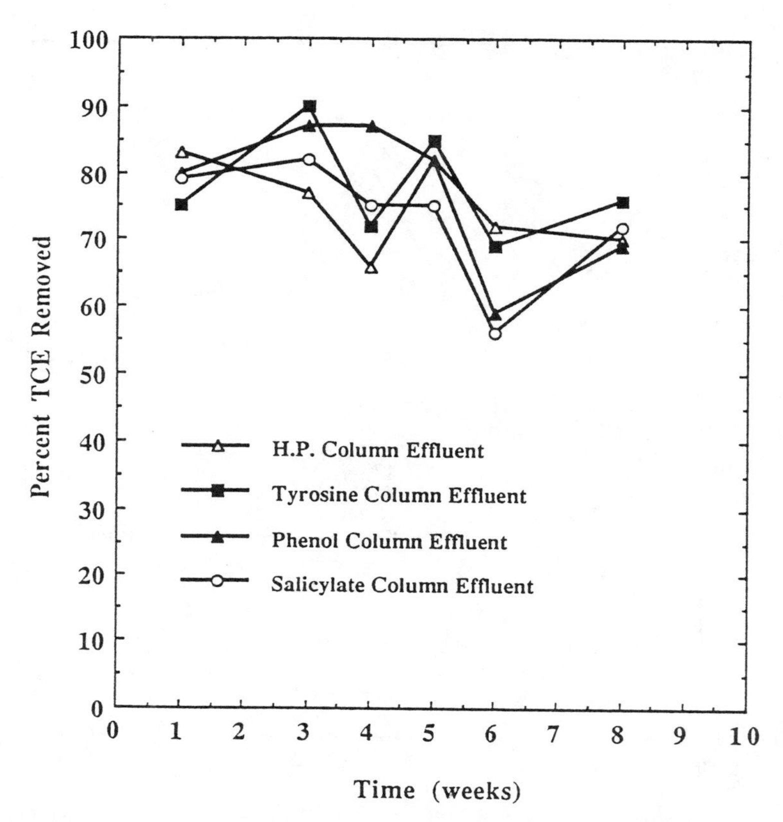

FIGURE 2. Percent TCE removed versus time in Column 2 – hydrogen peroxide (H.P.) column effluent; Column 4 – H.P., nutrients, and phenol; Column 5 – H.P., nutrients, and salicylate; and Column 6 – H.P., nutrients, and tyrosine.

tyrosine) showed similar TCE removal rates (between 54 and 89%) to Column 2, which received only hydrogen peroxide as an amendment. The variation between the effluent TCE concentrations for Columns 2, 4, 5, and 6 was less than 25%, which indicates that there were not significant differences between the columns amended with a potential oxygenase inducer and those without an inducer.

To further evaluate the groundwater amendments, hydrogen peroxide, and nutrients, Columns 2 and 3 were duplicated and studied for an additional 6 weeks. During Weeks 10 through 14 there was less than a 13% variation in the effluent TCE concentration between the corresponding duplicate columns, except for Week 11, where there was a 20% variation between the duplicate nutrient columns. The low level of variation between the duplicate hydrogen peroxide columns indicates that the treatment conditions can be successfully replicated. During Weeks 11 and 12, the percent TCE removed was at least 15% less in those columns receiving only hydrogen peroxide. These results indicate that nutrients may have become limiting in the soil and/or groundwater in those columns that received only hydrogen peroxide.

DISCUSSION

The 8-week and 14-week bench-scale studies indicated that the element critical to promoting in situ bioremediation of TCE-contaminated groundwater was oxygen. When oxygen was available in nonlimiting concentrations, more than 60 and as high as 87% TCE removals were observed in the groundwater effluent.

There were no significant differences in the percent TCE removed in the columns amended with hydrogen peroxide alone and those amended with hydrogen peroxide, nutrients, and the various potential oxygenase inducers. Because there was major TCE removal in the hydrogen peroxide and nutrient columns, this would indicate either that a natural oxygenase inducer already is present in the groundwater and/or soil or that a microbial species exists in the groundwater or soil that is able to directly degrade TCE without an inducer.

To further evaluate the possible biological mechanisms responsible for the degradation of TCE, an additional bench-scale study is being conducted in which half of the duplicate columns have the influent groundwater passed thorough a granular activated carbon (GAC) system to remove potential oxygenase inducers, and then are amended with hydrogen peroxide and nutrients, while the other half of the columns are not subjected to GAC treatment. This will provide information concerning possible inducers that already exist in the subsurface.

REFERENCES

Folsom, B. R., P. J. Chapman, and P. H. Pritchard. 1990. "Phenol and trichloroethylene degradation by *Pseudomonas cepacia* G4." *Appl. and Environ. Microbiol. 56*: 1279-1285.

Hopkins, G. D., L. Semprini, and P. L. McCarty. 1992. "Evaluation of enhanced *in situ* aerobic biodegradation of trichloroethylene and *cis-* and *trans*-1,2-dichloroethylene by phenol-utilizing bacteria." *Symposium on Bioremediation of Hazardous Wastes,* Chicago, IL.

Nelson, M. J. K., S. O. Montgomery, E. J. O'Neill, and P. W. Pritchard. 1986. "Aerobic metabolism of trichloroethylene by a bacterial isolate." *Appl. and Environ. Microbiol. 51*: 383-384.

Nelson, M. J. K., S. O. Montgomery, W. R. Mahaffey, and P. H. Pritchard. 1987. "Biodegradation of trichloroethylene and involvement of an aromatic biodegradative pathway." *Appl. Environ. Microbiol. 53*:949-954.

United States Patent Number: 4,925,802. 1990. "Methods for stimulating biodegradation of halogenated aliphatic hydrocarbons."

Wackett, L. P., and D. T. Gibson. 1988. "Degradation of trichloroethylene by toluene dioxygenase in whole-cell studies with *Pseudomonas putida* Fl." *Appl. Environ. Microbiol. 54*:1703-1708.

EVALUATING POLYCHLORINATED BIPHENYL BIOREMEDIATION PROCESSES: FROM LABORATORY FEASIBILITY TESTING TO PILOT DEMONSTRATIONS

M. J. R. Shannon, R. Rothmel, C. D. Chunn, and R. Unterman

BACKGROUND AND OVERVIEW

Current PCB research programs continue to emphasize (1) optimization of conditions for growth and activity of PCB-degrading bacteria, (2) isolation of superior and novel PCB degraders, (3) genetic engineering of PCB pathways, (4) development of various physical and chemical pretreatment and cotreatments for improving the effectiveness of biotreatment steps, (5) evaluation of alternative soil slurry bioreactors, (6) bench-scale modeling of an in situ biotreatment for dilute soils, and (7) design and testing of field pilot systems. Three major PCB research and development programs are described below.

IMPROVEMENTS IN MICROBIOLOGICAL ACTIVITIES

Microbial studies have been conducted using two complementary PCB-degrading bacterial strains, ENV 307 and ENV 360. These strains demonstrate exceptional ability to degrade higher-chlorinated PCB congeners (tetra-, penta-, hexa-) by using both 2,3- and 3,4- dioxygenase pathways, and are therefore substantially more effective than more commonly isolated PCB-degrading strains. The additive effect of using these two PCB-degrading cultures, in contrast to using one culture exclusively, is illustrated in Figure 1. Bacterial enrichments using soils from selected sites have resulted in the isolation of several new bacterial strains that possess unique PCB-degrading activity. One of these strains exhibits a PCB congener specificity that is distinct from other PCB-degrading organisms described to date (Unterman et al. 1991).

Using highly active ENV 307 and ENV 360 cultures, up to 80% destruction of the PCB in site soils containing 300 mg/kg Aroclor 1248™ has been achieved. Even greater levels of PCB destruction are observed when this aerobic biotreatment is preceded by other additives such as surfactant pretreatment which increase the bioavailability of PCBs. Additional research has been directed at developing techniques to increase the survivability of PCB-degrading bacteria after they have

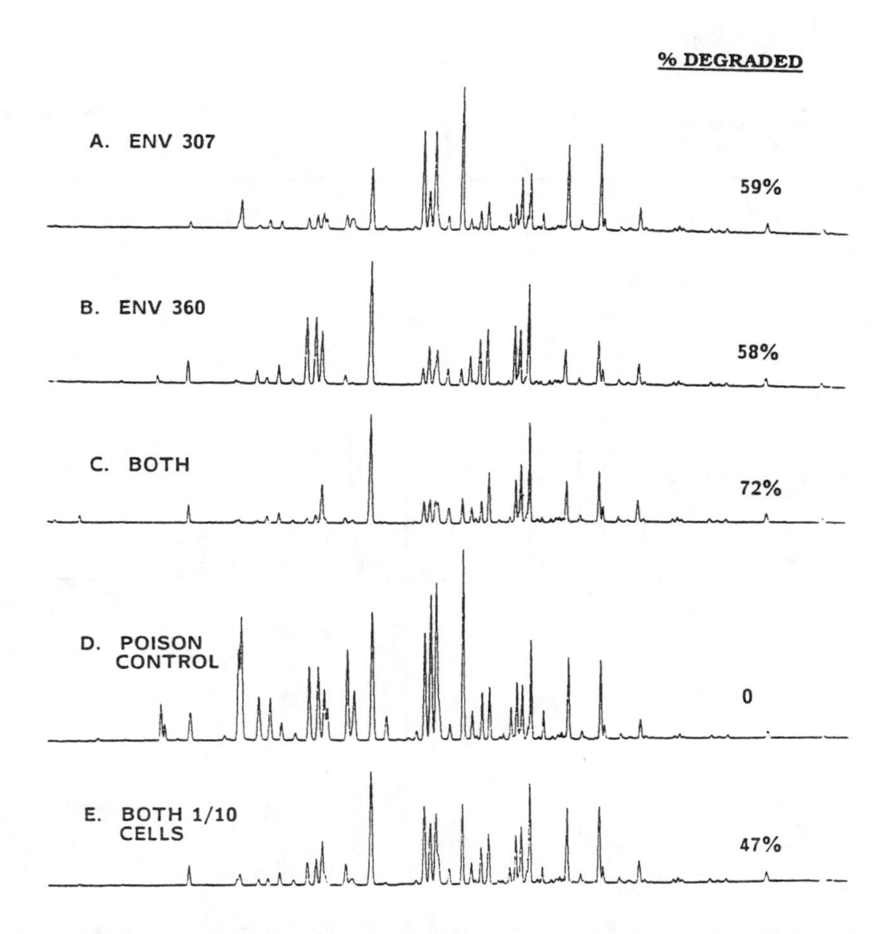

FIGURE 1. Complementary PCB-degrading activity of ENV 307 and ENV 360. The top three panels show that either ENV 307 or ENV 360 alone exhibits 59 or 58% destruction of the PCB; however, when both organisms are used together, even at one-half cell content for each, 72% of the PCB is destroyed. Whereas ENV 360 can degrade one set of PCB congeners, ENV 307 can degrade an additional group of specific congeners. It is this complementarity that is important for ultimately developing a full-scale bioremediation process for decontaminating soil and sludge.

been applied to soils, and a significant increase in the viable population of PCB-degrading microbes has been achieved in the treated soils with the addition of carbon sources such as biphenyl.

Other findings suggest that PCBs at several site locations are undergoing both aerobic and anaerobic transformation in situ. In the laboratory, we have demonstrated that a sequential anaerobic/aerobic process can remove more than 80% of the PCBs from a 1240 mg/kg Aroclor™-containing soil (see Figure 2).

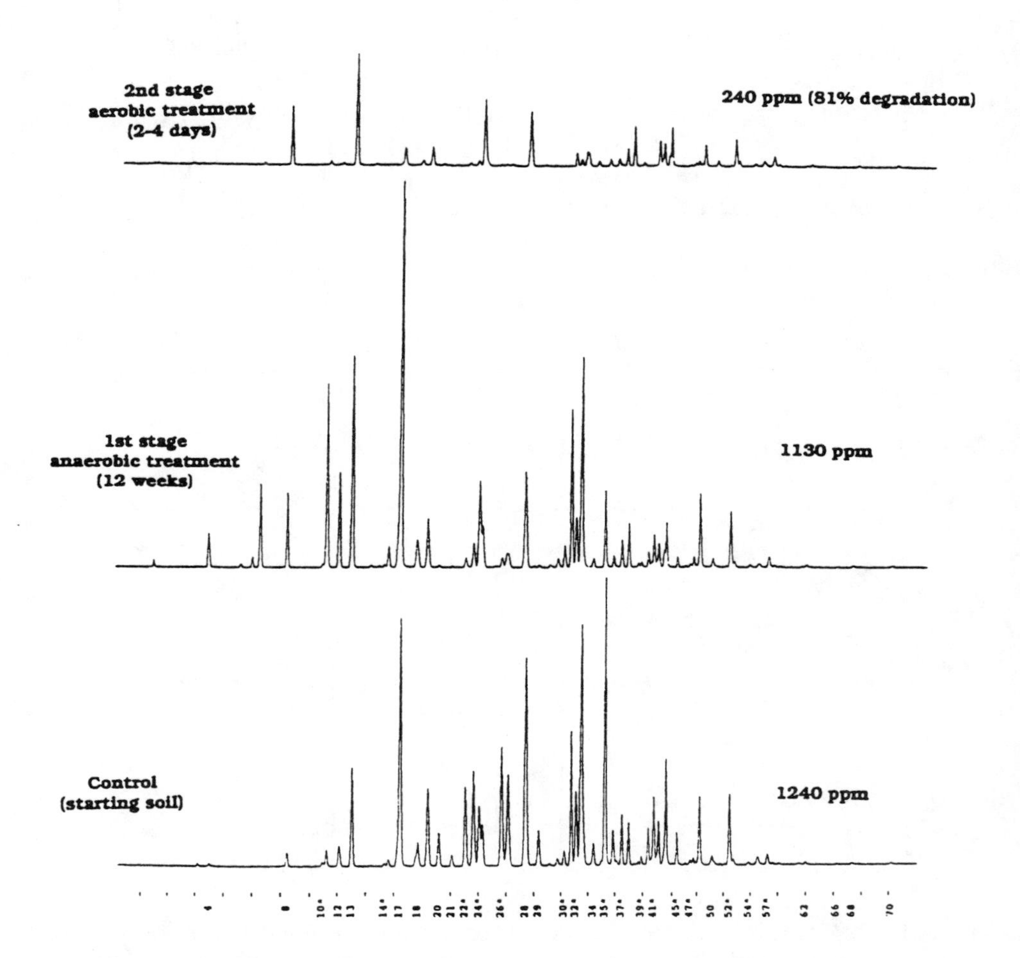

FIGURE 2. Sequential anaerobic/aerobic biodegradation of PCBs in soil. Substantial anaerobic degradation occurs in soil after being incubated under anaerobic conditions for 8 months. Compared to the PCB congener profile of an untreated sample it is apparent that the anaerobic treatment reduces the amount of higher chlorinated congeners and results in a corresponding increase of lower chlorinated congeners. This pattern of congener transformation is characteristic of the anaerobic dechlorination of PCBs. The subsequent aerobic biotreatment step resulted in a substantial 81% total degradation of PCBs.

This extent of removal was accomplished by using only unaugmented microbiological conditions for the anaerobic step; greater degrees of removal are expected as we optimize microbial conditions and employ any of several co- or pretreatments currently under development.

IMPROVING PCB BIOAVAILABILITY

The applicability of bioremediation to PCB-contaminated matrices is determined in part by the availability of the PCB molecules. The least problematic matrix to biotreat is composed of sandy materials having little to none of an organic component. Examples of this include sandy beaches and soils. Soils that are high in natural organic content (humates, etc.) are more problematic due to the binding of the PCB to the organic phase. The bioavailability of the PCB can be enhanced by any of several physical, chemical, or biological means (e.g., high-shear mixing, and chemical oxidation or biodegradation of co-contaminating non-PCB organics).

Problematic matrices include lagoons and landfilled materials that are high in non-PCB oils (e.g., cutting oils and hydraulic fluids). The concentration of these co-contaminating oils can vary significantly, and at high levels may decrease the bioavailability of the PCB. A matrix that contains nonbiodegradable oils at high concentrations is not a good candidate for bioremediation of PCB. Other matrices that contain low levels of nonbiodegradable oils or biodegradable oils are candidates for PCB bioremediation. One means of increasing PCB bioavailability is achieved by first biodegrading the co-contaminating oils. The subsequent step is to follow with a second-stage treatment with the PCB-degrading strains. This approach has proved to be effective with some difficult samples. Ultimately, the utility of this approach is realized only empirically, and as of this writing it cannot be easily predicted.

An alternative approach for increasing PCB bioavailability is to include a surfactant, solvent, or emulsifier to solubilize the PCB away from the surface of soil particles. Several commercially available surfactants can effectively augment PCB degradation. A new PCB-degrading strain has been isolated in our laboratories which produces its own surfactant (Rothmel et al. 1993). Experiments show that the surfactant stimulates the extent of PCB degradation well beyond the levels seen before with other strains. Currently under investigation is the possibility of using the strain in conjunction with other strains to achieve levels of degradation approaching 95%.

DEVELOPMENT OF ALTERNATIVE BIOTREATMENT SYSTEMS

Laboratory comparison of reactor-based versus in situ PCB processes has demonstrated significantly higher rates of PCB destruction in soil slurry reactors; however, for many sites the advantages of not excavating continues to favor the in situ process configuration as a very viable, albeit slower, alternative. Our reactor development program is currently assessing many system parameters including mixing mechanism (e.g., tumbler vs. stirred-tank) and rate, percent solids, bacterial dosing regime, the use of growing vs. stationary growth cells, and integrated co- and pretreatment steps.

TABLE 1. PCB bioremediation approaches/options – 1993.[a]

	In Situ		Reactor	
PCB	Aerobic	Anaerobic	Aerobic	Anaerobic
1242, 1248				
<100 mg/kg	++	+	+	–
>300 mg/kg	?	+	++	–
>1000 mg/kg	–	++	?	–
1254, 1260				
<100 mg/kg	–	+	–	–
>100 mg/kg	–	+	–	–

(a) The categorized options depicted are presented only to serve as a guide for those unfamiliar to the applicability of the biotreatment alternative. The applicability of PCB biotreatment will vary depending on the characteristics of the contaminated matrix.

Concurrently, a large bench-scale microcosm study is being conducted to optimize an in situ soil biotreatment process for a projected field trial. This work is focusing on dilute PCB contamination (less that 100 mg/kg) because of the current rate limitations of in situ treatments. Parameters being addressed include bacterial dosing and mixing regime, moisture content, nutrient addition, and additives.

SUMMARY

Many R&D programs throughout the United States, Canada, Japan, and Europe have produced microbial cultures capable of degrading many PCB congeners. The first commercial bioremediation systems to be developed should address Aroclor™-contaminated soils in the 25 mg/kg to 500 mg/kg range. However, by additional improvements in these aerobic strains as well as the characterization and further development of anaerobic cultures, bioremediation systems eventually could be developed to degrade PCBs in contaminated soils and sediments at concentrations as high as 5,000 mg/kg and perhaps higher. This overview of PCB bioremediation approaches and options is described in Table 1.

REFERENCES

Rothmel, R. K., M. J. R. Shannon, and R. Unterman. 1993. "Isolation and Characterization of a New PCB-degrading Bacterial Strain." *Abstract of the Annual Mtg. of the Amer. Soc. of Microbiology*, #Q-153. Atlanta, GA. In Press.

Unterman, R., C. D. Chunn, and M. J. R. Shannon. 1991. "Isolation and Characterization of a PCB-Degrading Bacterial Strain Exhibiting Novel Aerobic Congener Specificity." *Abstract of the Annual Mtg. of the Amer. Soc. of Microbiology*, #Q-49. Dallas, TX.

CONCEPTS IN IMPROVING POLYCHLORINATED BIPHENYL BIOAVAILABILITY TO BIOREMEDIATION STRATEGIES

P. J. Morris and P. H. Pritchard

INTRODUCTION

In situ bioremediation of polychlorinated biphenyls (PCBs) is an attractive disposal alternative due to the high cost of PCB transport, incineration, and other remedial procedures. However, PCBs are strongly sorbed onto soils and sediments, and this limits their availability to PCB-degrading microorganisms. This availability to the biotic community, referred to as bioavailability, is an increasingly important component in bioremediation strategies for hydrophobic compounds. Although microorganisms may be isolated or engineered to have enhanced enzymatic capabilities, these enzymes will be useless if the kinetics of desorption are much slower than the kinetics of biodegradation.

The soil environment has been shown to offer the microbial community sites for carrying out their respective activities. For example, Oldenhuis and colleagues (1989) observed enhanced biodegradation of a mixture of aromatic solvents in soil slurries that was not observed in liquid cultures. They suggested that soil (clay) had a protective effect on microorganisms with respect to toxicity of a combination of organic solvents. Whereas this is true for water-soluble compounds (e.g., xylenes and *o*-dichlorobenzene), for hydrophobic compounds such as PCBs and polycyclic aromatic hydrocarbons (PAHs), sorption to soil particles may actually inhibit biodegradation processes.

The objectives of this paper are to review information on abiotic and biotic interactions of PCBs in soil and sediment, and to suggest possible approaches to coupling enhanced availability to increased biodegradation. We are currently studying the biodegradation of PCBs found in a former drag strip soil in New York State that exemplifies the problems associated with bioavailability. The PCB concentration in this soil is roughly 500 mg/kg, and no longer resembles commercial Aroclor 1242™. Depletion of the di- and trichlorobiphenyls, probably due to the evaporation of these more volatile congeners, has resulted in a PCB congener profile that more closely resembles Aroclor 1248™. This drag strip soil is from the site of the first in situ PCB biodegradation field test ever conducted (McDermott et al. 1989).

In previous studies, when *Pseudomonas* sp. LB400 was incubated with the drag strip soil in the laboratory, it degraded 15% of the PCB in 1 day and 51%

in 3 days (McDermott et al. 1989). This is slower than that exhibited with laboratory-contaminated soil, but still demonstrates a significant reduction in PCB concentration. In the field, PCB biodegradation was first detectable after 8 to 10 weeks. After 13 weeks, the soil in the top 3 cm of the test plot showed approximately 20% biodegradation of the PCB; after 18 weeks, approximately 25% of the PCB had degraded. The rate of biodegradation at the drag strip was about 50% of the rate seen in the laboratory experiments. A large portion of the rate reduction can be attributed to the poor control of soil temperature and moisture content and the fragile nature of the added microorganism (i.e., *Pseudomonas* sp. strain LB400). However, even in optimized laboratory studies, approximately 50% of the total PCBs was unavailable to microbial degradation. This soil-bound, less-available PCB fraction is the focus of this discussion.

BACKGROUND

PCB mixtures (Aroclors™) have low solubility in water, and solubility decreases with an increase in the degree of chlorination. Although these properties offer excellent commercial applications (e.g., dielectric and industrial fluids, fire retardants, and plasticizers), they result in compounds resistant to most biological, chemical, and physical degradation processes. As compounds for bioremediation, PCBs represent an interesting challenge as both anaerobic and aerobic strategies are required to degrade heavily chlorinated PCB congeners. Anaerobically, reductive dechlorination of Aroclors™ has been observed in situ (Brown et al. 1984, Brown et al. 1987) and in laboratory studies (Quensen et al. 1988), and results in the accumulation of less-chlorinated PCB congeners. These less-chlorinated congeners are more susceptible to aerobic PCB degradation (Bedard et al. 1986, Bopp 1986, Furukawa et al. 1978). However, this biodegradation potential can be realized only if the PCBs are sufficiently bioavailable to microbial communities.

It is not currently known if sorption renders a compound completely unavailable for uptake and subsequent degradation by microorganisms. Several studies have shown that sorption decreases degradation rates by reducing bioavailability (Heitkamp et al. 1984, Subba-Rao & Alexander 1982, Wszolek & Alexander 1979). Ogram et al. (1985) demonstrated that sorbed (2,4-dichlorophenoxy)acetic acid (2,4-D) was unavailable to *Flavobacterium* sp, which is known to degrade 2,4-D in solution. Steinberg et al. (1987) demonstrated that freshly added ethylene dibromide was completely used by indigenous soil bacteria, whereas a residual soil fraction remained that was unavailable.

More recently, Guerin and Boyd (1992) developed bioavailability assays with two bacterial species (*Pseudomonas putida* ATCC 17484 and a Gram-negative soil isolate, NP-Alk) to examine naphthalene mineralization in soil-free and soil-containing systems. For strain 17484, the authors concluded that naphthalene mineralization exceeded that predicted if sorbed naphthalene was assumed to be unavailable, resulting in enhanced rates of naphthalene desorption from soil. This is in contrast to NP-Alk, where sorption limited both the rate and extent of naphthalene mineralization. Weissenfels et al. (1992) observed a decrease in biodegradability and biotoxicity of soil-bound PAHs. In a similar study, biodegradation of quinoline, an ionic, N-containing heterocycle associated with creosote wastes, was shown to be desorption-rate limited (Smith et al. 1992). For mixtures of hydrophobic compounds with a wide range of solubilities, such as Aroclors™, sorption will influence the distribution of individual congeners in the sorbed versus the aqueous phase. If microorganisms are able to degrade only aqueous-phase PCBs, the PCB congeners available in solution may differ considerably from those congeners sorbed onto soil/sediment. Depletion of the more water-soluble, lesser chlorinated PCB congeners by aerobic microorganisms will result in accumulation of the less-soluble, more tightly sorbed congeners. This accumulation clearly will influence bioremediation strategies.

BIOAVAILABILITY ENHANCEMENT STRATEGIES

Synthetic Surfactants. Most of the surfactant studies with PCBs have assessed the ability of surfactants to enhance removal of PCBs by soil-washing methods. Abdul and Gibson (1991) assessed the suitability of an alcohol ethoxylate surfactant for washing Aroclor 1248™ from a sandy soil. After 20 washings with the surfactant, about 66, 86, and 56% of the PCBs were removed by 5,000, 10,000, and 20,000 mg/L surfactant solutions, respectively. Their results show that as the aqueous surfactant concentration increases, a critical concentration is reached at which the surfactant molecules become arranged into structural units (micelles). Adsorption of the hydrophobic PCBs into the hydrophobic interior of the micelles increases the apparent solubility of the PCBs and thereby enhances their transport through the sand. Although this study met with some success in extracting PCBs, the ability to extract PCBs from a low-organic-matter, sandy soil will differ considerably from a soil with a higher silt, clay, and organic matter component where the PCBs may be more strongly integrated into the organic matrix.

In this regard, General Electric scientists have studied the efficiency of PCB solubilization by the anionic surfactant, Surco 233™ (formulation consisting primarily of sodium dodecylbenzenesulfonate [SDBS] plus lime soap-dispersing agents) compared to Triton X-100™ (a nonionic surfactant) (McDermott et al. 1989). A threefold increase of PCB solubilization was obtained by adding an electrolyte (5% Na2SO4 or 1% NaCl) to a Surco 233™ surfactant solution, largely due to an increase in the initial rate of PCB solubilization. Examination of solubilization kinetics with a model substrate (sand) shows that the majority of the PCB is found in the aqueous surfactant solution within the first hour of contact. The effectiveness of the surfactant solution varied with the soil type; Surco 233™ was more effective in removing PCBs from a soil with a high clay component, whereas Triton X-100™ was more effective with a sandy soil. This study is significant because it shows the differential behavior of surfactants based on soil type. Harkness and Bergeron (1990) extracted PCB-contaminated drag strip soil, a sandy loam, with SDBS; only 40% of the PCBs were removed after 2 h of extraction, suggesting that the extraction was not limited by surfactant loading.

Whereas surfactants may be useful for abiotic soil-washing techniques, integration of surfactants into biodegradation strategies may prove difficult. Laha and Luthy (1991) studied the biodegradation of phenanthrene in soil-water systems with nonionic surfactants. In the presence of surfactants at concentrations that resulted in micelle formation (above the critical micelle concentration), the mineralization of phenanthrene was substantially inhibited. Their studies suggested that the inhibition of biodegradation was not a toxicity effect or a result of the surfactant being the preferred substrate. Instead the release (or desorption) of phenanthrene from the micelles may be slow, resulting in a limited availability to the microbial degraders. Interestingly, surfactant concentrations below which micelles are formed in soil-water systems did not inhibit or enhance degradation of phenanthrene, compared to studies in the absence of surfactant. However, in a study by Liu (1980), emulsification of commercial PCB mixtures with sodium ligninsulfonate-enhanced PCB biodegradation by a *Pseudomonas* isolate in liquid cultures; a stable PCB-ligninsulfonate emulsion appeared to facilitate the bacterial degradation by overcoming substrate-surface area limitations.

Biosurfactants. Microbially-produced surfactants also may be exploited as agents to enhance bioavailability and possibly bioremediation. For example, an effective microbially produced surfactant can lower the surface tension of distilled water from 73 dyn/cm to a value less than 30 dyn/cm (Lang & Wagner 1987). Zhang and Miller (1992) studied a rhamnolipid produced by *Pseudomonas aeruginosa* ATCC 9027 and demonstrated that the addition of 300 mg/L rhamnolipid increased the biodegradation of octadecane from 5% (no biosurfactant added) to 20% (biosurfactant added). Thus, there is a need to more thoroughly study biologically-produced surfactants and to couple their ability to solubilize hydrophobic compounds with enhancement of biodegradation. The potential advantages of using microorganisms with biosurfactant-producing capabilities include possible localized effects, such as enhanced desorption from soil particles by attached microorganisms, and increased contaminant mobility into the cell. The biosurfactant, being part of the bacterial extracellular polysaccharide coat, could create a favorable microenvironment for enhanced microbial processes, including biodegradation.

Cosolvency. Whereas surfactants promote interfacial activity to enhance bioavailability, changes in the polarity of the soil/water environment also influence bioavailability. Sorption of low-polarity compounds, such as PCBs, has been modeled as partitioning between a polar solvent (e.g., water) and a sorbent (e.g., organic matter) (Chiou et al. 1979). With the soil acting as a flexible and amorphous polymer, changes in solvent polarity can result in expansion or contraction of the polymer, hence influencing the permeation of a sorbate (e.g., PCBs) through the polymer. Cosolvency has been defined as the effect of the addition of one or more organic cosolvents (e.g., methanol) on the solubility, sorption, and transport of organic chemicals (Nkedi-Kizza et al. 1987, Rao et al. 1991). This effect may be due, in many cases, to changes in polarity and subsequent changes in the permeation of soil polymers. A hydrophobic polymer (e.g., soil) will be condensed when water is the sole solvent; addition of an organic cosolvent will result in an expansion of the polymer and will result in increased diffusion (Brusseau et al. 1991). In fact, Griffin and Chou (1981) demonstrated that the mobility of PCBs, as determined by a soil-thin-layer chromatography technique, significantly increases with methanol as the elution solvent compared to water. The addition of a cosolvent to a site, whether intentional (e.g., landfill) or accidental (e.g., spill), could influence the mobility, hence the availability, of hydrophobic contaminants. The potential effect of cosolvency on bioremediation of PCBs must be weighed against the knowledge that enhanced mobility also will increase transfer of PCBs from unavailable, soil-bound sites, into the food chain.

SORPTION BEHAVIOR OF PCBs

A number of studies have been done on the kinetics and reversibility of the sorption-desorption of hydrophobic organic compounds such as PCBs, and this remains a controversial issue (Sklarew & Girvin 1987). The question revolves around whether all PCBs in soil/sediment eventually will be available to the biotic community or whether a portion of the PCBs remains irreversibly bound. For example, following rapid equilibration (20 h) of Aroclor 1254™ with soil, the majority of the sorbed PCB was shown to be resistant to further desorption (Fairbanks & O'Connor 1984). Di Toro and Horzempa (1982) observed that montmorillonite clay, kaolinite clay, and natural lake sediment showed linear isotherms, which effectively described the relationships between both adsorbed 2,4,5-2′,4′,5′-hexachlorobiphenyl (HCBP) and the aqueous HCBP concentration. Their laboratory equilibration studies indicate that sediment-adsorbed HCBP may comprise both reversibly and strongly bound or resistant components. The issue of reversibility becomes critical for in situ bioremediation strategies that must meet the stringent regulatory levels required for cleanup.

Coates and Elzerman (1986) conducted one of the first PCB studies to address the issue of the slow approach to sorption equilibrium. Desorption data obtained by the gas purge technique suggest that equilibration times for PCBs with low chlorine content (containing up to 4 chlorine atoms) are on the order of 6 weeks, and are months to years for PCBs with a significantly higher chlorine content (greater than 6 chlorine atoms). Slow release of PCBs support Karickhoff's recommendation that sorption of hydrophobic sorbates onto hydrophobic surfaces can be modeled as a two-component sorption system consisting of a component desorbing within minutes to hours and a much slower component that may take months to years to completely desorb (Karickhoff & Morris 1985).

In a later study by Girvin et al. (1990), the time to reach sorption equilibrium (i.e., sorption-diffusion equilibrium for a highly aggregated montmorillonitic soil containing 0.9% organic carbon) for several PCB congeners (e.g., 2,4′-CB; 2,5-2′,5′-CB; and 2,4,5-2′,4′,5′-CB) was determined to be 3, 5, and 10 months, respectively, using a gas-purge technique. This finding suggests that the diffusion of hydrophobic compounds into the interior porous structure of soil and sediment particles extends the time required for sorption equilibrium to time scales of weeks to years, depending on the hydrophobicity of the compound. It has been the use of the precise gas-purge technique to measure both the rapid and slower components of the sorption process for hydrophobic compounds that has cast doubt on earlier reports with the batch method that PCB sorption equilibration occurs within hours to days (Horzempa & Di Toro 1983, Nau-Ritter et al. 1982, Nau-Ritter & Wurster 1983).

Another factor contributing to the confusion surrounding the sorption/desorption of PCB was identified by Gschwend and Wu (1985) in their study of the fast sorption component. These authors showed in batch experiments that incomplete phase separation occurred between 2,4,5-2′,4′,5′-hexachlorobiphenyl in "true" aqueous solution and, associated with microparticles and/or organic macromolecules in the aqueous phase, introduced an experimental artifact. If in typical batch studies the soil/aqueous phase is not centrifuged sufficiently, the aqueous phase will contain nonsettling microparticulates to which PCBs are bound, thus giving erroneous measures of aqueous-phase PCB concentrations. The effect of these microparticles and dissolved organic matter (DOM) on PCB solubility was investigated by Chiou et al. (1986); solubility enhancements of relatively water-insoluble solutes (e.g., 2,4,5-2′,5′-CB and 2,4-4′-CB) was observed upon the addition of DOM of soil and aquatic origins. This effect may be described in terms of partition-like interaction of the solutes with the microscopic organic environment of the high-molecular-weight DOM species. Whether this enhanced solubility due to sorption onto microparticles or DOM results in increased bioavailability to the microbial community is as yet unknown, but it could be a crucial factor in the extent to which bioremediation can remove hydrophobic compounds from soil.

EFFECT OF OIL AND GREASE

Another factor affecting the bioavailability of PCBs for bioremediation involves co-contaminants. Many PCB-contaminated sites contain significant quantities of oil and grease, a general "class" of compounds encompassing a wide variety of petroleum products, including transformer oils. In general, this oil and grease component will be weathered, resulting in high-molecular-weight, stable (slowly biodegraded) compounds. For example, the drag strip soil we are studying contains roughly 1.5% oil and grease by weight, and a gas chromatography/flame ionization detector (GC-FID) trace of this fraction suggests that the oil is highly weathered.

The presence of this oil and grease component has been shown to affect the sorption characteristics of PCBs. Sayler and Colwell (1976) showed that the partitioning of PCBs into crude oil was approximately three times greater than partitioning onto suspended sediments. More recently, Boyd and Sun (1990) have shown that both natural organic matter and residual oil components of soils act as partition media for organic solutes, with the latter being roughly 10 times more effective as a sorptive phase. Sun and Boyd (1991) showed that residual PCB oil present in soils heavily contaminated with PCB mixtures (Aroclors™) functions as a highly effective partitioning medium. When the PCB-oil content in soil reaches a critical level (0.1%), the discrete PCB solute molecules start forming a separate anthropogenic sorptive phase that is highly effective in removing individual PCBs from aqueous solution. Harkness and Bergeron (1990) conducted resting cell assays with a PCB-degrading isolate and Aroclor 1242™ in the presence of various ratios of mineral oil. They observed inhibition of biodegradation of Aroclor 1242™ at all levels of mineral oil assayed (i.e., from 10:1 oil:PCB by weight to 200:1), suggesting that PCBs were partitioning into the mineral oil, resulting in a decrease in availability to the PCB-degrader.

SUMMARY

We are currently coupling studies on desorption of historically PCB-contaminated soil materials, particularly those co-contaminated with oil and grease, with biodegradation kinetics. To increase desorption of PCBs from historically contaminated PCB soils, developing methods that are compatible with PCB biodegradation is essential if in situ bioremediation of PCBs is to be successful. The integration of surfactants and cosolvents into PCB bioremediation strategies in soils needs to be further examined, but in a careful and thorough manner because, as we have attempted to show, PCB bioavailability is controlled by a very complex set of processes that we are only now beginning to more fully understand. The reality of PCB bioremediation strategies is one of regulatory levels and constraints. Understanding and enhancing bioavailability ultimately may be essential in using PCB biodegradation to remediate sites.

ACKNOWLEDGMENTS

Financial support for this work was provided by the General Electric Company. This work was performed as part of a Cooperative Research and Development Agreement between the U.S. Environmental Protection Agency Environmental Research Laboratory (Gulf Breeze, Florida) and SBP Technologies, Inc. (Stone Mountain, Georgia) as defined under the Federal Technology Transfer Act of 1986.

REFERENCES

Abdul, A. S., and T. L. Gibson. 1991. "Laboratory studies of surfactant-enhanced washing of polychlorinated biphenyl from sandy material." *Environ. Sci. Technol. 25*:665-671.

Bedard, D. L., R. Unterman, L. H. Bopp, M. J. Brennan, M. L. Haberl, and C. Johnson. 1986. "Rapid assay for screening and characterizing microorganisms for the ability to degrade polychlorinated biphenyls." *Appl. Environ. Microbiol. 51*:761-768.

Bopp, L. H. 1986. "Degradation of highly chlorinated PCBs by *Pseudomonas* strain LB400." *J. Ind. Microbiol. 1*:23-29.

Boyd, S. A., and S. Sun. 1990. "Residual petroleum and polychlorobiphenyl oils as sorptive phases for organic contaminants in soils." *Environ. Sci. Technol. 24*:142-144.

Brown, J. F., Jr., R. E. Wagner, D. L. Bedard, M. J. Brennan, J. C. Carnahan, and R. J. May. 1984. "PCB transformations in upper Hudson sediments." *Northeast. Environ. Sci. 3:*167-179.

Brown, J. F., Jr., D. L. Bedard, M. J. Brennan, J. C. Carnahan, H. Feng, and R. E. Wagner. 1987. "Polychlorinated biphenyl dechlorination in aquatic sediments." *Science 236:*709-712.

Brusseau, M. L., A. L. Wood, and P. S. C. Rao. 1991. "Influence of organic cosolvents on the sorption kinetics of hydrophobic organic chemicals." *Environ. Sci. Technol. 25:*903-910.

Chiou, C. T., L. J. Peters, and V. H. Freed. 1979. "A physical concept of soil-water equilibria for nonionic organic compounds." *Science 206:*831.

Chiou, C. T., R. L. Malcolm, T. I. Brinton, and D. E. Kile. 1986. "Water solubility enhancement of some organic pollutants and pesticides by dissolved humic and fulvic acids." *Environ. Sci. Technol. 20:*502-508.

Coates, J. T., and A. W. Elzerman. 1986. "Desorption kinetics for selected PCB congeners from river sediments." *J. Contam. Hydrol. 1:*191-210.

Di Toro, D. M., and L. M. Horzempa. 1982. "Reversible and resistant components of PCB adsorption-desorption: Isotherms." *Environ. Sci. Technol. 16:*594-602.

Fairbanks, B. C., and D. J. O'Connor. 1984. "Effect of sewage-sludge on the adsorption of polychlorinated biphenyls by three New Mexico soils." *J. Environ. Qual. 13:*297-300.

Furukawa, K., N. Tomikuza, and A. Kamibayashi. 1978. "Effect of chlorine substitution on the biodegradability of polychlorinated biphenyls." *Appl. Environ. Microbiol. 35:*223-227.

Girvin, D. C., D. S. Sklarew, A. J. Scott, and J. P. Zipperer. 1990. *Release and Attenuation of PCB Congeners: Measurement of Desorption Kinetics and Equilibration Sorption Partition Coefficients.* EPRI GS-6875, Electric Power Research Institute Report, Palo Alto, CA.

Griffin, R. A., and S. F. J. Chou. 1981. "Movement of PCBs and other persistent compounds through soil." *Wat. Sci. Tech. 13:*1153-1163.

Gschwend, P. M., and S-C. Wu. 1985. "On the constancy of sediment-water partition coefficients of hydrophobic organic pollutants." *Environ. Sci. Technol. 19:*90-96.

Guerin, W. F., and S. A. Boyd. 1992. "Differential bioavailability of soil-sorbed naphthalene to two bacterial species." *Appl. Environ. Microbiol. 58:*1142-1152.

Harkness, M. R., and J. A. Bergeron. 1990. "Availability of PCBs in soils and sediments to surfactant extraction and aerobic biodegradation." *In Ninth Progress Report for the Research and Development Program for the Destruction of PCBs*, pp. 109-120. General Electric Co. Corporate Research and Development, Schenectady, NY.

Heitkamp, M. A., J. N. Huckins, J. D. Petty, and J. L. Johnson. 1984. "Fate and metabolism of isopropylphenyl diphenyl phosphate in freshwater sediments." *Environ. Sci. Technol. 18:*434-439.

Horzempa, L. M. and D. M. Di Toro. 1983. "The extent and reversibility of polychlorinated biphenyl adsorption." *Water Res. 17:*851-859.

Karickhoff, S. W., and K. R. Morris. 1985. "Sorption dynamics of hydrophobic pollutants in sediment suspensions." *Environ. Toxicol. Chem. 4:*462-479.

Laha, S. and R. G. Luthy. 1991. "Inhibition of phenanthrene mineralization by nonionic surfactants in soil-water systems." *Environ. Sci. Technol. 25:*1920-1930.

Lang, S., and F. Wagner. 1987. "Structure and properties of biosurfactants." *Surfactant Sci. Ser. 25:*21-45.

Liu, D. 1980. "Enhancement of PCBs biodegradation by sodium ligninsulfonate." *Water Res. 14:*1467-1475.

McDermott, J. B., R. Unterman, M. J. Brennan, R. E. Brooks, D. P. Mobley, C. C. Schwartz, and D. K. Dietrich. 1989. "Two strategies for PCB soil remediation: Biodegradation and surfactant extraction." *Experimental Progress 8:*46-51.

Nau-Ritter, G. M., and C. F. Wurster. 1983. "Sorption of polychlorinated biphenyls (PCB) to clay particulates and effects of desorption on phytoplankton." *Water Res. 17:*383-387.

Nau-Ritter, G. M., C. F. Wurster, and R. G. Rowland. 1982. "Partitioning of [14C] PCB between water and particulates with various organic contents." *Water Res. 16:*1615-1618.

Nkedi-Kizza, P., P. S. C. Rao, and A. G. Hornsby. 1987. "Influence of organic cosolvents on leaching of hydrophobic organic chemicals through soils." *Environ. Sci. Technol. 21*:1107-1111.

Ogram, A. Y., R. E. Jessup, L. T. Ou, and P. S. C. Rao. 1985. "Effect of sorption on biological degradation rates of (2,4-dichlorophenoxy) acetic acid in soils." *Appl. Environ. Microbiol. 49*:582-587.

Oldenhuis, R., L. Kuijk, A. Lammers, D. B. Janssen, and B. Witholt. 1989. "Degradation of chlorinated and non-chlorinated aromatic solvents in soil suspensions by pure bacterial cultures." *Appl. Microbiol. Biotechnol. 30*:211-217.

Quensen, J. F., III, S. A. Boyd, and J. M. Tiedje. 1988. "Reductive dechlorination of polychlorinated biphenyls by anaerobic microorganisms from sediments." *Science* 242:752-754.

Rao, P. S. C., L. S. Lee, and A. L. Wood. 1991. *Solubility, Sorption and Transport of Hydrophobic Chemicals in Complex Mixtures.* EPA/600/M-91/009. U.S. Environmental Protection Agency, Ada, OK.

Sayler, G. S., and R. R. Colwell. 1976. "Partitioning of mercury and polychlorinated biphenyl by oil, water, and suspended sediment." *Environ. Sci. Technol. 10*:1142-1145.

Sklarew, D. S., and D. C. Girvin. 1987. "Attenuation of polychlorinated biphenyls in soils." *Rev. Environ. Contamin. Toxicol. 98*:1-41.

Smith, S. C., C. C. Ainsworth, S. J. Traina, and R. J. Hicks. 1992. "Effect of sorption on the biodegradation of quinoline." *Soil Sci. Soc. Am. J. 56:737-746.*

Steinberg, S. M., J. J. Pignatello, and B. L. Sawhney. 1987. "Persistence of 1,2-dibromoethane in soils: Entrapment in intraparticle micropores." *Environ. Sci. Technol. 21*:1201-1208.

Subba-Rao, R. V., and M. Alexander. 1982. "Effect of sorption on mineralization of low concentrations of aromatic compounds in lake water samples." *Appl. Environ. Microbiol. 44*:659-668.

Sun, S., and S. A. Boyd. 1991. "Sorption of polychlorobiphenyl (PCB) congeners by residual PCB-oil phases in soils." *J. Environ. Qual.* 20:557-561.

Weissenfels, W. D., H-J. Klewer, and J. Langhoff. 1992. "Adsorption of polycyclic aromatic hydrocarbons (PAHs) by soil particles: Influence on biodegradability and biotoxicity." *Appl. Microbiol. Biotechnol. 36*:689-696.

Wszolek, P. C., and M. Alexander. 1979. "Effect of desorption rate on the biodegradation of n-alkylamines bound to clay." *J. Agric. Food Chem.* 27:410-414.

Zhang, Y., and R. M. Miller. 1992. "Enhanced octadecane dispersion and biodegradation by a *Pseudomonas* rhamnolipid surfactant (biosurfactant)." *Appl. Environ. Microbiol. 58*:3276-3282.

FIELD STUDY OF AEROBIC POLYCHLORINATED BIPHENYL BIODEGRADATION IN HUDSON RIVER SEDIMENTS

M. R. Harkness, J. B. McDermott, D. A. Abramowicz, J. J. Salvo, W. P. Flanagan, M. L. Stephens, F. J. Mondello, R. J. May, J. H. Lobos, K. M. Carroll, M. J. Brennan, A. A. Bracco, K. M. Fish, G. L. Warner, P. R. Wilson, D. K. Dietrich, D. T. Lin, C. B. Morgan, and W. L. Gately

INTRODUCTION

Microbially mediated reductive dechlorination of polychlorinated biphenyls (PCBs) in Hudson River sediments has been extensively documented (Abramowicz 1990, Bedard 1990). This natural anaerobic process has reduced the average number of chlorines per PCB molecule in the sediment from 3.5 for the Aroclor™ 1242 commercial mixture originally released into the river to ~2.4 at present. The less-chlorinated PCB congeners produced by anaerobic dechlorination are suitable substrates for oxidative degradation by a wide range of aerobic organisms (Abramowicz 1990, Bedard 1990). Products of this oxidative attack include the corresponding chlorobenzoic acids, which are readily degraded by other aerobic bacteria (Hickey & Focht 1990).

Oxidative degradation of low-chlorinated PCBs by aerobic bacteria is well documented from laboratory experiments, most often using purified PCB-degrading bacteria acting on defined PCB congener mixtures in the absence of soil or sediment. Little is known about this process in the field. Therefore, laboratory experiments were performed under conditions that more closely resembled those encountered in the field to assess the potential for aerobic PCB biodegradation in Hudson River sediments. Based on those results, a 73-day study was conducted in the Hudson River to (1) demonstrate that PCBs could be biodegraded aerobically under field conditions, and (2) identify the key variables that influence the rate and extent of PCB biodegradation in these sediments.

LABORATORY RESULTS

PCB biodegradation studies were performed in 4-week experiments at 24°C using 250-mL shake flasks containing 50 g of PCB-contaminated Hudson River sediment and 50 mL of river water. Non-control flasks were amended with 130 mg ammonium sulfate, 70 mg potassium phosphate, and 20 mg biphenyl with and without an equal amount of sodium succinate as carbon sources. Biphenyl is known to induce PCB-degrading activity in selected bacterial strains (Focht & Brunner 1985). Several flasks were inoculated with *Alcaligenes eutrophus* H850, a purified bacterium able to degrade an unusually broad spectrum of PCB congeners, including many tetra- and penta-, and some hexachlorobiphenyls (Bedard et al. 1987). PCBs were extracted from the sediment and glassware/water with 1:1 mixtures of hexane/acetone and hexane/ether, respectively, and analyzed by high-resolution capillary gas chromatography after cleanup (Brown et al. 1987). Foam plug stoppers also were analyzed for volatilized PCBs. The results were compared to those from unamended controls flasks containing only sediment and water and to mercuric chloride-killed controls.

Maximal PCB biodegradation occurred in the presence of inorganic nutrients and biphenyl, with losses ranging from 30 to 40% over 4 weeks. PCB losses were highest among mono- and dichlorobiphenyl congener groups, whereas losses of higher congeners also were evident (Figure 1a,b). Enhanced biological nitrifying activity, resulting in shifts to lower pH (<5.0) over time, also was observed in these

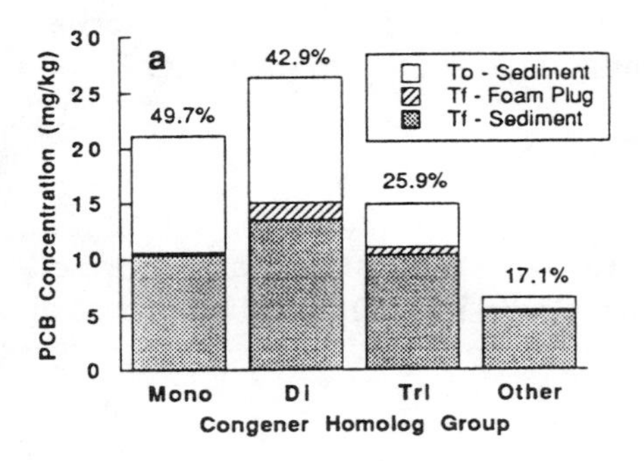

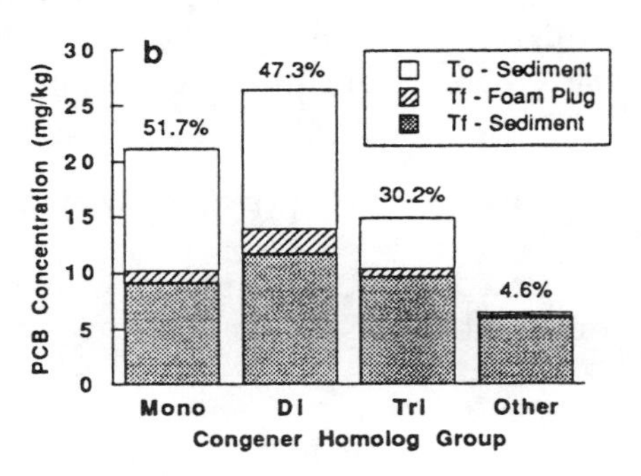

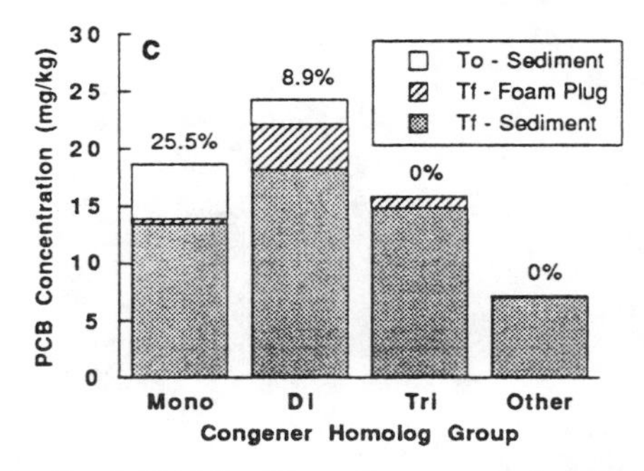

FIGURE 1. PCB biodegradation by homolog group in 28-day laboratory experiments using amended Hudson River sediments (a) with or (b) without H850 added and (c) for an unamended control. Biodegradation losses were calculated from the difference in PCB concentration measured in untreated sediment (T_o) and that measured in the sediment and foam plug after 28 days of treatment (T_f).

flasks. Neither the addition of sodium succinate nor inoculation with H850 enhanced the degradation results. Plate counts of H850 declined from 5×10^7 to $\sim 10^6$ colony-forming units (cfu)/mL over 28 days, whereas counts of indigenous biphenyl-metabolizing bacteria increased from 5×10^3 to $\sim 10^8$ cfu/mL in 2 weeks in flasks not inoculated with H850. Some volatilization losses were apparent in all the flasks. Biological PCB losses of 10 to 15% were observed in unamended control flasks (Figure 1c), whereas complete recovery of PCBs was achieved in killed controls.

The extent of biodegradation in these experiments did not exceed 50% of the PCBs in Hudson River sediments, despite the fact that most (>90%) of this material is biodegradable in the absence of sediment (Bedard et al. 1987). The rate of PCB biodegradation appears to be limited by the desorption kinetics of the PCBs from the polymeric, natural organic matrix of the sediments. In laboratory experiments performed by contacting Hudson River sediment with a PCB-absorbing resin, a labile PCB fraction desorbed readily, whereas a resistant fraction desorbed orders of magnitude more slowly, with ~45% of the PCBs residing in the resistant component (Figure 2). Over 6 months, an additional 50% of the PCBs making up the resistant fraction slowly desorbed from the sediment. This resistant PCB fraction is not likely to be bioavailable in its sorbed state, thereby constituting the primary limitation to short-term biodegradation.

FIELD STUDY DESCRIPTION

Based on these laboratory results, a large-scale field study was performed immediately offshore in the upper Hudson River, in six 1.8-m-diameter steel

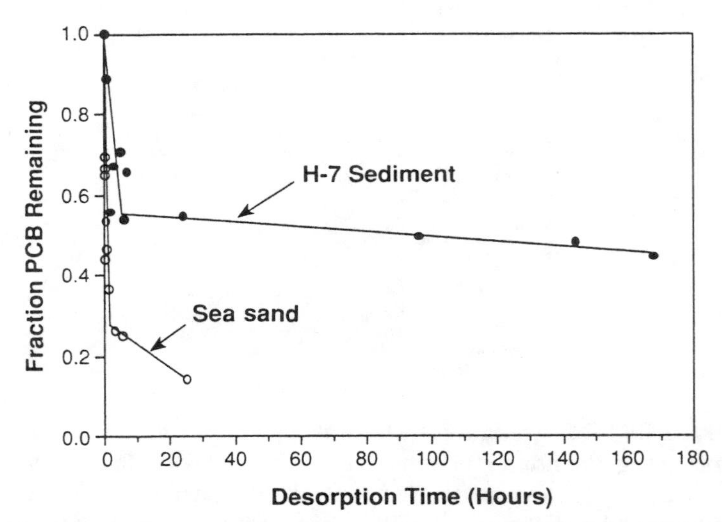

FIGURE 2. Desorption of PCBs from environmentally contaminated Hudson River sediment and Aroclor 1242™ spiked sea sand to XAD-4 resin.

caisson reactors (R101 to R106) driven into the river bottom to isolate the natural biota and sediment from the river environment (Figure 3). The field study began on August 9, 1991 and was run at ambient river temperatures (10 to 28°C). In it we investigated the effects of mixing mode and the addition of oxygen, inorganic nutrients, biphenyl, and PCB-degrading bacteria on aerobic PCB biodegradation (Figure 3). The sediments were mixed using two types of agitators; high-mix turbines turning at 40 rpm and low-mix rakes that rotated at 3 rpm. Aqueous-phase dissolved oxygen levels of 6.0 to 6.5 mg/L were maintained in four caissons (R102, R103, R105, R106) by automatically controlled addition of a 10% hydrogen peroxide solution. Diammonium phosphate, ammonium sulfate, potassium phosphate, and biphenyl were added to these caissons at the start or at intervals throughout the study. Two caissons (R102, R103) were inoculated with H850, and two caissons (R101, R104) served as controls and received no nutrient, oxygen, or microbial amendments to limit aerobic biological activity. All the caissons

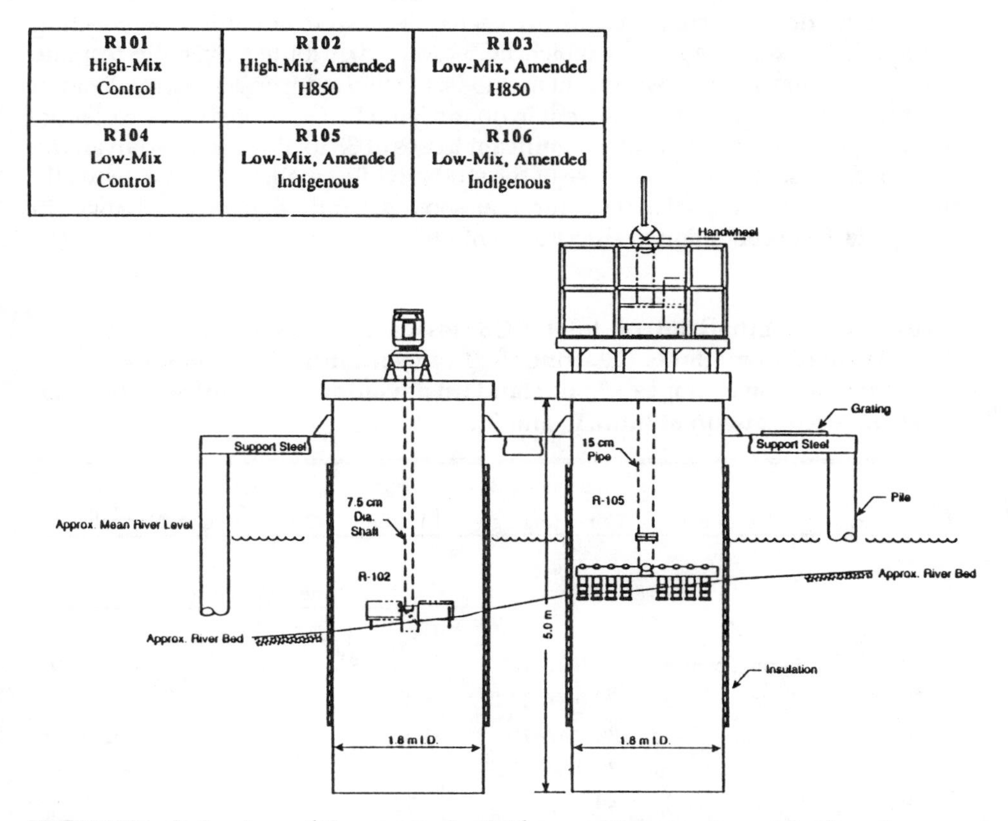

R101 High-Mix Control	R102 High-Mix, Amended H850	R103 Low-Mix, Amended H850
R104 Low-Mix Control	R105 Low-Mix, Amended Indigenous	R106 Low-Mix, Amended Indigenous

FIGURE 3. Side view of the research platform with experiment design shown. Amended caissons received nitrogen, phosphate, biphenyl, and hydrogen peroxide. (Used with permission, Harkness, M. R., et al., 1993. "In Situ Stimulation of Aerobic PCB Biodegradation in Hudson River Sediments." *Science* 259:503-507, copyright AAAS.)

were vented to the atmosphere through vapor traps containing XAD-2 resin to retain any volatilized PCBs.

Sediment cores for PCB analysis were taken from the caissons throughout the study. In each case, the mixers were shut down prior to sampling to allow the sediment to settle. In addition, aqueous and sediment samples also were taken periodically for bacterial counts and for chlorobenzoate and biphenyl analyses.

FIELD STUDY RESULTS

Three methods were used to calculate PCB concentration changes within the caissons: (1) direct concentration measurements, (2) concentration measurements normalized to a recalcitrant reference congener peak (peak 61 — 34-34-/236-34-chlorobiphenyl), and (3) concentration measurements normalized to the total organic carbon (TOC) content in each core sample.

The starting PCB concentrations in each caisson were not uniform, reflecting the heterogeneous manner in which PCBs were distributed over the test site (Table 1). Of the three measurement methods, the two independent normalization techniques were the most consistent in quantifying PCB changes in the caissons, demonstrating that statistically significant losses of 38 to 55% occurred in all the amended caissons over the course of the study, while smaller or nonstatistically significant losses occurred in the control caissons. Attempts to interpret changes in average PCB concentration by direct concentration measurement were complicated

TABLE 1. Comparison of total PCB losses after 73 days by different quantitation methods. Starting PCB concentrations are shown as direct concentration averages +/- one standard deviation. 12 samples were taken from each caisson at both T_0 and T_f.

Caisson	Treatment	Starting PCB Conc. (mg/kg)	Direct Measure	Peak 61 Normalized	TOC Normalized
			Percent Change (%)		
R101	hi-mix, control	6.0 +/- 1.9	+ 8.7 (a)	- 14.4	- 30.7 (a)
R102	hi-mix, H850	20.0 +/- 11.0	- 41.0	- 42.4	- 44.7
R103	low-mix, H850	30.2 +/- 10.6	- 36.8	- 37.8	- 55.5
R104	low-mix, control	39.9 +/- 15.6	- 41.8	- 4.3 (a)	+ 8.4 (a)
R105	low-mix, indig.	49.7 +/- 27.8	- 72.6	- 40.5	- 53.1
R106	low-mix, indig.	39.1 +/- 17.5	- 68.5	- 38.7	- 46.0

(a) Indicates changes were not statistically significant at the 95% confidence level by a two-sample t-test.

by sampling difficulties encountered in the low-mix caissons (Harkness et al. 1993).

The PCB losses observed in the amended caissons were highly congener-specific, with substantial reductions of primarily mono- and dichlorobiphenyls evident in the low-mix caissons (Figure 4c,e,f) and additional losses of trichlorobiphenyl congeners apparent in high-mix caisson R102 (Figure 4b). The performance of R102 was most similar to that of the laboratory shake flasks. R102 had

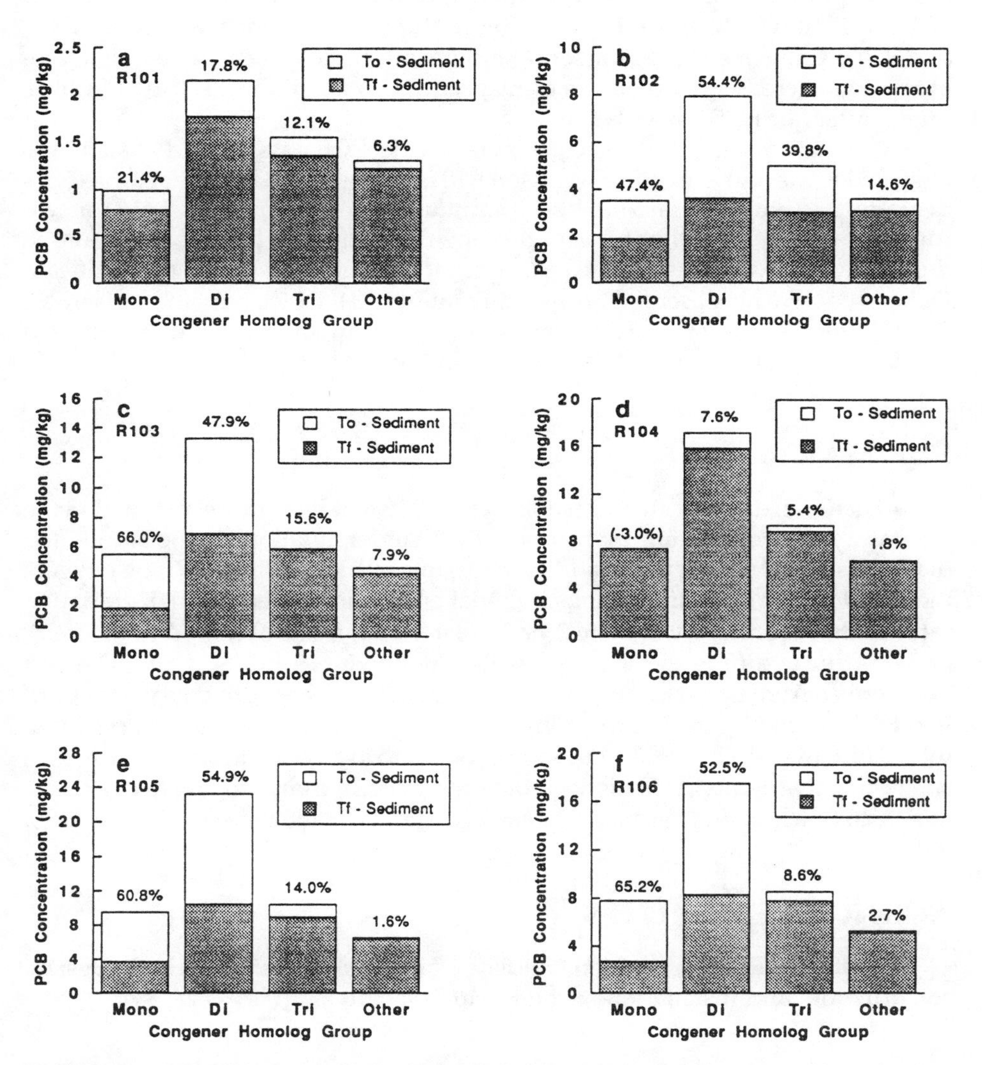

FIGURE 4. PCB biodegradation after 73 days by homolog group in caissons (a) R101 to (f) R106. Biodegradation losses were calculated from the difference in PCB concentration measured at the beginning of the study (T_o) and after 73 days of treatment (T_f) using data normalized to peak 61.

the highest peroxide demand, exhibited the highest rate of biological nitrifying activity, and operated at the lowest pH (<5.0) of any of the amended caissons. Although the broader PCB-degradative competence observed in this caisson may have been due to the presence of H850, it is more likely that the unique environment created there stimulated an indigenous PCB-degrading population with broader congener specificities.

The survival of H850 was quite poor during the study. Population counts of H850 dropped at least three orders of magnitude from starting concentrations of 10^7 to 10^8 cfu/mL within 10 days after multiple inoculations in caissons R102 and R103. Conversely, the numbers of indigenous biphenyl-metabolizing microorganisms increased by 6 orders of magnitude, to a maximum of 10^8 cfu/mL, independent of H850 inoculation.

A statistically significant congener-specific PCB loss of 14.4% also was observed in the high-mix control caisson R101, suggesting that some aerobic PCB-degrading activity may also have been stimulated there (Figure 4a). Low concentrations of oxygen (<2 mg/L) were present in the aqueous phase of this caisson due to oxygen exchange with the headspace at the high rate of mixing. Significant PCB losses were not observed in low-mix control R104, which remained anaerobic throughout the study (Figure 4d), nor were significant PCB losses detected from any caisson via monitored abiotic routes in the field test.

SUMMARY

In general, the field study results were consistent with those obtained in the laboratory, demonstrating that aerobic PCB biodegradation can be stimulated in Hudson River sediments by adding inorganic nutrients, biphenyl, and oxygen, mediated primarily by indigenous bacterial populations. The initial degradation rates in the laboratory were 2 to 3 times faster than those observed in the field, perhaps due to differences in length scales and mixing. Little additional benefit was derived from increased mixing or by inoculation of the sediments with a purified PCB-degrading bacterium. Limited biodegradation also was observed in a mixed but unamended control caisson exposed to low concentrations of oxygen, suggesting that aerobic PCB biodegradation may occur more broadly in river sediments, complementing the ongoing anaerobic dechlorination observed previously.

ACKNOWLEDGMENTS

We thank OHM Corporation (Findlay, Ohio) for their assistance in the design, construction, and installation of the research facility.

REFERENCES

Abramowicz, D. A. 1990. "Aerobic and Anaerobic Biodegradation of PCBs: A Review." *CRC Crit. Rev. Biotechnol.* *10*(3): 241-251.

Bedard, D. L. 1990. "Bacterial Transformations of Polychlorinated Biphenyls." In D. Kamely, A. Chakrabarty, and G.S. Omenn (Eds.), *Biotechnology and Biodegradation, Advances in Applied Biotechnology Series*, pp. 369-388. Portfolio Publishing Co., Woodlands, TX.

Bedard, D. L., R. E. Wagner, M. J. Brennan, M. L. Haberl, and J. R. Brown, Jr. 1987. "Extensive Degradation of Aroclors and Environmentally Transformed Polychlorinated Biphenyls by *Alcaligenes Eutrophus* H850." *Appl. Environ. Microbiol. 53*(5):1094-1102.

Brown J. F., Jr., R. E. Wagner, H. Feng, D. L. Bedard, M. J. Brennan, J. C. Carnahan, and R. J. May. 1987. "Environmental Dechlorination of PCBs." *Environ. Toxicol. Chem. 6*:579-593.

Focht, D. D., and W. Brunner. 1985. "Kinetics of Biphenyl and Polychlorinated Biphenyl Metabolism in Soil." *Appl. Environ. Microbiol. 50*(4):1058-1063.

Harkness, M. R., et al. 1993. "In Situ Stimulation of Aerobic PCB Biodegradation in Hudson River Sediments." *Science 259*:503-507.

Hickey, W. J., and D. D. Focht. 1990. "Degradation of Mono-, Di-, and Trihalogenated Benzoic Acids by *Pseudomonas Aeruginosa* JB2." *Appl. Environ. Microbiol. 56*(12):3842-3850.

BIOREMEDIATION OF SOIL CONTAMINATED WITH THE HERBICIDES 2,4-DICHLOROPHENOXYACETIC ACID (2,4-D) AND 4-CHLORO-2-METHYLPHENOXYACETIC ACID (MCPA)

I. McGhee and R. G. Burns

INTRODUCTION

Soil contamination by xenobiotic chemicals has become a serious worldwide problem not only because it reduces the value of land for agricultural and recreational use and habitation, but also because it is a source of water pollution. Soils may be decontaminated by physical, chemical, and biological methods, and these may be applied in situ or following excavation and removal (Bewley 1992). Economic considerations suggest that bioremediation in situ may be the most cost-effective method (Seech & Trevors 1991).

Many pesticides persist in soil because either the physical and chemical environment is unsuitable or the appropriate microbial strains are absent. For example, adsorption of herbicides to clays and humates is well known to influence their bioavailability and persistence, and compacted soils will not be conducive to aerobic microbial activity. The aim of soil bioremediation is to provide conditions to enable degradation to proceed at a maximum sustainable rate (Morgan & Watkinson 1989).

The purpose of the research reported here is to evaluate microbial inoculation and changes to soil pH, nutrient levels, and aeration as methods for the removal of 2,4-dichlorophenoxyacetic acid (2,4-D) and 4-chloro-2-methylphenoxyacetic acid (MCPA) from contaminated soil.

METHODS

The ability of three Gram-negative bacterial species isolated from a soil contaminated with pesticides (SB_5, SB_9, and SA_2) to degrade 2,4-D and MCPA was measured in an uncontaminated soil (41% sand, 31% clay, 28% silt, 11.8% organic matter [O.M.]; pH 5.7) spiked with 17 mg kg^{-1} and 60 mg kg^{-1} of herbicide, respectively. Soil (25 g) was surface-inoculated with 1 x 10^8 bacteria g^{-1} soil in Erlenmeyer flasks (125 mL), and incubated at 12.5 ± 2.5°C for 28 days. A second experiment used an industrially contaminated soil (37% sand, 25% clay, 38% silt, 9.8% O.M.; pH 6.7) containing 16.89 mg kg^{-1} 2,4-D and 14.22 mg kg^{-1} MCPA (and other pesticide residues). Soil was inoculated with isolate SB_5 or treated with nitrogen, phosphorus, and potassium fertilizer (NPK; 14 g kg^{-1}) or lime (225 g kg^{-1}) to raise the soil to pH 7.5 (previously established as optimal for SB_5 growth). In both experiments the disappearance of 2,4-D and MCPA was monitored by gas chromatograph/mass spectrophotometry (GC-MS) following Soxhlet extraction and methylation.

RESULTS AND DISCUSSION

At day 14, degradation in all the inoculated spiked soils was significantly higher than in the noninoculated controls (Table 1). However by day 28, percent degradation in the noninoculated soils was similar to that in the inoculated soils, probably through acclimation of indigenous microbial species (Smith & Lafond 1990).

The effect of bacterial inoculation, pH adjustment, and the addition of NPK fertilizer on the degradation of 2,4-D and MCPA in a contaminated soil was studied (Tables 2 and 3). The degradation of 2,4-D and MCPA by SB_5 after 28 days was 48% and 61%, respectively, in contrast to 96% and 99% in the spiked soil. The difference may be due to humic-pesticide interactions reducing the bioavailability of the compounds (Ogram et al. 1985) or the presence of other pesticides toxic to SB_5.

By day 14, similar amounts of 2,4-D and MCPA (15% and 21% respectively) had been degraded in the noninoculated soil, but by day 28, 34% of 2,4-D had been degraded in contrast to 70% of MCPA. The contaminated soil possesses indigenous microorganisms, some of which, under ideal laboratory conditions (homogenized, aerated, elevated temperature, constant moisture content), are able to degrade the target compounds.

The addition of NPK to noninoculated soil increased degradation of both 2,4-D and MCPA. By day 28, 75% (2,4-D) and 84% (MCPA) had been degraded compared with 34% and 70% in the controls (no NPK). The addition of nutrients to contaminated soil has been shown to stimulate the growth of indigenous microorganisms resulting in increased degradation of pesticides (Mueller et al. 1989). The addition of lime to noninoculated soil completely inhibited degradation of 2,4-D and MCPA up to day 14. However by day 28, degradation of 2,4-D was 29% (a value similar to that of the control) in contrast to 45% (control 70%) for MCPA. Whilst the elevated pH may be optimal for SB_5 growth in vitro, it may alter the bioavailability of the target pesticides in soil.

By day 14, 28% (2,4-D) and 58% (MCPA) had been degraded in inoculated soil compared with 58% and 83% in the SB_5 plus NPK soil. In soil plus NPK alone, 46% (2,4-D) and 75% (MCPA) degradation had occurred by day 14. However, by day 28 there was no significant difference between SB_5 and SB_5 plus NPK treatments. The addition of lime to inoculated soil resulted in a decreased rate of degradation of both herbicides: by day 28 degradation of 2,4-D was reduced from

TABLE 1. Effect of inoculation with bacterial isolates SB_5, SB_9, and SA_2 on the degradation of 2,4-D and MCPA in a spiked soil.

Pesticide	Isolate	Degradation				
		Day 0	Day 14		Day 28	
		mg kg^{-1}	mg kg^{-1} soil	% degradation	mg kg^{-1} soil	% degradation
2,4-D	SB_5	17.01 ± 0.68	$3.74 \pm 0.19^{a\alpha}$	78	$0.68 \pm 0.05^{a\beta}$	96
	SB_9		$5.61 \pm 0.34^{b\alpha}$	67	$1.53 \pm 0.11^{c\beta}$	91
	SA_2		$5.10 \pm 0.21^{b\alpha}$	70	$1.02 \pm 0.06^{b\beta}$	94
	Noninoculated		$12.41 \pm 0.87^{c\alpha}$	27	$1.19 \pm 0.11^{bc\beta}$	93
MCPA	SB_5	60.31 ± 3.60	$9.60 \pm 0.63^{a\alpha}$	84	$0.06 \pm 0.03^{a\beta}$	99
	SB_9		$1.20 \pm 0.05^{b\alpha}$	98	$0.06 \pm 0.04^{a\beta}$	99
	SA_2		$1.80 \pm 0.07^{c\alpha}$	97	$0.06 \pm 0.03^{a\beta}$	99
	Noninoculated		$46.20 \pm 1.21^{d\alpha}$	23	$3.60 \pm 0.22^{b\beta}$	94

Values within a column (latin) or row (greek) followed by the same letter do not differ significantly ($P<0.05$).

TABLE 2. Effect of inoculation with bacterial isolate SB_5, NPK, and lime on the degradation of 2,4-D in a contaminated soil.

Treatment	Degradation Day 14 mg kg^{-1} soil	Day 14 % degradation	Day 28 mg kg^{-1} soil	Day 28 % degradation
SB_5	12.23 ± 2.20^{abc}	28 ± 13	8.71 ± 1.42^{a}	48 ± 8
SA_5 + NPK	7.15 ± 1.80^{a}	58 ± 11	3.55 ± 0.57^{b}	79 ± 4
SB_5 + Lime	16.39 ± 1.88^{bc}	3 ± 11	12.91 ± 1.22^{c}	24 ± 7
NPK	9.18 ± 1.23^{ad}	46 ± 7	4.25 ± 0.91^{b}	75 ± 5
Lime	17.99 ± 1.20^{c}	0	12.03 ± 2.43^{ac}	29 ± 14
NPK + Lime	13.04 ± 1.60^{bd}	23 ± 9	10.52 ± 1.32^{ac}	38 ± 8
Control	14.37 ± 1.10^{bc}	15 ± 7	11.12 ± 2.10^{ac}	34 ± 12

Initial concentration of 2,4-D is 16.89 ± 1.53 mg kg^{-1}.

Values within a column followed by the same letter do not differ significantly ($P<0.05$).

TABLE 3. Effect of inoculation with bacterial isolate SB_5, NPK, and lime on the degradation of MCPA in a contaminated soil.

Treatment	Degradation			
	Day 14		Day 28	
	mg kg^{-1} soil	% degradation	mg kg^{-1} soil	% degradation
SB_5	6.05 ± 1.07^{ad}	58 ± 7.5	5.56 ± 0.51^{a}	61 ± 3.5
SA_5 + NPK	2.38 ± 0.34^{b}	83 ± 2.4	2.64 ± 0.55^{b}	81 ± 3.9
SB_5 + Lime	8.45 ± 1.70^{ac}	41 ± 11.9	8.43 ± 1.37^{ac}	41 ± 9.6
NPK	3.51 ± 0.26^{d}	75 ± 1.8	2.21 ± 1.19^{bd}	84 ± 8.4
Lime	14.54 ± 2.00^{c}	0	7.82 ± 0.89^{c}	45 ± 6.3
NPK + Lime	5.34 ± 0.17^{a}	62 ± 1.2	3.35 ± 0.08^{b}	76 ± 0.6
Control	11.26 ± 0.98^{c}	21 ± 6.9	4.25 ± 0.26^{d}	70 ± 1.8

Initial concentration of MCPA is 14.22 ± 1.51 mg kg^{-1}.

Values within a column followed by the same letter do not differ significantly ($P<0.05$).

48% (SB_5) to 24% (SB_5 plus lime) and MCPA from 61% (SB_5) and 41% (SB_5 plus lime).

Experiments are under way to study the individual effects of nitrogen, phosphorus, and potassium on the degradation of the target compounds, and a field experiment is planned.

ACKNOWLEDGMENT

This project is funded by the United Kingdom Science and Engineering Research Council.

REFERENCES

Bewley, R. J. F. 1992. "Bioremediation of Contaminated Ground." In J. F. Rees (Ed.), *Contaminated Land Treatment Technologies*, pp. 270-284. Elsevier Applied Science, London.

Morgan, P., and R. J. Watkinson. 1989. "Microbiological Methods for the Cleanup of Soil and Groundwater Contaminated with Halogenated Organic Compounds." *FEMS Microbiology Reviews 63*: 277.

Mueller, S. B., R. J. Chapman, and P. H. Pritchard. 1989. "Creosote-Contaminated Sites, Their Potential For Bioreclamation." *Environmental Science and Technology 23:* 1197.

Ogram, A. V., R. E. Jessup, L. T. Ou, and P. S. C. Rao. 1985. "Effects of Sorption on Biological Degradation Rates of 2,4-Dichlorophenoxyacetic Acids in Soils." *Applied and Environmental Microbiology 49:* 582-587.

Seech, A. G. B., and J. T. Trevors. 1991. "Environmental Variables and Evolution of Xenobiotic Catabolism in Bacteria." *Trends in Ecology and Evolution 6:* 79.

Smith, A. E., and G. P. Lafond. 1990. "Effect of Long Term Phenoxyalkanoic Acid Field Applications on the Rate of Microbial Degradation." *ACS Symposium Series 426:* 14.

FLUIDIZED-BED BIORECLAMATION OF GROUNDWATER CONTAMINATED WITH CHLOROPHENOLS

K. T. Järvinen and J. A. Puhakka

INTRODUCTION

In 1987, potable water in Kärkölä municipality, Finland, was found to be contaminated by chlorophenols (CPs); concentrations ranging from 56 to 190 mg/L were detected in various sites of this drinking water aquifer (Lampi et al. 1990). The CP level has remained stable (44 to 55 mg/L) in the groundwater during the period from 1990 to 1992 (Järvinen & Puhakka 1993).

This study is part of a larger project to develop a bioreclamation process for chlorophenol-contaminated groundwater. The project was initiated in 1988 with studies on aerobic degradation of model chlorophenols in fluidized-bed reactors (Mäkinen et al. 1993; Melin et al. 1993; Puhakka & Järvinen 1991, 1992; Shieh et al. 1990). The most recent experiments have been conducted at groundwater temperature using actual polluted groundwater from the Kärkölä aquifer.

MATERIALS AND METHODS

Experiments were carried out in 1-L glass fluidized-bed reactors using porous silica-based carriers (surface area 1.2 m^2/g, mean pore diameter 6.5 µm). The hydraulic retention time was held constant (5 h), and the bed expansion was maintained at 50% during the experiments as reported earlier (Shieh et al. 1990).

In the aerobic reactors, CPs were used as only carbon and energy sources. Nutrients and a phosphate buffer were added to feed solutions as described previously (Puhakka & Järvinen 1992). The reactor performance was monitored as inorganic chloride (ICl^-) release by ion-selective electrode and CP removal by gas chromatography (GC) from nonfiltered samples.

RESULTS

Dichlorophenol degradation was studied in a stable fluidized-bed reactor degrading the mixture of 2,4,6-TCP; 2,3,4,6-TeCP; and PCP. The feed was changed to 2,5-DCP feed (29.4±1.0 mg/L). For a period of 8 weeks the mean ICl^- release remained at 0.3±0.4 mg/L (12 samples), whereas complete degradation would have released 12.7 mg/L ICl^-. When the feed was changed to 2,4-DCP (45 mg/L), complete degradation was observed within 2 weeks (ICl⁻ release was 20.6 ±0.2 mg/L; 4 samples), indicating that 2,5-DCP was not bacteriocidic although it was recalcitrant. These results show that the biodegradation of dichlorophenols depended on the position of the chloride substituents in the aromatic ring.

PCP degradation (40 mg/L) was studied in a reactor degrading 2,3,4,6-TeCP (which included 11% PCP as an impurity). PCP degradation and ICl^- releases during startup are presented in Figure 1. PCP degradation started slowly and stabilized after 5 weeks of continuous operation. After 5 weeks, PCP degradation varied between 78 and 96% with the corresponding effluent PCP concentrations of 1.1 to 6.7 mg/L. On day 65, the PCP feed concentration was increased to 58 mg/L. After the reactor performance stabilized, the process was monitored from day 83 to day 94 (9 samples). PCP feed concentrations of 57.7 ±3.4 mg/L achieved effluent concentrations of 0.3 ±0.1 mg/L. The ICl^- release calculated from the GC results was 38.2 ±2.3 mg/L, whereas 44.4 ±2.1 mg/L of ICl^- was measured. The results indicate that more than 99.5% of the PCP was degraded at the PCP loading rate of 277 mg/L*d.

Degradation of a CP mixture was studied with 2,4-DCP acclimated fluidized-bed culture. The feed contained (mg/L) 2,4-DCP (8.1 ±0.4); 2,6-DCP (9.8 ±0.4);

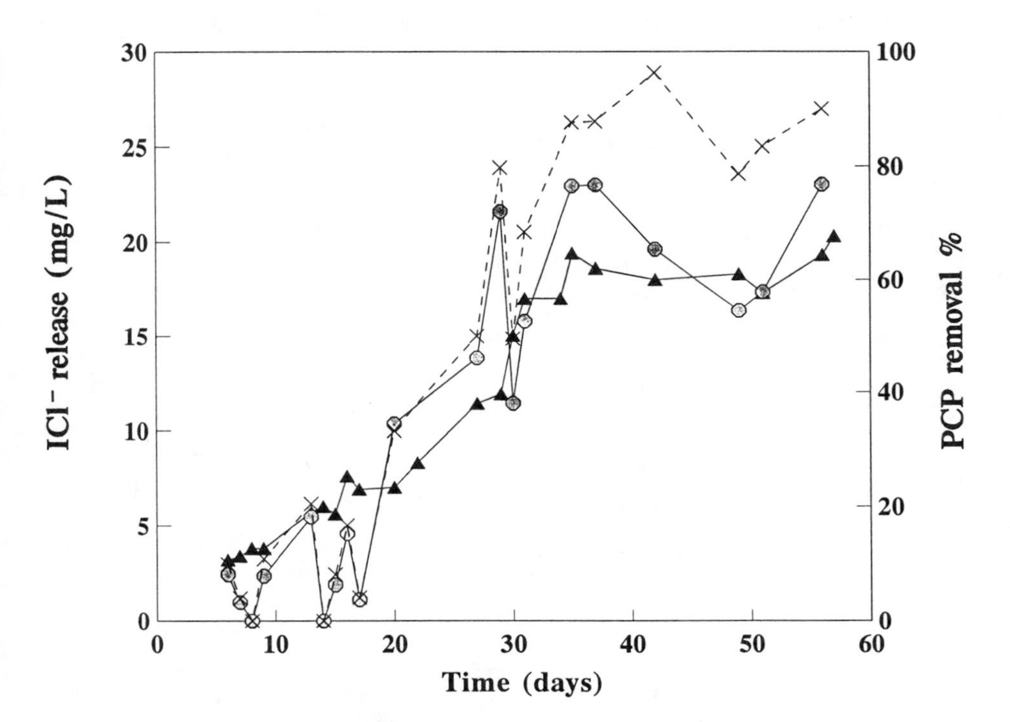

FIGURE 1. Startup of PCP degradation (40 mg/L) in aerobic continuous fluidized-bed reactor measured as inorganic chloride release (▲; solid line), chloride release calculated from gas chromatography results (o; solid line) and as PCP removal % (×; dashed line).

2,4,6-TCP (13.8 ±0.3); and PCP (9.0 ±0.4). More than 99% of 2,4-DCP degradation was apparent from the beginning (Figure 2). More than 99% degradation was achieved at day 18 for 2,4,6-TCP and at day 26 for 2,6-DCP. The lag phase for PCP degradation was about 2 weeks, and the degradation remained partial throughout the 6-week experiment.

After 60 days, the concentrations of CPs were changed to 9.1 ±0.2 mg/L for 2,4-DCP; 8.5 ±0.4 mg/L for 2,6-DCP; 11.2 ±0.4 mg/L for 2,4,6,-TCP; and 14.9 ±2.0 mg/L for PCP. The total feed concentration was 43.7 mg/L, which was equal to the CP loading rate of 210 mg/L*d. During 12 days of continuous operation, the degradation of CPs exceeded 99%, except for PCP which was 83%. The ICl^- release was 21.5 mg/L, although 22.1 mg/L had been expected from the GC measurements. The removal efficiency was not affected when the feed CP concentrations were doubled, indicating that the loading of the bioprocess was not yet near its upper capacity.

CP degradation in actual contaminated groundwater was studied in an aerobic fluidized-bed reactor at 14°C. After 10 days of groundwater feed, the process was monitored for 12 days. The results are presented in Table 1. Degradation of 2,4,6-TCP and 2,3,4,6-TeCP exceeded 99.9%, whereas the mean PCP degradation was 83.5%. The total CP degradation was 99.2%. Observed ICl^- releases agreed well with those calculated from GC results. These results showed that high-rate

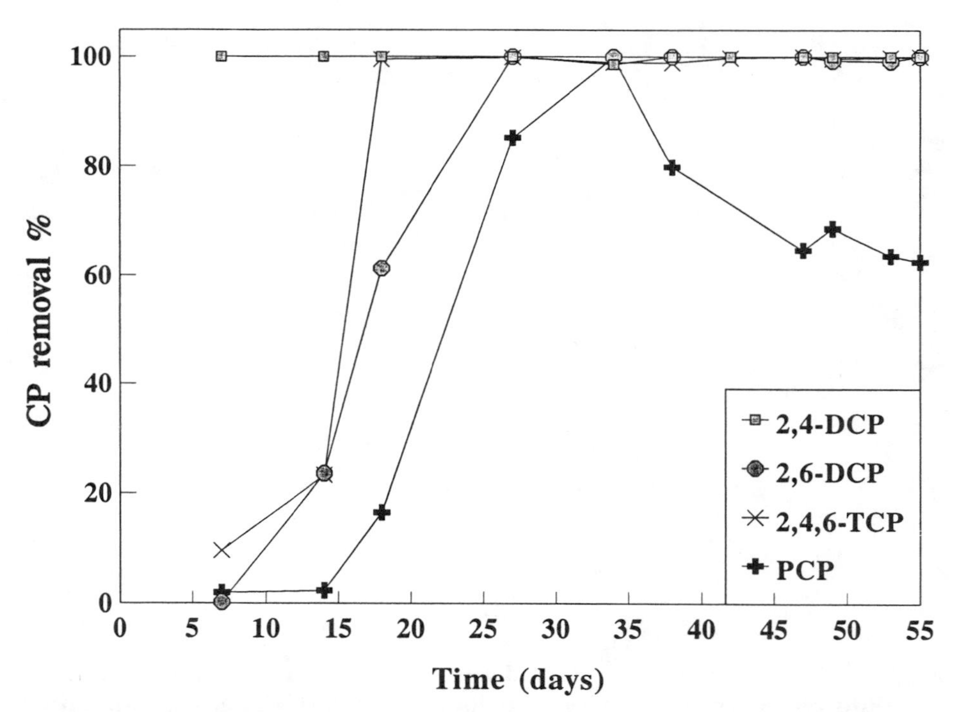

FIGURE 2. Startup of chlorophenol removal in aerobic fluidized-bed reactor. Prior to these experiments, the reactor was fed only 2,4-DCP.

TABLE 1. Steady-state chlorophenol (CP) degradation and inorganic chloride release (ICl^-) from contaminated groundwater in aerobic fluidized-bed process at the total CP loading rate of 217 mg/L*d. (Number of samples was 10.)

	CP-conc. (mg/L)		ICl^- (mg/L)	
	Influent	Effluent	Observed	Calc.
2,4,6-TCP	9.8±0.2	<0.01±0.00		
2,3,4,6-TeCP	33.3±1.4	<0.01±0.00		
PCP	2.2±0.2	0.4±0.3		
Total CP	45.3±1.7	0.4±0.3		
ICl^-			26.2±2.3	26.8±1.1

bioremediation of contaminated groundwater at suboptimal temperature was successful in the aerobic fluidized-bed process.

DISCUSSION

Our results demonstrated that aerobic fluidized-bed treatment was an effective method for biodegradation of chlorophenolic wood preservative constituents. In the contaminated groundwater 2,4,6-TCP and 2,3,4,6-TeCP were degraded more than 99%, whereas 83.5% of PCP was degraded. These results compare favorably with other methods tested for groundwater decontamination. In Finland, these include biofilter studies, with immobilized *Rhodococci* (Valo et al. 1990), treatment of groundwater in a municipal activated sludge treatment plant (Ettala et al. 1992), and adsorption to activated carbon (Ettala 1992). The *Rhodococcus* process did not degrade CPs in temperatures below 20°C and only 30 to 60% of the ICl^- release was apparent. In activated sludge treatment, the CP removals averaged 71%. In activated carbon treatment, more than 99% of CPs were adsorbed in the beginning of the experiment, but after a few weeks the adsorption capacity was reached.

Although complete PCP degradation required an acclimation period of a few months, biodegradation at high concentrations (58 mg/L) was achievable. DCPs with chlorine substituent in *meta*-position in the aromatic ring were recalcitrant. In our earlier study, 2,4-DCP and 2,6-DCP biodegraded, but 3,5-DCP did not (Puhakka et al. 1991). Poor degradation of certain mono- and dichlorophenols including 3,5-DCP has been observed in many pentachlorophenol-degrading strains (Apajalahti & Salkinoja-Salonen 1986, Karns et al. 1983, Steiert et al. 1987). However, these recalcitrant *meta*-substituted DCPs are not present in the wood preservative Ky-5 (Järvinen & Puhakka 1993).

We have shown that aerobic fluidized-bed treatment is an attractive process for high-rate, on-site bioreclamation of chlorophenol-contaminated groundwater even at low temperatures.

REFERENCES

Apajalahti, J. H. A., and M. S. Salkinoja-Salonen. 1986. "Degradation of polychlorinated phenols by *Rhodococcus chlorophenolicus*." *Appl. Microbiol. Biotechnol.* *25*(1): 62-67.

Ettala, M. 1992. "Cleanup of contaminated groundwater" (in Finnish). *Ympäristö ja terveys* *23*(2-3): 165-169.

Ettala, M., J. Koskela, and A. Kiesilä. 1992. "Removal of chlorophenols in a municipal sewage treatment plant using activated sludge." *Wat. Res.* *26*(6): 797-804.

Järvinen, K. T., and J. A. Puhakka. 1993. "Bioremediation of chlorophenol contaminated groundwater." *Environ. Technol. Lett.* (accepted for publication).

Karns, J. S., J. J. Kilbane, S. Duttagupta, and A. M. Chakrabarty. 1983. "Metabolism of halophenols by 2,4,5-trichlorophenoxyacetic acid-degrading *Pseudomonas cepatia*." *Appl. Environ. Microbiol.* *46*(5): 1176-1181.

Lampi, R., T. Vartiainen, and J. Tuomisto. 1990. "Population exposure to chlorophenols, dibenzo-*p*-dioxins and dibenzofurans after a prolonged groundwater pollution of chlorophenols." *Chemosphere* *20*(6): 625-634.

Mäkinen, P. M., T. J. Theno, J. F. Ferguson, J. E. Ongerth, and J. A. Puhakka. 1993. "Chlorophenol toxicity removal and monitoring in aerobic treatment: Recovery from process upsets." *Environ. Sci. Technol.*, *27*(7):1434-1439.

Melin, E., J. A. Puhakka, and W. K. Shieh. 1993. "Degradation of 4-chlorophenol in denitryfying fluidized-bed reactor." *J. Environ. Sci. Health, A28*(8):1801-1811.

Puhakka, J. A., and K. Järvinen. 1992. "Aerobic fluidized-bed treatment of polychlorinated phenolic wood preservative constituents." *Wat. Res.* *26*(6): 765-770.

Puhakka, J. A., E. Melin, K. Järvinen, T. Tuhkanen, and W. K. Shieh. 1991. "Oxic fluidized-bed treatment of dichlorophenols." *Wat. Sci. Tech.* *24*(3/4): 171-177.

Puhakka, J. A., W. K. Shieh, K. Järvinen, and E. Melin. 1992. "Chlorophenol degradation under oxic and anoxic conditions." *Wat. Sci. Tech.* *25*(1): 147-152.

Shieh, W. K., J. A, Puhakka, E. Melin and T. Tuhkanen. 1990. "Immobilized-cell degradation of chlorophenols." *J. Environ. Eng.* *116*(4): 683-697.

Steiert, J. G., J. J. Pignatello, and R. L. Crawford. 1987. "Degradation of chlorinated phenols by a pentachlorophenol-degrading bacterium." *Appl. Environ. Microbiol.* *53*(5): 907-910.

Valo, R. J., M. M. Häggblom, and M. S. Salkinoja-Salonen. 1990. "Bioremediation of chlorophenol containing simulated groundwater by immobilized bacteria." *Wat. Res.* *24*(2): 253-258.

APPLICATION OF A STRUCTURED KINETIC MODEL TO THE BIOREMEDIATION OF HANFORD GROUNDWATER

B. S. Hooker, R. S. Skeen, and J. N. Petersen

INTRODUCTION

Liquid wastes containing radioactive, hazardous, and regulated chemicals have been generated throughout the 40+ years of operations at the U.S. Department of Energy (DOE) Hanford Site. Some of these wastes were discharged to the soil column and many of the waste components, including nitrate, carbon tetrachloride (CCl_4), and several radionuclides, have been detected in the Hanford groundwater. Current DOE policy prohibits the disposal of contaminated liquids directly to the environment, and may require the remediation of existing contaminated groundwaters. In situ bioremediation is one technology currently being developed at Hanford to meet the need for cost-effective methods to remediate groundwater contaminated with CCl_4, nitrate, and other organic and inorganic contaminants.

A structured mathematical model has been developed that describes the kinetics of CCl_4 degradation by the Hanford groundwater microbial consortium. This model relates the destruction of CCl_4 to both assimilatory and dissimilatory denitrification reactions through a cometabolic pathway. Other researchers have shown that structured kinetic models can aid in understanding cellular processes (Frazier 1989, Hooker & Lee 1992). Once finalized, this model will be incorporated into a mathematical description of in situ bioremediation.

MATERIALS AND METHODS

Batch CCl_4 degradation experiments were perform using a balanced 2^{3-1} fractional factorial design. Experimental measurements of biomass growth, denitrification, and CCl_4 destruction, were obtained as described by Skeen et al. (1992). Values of kinetic parameters were obtained using the SimuSolv® package (Dow Chemical Company, Midland, Michigan). Kinetic parameters were formulated based on a lack of statistically valid interactions between experimental conditions and parameters values. Experimental responses considered in the model are concentrations of CCl_4, acetate, nitrate, nitrite, and biomass. Initial conditions in separate batch studies were varied for acetate, nitrate, and nitrite levels.

RESULTS AND DISCUSSION

To describe the growth, denitrification, and CCl_4 destruction properties of the Hanford consortium in a simulated groundwater system with an acetate substrate, the following reactions have been used.

Dissimilatory Denitrification:

$$8NO_3^- + 2CH_3COO^- + 2H^+ \Rightarrow 4CO_2 + 8NO_2^- + 4H_2O \tag{1}$$

$$8NO_2^- + 3CH_3COO^- + 11H^+ \Rightarrow 6CO_2 + 4N_2 + 10H_2O \tag{2}$$

Assimilatory Denitrification:

$$3.5CH_3COO^- + NO_3^- + 4.5H^+ \Rightarrow 2CO_2 + 3H_2O + C_5H_9O_3N \tag{3}$$

$$3.25CH_3COO^- + NO_2^- + 4.25H^+ \Rightarrow 1.5CO_2 + 2.5H_2O + C_5H_9O_3N \tag{4}$$

Equations 1 through 4 were formulated using the approach outlined by McCarty et al. (1969). Separate reactions were written for nitrate and nitrite rather than one overall denitrification expression since both species could be measured. Thus, an additional, experimentally determined variable was provided for comparison with the model. These equations reflect the approximate biomass composition of the Hanford consortium, measured at $C_5H_9O_3N$. The four rate expressions for dissimilatory and assimilatory denitrification reactions have been assumed to be first order in the appropriate electron acceptor concentration, to be first order in biomass concentration, and to follow Monod kinetics in acetate concentration. The rate expressions, coupled with the stoichiometric relationships outlined in equations 1 through 4, provide a description of the consumption of electron donor and acceptor, and subsequent production of biomass.

It has been demonstrated by Skeen et al. (1992) that the Hanford microbial consortium degrades CCl_4 only when the biomass is cycled between an electron acceptor-rich and electron acceptor-depleted environment. The observed phenomenon has been described mathematically by assuming that when all the nitrate and nitrite are depleted, the biomass changes state metabolically. Details of cell physiology causing this state change at this point are hypothetical. But, this may represent the point where the cells begin active assimilation of cell storage polymer. The change of state is denoted by the reaction:

$$X \Rightarrow X^* \tag{5}$$

where X is the normal cell state and X* is the new, substrate-depleted state. When nitrate and nitrite are reintroduced, X* is capable of destroying CCl_4. However, X* is transient, and as it metabolizes the nitrate and nitrite, it reverts to X, as shown by equations 6 and 7.

$$aX^{*}+8NO_3^{-}+2CH_3COO^{-}+2H^{+} \Rightarrow aX+4CO_2+8NO_2^{-}+4H_2O \quad (6)$$

$$bX^{*}+8NO_2^{-}+3CH_3COO^{-}+11H^{+} \Rightarrow bX+6CO_2+4N_2+10H_2O \quad (7)$$

The constants a and b represent the stoichiometric equivalent of cells changed in state by eight molecules of NO_3^- or NO_2^-, respectively. This transition is coupled with dissimilatory denitrification, the pathway that can provide metabolic energy for the conversion. The reaction from X to X*, represented by equation 5, was described using Monod kinetics for X with inhibition by the sum of nitrate and nitrite concentrations. In the reverse direction, the two reactions were described as first order in the concentration of X*, nitrate or nitrite, and acetate. The destruction of carbon tetrachloride was assumed to be directly dependent on the rate of the conversion of X* to X and first order in CCl_4 concentration.

The differential equations describing these reaction kinetics were solved numerically using SimuSolv®. Representative predictions demonstrating the link between X*, depletion of electron acceptors, and CCl_4 degradation are shown in Figure 1. Here we display X* concentration, the sum of the nitrate and nitrite concentrations, and CCl_4 concentration as functions of time. When the combined level of nitrate and nitrite reaches zero, the concentration of X* begins to increase. When either of the electron acceptors is reintroduced, CCl_4 is consumed and X* is converted back to X.

The predictions provided by the mathematical model described here were compared to experimental data obtained as reported by Skeen et al. (1992). Experimental data and model responses for one representative batch experiment from the factorial design are shown on Figure 2. It can be seen that the model

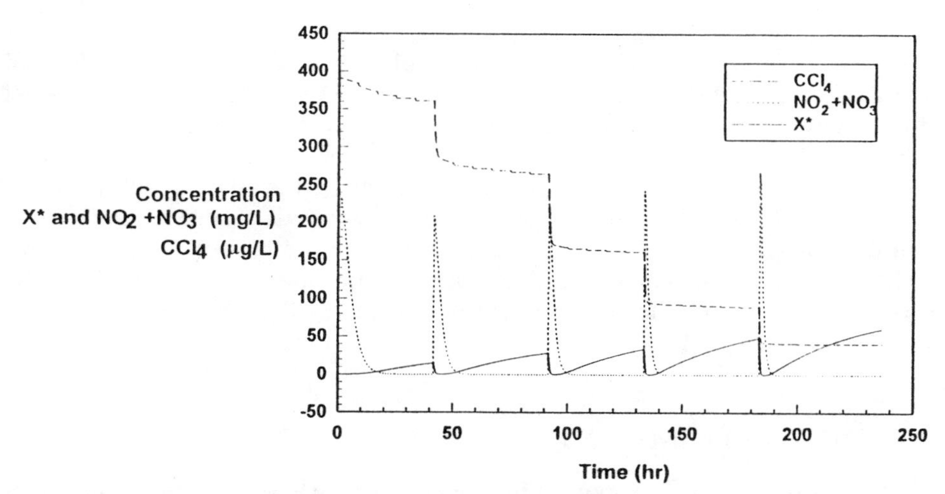

FIGURE 1. Model predictions for CCl_4 concentration, combined nitrate and nitrite concentrations, and X* concentration showing the relationship between the conversion of X* to X and the destruction of CCl_4.

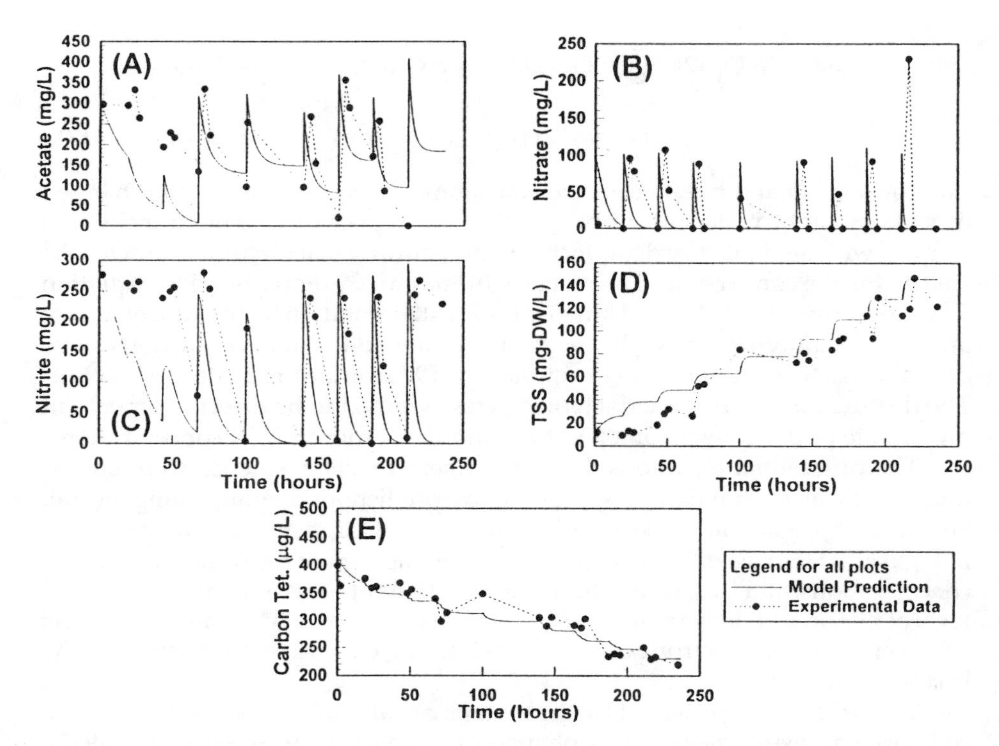

FIGURE 2. Comparison of model predictions and experimental data for (A) acetate, (B) nitrate, (C) nitrite, (D) total suspended solids, and (E) CCl_4 concentrations.

describes the experimental behavior of this pulse-fed reactor system. Additional work is being undertaken to ensure that the parameters employed in the model adequately represent the data obtained from a wide range of experimental conditions. This will include the evaluation of other published models, such as the description of CCl_4 destruction formulated by Semprini et al. (1991). Also, other kinetic forms which may reflect the current sets of data, such as competitive nitrate inhibition, will be further tested to confirm the proposed model. This will require additional batch CCl_4 destruction experiments. Once finalized, the model will be incorporated into a mathematical description of in situ bioremediation to aid in design.

ACKNOWLEDGMENTS

This research was supported by the U. S. Department of Energy Office of Technology Development, VOC-Arid Integrated Demonstration. Pacific Northwest Laboratory is operated by Battelle Memorial Institute for the U.S. Department of Energy under contract DE-AC06-76RLO 1830. Dr. Hooker and Dr. Petersen were supported by the Northwest College and University Association of Science (Washington State University) under Grant DE-FG06-89ER-75522 with the U.S. Department of Energy. The portion of the work at Washington State University was supported by a contract from the Pacific Northwest Laboratory.

REFERENCES

Frazier, G. C. 1989. "A Simple Leaky Cell Growth Model for Plant Cell Aggregates." *Biotech. Bioeng. 33*:313-320.

Hooker, B. S., and J. M. Lee. 1992. "Application of a New Structured Model to Tobacco Cell Culture." *Biotech. Bioeng. 39*:765-774.

McCarty, P. L., L. Beck, and P. S. Amant. 1969. "Biological Denitrification of Wastewaters by Addition of Organic Materials." In *Proceedings of the 24th Industrial Waste Conference*, pp. 1271-1285. Purdue University, West Lafayette, IN.

Semprini, L., G. D. Hopkins, D. B. Janssen, M. Lang, P. V. Roberts, and P. L. McCarty (Eds.). 1991. *In Situ Biotransformation of Carbon Tetrachloride Under Anoxic Conditions.* U.S. Environmental Protection Agency Technical Report, EPA/2-90/060, R. S. Kerr Environmental Research Laboratory, Ada, OK.

Skeen, R. S., S. M. Cote, M. J. Truex, and J. N. Petersen. 1992. "Kinetics of In Situ Bioremediation of Hanford Groundwater." In *Proceedings of the Spectrum '92 Conference*, pp. 1209-1213. Boise, ID.

MOLECULAR ENVIRONMENTAL DIAGNOSTICS OF TRICHLORO-ETHYLENE (TCE) CONTAMINATED SUBSURFACE ENVIRONMENTS

L. Jimenez, I. Rosario, J. Bowman, S. Koh, and G. S. Sayler

This investigation was undertaken to isolate DNA from trichloroethylene (TCE) contaminated subsurface sites at the Savannah River Site (SRS) for molecular analysis and diagnosis, to study the distribution of catabolic genotypes in subsurface sites, and to use this knowledge for eventual optimization of the bioremediation process.

Before the injection of methane into the subsurface, 161 subsurface sediment samples were obtained across the depth profile from 12 different boreholes (Hazen 1992). Microbial DNA extracted from these sediments was hybridized with three different gene probes, including a 410-bp single-stranded probe encoding the B protein of the soluble methane monooxygenase (sMMO) of *Methylosinus trichosporium* OB3b (Tsien & Hanson 1992); a 2.5-kb *Pst*I fragment of the toluene dioxygenase (*todC1C2BA*) from *Pseudomonas putida* F1 (Zylstra et al. 1989); and a 2.5-kb *Sma*I fragment of the methanol dehydrogenase (MDH) enzyme from *Methylobacterium organophilum* XX (Machlin & Hanson 1988). The DNA extraction and hybridization procedures are described elsewhere (Sayler & Layton 1990).

Table 1 shows the distribution of the three genes across the depth profile in site MHB 7T. No particular pattern was found for the three genes in the shallow and deep sediments of this borehole. The overall distribution of the genes through the depth profile appears chaotic. This also seems to be the case in the other 11 sites (data not shown). Figure 1 shows the frequencies of the three genes in the 12 sites analyzed; sMMO genes were detected in all sites, whereas *tod* genes were found in 10 sites across the depth profile. MDH genes were found only in 7 sites. The highest frequency for MDH was found in site MHT 1C. The relatively low frequency of the MDH genes is probably related to a lack of general specificity to the genes of a wide range of microorganisms possessing the enzyme MDH. Studies suggest only a segment of the methylotrophic bacterial community hybridizes to this particular probe (Bowman et al. 1993) The highest frequencies for sMMO were found in sites MHB 7T and MHT 9B with 65 and 63% of the sediments across the depth profile showing the presence of sMMO genes (Figure 1). On the other hand, the highest frequencies for *tod* were found in sites MHB 5V and MHB 7T with 67 and 65% of the sediments, respectively. The average frequencies for sMMO and *tod* in the 161 sediments analyzed were 37% and 40%, respectively.

TABLE 1. Distribution of sMMO, MDH, and *tod* genes across the depth profile in site MHB 7T.

Depth (meters)	sMMO	MDH	*tod*
0.9	–[a]	–	+
4.6	–	–	+
8.2	+	–	–
11.8	–	–	–
15.5	+	–	–
19.1	+	–	–
22.7	–	–	+
26.4	+	–	–
30.0	+	+	+
33.6	+	+	+
36.1	–	–	+
39.7	+	–	+
44.6	–	–	+
47.0	+	+	+
50.6	+	–	+
57.3	+	–	+
59.1	+	–	–
Positive	11	3	11
Percentage	65	18	65

(a) +, positive hybridization; –, negative hybridization.

Based on these results it seems that the potential to degrade TCE is widespread across all sites at the SRS. Furthermore, methane enrichments from adjacent groundwater wells yielded 25 type II methanotrophic bacteria which showed high sMMO activity and high TCE degradation rates (Table 2). The majority of the isolates showed sMMO activity and the ability to degrade TCE (Bowman et al. 1993).

On the basis of the gene distribution, high TCE degradation rates, and high sMMO activity, we can conclude that subsurface methanotrophic bacterial communities at the SRS have the potential to be useful for in situ bioremediation of TCE-contaminated sites. Gene probe technologies can be very useful for the initial diagnosis, monitoring, and optimization of the biological cleanup of a given contaminated subsurface site. Further studies will determine the effect of methane injection on the distribution of the cometabolic genotypes in subsurface environments.

ACKNOWLEDGMENTS

The information contained in this article was obtained under contract RO1-1015-04 between the University of Tennessee and Westinghouse Savannah River Company through the Martin Marietta Corporation.

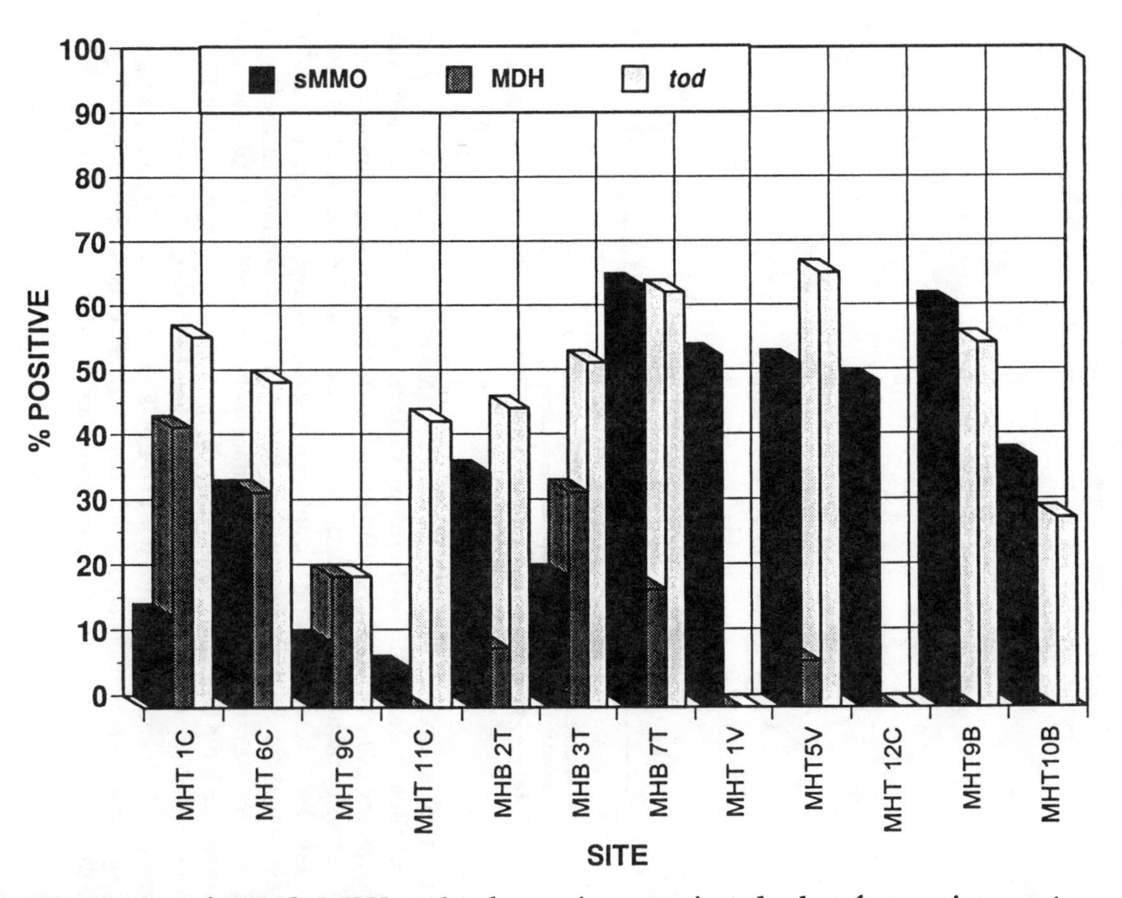

FIGURE 1. Distribution of sMMO, MDH, and *tod* genes in contaminated subsurface environments. Percentages are based on the number of sediments probing positive.

TABLE 2. Soluble MMO activity and TCE degradation by methanotrophic bacteria isolated from TCE-contaminated groundwater sites.

Site MHT	Isolate	Naphthalene oxidation (nmol/h/mg protein)[a]	TCE oxidation (nmol/h/mg protein)[b]	sMMO gene probe	MDH gene probe
1C	A	<0.1	N.D.	–	–
2C	A	3±1	N.D.	–	–
	B	6±2	N.D.	+	–
	C	307±13	58 ± 9	+	–
	D	328±22	100 ± 21	+	–
3C	A	N.D.	N.D.	–	–
4C	A	247±14	106 ± 17	+	–
	B	217±3	86 ± 10	+	–
	C	69±5	7 ± 2	+	–
5C	A	8±2	N.D.	–	–
	B	6±1	N.D.	–	–
	C	536±14	190 ± 8	+	–
	D	678±14	210 ± 12	+	–
6C	A	594±20	184 ± 21	+	–
	B	422±19	134 ± 23	+	–
7C	A	256±30	97 ± 15	+	–
	B	434 ± 25	129 ± 5	+	–
8C	A	173 ± 16	53 ± 9	+	–
	B	N.D.	N.D.	–	–
9B	A	137 ± 8	67 ± 5	–	–
	B	48 ± 6	6 ± 2	–	–
	SPa	37 ± 4	2 ± 1	+	+
9C	A	41 ± 1	60 ± 11	–	+
	C	12 ± 2	N.D.	+	+
10C	A	21 ± 3	8 ± 3	+	+
M trichosporium OB3b		315 ± 7	72 ± 6	+	+

(a) Naphthalene oxidation and TCE degradation rate analyses performed according to Koh et al. (1993). Strains tested in identical conditions. TCE degradation rates calculated from 1 h incubation period.

(b) N.D., not detected; +, positive hybridization; –, negative hybridization.

REFERENCES

Bowman, J. P., L. Jiménez, I. Rosario, T. C. Hazen, and G. S. Sayler. 1993. "Characterization of methanotrophic bacterial communities present in a trichloroethylene contaminated subsurface groundwater site." *Appl. Environ. Microbiol.* In press.

Hazen, T. C. 1992. *Test Plan for In Situ Bioremediation Demonstration of the Savannah River Integrated Technology Demonstration Project.* WSRC-RD-91-3. Westinghouse Savannah River Company, Aiken, SC.

Koh, S.-C., J. P. Bowman, and G. S. Sayler. 1993. "Soluble methane monooxygenase production and trichloroethylene degradation by a type I methanotroph, *Methylomonas methanica* 68-1." *Appl. Environ. Microbiol. 59*:960-967.

Machlin, S. M., and R. S. Hanson. 1988. "Nucleotide sequence and transcriptional start site of the *Methylobacterium organophillum* XX methanol dehydrogenase structural gene." *J. Bact. 170*:4739-4747.

Sayler, G. S., and A. C. Layton. 1990. "Environmental application of nucleic acid hybridization." *Ann. Rev. Microbiol. 44*:625-648.

Tsien, H. C., and R. S. Hanson. 1992. "Soluble methane monooxygenase component B gene probe for identification of methanotrophs that rapidly degrade TCE." *Appl. Environ. Microbiol. 58*:953-960.

Zylstra, G. J., L. P. Wackett, and D. T. Gibson. 1989. "Trichloroethylene degradation by *Escherichia coli* containing the cloned *Pseudomonas putida* F1 toluene dioxygenase structural gene." *Appl. Environ. Microbiol. 55*:3162-3166.

BASELINE CHARACTERIZATION AND REMEDIATION-INDUCED CHANGES IN TCE DEGRADATIVE POTENTIAL USING ENRICHMENT TECHNIQUES AND DNA PROBE ANALYSIS

F. J. Brockman, W. Sun, A. Ogram, W. Payne, and D. Workman

INTRODUCTION

In situ remediation of trichloroethylene (TCE) in deep unsaturated and saturated sediments is being conducted at the Savannah River Site, South Carolina, by injection of air and methane to stimulate the growth and activity of indigenous methanotrophs (Phelps et al. 1988), which can degrade TCE via the methane monooxygenase enzyme. The objective of our research is to characterize the response of the microbial community to bioremediation activities using cultural approaches for select physiological groups that possess TCE-degradative ability, and using DNA probes specific for genes that identify TCE-degrading microorganisms. Of particular interest is an assessment of how well the available probes correlate with degradation of TCE as observed in enrichment cultures and in field measurements of contaminant concentrations in sediment and groundwater. The use of both approaches allows a more comprehensive and defensible analysis of the success of bioremediation.

SAMPLING LOCATIONS AND EXPERIMENTAL METHODS

A cross-section of the site is shown in Figure 1. Concentrations of TCE between 100 ppb and 5 ppm are found primarily between 90 to 150 ft (27.4 to 45.7 m) and correspond closely to the "tan clay zone" which consists of thin, discontinuous, interlayered sands and clays (Eddy et al. 1991). From July 27, 1990 to December 13, 1990, air was injected into the saturated zone via a slotted horizontal well, and a vacuum was established on a second slotted horizontal well in the unsaturated zone. At the completion of the air injection campaign, sediment samples were taken at 12-ft (3.7-m) intervals to 200 ft (61.0 m) from boreholes MHB-5T and MHT-12C (Figure 2). Air and methane (1% of air) was injected from April 20, 1992 to July 24, 1992. At the completion of the 1% methane injection campaign, sediment samples were taken at 10-ft (3.3-m) intervals to 130 or 140 ft (39.6 or 42.7 m) from boreholes MHC-1, MHC-2, MHC-3, and MHV-11

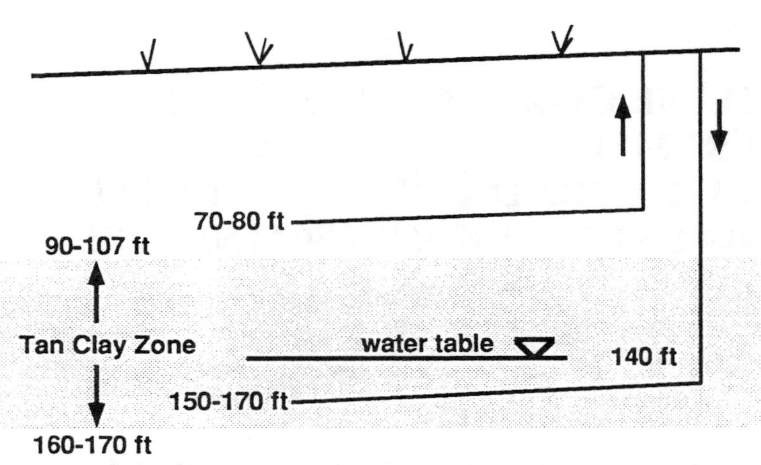

FIGURE 1. Cross-section of the subsurface at the Savannah River bioremediation site, showing the location of horizontal injection and extraction wells and the geological unit which contains the majority of the TCE contamination. Figure is not drawn to scale. (To convert ft to m, divide by 3.28.)

(Figure 2). Boreholes MHC-1 and MHC-2 were located near groundwater monitoring wells that had shown small increases in methanotroph populations during 1% methane injection (i.e., were weakly influenced), whereas boreholes MHV-11 and MHC-3 were located near groundwater monitoring wells that had shown large increases in methanotroph populations during 1% methane injection (i.e., were strongly influenced) (T. Hazen and T. Phelps, personal communication).

Enrichments were carried out (in triplicate) for aerobes, denitrifiers, Fe(III)/Mn(IV)-reducers, sulfate-reducers, methane-oxidizers, propane-oxidizers, and ammonia-reducers in samples from the air injection campaign. Enrichments received 1 g of sediment per vial and were spiked with a mixture of 0.8 ppm each of TCE and PCE. Vials were analyzed at 22 weeks for contaminant removal by headspace gas chromatography (GC). A combined enrichment/most probable number (MPN) approach consisting of 12 vials per medium per sample was used for samples from the 1% methane injection campaign. At the low dilution, three vials were spiked with 0.8 ppm TCE and three vials with 0.8 ppm PCE (excluding the enrichment for dinitrogen-fixers). The intermediate and high dilutions (three vials each) contained one-tenth and one-hundredth the mass of sediment, respectively, of the low dilution; these dilutions did not receive TCE or PCE. Enrichments were carried out for methane-oxidizers and dinitrogen-fixers using an equivalent of 0.01 g of sediment per vial at the low dilution, and for Fe(III)-reducers, sulfate-reducers, methanogens, propane-oxidizers, and ammonia-oxidizers using an equivalent of 0.1 g sediment per vial at the low dilution. An exception was that enrichments for methane-oxidizers at the 120 to 140 ft (36.6 to 42.7 m) depths used an equivalent of 0.001 g (instead of 0.01 g) of sediment at

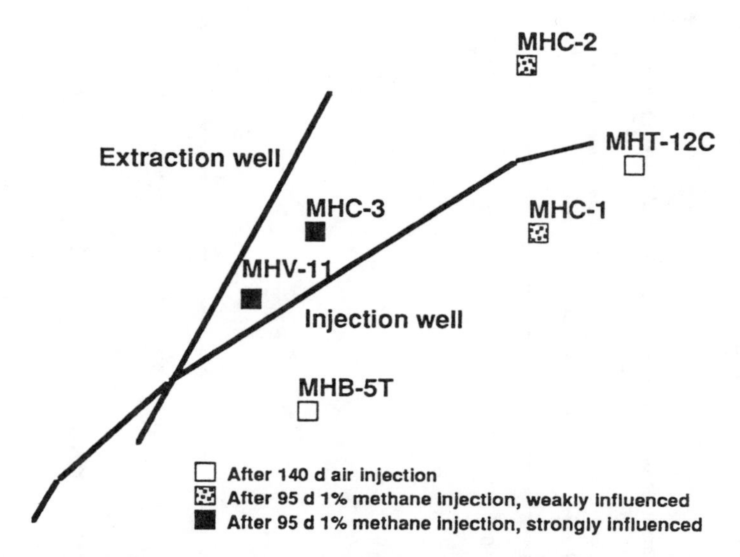

FIGURE 2. Plan view of the Savannah River bioremediation site, showing the location of boreholes from which sediment samples were analyzed.

the low dilution. Preliminary assessment of nitrogen-fixer MPNs (presumptive since acetylene reduction assays are not complete) were made by recording the development of a submerged band of growth in the semisolid medium. Headspace GC analysis for contaminant removal was at 10 weeks for aerobic incubations and at 14 weeks for anaerobic incubations.

Samples for DNA extraction were frozen at –70° C and shipped overnight in a frozen condition. DNA was extracted from 3 to 4 g of sediment and fixed to hybridization membranes. Sediment DNA was analyzed with gene probes corresponding to the soluble methane monooxygenase (MMO) (Tsien & Hanson, 1992), methanol dehydrogenase (MDH) (Bastien et al. 1989), toluene dioxygenase (TOD) (Zylstra et al. 1989), toluene monooxygenase (TMO) (Winter et al. 1989), and haloalkane dehalogenase (DH) (Janssen et al. 1989) genes. The genes were received in, or subcloned into, in vitro transcription systems for production of radiolabeled mRNA probes. Prehybridization and hybridization was at 42°C in 50% formamide, with final washes at 65°C in 0.1X saline sodium citrate and 0.1% sodium dodecyl sulfate.

RESULTS AND DISCUSSION

In this report, we will focus on the "tan clay zone" sediments present above the water table. Table 1 summarizes the cultural data for the air injection and 1% methane injection campaigns. Culturable aerobic heterotrophs on peptone-tryptone-yeast-glucose agar and actinomycete isolation agar were highest in the samples strongly influenced by the 1% methane injection campaign, and lowest

TABLE 1. Cultural characteristics of sediments at the end of the air injection and 1% methane injection campaigns.

Campaign, Sediment Type, Number of Samples	log Colony-Forming Units on		log MPN dinitrogen fixers (b)	Percent of Sediment Samples Showing TCE Removal (c) in Enrichments					
	10% PTYGA (a)	10% AIA (a)		methane-oxidizers	propane-oxidizers	ammonia-oxidizers	Fe(III)-reducers	sulfate-reducers	methanogens
air, n=4	1,1,1,1,0,0	1,2,1,0,0,0	nd (d)	50	0	0	0	0	nd (d)
1% methane, weakly influenced, n=6 or 7	0,0,1,3,2,0	0,3,2,1,0,0	0,2,3,2	0	0	0	14	43	43
1% methane, strongly influenced, n=7 or 8	1,0,0,1,3,2	1,1,1,2,2,0	2,5,1,0	0	11	37	0	0	0

(a) Expressed by class: <2.0, 2.0-2.9, 3.0-3.9, 4.0-4.9, 5.0-5.9, and 6.0-6.9, respectively. PTYG, peptone-tryptone-yeast extract-glucose agar. AIA, actinomycete isolation agar.
(b) Expressed by class: <2.5, 2.5-3.4, 3.5-4.4, and >4.4, respectively. MPN, most probable number.
(c) Positive result was defined as one or more vials with >90% removal compared to sterile controls.
(d) Not determined

in the samples from the end of the air injection campaign. Populations of dinitrogen fixers at the end of the 1% methane injection campaign were greater in samples from the weakly influenced region than in samples from the strongly influenced region, suggesting that greater air flux may have reduced populations of dinitrogen-fixers.

The level of detection in the enrichments was one (iron-reducers, sulfate-reducers, methanogens, propane-oxidizers, and ammonia-oxidizers) and two or three (methane-oxidizers) orders of magnitude higher at the end of the air injection campaign than at the end of the 1% methane injection campaign. With this in mind, TCE was removed from 50% of the methane-oxidizer enrichments at the end of the air injection campaign, however TCE removal was not detected in methane-oxidizer enrichments from the 1% methane injection campaign. Higher dilutions were carried out for methane-oxidizers in the 1% methane injection campaign than in the air injection campaign because methane-oxidizer MPNs of log 4-log 5 cells per milliliter were present in nearby groundwater monitoring wells during the 1% methane injection campaign. Thus, these data demonstrate that the chemical and/or physical conditions at the groundwater monitoring wells (versus the bulk sediment) exert a strong positive effect on methane-oxidizer populations, and that methane-oxidizer populations in the bulk sediment (as determined by cultural analysis) could not have increased more than two to three orders of magnitude.

Ammonia-oxidizer enrichments from three of the four MHC-3 sediments (borehole strongly influenced by 1% methane injection) removed TCE, whereas TCE was not removed from ammonia-oxidizer enrichments from the air injection campaign or from the weakly effected region in the 1% methane injection campaign (Table 1). This result suggests that TCE removal by ammonia-oxidizers may be an important process at the site, and that this physiological group could be targeted for further stimulation. Some anaerobic Fe(III)-reducer, sulfate-reducer, and methanogen enrichments from the weakly influenced region in the 1% methane injection campaign removed TCE, but enrichments from the air injection campaign and from the strongly influenced region in the 1% methane injection campaign did not remove TCE. Apparently, the development of increased biomass combined with reduced oxygen delivery to the weakly influenced regions compared to the strongly effected regions during the 1% methane injection campaign provided favorable conditions for the growth of these anaerobes.

The number of sediment samples hybridizing to the MMO, MDH, and TMO probes was lowest for the air injection campaign and highest for the strongly influenced boreholes in the 1% methane injection campaign (Table 2). In addition, the number of picograms of probe (for the MMO, MDH, TMO, and DH probes) hybridizing to sediment DNA increased in the order air injection campaign $<$ weakly influenced boreholes in the 1% methane injection campaign $<$ strongly influenced boreholes in the 1% methane injection campaign. Thus, the number and intensity of hybridization of probes corresponding to methanotrophs (i.e., MMO and MDH) increased in response to remediation. Of particular importance is the observation that although enrichments did not provide evidence for increased methane-oxidizer populations in the bulk sediment after the 1% methane

TABLE 2. Gene probe analysis of DNA extracted from sediment samples. Picograms of probe per g sediment hybridizing to sediment DNA: (-) no hybridization; (+) 0.2 - 1.0 (++) >1 - <10; (+++) <10 - <100. MMO, soluble methane monooxygenase; MDH, methanol dehydrogenase; TMO, toluene monooxygenase; TOD, toluene dioxygenase; HD, haloalkane dehalogenase.

Borehole	Depth (ft)	MMO	MDH	TMO	TOD	DH
End of Air Injection Campaign						
MHB-5T	99	-	+	-	-	+
	111	-	-	-	-	+
	123	-	+	+	+	-
	135	-	-	-	+	+
MHT-12C	109	-	-	-	+	+
	120	-	-	-	+	+
	129	-	-	-	+	+
End of 1% Methane Injection, Weakly Influenced Boreholes						
MHC-1	110	++	++	-	+	+
	120	++	-	-	+	+
	130	++	++	+	+	+
MHC-2	110	-	-	-	-	+
	120	+	-	+	-	+
	130	+	+	-	-	++
	140	+	-	+	-	-
End of 1% Methane Injection, Strongly Influenced Boreholes						
MHC-3	110	-	+	-	+	+
	120	+	++	+	+	+++
	130	+	++	+	+	+
	140	+++	++++	+++	+	+++
MHV-11	100	-	-	+	-	+
	110	+	-	+	-	+
	120	+	+	++	-	+
	130	-	+	-	-	+

injection campaign, the MMO and MDH gene probes provided strong evidence that methane-oxidizer populations did increase in the bulk sediment during the 1% methane injection campaign. In addition, the MMO and MDH gene probes show that sediment samples from the MHC-1 borehole (weakly influenced as defined by cultural data from a nearby groundwater monitoring well) were stimulated more than those from the MHV-11 borehole (strongly influenced as defined by cultural data from a nearby groundwater monitoring well). These are examples of how the combined use of cultural and nucleic acid methods allows a more comprehensive and defensible analysis of the success of bioremediation.

REFERENCES

Bastien, C., S. Machlin, Y. Zhang, K. Donaldson, and R. S. Hanson. 1989. "Organization of genes required for the oxidation of methanol to formaldehyde in three Type II methanotrophs." *Appl. Environ. Microbiol. 55*:3124-3130.

Eddy, C. A., B. B. Looney, J. M. Dougherty, T. C. Hazen, and D. S. Kaback. 1991. *Characterization of the Geology, Geochemistry, Hydrology and Microbiology of the In Situ Air Stripping Demonstration Site at the Savannah River Site (U).* Westinghouse Savannah River Company, WSRC-RD-91-21. Aiken, SC.

Janssen, D. B., F. Pries, J van der Ploeg, B. Kazemier, P. Terpstra, and B. Witholt. 1989. "Cloning of 1,2-dichloroethane degradation genes of *Xanthobacter autotrophicus* GJ10 and expression and sequencing of the *dhlA* gene." *J. Bacteriol. 171*:6791-6799.

Phelps, T. J., D. Ringelberg, D. Hendrick, J. Davis, C. B. Fliermans, and D. C. White. 1988. "Microbial biomass and activities associated with subsurface environments contaminated with chlorinated hydrocarbons." *Geomicrobiol. J. 6*:157-170.

Tsien, H.-C., and R. S. Hanson. 1992. "Soluble methane monooxygenase component B gene probe for identification of methanotrophs that rapidly degrade trichloroethylene." *Appl. Environ. Microbiol. 58*:953-960.

Winter, R. B., K-M. Yen, and B. D. Ensley. 1989. "Efficient degradation of trichloroethylene by a recombinant *Escherichia coli*." *Bio/Tech. 7*:282-285.

Zylstra, G. J., L. P. Wackett, and D. T. Gibson. 1989. "Trichloroethylene degradation by *Escherichia coli* containing the cloned *Pseudomonas putida* F1 toluene dioxygenase genes." *Appl. Env. Microbiol. 55*:3162-3166.

APPLICATION OF MICROBIAL BIOMASS AND ACTIVITY MEASURES TO ASSESS IN SITU BIOREMEDIATION OF CHLORINATED SOLVENTS

T. J. Phelps, S. M. Pfiffner, R. Mackowski, D. Ringelberg, D. C. White, S. E. Herbes, and A. V. Palumbo

INTRODUCTION

Evaluating the effectiveness of chlorinated solvent remediation in the subsurface can be a significant problem given uncertainties in estimating the total mass of contaminants present. If the remediation technique is a biological activity, information on the progress and success of the remediation may be gained by monitoring changes in the mass and activities of microbial populations.

The in situ bioremediation demonstration at the U.S. Department of Energy (DOE) Savannah River Site (SRS) is designed to test the effectiveness of methane injection for the stimulation of trichloroethylene (TCE) degradation in sediments. Past studies have shown the potential for TCE degradation by native microbial populations (Fliermans et al. 1988). A control phase without treatment was followed by a phase withdrawing air. The next phase included vacuum extraction plus air injection into the lower horizontal well located below the water table. The next period included the injection of 1% methane in air followed by injection of 4% methane in air. Based on the literature (Little et al. 1988, Wilson & Wilson 1985), it was hypothesized that the injection of methane would stimulate methanotrophic populations and thus accelerate biological degradation of TCE. Measuring the success of bioremediation is a complex effort that includes monitoring of changes in microbial populations associated with TCE degradation. These monitoring efforts are described in this paper and in related papers (Brockman et al. 1994, etc.) in this volume (p. 397).

METHODS

The primary source of samples for monitoring bioremediation was groundwater obtained from a series of 12 wells located at and around the site at Savannah River. Wells were sampled twice monthly and experiments were initiated on site or samples were stored on ice until processed at the University of Tennessee (UT) or Oak Ridge National Laboratory (ORNL).

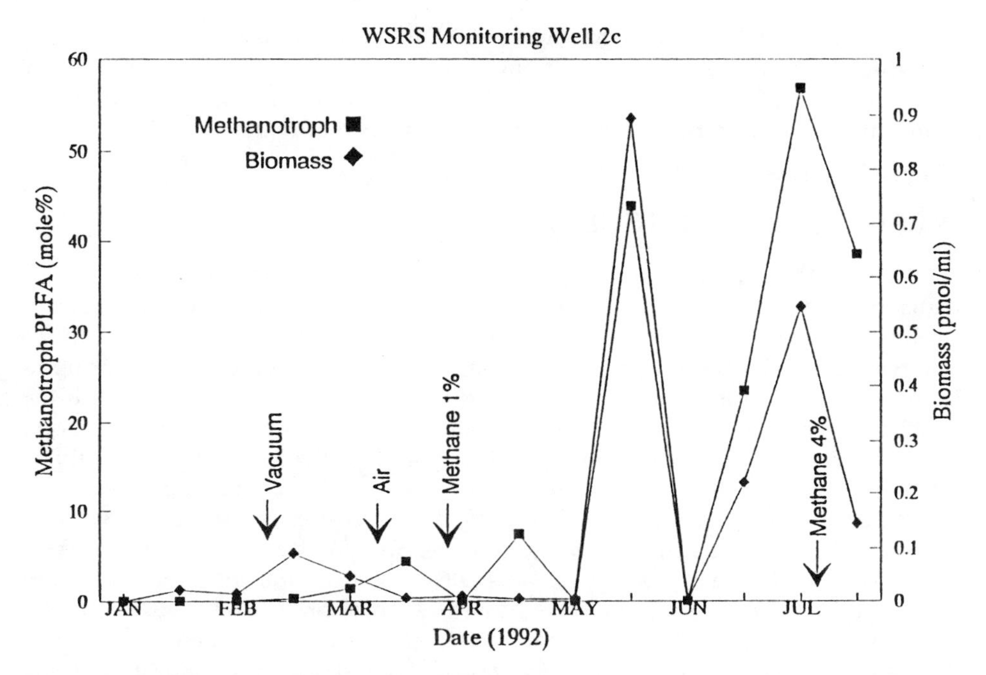

FIGURE 1. Methanotrophic PLFA and total biomass over time. Mole percent of methanotrophic PLFA (squares) was compared to biomass (diamonds) in well MHT-2C. Type II methanotrophic PLFA included 18:1w8c, 18:1w7c, 18:2a, and 18:2b.

Microbial biomass was evaluated using turbidimetric most probable number (MPN) techniques for methanotrophic and methylotrophic populations. A phosphate- and bicarbonate-buffered mineral salts medium, supplemented with 5% methane (vol/vol, headspace), was used to enumerate methanotrophs exhibiting turbidity. The methylotrophic medium was supplemented with 3 mg L^{-1} of yeast extract, 5 mM methanol, and 3% headspace propane. Microbial activity was assessed by measuring TCE mineralization in time course experiments and enrichments as described previously (Phelps et al. 1989). Assays used 10 mL groundwater, 1.0 mL sterile water and 0.5 µCi of carrier-free [1,2-^{14}C]TCE and were incubated 30 days. Methane and CO_2 were assayed using a Shimadzu GC-8A gas chromatograph (GC) with a thermal conductivity detector (TCD). Radioactive $C0_2$, VC, TCE, and PCE were quantified using a gas proportional counter (GPC) analyzing the effluents from the gas chromatograph (GC-GPC). Groundwater samples (1 L) were filtered through 0.2µm inorganic filters and analyzed for biomass and community structure. Microbial phospholipids were extracted (Phelps et al. 1989) and methyl esters of the phospholipid fatty acids (PLFAs) were analyzed by GC.

Specific nitrogen transformations were evaluated by measuring urease activity (adding urea and measuring ammonia production) and ammonia uptake (by measurement of the loss of ammonia after addition) in well water. Ammonia concentrations were measured using a Technicon™ autoanalyzer.

RESULTS AND DISCUSSION

The numbers of methanotrophs and methylotrophs increased significantly during the treatments (Table 1). During the control period, 19% of the analyses indicated presence of methanotrophs in 10-mL samples. None of the analyses from the control, vacuum extraction or air injection periods indicated $>10^3$ methanotrophs/mL. Methanotrophic populations increased dramatically with the initiation of methane injection. Wells exhibiting $>10^1$ methanotrophs/mL increased from <10% to 47% with 1% methane additions and to 85% during the 4% methane injections. During the final 2 months of 1% methane injection, 53% of the samples indicated $>10^3$ methanotrophs/mL. During the control period, 61% of the samples indicated the presence of $>10^1$ methylotrophs/mL. Between the control and air injection periods, the percentage of wells with $>10^3$ methylotrophs increased from 11% to 81%. These results indicated that methanotrophic and methylotrophic populations increased >100-fold during the treatments. Increased methanotrophic densities within several individual wells were 4 to 5 orders of magnitude (data not shown). Interestingly, both methanotrophic and methylotrophic densities decreased in many wells during the 4% methane injection phase, suggesting that nutrients or other constraints may have become limiting.

Further evidence of increased methanotrophic densities was shown by the PLFA analyses (i.e., well MHT-2C, Figure 1). Well MHT-2C showed an increase in lipid biomass from <5 pmol PLFA/mL to values often >30 pmol/mL shortly after the onset of methane injection. Increased methanotrophic populations were

TABLE 1. Average percent of wells with presumptive methanotrophs (methane) and methylotrophs (methy.) at greater than 0.1/mL, 10^1/mL, and 10^3/mL.

	>0.1/mL Presumptive		$>10^1$/mL Presumptive		$>10^3$/mL Presumptive	
Treatment	Methy.	Methane	Methy.	Methane	Methy.	Methane
Control	100	19	61	3	11	0
Vacuum Extraction	100	13	100	8	42	0
Air Injection	100	67	100	8	83	0
1% Methane	100	82	100	47	79	32
4% Methane	100	92	100	85	81	38

detected by signature PLFAs of type II methanotrophs resembling *Methylosinus trichosporium* OB3b. When total lipid biomass was compared to mole percent PLFA for type II methanotrophs, the increases in methanotrophic PLFA corresponded to the increases in lipid biomass. By the end of the 1% methane injection, lipids typical of type II methanotrophs, including 18:1w8c, 18:1w7c, 18:2a, and 18:2b, represented more than half of the total PLFA present.

Enrichment experiments examining $^{14}CO_2$ production from [1,2-^{14}C]TCE showed potential for TCE degradation throughout the operations at the SRS. For example, during the 1% methane injection, the amount of TCE degradation in the enrichments typically was <10% (Figure 2) but there was substantially more TCE degradation (>20%) in some of the analyses (6C, 7C, 8C). Maximum TCE degradation in the enrichments was 56% after 30 days. If bioremediation were optimized, most analyses should reveal mineralization within 10 days without the need for enriching nutrients, a criterion that fewer than 5% of the analyses met.

SUMMARY

There have been substantial changes in biological activity and biomass with the increasingly aggressive measures to promote TCE degradation. The data indicate the success in stimulating TCE-degrading populations. Other nutrients may be limiting, and a further phase (methane injection with nutrient addition)

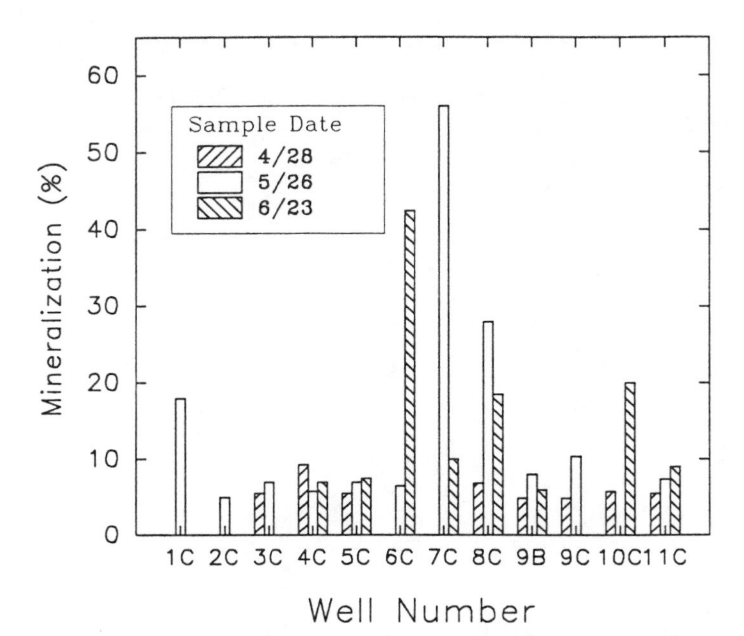

FIGURE 2. Mineralization of [1,2-^{14}C]TCE to $^{14}CO_2$ by microorganisms enriched from groundwater. $^{14}CO_2$ in blanks and control tubes after 30-d experiments were 1.4 ± 1.0%.

is currently being implemented to further increase critical population levels and degradative activity.

ACKNOWLEDGMENT

Oak Ridge National Laboratory is managed by Martin Marietta Energy Systems, Inc. for the U. S. Department of Energy under contract DE-AC05-84-OR21400. This research was funded by the DOE Office of Technology Development.

REFERENCES

Brockman, F., D. Workman, W. Sun, A. Ogram, and L. Bucey. 1994. "Baseline characterization of TCE degradative potential by enrichment techniques and DNA probe analysis." In R. E. Hinchee, A. Leeson, L. Semprini, and S. K. Ong (Eds.), *Bioremediation of Chlorinated and Polycyclic Aromatic Hydrocarbons*. Lewis Publishers, Ann Arbor, MI.

Fliermans, C. B., T. J. Phelps, D. Ringelberg, A. T. Mikell, and D. C. White. 1988. "Mineralization of trichloroethylene by heterotrophic enrichment cultures." *Appl. Environ. Microbiol. 54*: 1709-1716.

Little, C. D., A. V. Palumbo, S. E. Herbes, M. E. Lindstrom, R. L. Tyndall, and P. J. Gilmer. 1988. "Trichloroethylene biodegradation by pure cultures of a methane-oxidizing bacterium." *Appl. Environ. Microbiol. 54*:951-956.

Phelps, T. J., D. Ringelberg, D. Hedrick, J. Davis, C. B. Fliermans, and D.C. White. 1989. "Microbial activities and biomass associated with subsurface environments contaminated with chlorinated hydrocarbons." *Geomicrobiology J. 6*: 157-170.

Wilson, J. T., and B. H. Wilson. 1985. "Biotransformation of trichloroethylene in soil." *Appl. Environ. Microbiol. 49*:242-243.

SPATIAL VARIABILITY OF AEROBIC DEGRADATION POTENTIAL FOR ORGANIC POLLUTANTS

P. H. Nielsen and T. H. Christensen

INTRODUCTION

During recent years, comprehensive research has been performed to evaluate the biological degradation of specific organic compounds in groundwater (Ghiorse & Wilson 1988, Van Beelen 1990). Appreciation of the small-scale heterogeneities of groundwater and aquifers in terms of hydrogeology (Bjerg et al. 1992, Sudicky 1986), hydrogeochemistry (Bjerg & Christensen 1992, Pedersen et al. 1991), and microbiology (Albrechtsen & Winding 1992) is emerging. However, variations as to the biological degradation of organic contaminants have so far gained very little attention. Most of the studies have reported on one or only a few samples representing the groundwater or the aquifer. The hydrogeological, hydrogeochemical, and microbial variations could potentially support variations in degradation of organic compounds due to variations in transport and availability of bacteria, nutrients, and electron acceptors. The purpose of this project is to study the variability in biological degradation of aromatic hydrocarbons within an aerobic section of a landfill pollution plume.

MATERIALS AND METHODS

Location. The study was performed with sediment and groundwater from a shallow, unconfined, glaciofluvial sandy aquifer. The part of the aquifer we studied is aerobic and is influenced by the pollution plume from the Vejen landfill as seen by slightly increased concentrations of chloride. Groundwater and sediment samples were taken in 8 localities within an area of 15 m × 30 m.

Sampling. Groundwater samples were taken 2 m below the groundwater table in iron pipes (2.6 cm, I.D.) equipped with an iron tip and a 10-cm screen as described by Lyngkilde and Christensen (1992). Sediment samples were taken by manual equipment (Eikelkamp®) in a 10-cm cased borehole with a stainless steel bailer in the same points as the groundwater samples.

Preparation and Loading. Laboratory batch microcosms were made in 2.5-L glass bottles equipped with a glass valve used in sampling. Groundwater was saturated with oxygen by bubbling atmospheric air through a diffuser for approximately 1 h. The fine fraction (particle size < 75 µm) of 1 kg of sediment was suspended in 2 L of groundwater in a dry-sterilized 5-L bottle. The procedure is described in detail by Holm et al. (1992). After preparation of the microcosm, 19 specific organic contaminants were added to the suspension. The specific organic compounds included approximately 150 µg/L each of phenolic hydrocarbons, chlorinated aliphatic hydrocarbons, and aromatic hydrocarbons. The aliphatic hydrocarbons were not degraded and the phenolic hydrocarbons have yet not been analyzed. Therefore only aromatic hydrocarbons are discussed in this article. All microcosms were made in replicates and incubated in the dark in a slowly rotating box at 10°C.

Sampling. The microcosms were incubated for 149 days. During this period, samples for specific organic compounds, O_2, NO_3^-, and dissolved organic carbon (DOC) were taken. A slight overpressure was maintained in the microcosm by forcing atmospheric air into the microcosm with a syringe. Samples were taken 6 times per week in the first week of the experiment and once per 2 weeks by the end. When opening the valve, a sample was pushed out of the microcosms into a sample bottle by the overpressure. Samples for specific organic compounds were analyzed by a gas chromatograph (GC) with a flame ionization detector (FID). Biologically deactivated control experiments were made by poisoning two microcosms with 250 mg/L of formaldehyde.

Statistical Analysis. The statistical analysis of degradation rates was performed as a one-way analysis of variance (ANOVA) assuming that the degradation rates are normally and independently distributed, have the same variance, and are homogeneous within replicates.

RESULTS AND DISCUSSION

Groundwater Characteristics. Groundwater samples were clearly aerobic because concentrations of oxygen were more than 2 mg/L and concentrations of reduced species such as NH_4^+, Fe^{2+}, and Mn^{2+} were very low. The concentrations of NO_3^- were quite high, 17 to 44 mg/L. All parameters measured in the groundwater in the eight sampling points were in the same range, and none of the sampling points differed significantly from the others. The addition of atmospheric air to the microcosms as part of the sampling procedure ensured at least 9 mg/L of O_2 in the microcosms during the experiment.

Examples of Degradation Curves. Figure 1 presents biphenyl degradation curves for microcosms loaded with groundwater and sediment from Localities 3 and 7, including replicates (A and B) and a biologically deactivated control. Biphenyl was degraded within 35 days. Therefore results from only the first 45 days of the experimental period are shown for the biologically active microcosms. The 5 degradation curves presented are representative of the 112 total degradation

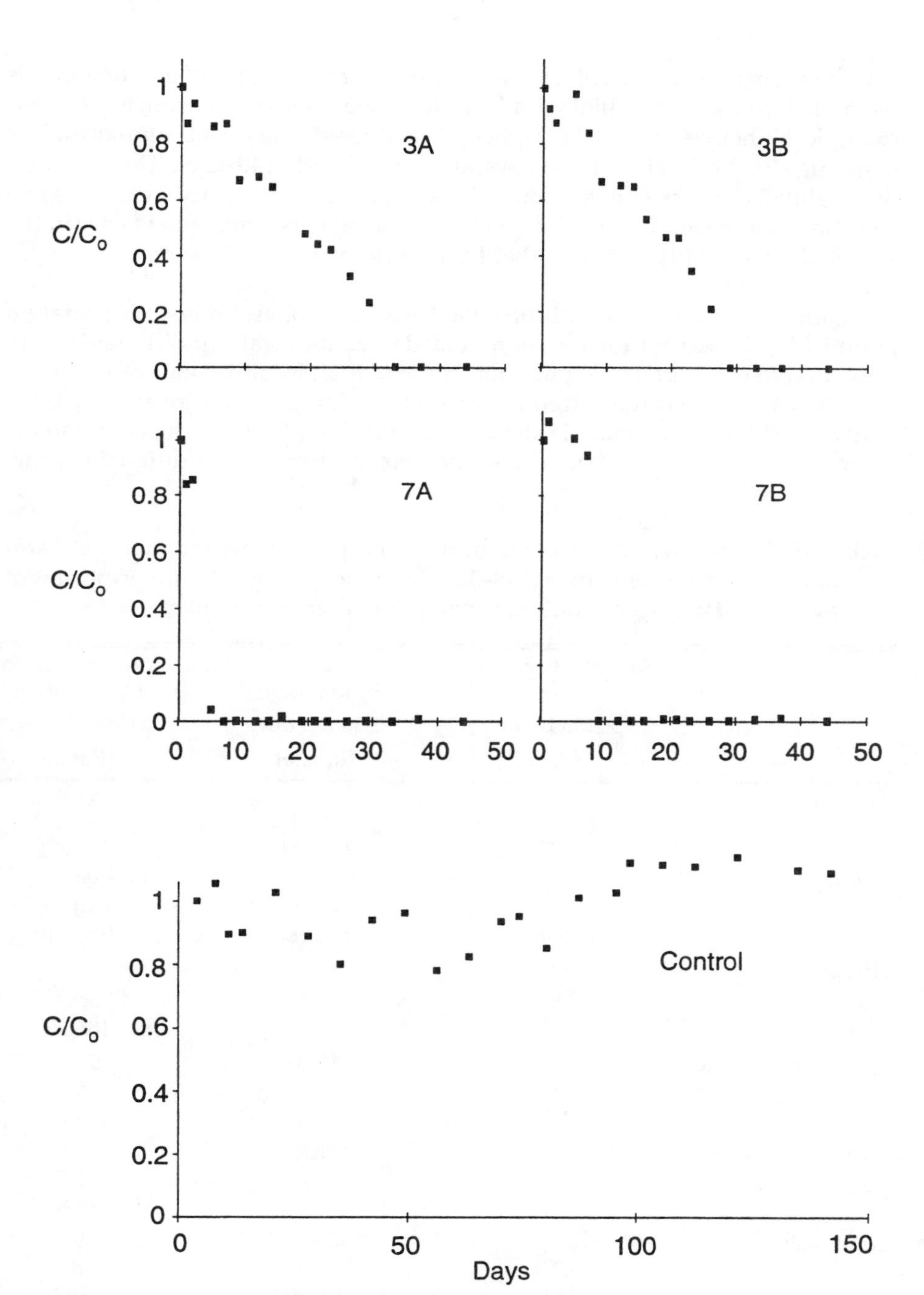

FIGURE 1. Normalized concentrations of biphenyl as a function of time in batch degradation experiments with sediment and groundwater from Localities 3 and 7 (A and B are replicates). The concentrations are normalized by the initial concentration.

curves (7 compounds, 8 localities, 2 replicates). Concentrations of the compounds did not decrease in the biologically deactivated control experiment (see, for example, biphenyl in Figure 1). Biphenyl was degraded to a concentration of less than 2 µg/L in both replicates of Localities 3 and 7 within 40 days. The replicates of Locality 3 are very similar, whereas the replicates of Locality 7 exhibit some variation. However, biphenyl degradation seems to be somewhat faster in the sample from Locality 7 than in that from Locality 3.

Quantification of Degradation Patterns. Lag phase (days), degradation period (days), percent degradation, and degradation rate (µg/[L·day]) were estimated to compare all degradation curves for all localities and replicates.

The results are summarized in Table 1. The lag phases were all very short (2 to 7 days) and very similar in all localities and all replicates for all compounds, whereas the degradation period and the percent degradation differed for the

TABLE 1. Lag phases, degradation period, and percent degradation for 7 aromatic hydrocarbons in a 149-day laboratory degradation experiment (average, standard deviation, range). Number of samples, n = 16.

	Average Lag Phase (days) ±Stand. dev. [Range]	Average Degradation Period (days) ±Stand. dev. [Range]	Average Percent Degradation ±Stand. dev. [Range]
Benzene	4.6 ±1.4 [2-7]	31.5 ±12.9 [16-61]	97.8 ±0.7 [97-98]
Toluene	4.1 ±1.8 [2-7]	26.6 ±11.0 [16-44]	99.8 ±0.2 [99-100]
o-Xylene	4.7 ±1.9 [2-7]	82.0 0.0 82	86.9 ±5.3 [83-95]
p-Dichlorobenzene	4.9 ±3.3 [0-7]	82.0 0.0 82	78.3 ±5.9 [75-78]
o-Dichlorobenzene	4.5 ±2.6 [0-7]	82.0 0.0 82	81.0 ±5.2 [79-82]
Naphthalene	4.5 ±2.3 [2-9]	15.2 ±8.4 [9-44]	99.9 ±0.1 100
Biphenyl	4.5 ±2.2 [2-7]	29.7 ±34.3 [5-149]	100.0 0.0 100

different compounds. Benzene, toluene, naphthalene, and biphenyl all were degraded to less than 2 μg/L within 1 month, whereas the degradation of *o*-xylene, *p*-dichlorobenzene, and *o*-dichlorobenzene continued for a period of approximately 3 months when the compounds were degraded approximately 80%. The three compounds were degraded to approximately the same degree. Sampling was once per week in this period and degradation stopped between day 75 and day 82 in all microcosms. Biphenyl was degraded fast in some localities (e.g., Locality 7) and slower in other localities (e.g., Locality 3), but less than 2 μg/L was left in all experiments.

In general, the variations among the 8 sampling localities seem relatively moderate. Where the largest variation is observed (biphenyl), degradation is fairly rapid, thus reducing the environmental significance of the variation. However, closer examination of the degradation rates reveals some statistically significant variations. The degradation rates for benzene and biphenyl are shown as bar diagrams in Figure 2.

Variations Among Localities. Figure 2 shows two features with respect to variation in degradation rates. (1) For benzene, the experimental variation among replicates is very small, although the variation among localities with respect to

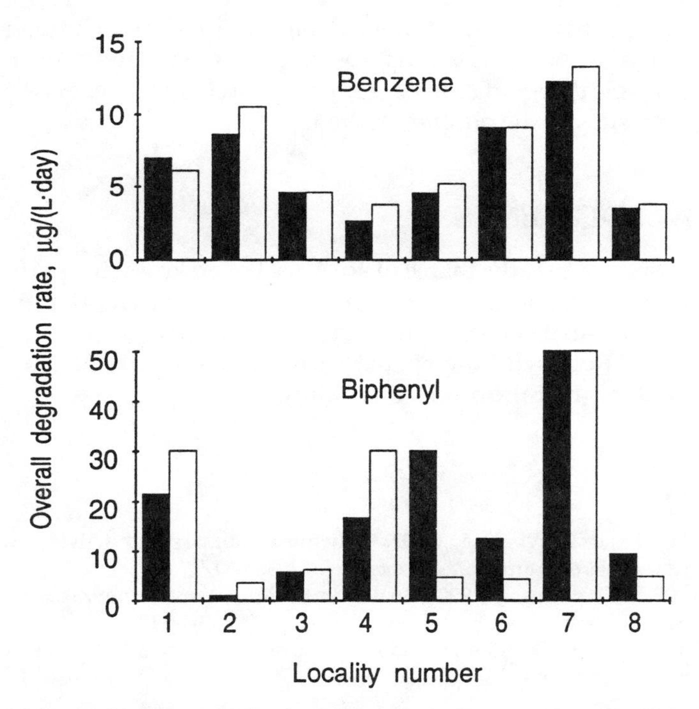

FIGURE 2. Degradation rates for benzene and biphenyl in 8 localities in a 15 m × 30 m area of an aerobic landfill pollution plume. Black bars: replicate A; white bars: replicate B.

degradation rates is both substantial (up to a factor of 4) and statistically significant (5%). (2) For biphenyl, the experimental variation is substantial, but the differences (up to a factor of 15) among localities are statistically significant (5%).

The localities studied are too few to allow for a geostatistical analysis of the spatial variations in the degradation rates observed for the studied compounds. A general comparison of localities showing, for example, high degradation rates and their physical location does not indicate much of a scale correlation.

CONCLUSIONS

Based on the results of the 149-day-long laboratory degradation experiment, we can conclude that benzene, toluene, *o*-xylene, naphthalene, biphenyl, *p*-dichlorobenzene, and *o*-dichlorobenzene were biologically degraded in 8 localities representing a 15 m × 30 m section of the aerobic zone of a landfill pollution plume. All compounds showed short lag phases (max. 7 days). Benzene, toluene, naphthalene, and biphenyl were degraded to less than 2 µg/L in all replicates of the 8 localities, and *p*- and *o*-dichlorobenzene were degraded to approximately the same degree (80%) in all replicates for the 8 different localities.

The laboratory batch technique was very reproducible for some compounds (e.g., benzene) and less reproducible for other compounds (e.g., biphenyl). The degradation rate of benzene and biphenyl varied significantly among localities (up to a factor of 4 for benzene and 15 for biphenyl). However, the spatial variation in biological degradation of aromatic hydrocarbons seems to be modest in the sampling sites of the aquifer studied.

ACKNOWLEDGMENTS

The authors are greatly thankful to Anja Foverskov who performed the chemical analysis. This study was a part of a major research program focusing on the effects of waste disposal on groundwater. The program is funded by the Danish Technical Research Council, the Technical University of Denmark, and the Commission of the European Community.

REFERENCES

Albrechtsen, H.-J., and A. Winding. 1992. "Microbial biomass and activity in subsurface sediments from Vejen, Denmark." *Microb. Ecol. 23*: 303-317.

Bjerg, P. L., and T. H. Christensen. 1992. "Spatial and temporal small scale variation in groundwater quality of a shallow sandy aquifer." *J. Hydrol. 138*: 133-149.

Bjerg, P. L., K. Hinsby, T. H. Christensen, and P. Gravesen. 1992. "Spatial variability of hydraulic conductivity of an unconfined sandy aquifer determined by a mini slug test." *J. Hydrol. 136*: 107-122.

Ghiorse, W. C., and J. T. Wilson. 1988. "Microbial ecology of the terrestrial subsurface." *Adv. Appl. Microbiol. 33*: 107-172.

Holm, P. E., P. H. Nielsen, H.-J. Albrechtsen, and T. H. Christensen. 1992. "Importance of unattached bacteria and bacteria attached to sediment in determining potentials for degradation of xenobiotic organic contaminants in an aerobic aquifer." *Appl. Environ. Microbiol. 58*(9): 3020-3026.

Lyngkilde, J., and T. H. Christensen. 1992. "Redox zones of a landfill leachate pollution plume (Vejen, Denmark)." *J. Contam. Hydrol. 10*: 273-289.

Pedersen, J. K., P. L. Bjerg, and T. H. Christensen. 1991. "Correlation of nitrate profiles with groundwater and sediment characteristics in a shallow sandy aquifer." *J. Hydrol. 124*: 263-277.

Sudicky, E. A. 1986. "A natural gradient experiment on solute transport in sand aquifer: Spatial variability of hydraulic conductivity and its role in the dispersion process." *Wat. Resour. Res. 22*(13): 2069-2082.

Van Beelen, P. 1990. "Degradation of organic pollutants in groundwater." *Stygologia 5*(4): 199-212.

IN SITU MEASUREMENT OF DEGRADATION OF SPECIFIC ORGANIC COMPOUNDS UNDER AEROBIC, DENITRIFYING, IRON(III)-REDUCING, AND METHANOGENIC GROUNDWATER CONDITIONS

P. H. Nielsen and T. H. Christensen

INTRODUCTION

Degradation of specific organic contaminants in aquifers usually is studied in the laboratory due to the simplicity and controllability of laboratory experiments compared to field experiments. However, laboratory experiments require collection, transport, storage, and handling of aquifer samples (groundwater and/or aquifer sediment) that may affect parameters such as pH, Eh, dissolved oxygen, and temperature. In addition, laboratory experiments usually are performed at a much higher groundwater-to-sediment ratio than found in the field. Redox conditions are important with respect to degradation of specific organic compounds, and laboratory experiments often may need substantial manipulation to ensure proper control of the redox level, particularly under anaerobic conditions. All these aspects may bring into question the field applicability of results from laboratory studies and may require measurements of degradation potentials directly in the field.

This paper presents selected in situ measurements of the degradation potential of a mixture of 19 specific organic compounds under methanogenic, iron(III) — reducing, denitrifying, and aerobic conditions in a landfill pollution plume using in situ microcosms. The pollution plume was selected because the plume had induced these different redox conditions to develop in the same aquifer during the last 15 years.

EXPERIMENTAL DESIGN AND METHODS

Location. Degradation of specific organic compounds was investigated using in situ microcosms installed in 4 different redox zones of the Vejen landfill (Denmark) pollution plume. Figure 1 presents a vertical transect of the pollution plume, showing the redox zones of the aquifer and in situ microcosms installed in methanogenic, iron(III)-reducing, denitrifying, and aerobic conditions. The aquifer is a glaciofluvial sandy deposit with a pore flow velocity of 150 to 200 m

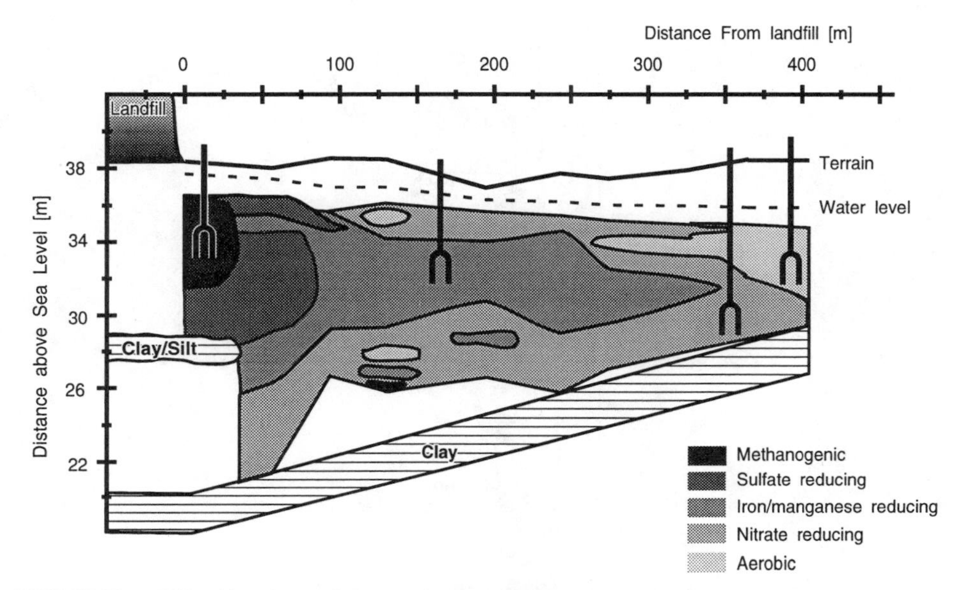

FIGURE 1. Distribution of the redox zones in the groundwater downgradient of Vejen landfill, Denmark. In situ microcosms were installed in four different redox zones in the leachate pollution plume.

per year. The redox zones of the pollution plume have been described by Lyngkilde and Christensen (1992).

In Situ Microcosm. The in situ microcosm (Figure 2) originally was developed at the University of Waterloo, Canada (Gillham et al. 1990) and has been used for different purposes by different authors (e.g., Acton & Barker 1992, Holm et al. 1992, and Nielsen et al. 1992). The in situ microcosm is installed in the groundwater zone and isolates 2 L of aquifer material including both sediment and groundwater. The in situ microcosm is open in the bottom and screened in the top. The in situ microcosm is loaded from the surface with a mixture of specific organic compounds.

Installation. The in situ microcosms were installed in cased boreholes drilled by hand equipment (Eikelkamp®). Aquifer material was forced into the microcosms when driving the microcosms into the ground using a Cobra® jackhammer. During the installation, the casing was withdrawn. Five pore volumes of groundwater were pumped from the microcosms for development prior to loading. All in situ microcosms were installed with the open end approximately 3 m below the groundwater table.

Loading. Groundwater from the microcosm was pumped into a 5-L Tedlar® bag and spiked with a mixture of 19 specific organic compounds including aromatic hydrocarbons (benzene, toluene, *o*-xylene, *p*-dichlorobenzene, *o*-dichlorobenzene, naphthalene, and biphenyl, nitrobenzene); phenolic hydrocarbons (phenol, *o*-cresol,

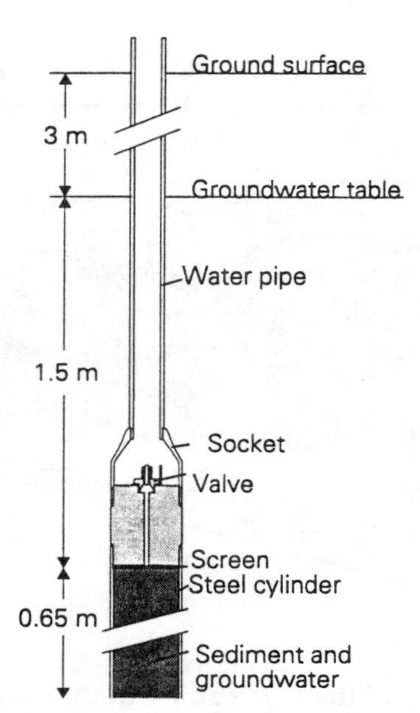

FIGURE 2. Sketch of an in situ microcosm installed in the aquifer.

o-nitrophenol, *p*-nitrophenol, 2,6-dichlorophenol, 2,4-dichlorophenol, and 4,6-*o*-dichlorocresol); and chlorinated aliphatic hydrocarbons (1,1,1-trichloroethane, trichloroethene, tetrachloromethane, and tetrachloroethene). Several of these compounds frequently have been found in groundwater polluted by waste disposal (Arneth et al. 1989). Tritiated water was used as a tracer and formaldehyde (250 mg/L) was used as a biocide in biologically deactivated control experiments. The concentration of each specific organic compound in the 5-L Tedlar® bag was approximately 150 µg/L. Water was pumped slowly (2.5 L/h) from the Tedlar® bag into the in situ microcosm through a 4-mm (I.D.) stainless steel tube using a Watson-Marlow® peristaltic pump.

Sampling. In situ microcosms were sampled using a stainless steel syringe that was handled from the ground surface (5 mL of groundwater was discarded before sampling), and 10-mL samples were transferred from the syringe to a measuring flask and extracted with 100 µL of pentane for gas chromatography (GC) analysis. Samples for inorganic analysis were transferred to plastic bottles and preserved for laboratory analysis. Oxygen was measured by Winkler titration in the field. The microcosms were sampled approximately once a week during a 3-month period.

RESULTS AND DISCUSSION

A constant level of the tracer in all the in situ microcosms described indicated that the microcosms were functioning hydraulically well.

Aerobic Conditions. The oxygen concentration in four in situ microcosms installed in the aerobic zone decreased from 10 mg/L in the beginning of the experiment to about 2 mg/L by the end of experiment, indicating aerobic conditions during the whole period. A high degradation potential was measured in the aerobic zone of the pollution plume. Many of the phenolic and aromatic hydrocarbons were degraded efficiently during a period of only a few months. Under aerobic conditions the chlorinated aliphatic compounds were not degraded. Figure 3A shows degradation curves for naphthalene, *p*-nitrophenol, and 1,1,1-trichloroethane. Naphthalene and *p*-nitrophenol were degraded completely within a 15-day period, whereas 1,1,1-trichloroethane was not degraded at all.

Denitrifying Conditions. No O_2 was detectable in the four microcosms installed in the denitrifying zone, and NO_3^- concentration decreased slowly during the experiment. This indicated that denitrification was taking place in the microcosm. The degradation potential was very low for the chlorinated aliphatic compounds and for the aromatic compounds in the denitrifying zone. Only nitrobenzene and some phenolic compounds were degraded. In Figure 3B the initial removal of naphthalene and *p*-nitrophenol probably was due to sorption during the first few days of the experiment. After this period naphthalene was not degraded, whereas *p*-nitrophenol was degraded over a period of approximately 50 days. Under denitrifying conditions, 1,1,1-trichloroethane was not degraded.

Iron(III)-Reducing Conditions. In the iron(III)-reducing zone, only one in situ microcosm was successfully installed. In this microcosm no O_2 was detectable and only traces of NO_3^- were found. The concentration of Fe^{2+} increased during the first 20 days of the experiment, indicating that Fe^{3+}-reduction was going on in the beginning of the experiment. The preliminary results from this zone show degradation of aromatic, phenolic, and chlorinated aliphatic compounds. In Figure 3C, naphthalene, *p*-nitrophenol, and 1,1,1-trichloroethane seem to be degraded in the iron(III)-reducing zone.

Methanogenic Conditions. In three in situ microcosms installed in the methanogenic zone of the pollution plume, elevated concentrations of CH_4 were measured in the groundwater. At the end of the experiment CH_4 was produced in water from the microcosm. In methanogenic conditions aromatic hydrocarbons were not degraded, whereas some chlorinated aliphatic hydrocarbons and phenolic hydrocarbons were degraded. Figure 3D shows the concentration of naphthalene, *p*-nitrophenol, and 1,1,1-trichloroethane. The three compounds seem to be sorbed partly to the sediment. This is probably due to a higher concentration of organic matter in the sediment close to the landfill. Naphthalene was not degraded within a period of 90 days, whereas 1,1,1-trichloroethane was degraded after a lag period of 40 days, and *p*-nitrophenol was degraded in a period of 80 days, beginning with a very fast degradation.

Control In Situ Microcosms. To distinguish between biological degradation of the specific organic compounds and abiotic removal of the compounds, biologically

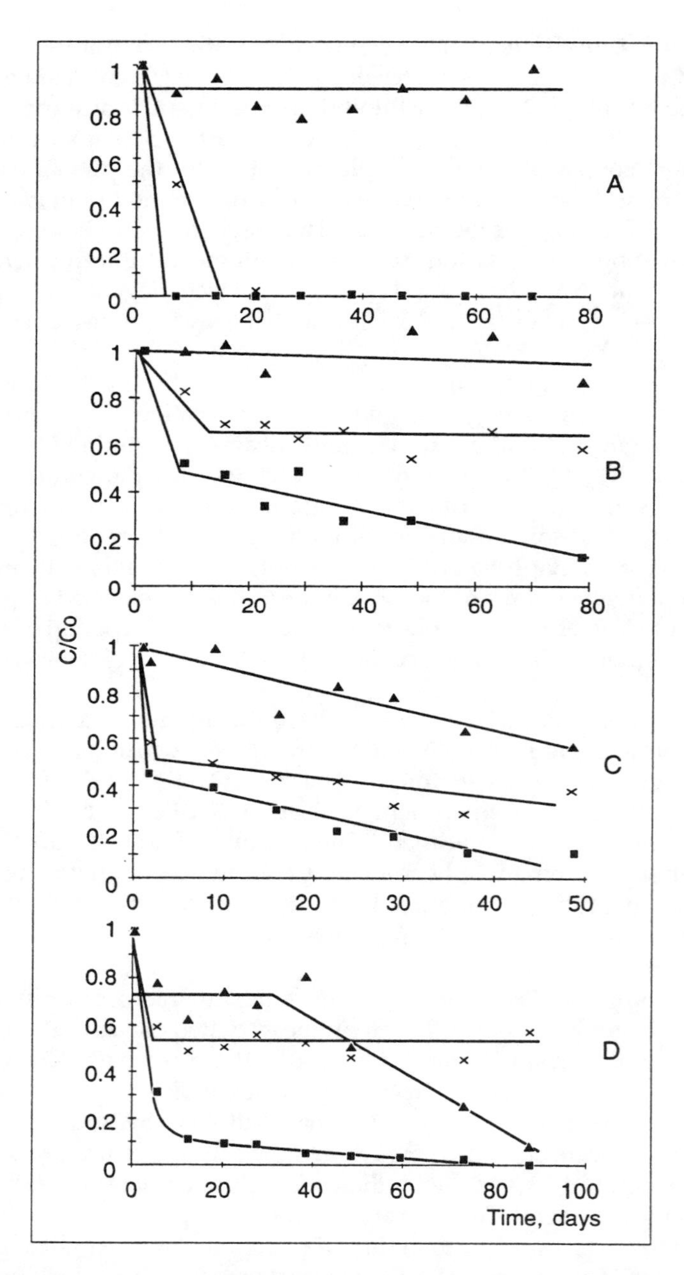

FIGURE 3. Concentration of 1,1,1-trichloroethane (▲), naphthalene (×), and *p*-nitrophenol (■) as a function of time using in situ microcosms installed in the four dominating redox zones in the Vejen landfill pollution plume. A: aerobic, B: denitrifying, C: iron(III)-reducing, D: methanogenic. The concentrations are normalized with respect to the initial concentrations.

deactivated control in situ microcosms were installed in the aerobic and denitrifying zones. These microcosms showed an initial nonbiological removal of some of the specific organic compounds, which was interpreted as due to sorption. Interpretation of degradation experiments using in situ microcosms should not include the initial period until sorption equilibria have become established.

CONCLUSION

In situ microcosms offer a unique means for measuring degradation potentials in groundwater for specific organic compounds under relatively undisturbed field conditions. It should, however, be kept in mind that the in situ microcosm partially isolates a small section of the aquifer and that certain chemical changes may occur during the experiment. This factor may lead to alterations in redox conditions in the microcosms. Therefore, it is important to measure the redox conditions in the microcosm from the beginning to the end of an experiment. Furthermore, the organic chemicals are sorbed to the sediment to different degrees depending on sediment and compound properties. Degradation experiments using in situ microcosms should not be interpreted before the sorption has reached equilibrium. In these experiments this period was about 1 week.

Several aromatic hydrocarbons (e.g., naphthalene) were degraded within in situ microcosms installed under aerobic conditions but not under more reduced conditions. Some chlorinated aliphatic hydrocarbons (e.g., 1,1,1-trichloroethane) were degraded under methanogenic and iron(III)-reducing conditions but not under more oxidized conditions. Several phenolic hydrocarbons (e.g., *p*-nitrophenol) were degraded under reduced as well as oxidized redox conditions. The redox gradient appearing in organic pollution plumes seems to be of importance for the degradation of the specific organic contaminants.

ACKNOWLEDGMENTS

Anja Foverskov performed the major part of the chemical analysis. Helga Bjarnadóttir, Pernille Nielsen, Pernille Smith, and Pia Winter performed several of the experiments as partial fulfillment for their master degrees.

REFERENCES

Acton, D. W., and J. F. Barker. 1992. "In Situ Biodegradation Potential of Aromatic Hydrocarbons in Anaerobic Groundwaters." *J. Contam. Hydrol. 9*: 325-352.

Arneth, J.-D., G. Milde, H. Kerndorff, and R. Schleyer. 1989. "Waste Deposit Influences on Groundwater Quality as a Tool for Waste Type and Site Selection for Final Storage Quality." In P. Baccini (Ed.), *Lecture Notes in Earth Sciences, The Landfill 20*: 399-424. Springer Verlag, Berlin, Germany.

Gillham, R. W., R. C. Starr, and D. J. Miller. 1990. "A Device for In Situ Determination of Geochemical Transport Parameters. 2. Biochemical Reactions." *Ground Water 28*(6): 858-862.

Holm, P. E., P. H. Nielsen, H.-J. Albrechtsen, and T. H. Christensen. 1992. "Importance of Unattached Bacteria and Bacteria Attached to Sediment in Determining Potentials for Degradation of Xenobiotic Organic Contaminants in an Aerobic Aquifer." *Appl. Environ. Microbiol. 58*(9): 3020-3026.

Lyngkilde, J., and T. H. Christensen. 1992. "Redox Zones of a Landfill Leachate Pollution Plume (Vejen, Denmark)." *J. Contam. Hydrol. 10*: 273-289.

Nielsen, P. H., P. E. Holm, and T. H. Christensen. 1992. "A Field Method for Determination of Groundwater and Groundwater Sediment Associated Potentials for Degradation of Xenobiotic Organic Compounds." *Chemosphere 25*(4): 449-462.

ANAEROBIC METABOLISM OF CHLORINATED BENZENES IN SOIL UNDER DIFFERENT REDOX POTENTIALS

K. Ramanand, M. T. Balba, and J. Duffy

INTRODUCTION

Halogenated aromatic chemicals, because of their widespread use, can contaminate groundwater reserves. Detoxification of such habitats by anaerobic bioprocesses seems appropriate because many subsurface sites contain little or no oxygen.

The anaerobic destruction of halobenzoates by reductive dehalogenation was first documented by Suflita et al. (1982). Later on, several research groups studied extensively the environmental fate of different groups of halogenated compounds under methanogenic conditions (Kuhn & Suflita 1989; Mohn & Tiedje 1992) or under denitrifying and sulfidogenic conditions (Genthner et al. 1989; Haggblom & Young 1990; Kohring et al. 1989; Schennen et al. 1985; Townsend et al. 1992), including a pure culture study with sulfate-reducing bacterium, *Desulfomonile tiedjei* (DeWeerd et al. 1990).

Many subsurface environments have substantial concentrations of dissolved organic carbon, and ions such as nitrate, sulfate, iron, manganese, and/or carbonate. These ions can serve as electron acceptors and influence the turnover of carbon in various ecosystems. In nature, these electron acceptors are sequentially utilized, eventually creating a highly reducing state where CO_2 will act as the ultimate electron acceptor. Chlorobenzenes are common groundwater contaminants and their fate in such transitional redox conditions is not well understood. The present study summarizes the results of the anaerobic metabolism of a mixture of chlorobenzenes in soil under different redox conditions and describes the dechlorination of selected tetra-, tri-, and dichlorobenzenes to monochlorobenzene.

METHODS AND MATERIALS

Soil Slurry Microcosm. Clean soil was obtained from 2 feet (0.6 m) below ground surface from a location in Niagara Falls, New York. The soil was characterized as a sandy clay loam and did not contain any sulfate or nitrate. The slurry microcosms were constructed by placing 50 g of soil and 50 mL of sterile anaerobic mineral salts medium (Shelton & Tiedje 1984) in 150-mL serum bottles as described previously (Ramanand & Suflita 1991). All bottles received filter-sterilized sodium sulfide (1 mM), and resazurin served as the anaerobic indicator. The bottles were sealed with Teflon™-lined butyl rubber stoppers, crimped, and removed from the glove box. The bottles incubated under sulfate-reducing (headspace, H_2/CO_2), and nitrate-reducing (headspace, N_2/CO_2) conditions were amended exogenously with 20 mM of Na_2SO_4 and $NaNO_3$, respectively. Methanogenic slurries (headspace, H_2/CO_2) did not receive an exogenous amendment of sulfate or nitrate. The soil slurries under different redox conditions were then amended with a mixture of pure chlorobenzenes to give 20 to 25 mg/L each of 1,2-dichlorobenzene (DCB); 1,2,4-trichlorobenzene (TCB); and 1,2,3,4-tetrachlorobenzene (TTCB). Experiments were performed in duplicates and compared to sodium azide (1,000 mg/L)-treated controls. At intervals, slurries were analyzed for sulfate and nitrate by high-pressure liquid chromatography (HPLC) or extracted with pentane and analyzed for the residues by gas chromatography (GC).

Extraction and Analysis. The organic compounds in soil slurries (100 mL) were extracted with pentane (40 mL) and the concentration of the substrates and metabolites in pentane extract was determined by a 5840A Hewlett Packard GC equipped with flame ionization detector and a DB-624 fused silica megabore column (30 m × 0.53 mm, 3.0 µm film thickness). The operating temperatures of the injector and detector were 250°C and 275°C, respectively. Nitrogen served as the makeup gas at a flowrate of 30 mL/min. Helium was used as the carrier gas at a flow rate of 6 mL/min. The column was operated under programmed conditions from 40°C initially for 5 min, increasing by 3°C/min until 18 min, 1°C/min until 22 min, and 10°C/min at 22 min with an isothermal temperature of 260°C during the last 10 min. The compounds were identified and quantified by comparing their retention times and integrator responses with those of authentic materials.

Sulfate and nitrate were analyzed by a Perkin Elmer HPLC (Model 250). The anions were separated on a PRP X-100 column (250 × 4.1 mm) using 10 mM potassium hydrogen phthalate buffer (pH 4.0) as the mobile phase at a flowrate of 1 to 1.5 mL/min and detected by indirect ultraviolet (UV) absorbance operated at 315 nm.

RESULTS AND DISCUSSION: METABOLISM OF CHLOROBENZENES IN SOIL SLURRY MICROCOSMS

Methanogenic Conditions. The anaerobic degradation of a mixture of 1,2,3,4-TTCB, 1,2,4-TCB, and 1,2-DCB was favored in methanogenic incubations among different redox conditions tested. After a typical lag period of 2 to 3 months, about 75% of the original 1,2,3,4-TTCB level was biodegraded within the next 2 months (Figure 1). Similarly, 1,2,4-TCB and 1,2-DCB declined to about 30% and 60% of the original concentration, respectively, within 4 months, and to negligible levels in 6 months. Thus, the degradation of substrates was faster for chemicals with more chlorine atoms compared to lesser chlorinated analogs.

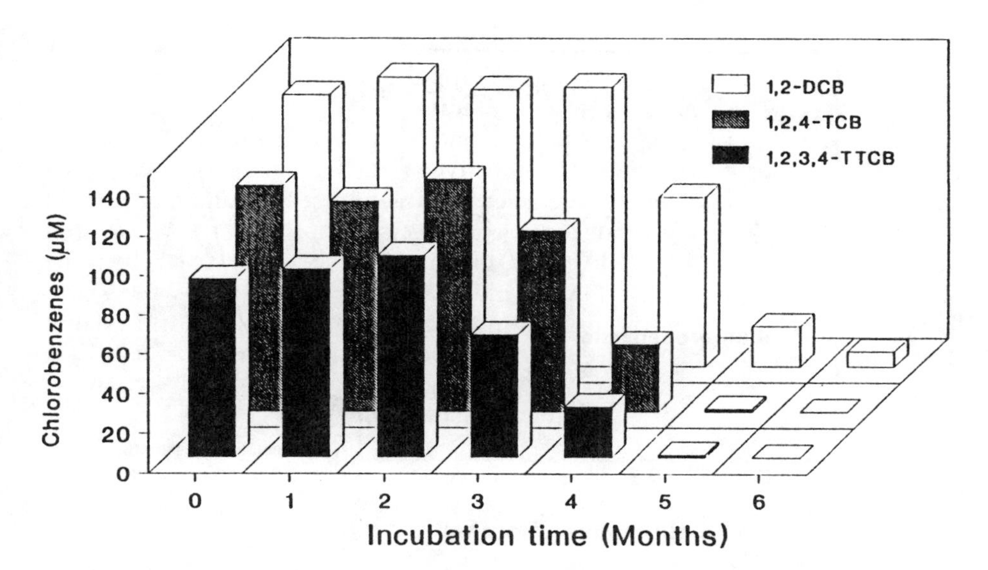

FIGURE 1. Anaerobic biodegradation of 1,2,3,4-TTCB, 1,2,4-TCB, and 1,2-DCB in soil slurry microcosms under methanogenic conditions.

Metabolism of the parent substrates (mixture of 1,2,3,4-TTCB, 1,2,4-TCB, and 1,2-DCB) resulted in the formation of monochlorobenzene (MCB) as the main metabolic product and the transient accumulation of the intermediates 1,2,3-TCB, 1,4-DCB, and 1,3-DCB (Figure 2). Stoichiometric analysis revealed that more than 82% of the added parent substrates was biotransformed to MCB. These results suggested that the tested substrates were catalyzed by sequential reductive dechlorination resulting in MCB formation as the dead-end product. No dechlorination of MCB occurred even 4 months after reaching its maximum concentration. These observations are consistent with reports described recently (Fathepure et al. 1988, Fathepure & Vogel 1991; Holliger et al. 1992). In these studies, the metabolism of hexachlorobenzene (HCB) and pentachlorobenzene (PCB) resulted in major accumulation of 1,3,5-TCB and trace amounts of DCB isomers. The 1,3,5-TCB and DCBs were not further transformed. In our studies, the slurries amended with a mixture of 1,2,3,4-TTCB, 1,2,4-TCB, and 1,2-DCB did not result in DCB accumulation, but instead resulted in near-stoichiometric amounts of MCB. This finding is consistent with the previous study by Bosma et al. (1988), which is the only cited literature on the dehalogenation of TCB isomers to MCB. A tentative pathway describing the biotransformation of chlorobenzenes based on the results of methanogenic studies is illustrated in Figure 3.

In an ongoing preliminary study using enriched soil cultures and cross-acclimation techniques, a mixture of HCB and PCB was successfully dechlorinated via the intermediates 1,2,3,4-TTCB, 1,2,3- and 1,2,4-TCB, DCBs, and MCB but not by, 1,3,5-TCB (data not shown). Such a complete reductive dehalogenation of HCB to MCB by a single biological system has not been demonstrated previously.

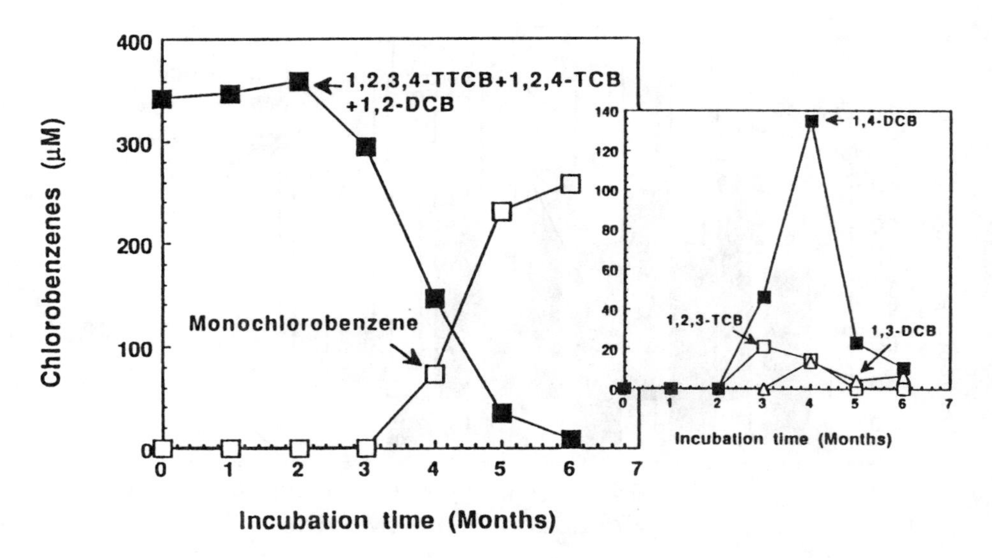

FIGURE 2. Biotransformation of highly chlorinated benzenes to monochlorobenzene in soil slurry microcosms under methanogenic conditions: production of transient metabolites.

These preliminary results suggest that the dechlorination pathway of HCB and microbiological species involved in this system may be different from those reported previously. Experiments are under way to further elucidate this pathway.

Sulfate-Reducing Conditions. In the soil slurries amended with sulfate to serve as an alternative electron acceptor, the parent substrates (mixture of 1,2,3,4-TTCB, 1,2,4-TCB, and 1,2-DCB) were not metabolized even after 6 months (Figure 4). The sulfate levels in the slurries were monitored periodically with time, and when necessary, sulfate was added to maintain sulfate levels at 15 to 20 mM. The sulfate concentration declined during the first 4 months, however, from 20 mM to about 5.23 mM. Substrate-unamended controls also exhibited similar sulfate losses. This indicated that sulfate decrease is caused by endogenous sulfate-reducing bacterial activity and the dissimilation of soil organic matter.

Previous studies suggested that the presence of sulfate inhibits the breakdown of chlorinated compounds (Mohn & Tiedje 1992). In the present studies, possible sulfate inhibitory affect was further examined by the addition of lactate, which is considered a preferred carbon source for sulfate-reducing bacteria (Postgate 1984). Therefore, 4 months after the sulfate concentration declined to 5.23 mM, one set of microcosms was amended with 10 mM of additional sulfate while the other set received 5 mM lactate. The microcosms that received sulfate showed no loss of parent substrates (1,2,3,4-TTCB, 1,2,4-TCB, and 1,2-DCB) during the next 5 months. Sulfate presence inhibited dehalogenation in a manner similar to comparable studies reported earlier (Genthner et al. 1989; Gibson & Suflita

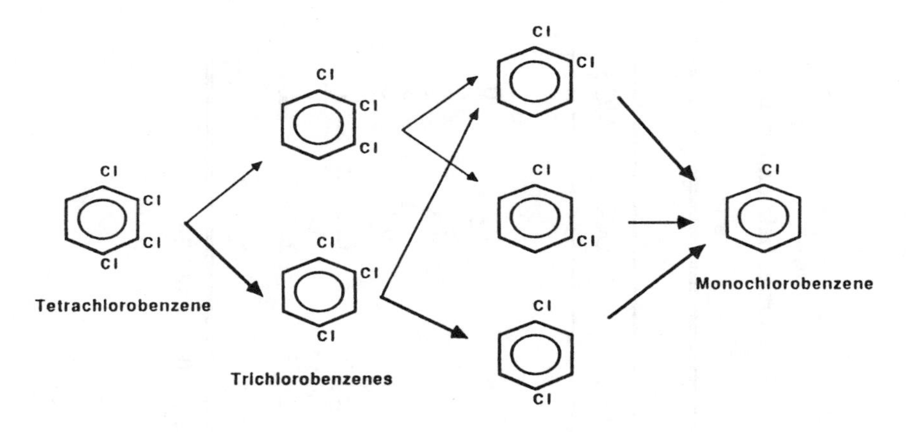

FIGURE 3. Proposed pathway for chlorobenzene metabolism in soil under methanogenic incubations.

1990). It is possible that sulfate is utilized as the preferred electron acceptor relative to the parent chlorobenzenes. The microcosms that received lactate, however, showed rapid loss of 1,2,3,4-TTCB, whereas the other two parent substrates (1,2,4-TCB, and 1,2-DCB) required longer adaptation before they were biodegraded (Table 1). The inhibitory effect of sulfate appeared to be reversible, because parent

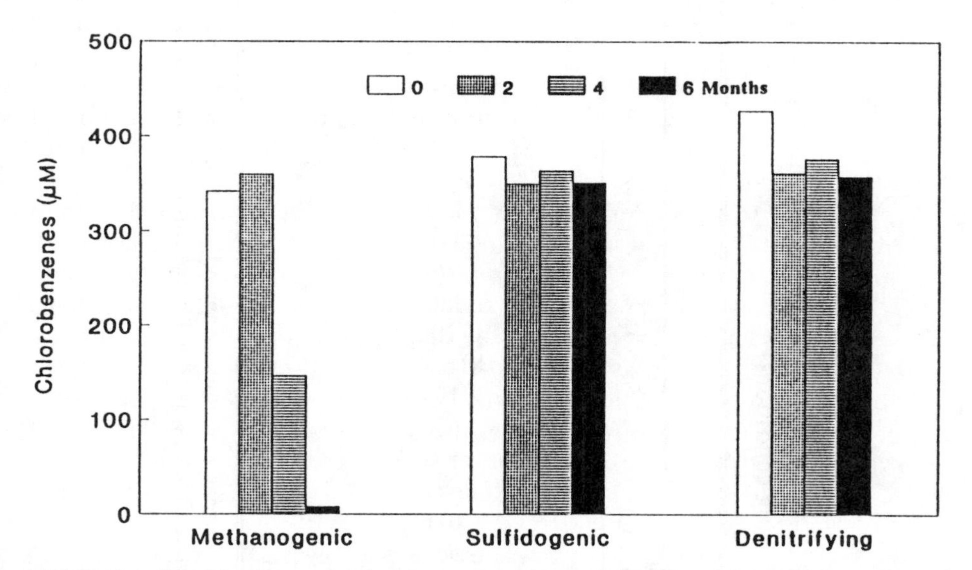

FIGURE 4. Metabolism of chlorobenzenes (1,2,3,4-TTCB + 1,2,4-TCB + 1,2-DCB) in soil slurry microcosms under methanogenic, sulfate-reducing, and nitrate-reducing conditions.

TABLE 1. Anaerobic degradation of chlorobenzenes in soil after the addition of lactate to the slurries held under sulfate-reducing conditions.

Time (Months)	Chlorobenzenes recovered in μM						
	1,2,3,4-TTCB	1,2,4-TCB	1,2-DCB	1,2,3-TCB	1,3-DCB	1,4-DCB	MCB
	(Initially [a] sulfate-reducing conditions)						
4[b]	86.2	125.1	152.4	0	0	0	0
5	57.9	140.5	132.6	0	4.6	5.8	0
6	3.3	219.8	155.8	0	4.9	6.7	0
8	0	113.0	112.9	0	6.7	10.9	0
9	0	27.0	44.9	0	32.0	25.2	160.7
9.4	0	23.5	33.8	0	35.2	23.5	205.5

(a) The slurries were initially amended with 115.8 μM of 1,2,3,4-TTCB; 137.8 μM of 1,2,4-TCB; and 170.1 μM of 1,2-DCB.
(b) Lactate (5 mM) was added to the slurries at the end of 4 months.

TABLE 2. Anaerobic degradation of chlorobenzenes in soil after the addition of lactate to the slurries held under nitrate-reducing conditions.

Time (Months)	Chlorobenzenes recovered in µM						
	1,2,3,4-TTCB	1,2,4-TCB	1,2-DCB	1,2,3-TCB	1,3-DCB	1,4-DCB	MCB
	(Initially [a] sulfate-reducing conditions)						
4[b]	92.2	133.9	150.3	0	0	0	0
6	91.2	124.6	142.2	0	0	0	0
9	43.5	44.1	191.4	39.7	11.6	90.6	20.4
9.4	15.6	14.4	35.0	18.9	15.8	55.9	339.6

(a) The slurries were initially amended with 115.8 µM of 1,2,3,4-TTCB; 137.8 µM of 1,2,4-TCB; and 170.1 µM of 1,2-DCB.
(b) Lactate (5 mM) was added to the slurries at the end of 4 months.

chlorobenzene concentration declined with time due to sulfate removal upon lactate addition. Based on the breakdown of substrates and pattern of metabolites produced (Table 1), a pathway similar to that observed in methanogenic incubations was noted.

Nitrate-Reducing Conditions. The anaerobic metabolism of chlorobenzenes tested was similarly inhibited in the presence of nitrate (Figure 4). Slurries containing nitrate showed no loss of parent substrates (1,2,3,4-TTCB, 1,2,4-TCB, and 1,2-DCB) even after prolonged incubation. However, the slurries that received lactate showed active dechlorination after 4 months of lactate addition. The substrate metabolic profiles and intermediates detected (Table 2) suggested a pathway similar to those observed in methanogenic incubations.

CONCLUSIONS

Anaerobic metabolism of chlorinated benzenes in soil slurries revealed that their biodegradation was favored under methanogenic conditions but not in the presence of sulfate or nitrate. Under the former incubation, 1,2,3,4-TTCB, 1,2,4-TCB, and 1,2-DCB declined to low levels with 1,2,3,4-TTCB degrading faster than 1,2,4-TCB, and 1,2-DCB. Metabolism of these substrates proceeded via reductive dechlorination and produced 1,2,3-TCB, 1,4-DCB, and 1,3-DCB as transient metabolites prior to transformation to stoichiometric amounts of MCB. The presence of sulfate and nitrate inhibited dehalogenation of parent substrates. However, restoring methanogenic conditions in these microbial systems by stimulating the removal of sulfate or nitrate with suitable electron donors (lactate) resulted in the breakdown of parent chlorobenzenes. This strongly indicated that dehalogenating microorganisms coexist with sulfate- or nitrate-reducers, and that their dehalogenating activity can be fully expressed once the environmental conditions are manipulated to remove these electron acceptors from the system. These findings help explain the metabolic fate of selected chlorobenzenes in anaerobic environments under transitional redox conditions and suggest that the manipulation of site conditions can significantly enhance the degradation of halogenated compounds.

REFERENCES

Bosma, T. N. P., J. R. Van der Meer, G. Schraa, M. E. Tros, and A. J. B. Zehnder. 1988. "Reductive Dechlorination of all Trichloro- and Dichlorobenzene Isomers." *FEMS Microbiol. Ecol. 53*: 223-229.

DeWeerd, K. A., L. Mandelco, R. S. Tanner, C. R. Woese, and J. M. Suflita. 1990. "*Desulfomonile tiedjei* Gen. Nov. and Sp. Nov., a Novel Dehalogenating, Sulfate-Reducing Bacterium." *Arch. Microbiol. 154*: 23-30.

Fathepure, B. Z., J. M. Tiedje, and S. A. Boyd. 1988. "Reductive Dechlorination of Hexachlorobenzene to Tri-and Dichlorobenzenes in Anaerobic Sewage Sludge." *Appl. Environ. Microbiol. 54*: 327-330.

Fathepure, B. Z., and T. M. Vogel. 1991. "Complete Degradation of Polychlorinated Hydrocarbons by a Two-Stage Biofilm Reactor." *Appl. Environ. Microbiol. 57*: 3418-3422.

Genthner, B. R. S, W. A. Price III, and P. H. Pritchard. 1989. "Characterization of Anaerobic Dechlorinating Consortia Derived from Aquatic Sediments." *Appl. Environ. Microbiol. 55*: 1472-1476.

Gibson, S. A., and J. M. Suflita. 1990. "Anaerobic Biodegradation of 2,4,5-Trichlorophenoxyacetic Acid in Samples from a Methanogenic Aquifer: Stimulation by Short-Chain Organic Acids and Alcohols." *Appl. Environ. Microbiol. 56*: 1825-1832.

Haggblom, M. M., and L. Y. Young. 1990. "Chlorophenol Degradation Coupled to Sulfate Reduction." *Appl. Environ. Microbiol. 56*: 3255-3260.

Holliger, C., G. Schraa, A. J. M. Stams, and A. J. B. Zehnder. 1992. "Enrichment and Properties of an Anaerobic Mixed Culture Reductively Dechlorinating 1,2,3-Trichlorobenzene to 1,3-Dichlorobenzene." *Appl. Environ. Microbiol. 58*: 1636-1644.

Kohring, G. -W., X. Zhang, and J. Wiegel. 1989. "Anaerobic Dechlorination of 2,4-Dichlorophenol in Freshwater Sediments in the Presence of Sulfate." *Appl. Environ. Microbiol. 55*: 2735-2737.

Kuhn, E. P., and J. M. Suflita. 1989. "Dehalogenation of Pesticides by Anaerobic Microorganisms in Soils and Groundwater—A Review." In B. L. Sawhney and K. Brown (Eds.), *Reactions and Movement of Organic Chemicals in Soils.* Soil Science Society of America Special Publication no. 22, pp. 111-180. Soil Science Society of America and American Society of Agronomy, Madison, WI.

Mohn, W. W., and J. M. Tiedje. 1992. "Microbial Reductive Dehalogenation." *Microbiol. Rev. 56*: 482-507.

Postgate, J. R. *The Sulfate-Reducing Bacteria,* 2nd ed., 1984, Cambridge University Press, Cambridge, England.

Ramanand, K., and J. M. Suflita. 1991. "Anaerobic Degradation of m-Cresol in Anoxic Aquifer Slurries: Carboxylation Reactions in a Sulfate-Reducing Bacterial Enrichment." *Appl. Environ. Microbiol. 57*: 1689-1695.

Schennen, U., K. Braun, and K. Knackmuss. 1985. "Anaerobic Degradation of 2-Fluorobenzoate by Benzoate-Degrading Denitrifying Bacteria." *Appl. Environ. Microbiol. 161*: 321-325.

Shelton, D. R., and J. M. Tiedje. 1984. "Isolation and Partial Characterization of Bacteria in an Anaerobic Consortium that Mineralizes 3-Chlorobenzoic Acid." *Appl. Environ. Microbiol. 48*: 840-848.

Suflita, J. M., A. Horowitz, D. R. Shelton, and J. M. Tiedje. 1982. "Dehalogenation: a Novel Pathway for the Anaerobic Biodegradation of Haloaromatic Compounds." *Science 218*: 1115-1116.

Townsend, G. T., K. Ramanand, and J. M. Suflita. "Reductive Dehalogenation of 3-Chlorobenzoate in the Presence of Sulfate." Presented at American Society for Microbiology Conference on Anaerobic Dehalogenation and its Environmental Implications, 1992, Aug. 31-Sept. 4, Atlanta, GA.

AN APPROACH TO THE REGULATION OF BIOREMEDIATION OF POLYCHLORINATED BIPHENYLS

J. Blake

INTRODUCTION

The U.S. Environmental Protection Agency's (EPA's) disposal approval program for polychlorinated biphenyls (PCBs) began with the passage of the Toxic Substances Control Act (TSCA) in 1976. TSCA Section 6(e) required that PCBs be disposed of using methods prescribed by EPA. Due to site-specific considerations concerning the disposal of PCBs, EPA chose to issue approvals for most disposal activities. The PCB disposal regulations subsequently written are found in 40 CFR Part 761.

Since 1983, PCB disposal approvals (i.e., permits), have been issued by each of the 10 EPA Regions and by EPA Headquarters in Washington, D.C. Regional approvals are valid in the Region in which they are issued and Headquarters' approvals are valid anywhere in the United States. However, whether TSCA approvals are issued by EPA Headquarters or by an EPA Region, the EPA Regions enforce compliance with the conditions of a company's approval.

Under the PCB regulations, incineration is the standard for PCB destruction, and the performance standards for PCB incinerators are provided in the regulations. Provision is also made in the regulations for approval by EPA of technologies other than incineration, or alternative to incineration, for destruction of PCBs if "equivalency to incineration" can be demonstrated.

The flexibility of TSCA makes possible a systematic, organized approval program for alternative methods of destruction of PCBs such as bioremediation. TSCA gives EPA the responsibility to develop guidance outside of the regulations for approval of alternative methods of PCB destruction. As a matter of policy, the EPA approval program for all technologies, including bioremediation, is based on performance. This means that the results of a successful demonstration to EPA by a company of their process determines the conditions under which the company will be allowed to operate.

REQUIREMENTS FOR COMMERCIAL APPROVAL

Under TSCA, EPA issues two types of PCB disposal approvals, the commercial operating permit and the Research and Development (R&D) permit. When evaluating the performance of bioremediation processes for possible commercial use, writers of TSCA approvals must draw upon the experience of approving nonbiological or mechanically engineered disposal processes.

To approve a bioremediation process for commercial operation, EPA must make the finding that the process is equivalent to incineration in its ability to destroy PCBs, that it produces no toxic by-products or toxic emissions, and that any microorganisms used as inoculum pose no unreasonable risk to human health or the environment. Equivalency to incineration for a bioremediation process is currently defined as destruction of PCBs to a soil residue of less than 2 parts per million (ppm) per congener peak quantitated with an individual congener standard, such as the DCMA (Dry Color Manufacturer's Association) standard, as opposed to a total Aroclor™ standard.

At the present time, incineration is the only PCB disposal process approved to remove PCBs from soils and sediments. And, because of the potential cost saving, there is great incentive to develop biological processes for commercial use instead of incineration or chemical waste landfilling for disposal of PCB soils.

To illustrate, consider the following example. Although EPA does not know of every site in the United States where PCBs have been spilled or disposed, EPA does have estimates of the number of sites contaminated with PCBs. In 1991, the EPA Office of Emergency and Remedial Response (Superfund) determined that PCBs were the "predominant waste type" at approximately 20% of the sites on the National Priorities List (NPL) alone. The NPL sites contain approximately 34,070,000 yd^3 (26,000,000 m^3) of material contaminated with PCBs and other substances. Some idea of the cost of cleanup is possible when incineration costs of around \$2,000/yd^3 (0.765 m^3) are considered.

REQUIREMENTS FOR EVALUATION OF PERFORMANCE

R&D Permit. Under TSCA, approvals are issued for R&D as well as for commercial operation. Because the EPA approval program is performance based, if a technology is very new and never has been used to destroy hazardous wastes, a company may be asked to apply for an R&D permit to generate the data necessary for a proper evaluation of their PCB disposal process by EPA before a demonstration for commercial operations is approved.

Under the current regulations, any research on the performance of a biological process for PCB destruction must be carried out under the terms and conditions of an R&D approval. EPA considers the R&D bioremediation approval to be a well-designed scientific experiment which can produce much needed data on the biodegradation of PCBs, prior to consideration of bioremediation for universal application under a commercial approval.

In the future, such experimental and analytical detail probably will not be required as the technology becomes better understood and perfected. However, thorough evaluation of the technology is necessary because approving a company's bioremediation process too early risks bringing public disillusionment and discredit to EPA's review and approval process.

Demonstration of Commercial Operation. During the demonstration to EPA of the performance of a mechanically engineered or nonbiological process, Agency engineers require that a company successfully destroy PCBs in three successive runs, or disposal treatments, to obtain an approval to operate commercially. The content of the approval subsequently written reflects the operating conditions and results of the successful demonstration. Because bioremediation requires a very long period of time to be effective, the exact requirements for a bioremediation process that would be equivalent to the three treatment runs have not yet been decided. It is certain, though, that the method chosen to evaluate bioremediation processes should be designed to show that the process will be safe and effective in any situation at any site if implemented on a commercial scale. One proposal is that 3 years of successful R&D should be required before a commercial operating permit is issued for a bioremediation PCB disposal process.

In addition, bioremediation may be successful in the environment at some PCB-contaminated sites and not at others. For this reason, the Commercial Operating Approval of a bioremediation company will always have a treatability study component. There will have to be a central laboratory in which soil and sediment samples from a site being considered for cleanup with a biodegradation process will be subjected to a specific testing protocol. Then, once screening tests in the laboratory demonstrate the potential for microbial metabolism of PCBs under controlled conditions, a biological process is ready for testing in field environments. However, from our experiences, extrapolation of laboratory data to the field often is not possible.

Validation of Biological Degradation. Many biodegradation processes that are successful in the laboratory either do not work well or do not work at all in the field for a variety of reasons. And, even if a bioremediation process appears to destroy PCBs in the field, data must show microbial involvement in the degradation process. To this end, EPA requires that a company devise a successful strategy to demonstrate as unequivocally as possible that biodegradation has taken place and that the PCB molecule has not simply volatilized, sorbed, transported, or attenuated by some other nonbiological or "abiotic" reaction.

Congener-Specific Analytical Method. As part of such a validation strategy, EPA requires that a company use a congener-specific method for PCB analysis. Analytical methods used for commercial formulations or mixtures of PCBs, for example Aroclors™, just are not suitable and their use can mask congener-specific or chemical-specific degradation.

PCB analysis is complicated because there are 209 possible congeners or different chemicals that are called polychlorinated biphenyls. They differ in the number and position of chlorines on the basic biphenyl molecule. In the microbial degradation process, each of these congeners differs in its susceptibility to breakdown.

Corroboration from Several Sources. In addition to a suitable analytical method, EPA requires corroborating data from several different sources to demonstrate biodegradation at a PCB-contaminated site. Such sources might include rough mass balances assembled from site data showing that biological losses of PCBs exceeded expected abiotic or nonbiological losses, the presence of characteristic PCB metabolites, decreasing ratios of biodegradable-to-nonbiodegradable PCB congeners, and the presence of metabolically adapted microbes in laboratory assays with environmental samples collected from a contaminated site. However, a company can propose any strategy validating biodegradation, and EPA will evaluate it.

Once several years of rigorous R&D have shown that a company's process effectively biodegrades PCBs in soils and sediments in spatially separated, heterogeneous field sites, the company can apply for approval from TSCA Headquarters to operate its process commercially anywhere in the United States.

BIOREMEDIATION R&D PERMITS ISSUED BY EPA HEADQUARTERS

As of this writing, nine R&D approvals for the study of the bioremediation of PCBs in soils and sediments have been issued under TSCA by EPA Headquarters in Washington, D.C. Most of these approvals are for large-scale, in situ processes. These R&D approvals are written by EPA Headquarters for 1 year to provide for the study of microbial degradation. A 1-year approval includes 1 growing season and producing a report.

Research has shown that PCBs do biodegrade in the environment, but at a very slow rate. However, no one yet has demonstrated a process to EPA that can accelerate PCB biodegradation to rates necessary to make such a process commercially viable.

Approvals for R&D on bioremediation have been issued by EPA Headquarters to General Electric Company, Texas Eastern Gas Pipeline Company, IT Corporation, Mycotech Corporation, Envirogen, Inc., Coastal Oil and Gas Corporation, PacifiCorp, and Safetec.

DEGRADATION AND MINERALIZATION OF POLYCHLORINATED BIPHENYLS BY WHITE-ROT FUNGI IN SOLID-PHASE AND SOIL INCUBATION EXPERIMENTS

A. Zeddel, A. Majcherczyk, and A. Hüttermann

INTRODUCTION

Polychlorinated biphenyls (PCBs) are recalcitrant anthropogenic substances that generally are not degradable by bacteria, with the exception of some specialized species that degrade mainly mono-, di-, and some trichlorinated PCBs under aerobic conditions (Abramowicz 1990). Addition of biphenyls as a source of carbon and frequent reinoculation are necessary to support otherwise unstable bacterial decontamination activity (Bedard 1990). Anaerobic PCB dechlorination by bacteria results in the environmental accumulation of tri- and tetrachlorinated biphenyls. The application of white-rot fungi to the degradation of xenobiotics such as DDT or Seveso dioxin has been demonstrated in liquid culture (Bumpus 1985) as well as to water decolorization (Van Ba et al. 1985); in soil remediation (Loske et al. 1990, Lamar & Dietrich 1990, Ryan & Bumpus 1989); and in the biofiltration of polluted air (Majcherczyk et al. 1990). The degradation of PCBs in liquid cultures has been demonstrated with the white-rot fungus *Phanerochaete chrysosporium* (Eaton 1985).

In this paper we report on the use of white-rot fungi for degradation of PCBs in solid phases, including soil under aerobic conditions. It could be shown that mono- to pentachlorinated PCBs are degradable when mixed with wood chips by incubation with *Pleurotus ostreatus* or *Trametes versicolor*. Addition of soil does not affect this degradation. Mixing the PCB extract with soil and incubating under the same conditions as before results in a decrease of the degradation rate. PCB congeners that remain relative to the mean reduction of the other congeners are for the most part 4,4′ substituted.

METHODS

PCBs were extracted from a sandy loam soil contaminated by transformer oil (Clophen A 30). The extract was evaporated on spruce wood chips and mixed until a homogeneous distribution had been achieved. The sum of all PCB isomers was 2,500 mg/kg of dry wood on wood chips and 450 mg/kg in soil. *Pleurotus ostreatus* ssp. *florida* Kummer, *Trametes versicolor* Lloyd, and *Phanerochaete chrysosporium* Burd (ATCC 24725) were obtained from the Institute of Forest Botanics, University of Göttingen, Germany. The stock cultures of all fungi were cultivated on 3% malt agar. The starting inoculum for degradation studies was prepared from millet cultures. Fresh potato pulp consisting of 20% dry weight solids was used without sterilization. Wood chips were mixed with pulp and water (1:4:1w/w). Soil homogenization was performed by adding pulp and liquid (water; Tween 80, Merck, Darmstadt, Germany; P3 Ferrolin 8643 and SP8648, Henkel, Düsseldorf, Germany) to contaminated sandy soil, mixing for 20 min, and solidifying with wood chips (chips:pulp:water:soil 2:2:5:1w/w). Experiments were performed in sealed 700-mL glass tubes (7 cm diameter) with Teflon™ caps. The seals were pierced allowing the tubes to be ventilated continuously with water-saturated air from the bottom to the top at a flowrate of 600 mL/h. The bottom was filled with porous clay media to maintain the relative humidity by way of a water sink. Volatile substances were controlled via an isooctane trap, but none were detected. Concentrations of O_2 and CO_2 in the culture atmospheres were measured in each experiment by gas chromatographic (GC) analyses of air samples with combined Porapack and molecular sieve columns and compared with standard mixed gases (Suppelco, Bad Homburg v.d.H., Germany). The oxygen concentration was maintained at about 15% during the entire cultivation period.

After incubation, the contents of the flasks and tubes were ground and an aliquot was extracted with acetone. PCBs were extracted with *n*-hexane from an aliquot of acetone extract after addition of isooctane and water (1:4:1) and analyzed by GC-mass spectrometric detection (MSD) (Hewlett Packard, Palo Alto, California). The PCBs were separated on an HP Ultra 2 Column (15 m × 0.2 mm, film thickness 33 μm). Substances including 2-chlorobiphenyl (CB), 2,3-diCB, 2,4,5- and 2,4,4′-triCB, 2,2′,5,5′- and 3,3′,4,4′-tetraCB, 2,2′,4,5,5′-pentaCB, 2,2′,3,4,4′,5-, 2,2′,4,4′,5,5′-, and 2,2′,3,4,4′,5,5′-hexaCB were identified by means of external standards. Other substances were identified by comparing their relative retention time with literature data. Losses during the analysis were corrected using the nondegradable PCB isomer 2,2′,4,4′,5,5′-hexaCB as an internal standard.

RESULTS

Fungal growth was observed by visual inspection. After 2 weeks of incubation, the substrates were covered completely with fungi in the case of the *Trametes* and *Pleurotus* cultures. *Phanerochaete chrysosporium* did not develop a thick mycelium but was found evenly distributed throughout all the wood chips. Contaminants (*Trichoderma* spec.) were found only in some flasks with *Phanerochaete chrysosporium* cultures and in control flasks after 7 weeks of incubation.

Optimal degradation was achieved by incubation of white-rot fungi directly on contaminated wood chips. Comparisons of the original chromatograms demonstrated the efficiency of, in this case, *Pleurotus ostreatus* ssp. *florida* to decontaminate a wide range of differently chlorinated compounds after incubation for 5 weeks (Figure 1). Table 1 presents an overview of the degradation potential

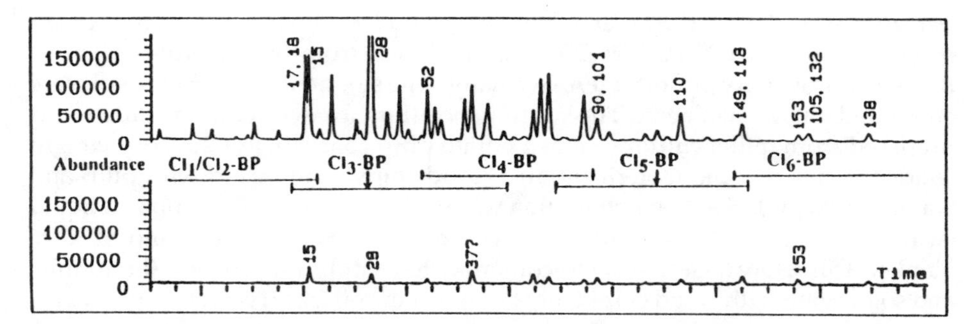

FIGURE 1. Comparison of GC-chromatograms of control (upper) and degradation result (lower) after treatment with fungi for 7 weeks. PCBs on wood chips mixed with potato pulp and soil. The numbers above the lower chromatogram follow an isomer-categorizing scheme from Ballschmiter and Zell (1980).

for PCB-contaminated wood chips in this experimental setup. Degradation of PCBs up to trichlorobiphenyls was nearly equal for all isomers. Mono- and dichlorobiphenyls were completely degraded; the only exception was 4,4′-dichlorobiphenyl, which was reduced only to 40% by *Trametes versicolor* and to 20% by *Pleurotus ostreatus* species. Tetrachlorobiphenyls differ from this pattern; their degradation was more dependent on the chlorine substitution pattern. *Phanerochaete chrysosporium* was not able to degrade PCBs higher than two chlorines per biphenyl.

A second experimental design combined a matrix of contaminated wood chips with garden soil. The results of this series corresponded with the degradation of the pure compounds described above. No disturbance by the soil matrix was detected.

TABLE 1. Degradation potential of white-rot fungi in relation to degree of PCB chlorination.

	T. versicolor	*P. ostreatus*	*P. chrysosporium*
Monochlorobiphenyl	over 97%	over 99%	40% to 90%
Dichlorobiphenyl	over 96%	over 98%	up to 80%
Trichlorobiphenyl	over 80%	85% to over 95%	up to 30%
Tetrachlorobiphenyl	20% to over 90%	20% to over 90%	no degradation
Pentachlorobiphenyl	up to 40%	up to 60%	no degradation
Hexachlorobiphenyl	up to 50%	up to 35%	no degradation
Heptachlorobiphenyl	no degradation	no degradation	no degradation

In contrast, direct coating of the same PCB-isomer mixture as above to the soil prior to mixing with wood chips greatly reduced the rate of their degradation. After 7 weeks, only 50 to 60% of the degradation obtained on wood chips was observed in this system, in spite of the fact that, except for the soil, the overall compositions of the two solid state systems were identical. The continuation of degradative activity following the first 4 weeks of rapid growth was observed in two incubation periods (Figure 2). Homogenization with surfactants was not useful for obtaining better bioavailability of PCBs in this case.

In all three experimental setups and with each fungal culture, individual PCB isomers were degraded in similar amounts in the 5-week incubation period. Degradation was slower in some cases, with relative enrichment of more recalcitrant isomers. After 5 weeks of incubation with *Pleurotus ostreatus*, the following compounds were only partially degraded (as seen in the lower chromatogram of Figure 1 with isomer-categorizing numbers from Ballschmiter and Zell [1980]):

4,4′-dichlorobiphenyl (15)	2,4,4′-trichlorobiphenyl (28)
2,2′,5,5′-tetrachlorobiphenyl (52)	3,4,4′-trichlorobiphenyl (37)
2,3′,4,4′-tetrachlorobiphenyl (66)	2,3,4,4′-tetrachlorobiphenyl (60)
2,2′,4,5,5′-pentachlorobiphenyl (101)	2,3′,4,4′,5-pentachlorobiphenyl (118)

Only 2,2′,4,4′,5,5′-hexachlorobiphenyl (153) was not degraded at all in either the controls or the fungus cultures.

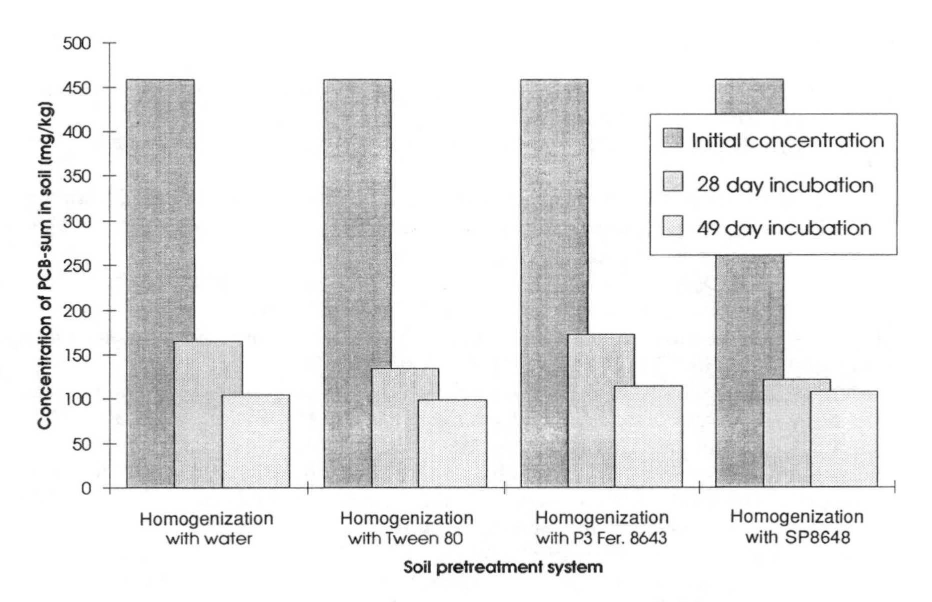

FIGURE 2. Degradation of PCB-contaminated soil by *Pleurotus ostreatus* ssp. *florida* after a 28- and 49-day incubation period. The soil was homogenized by adding pulp and liquid (water or 2% surfactant: Tween 80, P3 Ferrolin 8643, and SP8648) to the contaminated sandy soil. This mixture was stirred for 20 min and solidified with wood chips.

CONCLUSION

Incubation with white-rot fungi was found to be applicable for the degradation of PCBs. In contrast to bacteria, the fungi were observed to degrade a wide range of isomers of the environmentally important tri- to pentachlorinated biphenyls. Optimal reduction of PCB congeners was achieved by incubating the fungus with contaminated wood chips. This was an effective method for removing PCB-contaminated extracts by mixing them with wood. Solid wastes containing PCBs either are degradable in the original phase or can be extracted. The latter can be useful in case of low bioavailability of PCBs, e.g., in loamy or clay soils. PCBs in sandy soils are degradable without extraction but show a decreased loss of PCBs in the same incubation time. For degradation of higher chlorinated PCBs, it may be possible to perform an anaerobic pretreatment system for dechlorination following the aerobic biological combustion by white-rot fungi.

No formation of highly toxic chlorinated dioxins or furans was detected via measurement by GC-HRMS in the experiments described above, and ongoing experiments show mineralization of PCBs. Up to 10% of initial ^{14}C-activity was mineralized to ^{14}C-CO_2 during incubation of ^{14}C-PCB (mainly tetrachlorinated PCBs and total degradation of 50%) with *Pleurotus ostreatus*, and 33% of initial ^{36}Cl-PCB activity was found in chlorine ions after degradation of 58% of PCBs by *Pleurotus ostreatus* (data will be shown elsewhere).

REFERENCES

Abramowicz, D. A. 1990. "Aerobic and Anaerobic Biodegradation of PCBs — A Review." *Critical Reviews in Biotechnology 10*(3): 241-249.

Ballschmiter, K., and M. Zell. 1980. "Analysis of PCBs by GC." *Fresenius Z. Anal. Chem. 302*: 20-31.

Bedard, D. L. 1990. "Bacterial Transformation of Polychlorinated Biphenyls." In D. Kamely, A. Chakrabarty, and G. S. Omen (Eds.), *Advances in Applied Biotechnology Series*, Vol. 4. Portfolio Publ. Co., The Woodlands, TX.

Bumpus, J. A., M. Tien, D. Wright, and S. D. Aust. 1985. "Oxidation of Persistent Environmental Pollutants by a White-Rot Fungus." *Science 228/4706*: 1434-1436.

Eaton, D. C. 1985. "Mineralisation of Polychlorinated Biphenyls by *Phanerochaete chrysosporium*: A Lignolytic Fungus." *Enzyme Microb. Technol. 7/5*: 194-196.

Lamar, R. T., and D. M. Dietrich. 1990. "In Situ Depletion of Pentachlorophenol from Contaminated Soil by *Phanerochaete* ssp." *Applied and Environmental Microbiology 56/10*: 3093-3100.

Loske, D., A. Hüttermann, and A. Majcherczyk. 1990. "Use of White-Rot Fungi for the Clean-up of Contaminated Sites." In M. P. Coughlan and A. Collaco (Eds.), *Advances of Biological Treatment of Lignocellulosic Materials*, pp. 311-322. Elsevier Applied Science, London, UK.

Majcherczyk, A., A. Braun-Lüllemann, and A. Hüttermann. 1990. "Biofiltration of Polluted Air by a Complex Filter Based on White-Rot Fungi Growing on Lignocellulosic Substrate." In M. P. Coughlan and A. Collaco (Eds.), *Advances in Biological Treatment of Lignocellulosic Materials*, pp. 323-329. Elsevier Applied Science, London, UK.

Ryan, T. P., and J. A. Bumpus. 1989. "Biodegradation of 2,4,5-Trichlorphenoxyacetic Acid in Liquid Culture and in Soil by *Phanerochaete chrysosporium*." *Appl. Microb. Biotechnology 31*: 302-307.

Van Ba, H., H. Chang, T. W. Joyce, and T. K. Kirk. 1985. "Dechlorination of Chlor-Organics by a White-Rot Fungus." *Tappi Journal 68*: 98-102.

DEGRADATION OF ENVIRONMENTAL POLLUTANTS BY WHITE-ROT FUNGI

S. D. Aust, M. M. Shah, D. P. Barr, and N. Chung

INTRODUCTION

The technologies available for the removal of hazardous organopollutants can be divided into four groups:

1. Technology based on separation of contaminants;
2. Technology based on stabilization of contaminants;
3. Technology based on degradation of contaminants by chemical processes; and
4. Technology based on degradation of contaminants by biological processes.

The technologies belonging to the first two groups do not provide permanent solutions for the treatment of hazardous wastes. In general, technologies related to the third and fourth groups are valuable as they are targeted toward eliminating the chemicals. Between the chemical and biological technologies, degradation of contaminants via chemical treatment processes generally is more expensive than biological treatment processes. However, at present, application of biological treatment processes is limited only to the degradation of easily biodegradable chemicals.

Although chemicals such as DDT, lindane, dieldrin, PCB, benzo(a)pyrene, etc. generally are considered to be recalcitrant, the white-rot fungus *Phanerochaete chrysosporium* has been shown to mineralize these chemicals to CO_2 (Bumpus et al. 1985). The ability of this fungus to degrade these chemicals is related to its lignin-degrading enzyme system (Aust 1990, Tien 1987). Some of the most important components of its lignin-degrading enzyme systems are lignin peroxidases, H_2O_2, oxalate, and veratryl alcohol (Barr et al. 1992, Shah et al. 1992, Tien 1987).

This paper proposes a mechanism for the degradation of halogenated organics by the fungus. It is shown that the degradation of these chemicals would involve both oxidative and reductive reactions. For these reactions, lignin peroxidases are activated by H_2O_2; the activated enzyme oxidizes veratryl alcohol to its cation radical; this radical in turn oxidizes organic acid (EDTA/oxalate) to form organic acid anion radical. The organic acid anion radical is a powerful reductant ($E°_{½} \cong -1.9$ V). Reductions catalyzed by the organic acid radicals result in the production of CO_2. Using CCl_4 as a representative of halogenated organics, we showed that CCl_4 is reduced to the trichloromethyl radical by this sequence of reactions. The organic acid anion radical also can reduce molecular oxygen to superoxide, which can dehalogenate some chemicals. In the presence of iron, the hydroxyl radical can be generated by Haber-Weiss chemistry. The hydroxyl radical is a powerful oxidant that may be involved in the degradation of very recalcitrant chemicals.

RESULTS

Ethylenediaminetetraacetic acid (EDTA) is an effective inhibitor of the veratryl alcohol oxidase activity of lignin peroxidase H2 (LiPH2). However, EDTA was not found to inhibit H_2O_2 consumption by the enzyme. Electron spin resonance (ESR) spin-trapping techniques were used to show that EDTA was being oxidized to a radical in this system. Veratryl alcohol and H_2O_2 were required, along with the enzyme, to give the EDTA radical. Because the EDTA radical is known to be an excellent reductant, assays were conducted to show that various electron acceptors could be reduced in this system (Table 1). EDTA could be replaced by oxalate, which is produced by *P. chrysosporium* (Barr et al. 1992). Molecular oxygen also can be reduced by the organic acid radical to produce superoxide (Figure 1A). Upon addition of ferric iron, hydroxyl radicals were detected (Figure 1B). When CCl_4 was added to the reaction mixture, it was reduced to the trichloromethyl radical (Figure 2).

SUMMARY

Figure 3 show the proposed scheme for possible mechanisms for halogenated organic degradation by the lignin-degrading enzyme system of the fungus. Many highly halogenated pollutants require reduction before they can be oxidized.

TABLE 1. Reduction of various electron acceptor molecules by LiPH2 in the presence of EDTA and veratryl alcohol.

Electron Acceptor	Initial Rate of Reduction[a]
	nmol/mL/min
Cytochrome c	4.8 ± 0.2
NBT	2.8 ±0.1
Ferric iron	5.8 ± 0.1

[a] Reaction mixture contained 0.1 M sodium acetate buffer, pH 5.5, 0.5 µM LiPH2, 2 mM veratryl alcohol, 2 mM EDTA, and 100 µM H_2O_2. The concentration of cytochrome c, nitroblue tetrazolium (NBT), and ferric iron was 50 µM. For the ferric reduction assay, 1 mM 1,10-phenanthroline was used.

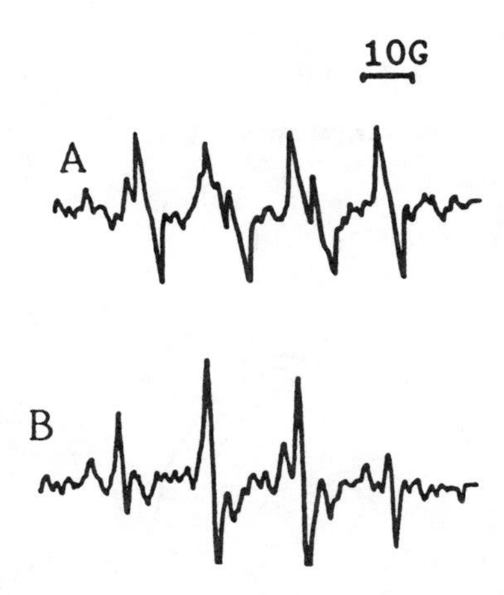

FIGURE 1. (A) ESR spectrum of DMPO-superoxide radical adduct. The reaction mixture contained 15 µM lignin peroxidase H2, 500 µM H_2O_2, 6 mM veratryl alcohol, 3 mM EDTA, 50 mM phosphate buffer (pH 6.5), and 60 mM DMPO (5,5-dimethyl-1-pyrroline N-oxide). No DMPO-superoxide radical adduct was observed on removal of enzyme or EDTA. (B) ESR spectrum of the DMPO-OH radical adduct formed by lignin peroxidase H2. The reaction mixture contained 100 mM sodium acetate (pH 5.0), 0.63 µM lignin peroxidase H2, 500 µM H_2O_2, 500 µM veratryl alcohol, 500 µM EDTA, 100 µM $FeCl_3$, and 40 mM DMPO. (Barr et al. 1992. Reproduced with permission of Academic Press.)

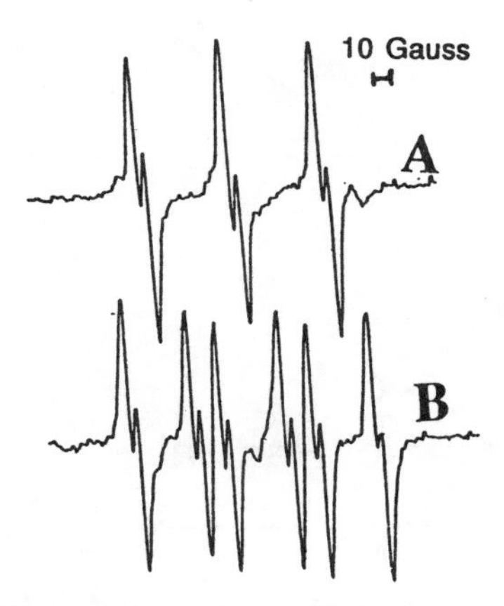

FIGURE 2. ESR spectra of α-phenyl-N-*tert*-butylnitrone (PBN)-CCl_3 radical adducts. (A) The reaction mixture contained 0.1 M phosphate buffer, 25 µM lignin peroxidase H2, 500 µM H_2O_2, 1 mM veratryl alcohol, 4 mM EDTA, 1% CCl_4, and 90 mM PBN. (B) The conditions were the same as described in (A), except that ^{13}C-labeled CCl_4 was substituted for CCl_4.

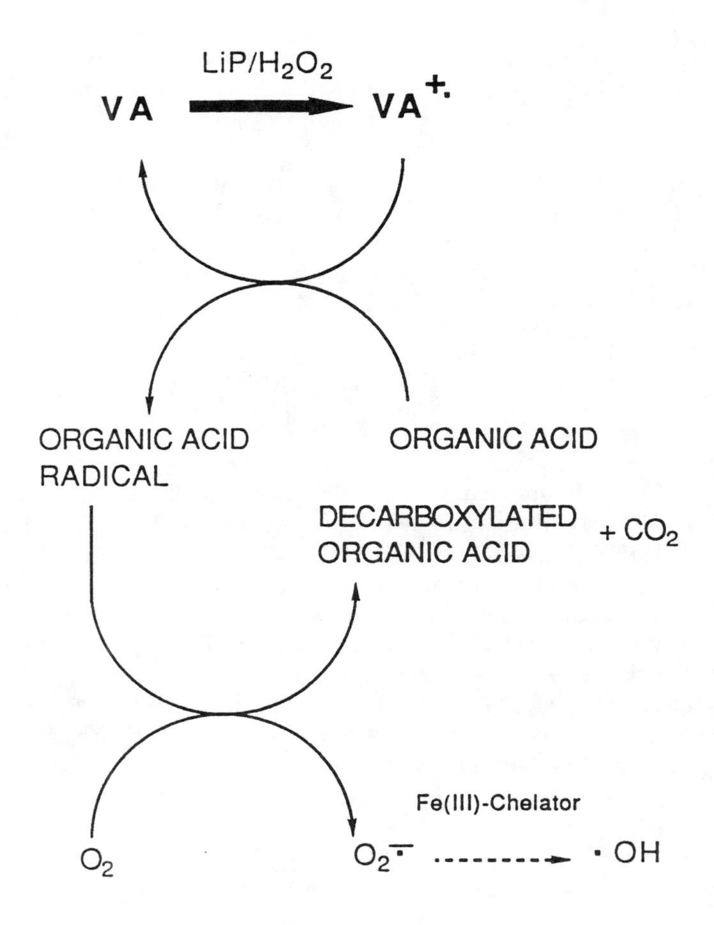

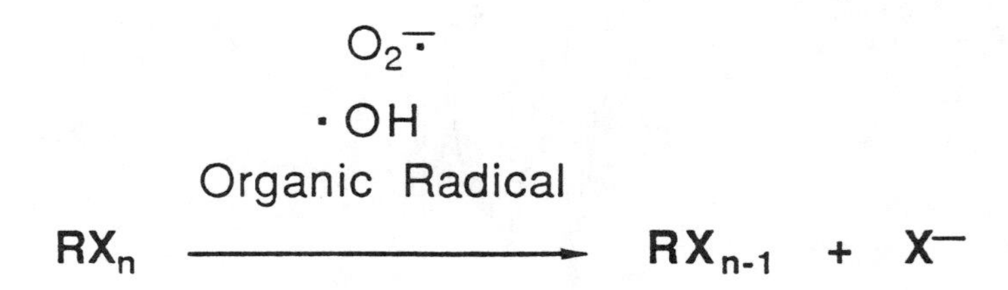

FIGURE 3. The proposed mechanism for the degradation of halogenated pollutants by lignin peroxidases. Organic acid is oxalate. RX_n, RX_{n-1}, and RX are halogenated organics with various degrees of halogenations. X^- represents halide. According to the proposed scheme, veratryl alcohol serves as the free radical mediator and organic acid is the reductant. The organic acid radicals, superoxide, and hydroxyl radicals are proposed to be reacting species in the degradation of halogenated organics.

This can be accomplished by the peroxidases secreted by enzymes produced by *P. chrysosporium*. Even though peroxidases generally are thought to oxidize chemicals, the fungus uses additional compounds, which it produces to catalyze reductions. The first is veratryl alcohol, considered a free radical mediator because it is readily oxidized by LiPH2 but is also readily reduced back to the alcohol by organic acids, including oxalate, which also is produced by the fungus. The oxalate radical is a very effective reductant, partly because its oxidation gives CO_2 and thus renders the reaction irreversible. The organic acid radicals can catalyze reductive dechlorination, as shown for CCl_4, or they can reduce molecular oxygen to superoxide.

Alternatively, in the presence of iron, the hydroxyl radical (a powerful oxidant) can be produced. Halogenated organics such as PCB and polychlorophenols can be hydroxylated via the electrophilic addition of the hydroxyl radical. Such a reaction lowers the oxidation potential of these halogenated compounds making them more easy to degrade by oxidative reactions. These mechanisms may be responsible for part of the reason that white-rot fungi are able to oxidize a wide variety of halogenated environmental pollutants.

REFERENCES

Aust, S. D. 1990. "Degradation of Environmental Pollutants by *Phanerochaete chrysosporium*." *Microb. Ecol. 20*: 197-209.

Barr, D. P., M. M. Shah, T. A. Grover, and S. D. Aust. 1992. "Production of Hydroxyl Radicals by Lignin Peroxidase from *Phanerochaete chrysosporium*." *Arch. Biochem. Biophys. 298*, 480-485.

Bumpus, J. A., M. Tien, D. Wright, and S. D. Aust. 1985. "Oxidation of Persistent Environmental Pollutants by a White Rot Fungus." *Science 228*: 1434-1436.

Shah, M. M., T. A. Grover, D. P. Barr, and S. D. Aust. 1992. "On the Mechanism of Inhibition of Veratryl Alcohol Oxidase Activity of Lignin Peroxidase H2 by EDTA." *J. Biol. Chem. 267*: 21564-21569.

Tien, M. 1987. "Properties of Ligninase from *Phanerochaete chrysosporium* and Possible Applications." *Crit. Rev. Microbiol. 15*: 141-168.

THE ACTIVATED SOIL PROCESS: PRODUCTION AND USE OF A BACTERIAL CONSORTIUM FOR THE BIOREMEDIATION OF CONTAMINATED SOIL FROM WOOD-PRESERVING INDUSTRIES

M. P. Otte, J. Gagnon, Y. Comeau,
N. Matte, C. W. Greer, and R. Samson

INTRODUCTION

The bioremediation of wood treatment plant soils contaminated with pentachlorophenol (PCP) and creosote can be performed using an aerobic biopile. Although the technology is simple and cost effective, inhibition of the growth and activity of the biomass can occur due to high concentrations of contaminants, low bioavailability of substrates, or lack of some essential nutrient(s) or oxygen (Kearney & Kellogg 1985). The observed decline in the survival and activity of pure laboratory cultures introduced into soil is caused by extreme changes in environmental conditions. The activity of pure cultures is often contaminant-specific and is less applicable to the complex mixture of pollutants encountered in the wood-preserving industry. A better approach to control the environmental conditions during treatment is to use a soil/slurry bioreactor (Mueller et al. 1991). Treatment in bioreactors is not always possible, so to make use of the advantages of a soil/slurry bioreactor, the indigenous microorganisms that possess pollutant biodegradation potential can be cultured initially in a reactor and the inoculum can then be used to enhance biopile treatment.

This research evaluated the activated soil process, a biotreatment concept in which the contaminant-degrading microorganisms present in contaminated soil are cultured in a soil/slurry bioreactor. The soil biopile is then bioaugmented with this acclimated, resistant, and active biomass to enhance the efficiency of biodegradation. The survival and adaptability of the acclimated consortium and its ability to degrade PCP and some polycyclic aromatic hydrocarbons (PAHs) were tested at different PCP loading rates in a fed-batch soil/slurry bioreactor.

MATERIALS AND METHODS

The contaminated soil was obtained from two different wood-preserving sites. These silty sand soils have been contaminated for several years with PCP

(100 to 1,500 mg/kg), PAHs (15 to 4,700 mg/kg), and copper/chromium/arsenate (CCA, 45/15/190 mg/kg).

Microcosm Studies. Mineralization studies were conducted in 125-mL microcosms containing 20 mL of a soil slurry (5% w/v) in minimal salts medium (MSM) (Greer et al. 1990). Flasks were sealed with rubber septa, incubated at 23°C, and agitated at 200 rpm. Nonlabeled and ^{14}C-labeled contaminants (PCP, PAHs) were added to the slurry, and the rate of mineralization was followed by measuring the production of $^{14}CO_2$. Identification of bacteria in the consortium was performed by plating the biomass on a PCP-selective medium (PCP, 100 mg/L; nutrient broth, 0.2 g/L; oxoid agar 10 g/L; MSM) and using the Analytical Profile Index (API) method.

Bioreactor Study. A 15-L Bioengineering™ reactor was used for biomass production. Oxygen (98% saturation), pH (7.0), temperature (25°C), and agitation (400 rpm) were controlled. Gas emissions were trapped by a NaOH (0.1 N) trap followed by an activated carbon trap. The reactor contained a 5% (w/v) soil slurry (500 g of contaminated soil in 10.0 L of MSM) and was fed for 23 days at increasing PCP loading rates (0, 50, 100, 300, 500, 700, and 800 mg PCP/L·d) with flowrates ranging from 144 to 316 mL/d. The feed solution was PCP in water adjusted to pH 10 with NaOH. Slurry samples were taken from the bioreactor before each rate increase to determine PCP-degrading activity in microcosms spiked with 100 mg/L of [U-^{14}C]-PCP. Mineralization was then monitored for 10 days.

Soxhlet extraction was used to analyze the PCP and PAHs in the soil. PCP and Cl^- were determined by high-performance liquid chromatography (HPLC). PAHs were analyzed using gas chromatography/mass spectrometry (GC/MS). The $^{14}CO_2$ production was analyzed with a Packard Tri-Carb 4530 scintillation counter.

RESULTS AND DISCUSSION

The influence of a solid support on consortium activity was tested. Monitoring of the ^{14}C-PCP mineralization rate in a soil slurry (5%w/v) and two subsequent dilutions of this slurry (1%, 0.01%) demonstrated that the consortium activity decreased significantly with the reduction of soil concentration in the slurry (Figure 1). The time for complete PCP mineralization increased from 3 days for the undiluted slurry to more than 28 days for the highest slurry dilution. In the highest slurry dilution, 92% (5 days for complete mineralization) of the consortium activity was recovered by the addition of sterile soil (1% w/v). These results suggested that a solid support was necessary to maintain consortium activity, although the requirement for some essential nutrient(s) from the soil cannot be excluded. Similar results were obtained for benzene, toluene, ethylbenzene, and xylenes (BTEX) and naphthalene biodegradation in aquifer sediments (Holm et al. 1992). Two PCP-degrading strains were isolated from the consortium and one was identified as *Pseudomonas luteola*.

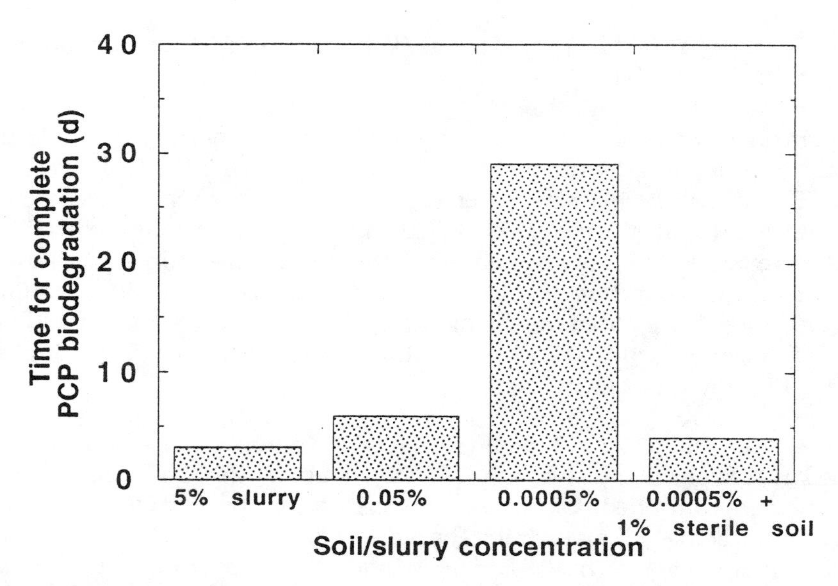

FIGURE 1. Effect of soil addition on the degradation of 100 mg/L PCP.

PCP degradation in batch culture was inhibited at concentrations greater than 400 mg/L, which compares favorably to the 160 mg/L inhibition level for *Pseudomonas* sp. observed by Radehaus and Schmidt (1992). In fed-batch culture, loaded at increasing PCP concentrations, as much as 800 mg/L·d was degraded. Chloride ion analysis demonstrated a stoichiometric production of Cl^- from the introduced PCP. In addition, a 3-fold increase in the rate of PCP mineralization occurred with increasing feed rates from 0 to 500 mg/L·d (Figure 2). The maximum rate of mineralization occurred after 40, 16, 10, and 9 hours when the bioreactor was fed at 0, 100, 300, and 500 mg/L·d, respectively. This maximum rate corresponded to the inflection point of the curves. Previous studies with 2,4-dichlorophenoxyacetic acid degradation showed that, at the inflection point, the substrate was no longer detectable (Comeau et al. 1993).

Culturing the biomass on PCP in the slurry bioreactor also stimulated PAH-degrading activity. A mass balance showed that the quantity of several PAHs decreased substantially after 500 h in the soil slurry bioreactor: acenaphthene, 99 to 10 mg; phenanthrene, 40 to 13 mg; fluoranthene, 112 to 29 mg; pyrene, 73 to 26 mg; and benzo(*a*)anthracene, 21 to 9 mg. These results take into account losses due to volatilization and adsorption to the soil and vessel. Acclimation of the consortium in the bioreactor increased the mineralization of pyrene and phenanthrene 58-fold and 26-fold, respectively, over the rates observed in the native soil.

Validation of the efficiency of the activated soil process was tested on contaminated soils. Inoculum from the soil/slurry in MSM (5%w/v) was introduced into static soil microcosms, and PCP mineralization was monitored. Preliminary results indicated a significant increase in the rate of PCP mineralization in the presence of the soil/slurry inoculum compared to uninoculated control microcosms.

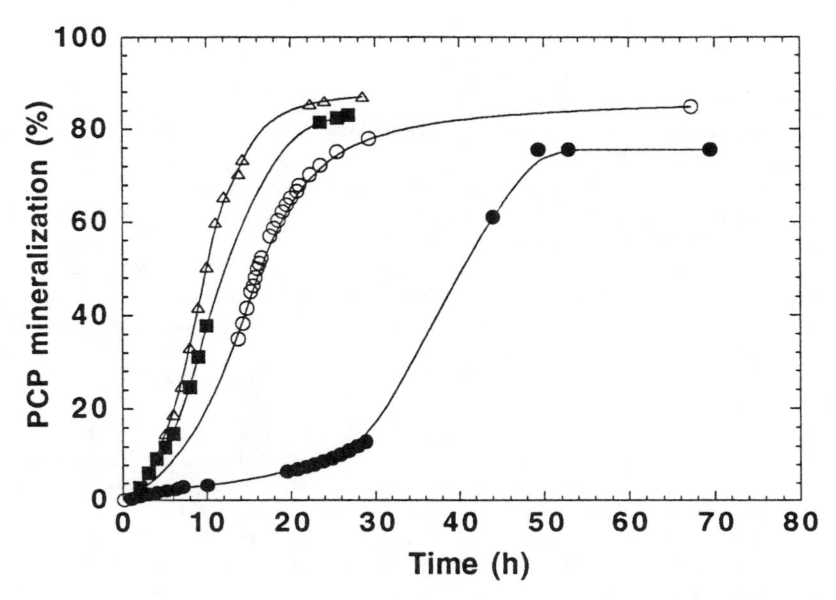

FIGURE 2. Effect of PCP loading rate on the mineralization of PCP. Samples were taken from the soil/slurry reactor (for determination of mineralization rates) when the PCP loading rates were 0 mg/L/d (•), 100 mg/L·d (○), 300 mg/L·d (■), or 500 mg/L·d (△). The PCP mineralization rate was determined in a liquid microcosm containing a 100 mg/L solution.

CONCLUSION

This study indicated that the use of activated soil, is a promising technique for biorestoration of contaminated soils. The activated soil concept is based on the cultivation of the biomass from a fraction of the contaminated soil to use as inocula for subsequent bioaugmentation on a large scale. A bacterial consortium able to degrade PCP and PAHs was characterized and identified in contaminated soil from a wood-preserving facility. To obtain an active consortium from the contaminated soil it was necessary to maintain the soil as a support for the bacteria and contaminants. Production of consortia in a fed-batch slurry reactor highly enhanced its activity both for PCP and PAHs biodegradation. The results indicated that a significant reduction in the time required for remediation of contaminated soil could be achieved at the laboratory scale.

ACKNOWLEDGMENTS

The authors thank Chantale Beaulieu, Stéphane Deschamps, Louise Deschênes, Jalal Hawari, Denis Rho, and Sylvie Sanschagrin for their valued support in this study.

REFERENCES

Comeau, Y., C. W. Greer, and R. Samson. 1993. "Role of inoculum preparation and density on the bioremediation of 2,4-D contaminated soil by bioaugmentation." *Appl. Microbiol. Biotechnol. 5*: 681-687.

Greer, C. W., J. Hawari, and R. Samson. 1990. "Influence of environmental factors on 2,4-dichlorophenoxyacetic acid degradation by *Pseudomonas cepacia* isolated from peat." *Arch. Microbiol. 154*: 317-322.

Holm, P. E., P. H. Nielsen, H.-J. Albrechtsen, and T. H. Christensen. 1992. "Importance of unattached bacteria and bacteria attached to sediment in determining potentials for degradation of xenobiotic organic contaminants in an aerobic aquifer." *Appl. Environ. Microbiol. 58*: 3020-3026.

Kearney, P. C., and S. T. Kellogg. 1985. "Microbial adaptation to pesticides." *Pure Appl. Chem. 57*: 389-403.

Mueller, J. G., S. E. Lantz, B. O. Blattmann, and P. J. Chapman. 1991. "Bench-scale evaluation of alternative biological treatment processes for the remediation of pentachlorophenol and creosote contaminated materials: Slurry-phase bioremediation." *Environ. Sci. Technol. 25*: 1055-1061.

Radehaus, P. M., and S. K. Schmidt. 1992. "Characterization of a novel *Pseudomonas* sp. that mineralizes high concentrations of pentachlorophenol." *Appl. Environ. Microbiol. 58*(9): 2879-2885.

ON-SITE/EX SITU BIOREMEDIATION OF INDUSTRIAL SOILS CONTAINING CHLORINATED PHENOLS AND POLYCYCLIC AROMATIC HYDROCARBONS

A. G. Seech, I. J. Marvan, and J. T. Trevors

INTRODUCTION

Pentachlorophenol (PCP), and its sodium salt, are toxic xenobiotics. The use of these compounds in North America has been associated primarily with the wood treatment industry (Rao 1978). This use has resulted in significant deposition of PCP to soil, and its subsequent dispersal throughout the environment. Similarly, the deposition of polycyclic aromatic hydrocarbons (PAHs) to soil is of great concern, as the environmental fate of such pollutants often includes leaching to groundwater.

Biodegradation is the ultimate fate of most organic compounds in terrestrial systems. Microbial degradation of PCP in soil has been reported (Edgehill & Finn 1983), and biodegradation of PAHs does occur under natural conditions. However, the persistence of PCP (Kituenen et al. 1987), and of PAHs in the environment suggests that the biodegradation of both these pollutants is exceeded by their rates of deposition. It becomes apparent that the rate at which the biodegradation process occurs is of great importance, as it can affect the partitioning of chemicals between biodegradation and other less desirable fates, such as bioaccumulation and migration to groundwater.

GRACE Dearborn's organic amendment bioremediation technology involves the addition of solid-phase, biodegradable organic amendments that have been prepared to specific particle-size ranges (e.g., D6380). The organic amendment technology increases the soil's water-holding capacity and increases macroporosity (and hence oxygen diffusion) and the ability of hydrophobic soils to support a xenobiotic-degrading microbial biomass.

The purpose of this study was to demonstrate the influence of GRACE Dearborn's patent-pending organic amendment bioremediation technology on the rate and extent of PCP and PAH degradation, as applied to the on-site/ex situ bioremediation of industrial soils containing chlorinated phenols and PAHs.

MATERIALS AND METHODS

Site Selection. The pilot-scale demonstration was performed at an industrial site where several decades of wood treatment had resulted in deposition of chlorinated phenols, creosote, and petroleum hydrocarbons to the soil. Analyses of the soil prior to treatment revealed chlorinated phenol concentrations of approximately 680 mg/kg, and total PAH concentrations of more than 1,400 mg/kg. The total petroleum hydrocarbon (TPH) concentration of the soil ranged from 5,700 to 7,000 mg/kg. Textural analysis (Day 1965) classified the soil as a fine sandy loam (72.3% sand, 23.5% silt, 4.2% clay). The soil pH was 7.4 (Peech 1965), and the organic carbon content was 1.8% (Nelson & Sommers 1982).

Soil Treatment. The on-site/ex situ treatment area was constructed using a high-density polyethylene liner, and was enclosed with a steel/transparent polyethylene structure. Electric heating allowed continuation of the work through the fall and winter months. Approximately 12 tonnes of soil was excavated from an area near the wood treatment apparatus, transported to the treatment area, and spread to a depth of approximately 0.5 m. The treatment applied to the soil consisted of a low-nutrient-content organic amendment (D6380), and inoculation with a strain of *Pseudomonas resinovorans*. The soil moisture was set to 80% of water-holding capacity by trickle irrigation. The soil temperature was maintained at 11 to 28°C in the winter months. During the summer months, soil temperature was not controlled and ranged from 18 to 34°C.

Microbiology. An organism capable of degrading PCP was isolated on site by enrichment culture using soil collected from the treatment area. The biological identification system (Biolog Inc., Hayward, California) was used to identify the culture as a strain of *Pseudomonas resinovorans*. The ex situ plots were inoculated with the PCP-degrading bacteria at a density of 7×10^5 cells/g. The ability of the culture to degrade PCP was verified by the evolution of $^{14}C-CO_2$ from aqueous cultures supplemented with ^{14}C-PCP.

RESULTS AND DISCUSSION

Figure 1 presents data on the biodegradation of PCP in the treated soil. The data illustrate rapid biodegradation of PCP, with a 99% reduction in the soil's initial PCP concentration. After 207 days of treatment, the PCP concentration reduced from approximately 680 mg/kg to 6 mg/kg. In order to verify that observed reductions in chlorinated phenols were caused by biodegradation and not by other processes such as volatilization or adsorption, $^{14}C-CO_2$ evolution was monitored from treatment area soil that was supplemented with ^{14}C-PCP in the laboratory. In the first 48 hours, more than 12% of the added ^{14}C-PCP was recovered as $^{14}C-CO_2$ (data not presented), verifying ring-cleavage and biodegradation of the chemical.

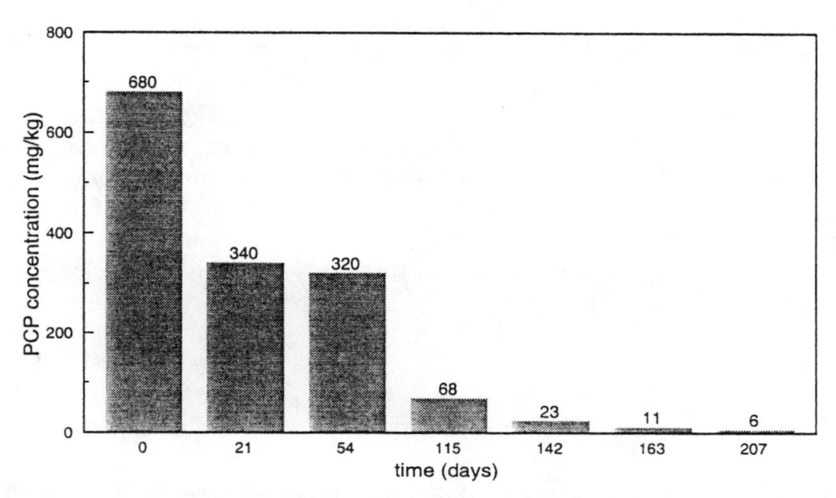

FIGURE 1. Residual PCP concentrations in wood treatment soil treated on-site with Dearborn organic amendment technology (preliminary data for Ontario soil).

Rapid biodegradation of PAHs also was observed. The total PAH concentration was reduced from more than 1,485 mg/kg to 35 mg/kg, after 207 days of treatment (Figure 2). Concentrations of specific PAHs, including the most recalcitrant, carcinogenic isomers (e.g., benzo(a)pyrene), were reduced to below the cleanup guidelines for industrial soils in Canada. Concentrations of chlorinated phenols and PAHs in the untilled control portion of the treatment area, where no treatments were applied, did not change significantly during the demonstration (data not presented).

During the same time period the TPH concentration was reduced to 34 mg/kg from 6,325 mg/kg (Figure 3). TPH concentrations in the control section did not change significantly.

Performance data collected during this demonstration indicate that the organic amendment technology can provide a cost-effective means of remediating industrial soils containing high concentrations of chlorinated phenols and PAHs.

ACKNOWLEDGMENTS

GRACE Dearborn Inc. expresses its appreciation to Environment Canada and to the Ontario Ministry of the Environment for funding the on-site portion of this study, and to the Wastewater Technology Centre for scientific and technical assistance.

REFERENCES

Day, P. R. 1965. "Particle fractionation and particle size analysis." In C. A. Black (Ed.), *Methods of Soil Analysis*. Agronomy No. 9, Part 2, pp. 545-567. Am. Soc. Agron., Madison, WI.

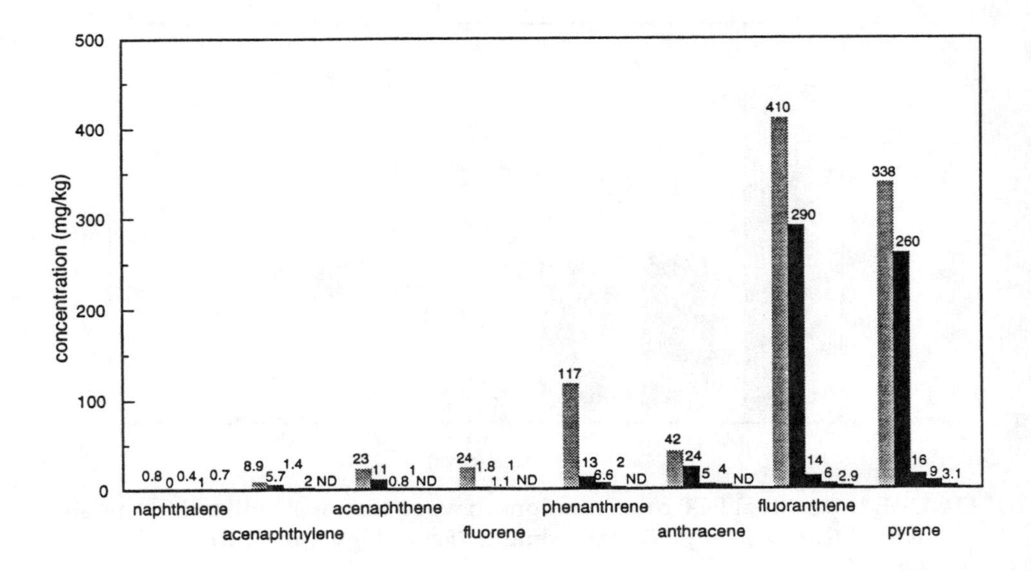

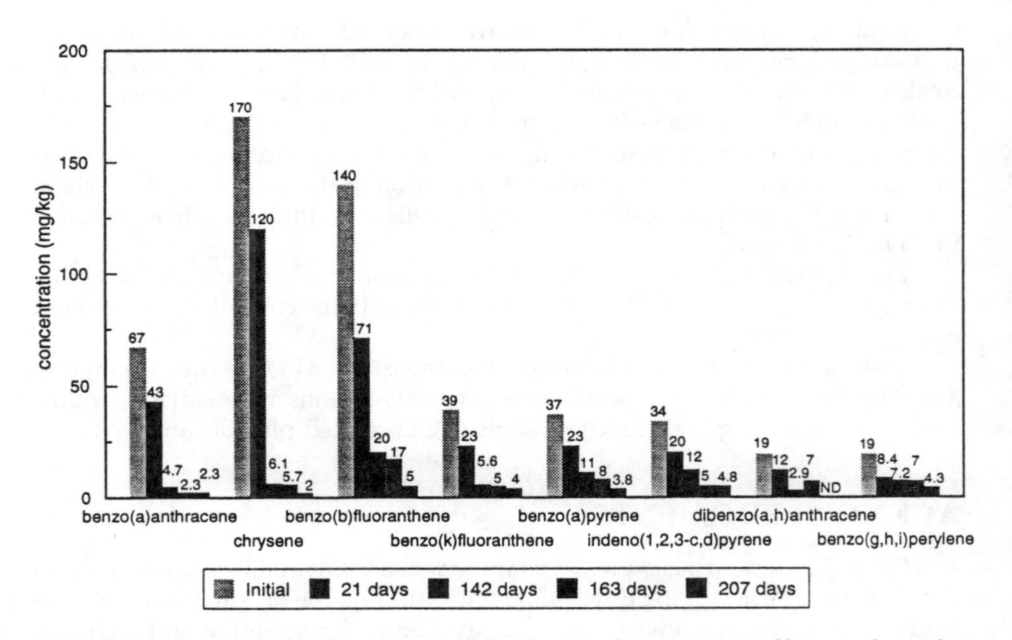

FIGURE 2. Residual PAH concentrations in wood treatment soil treated on-site with Dearborn organic amendment technology (preliminary data for Ontario soil).

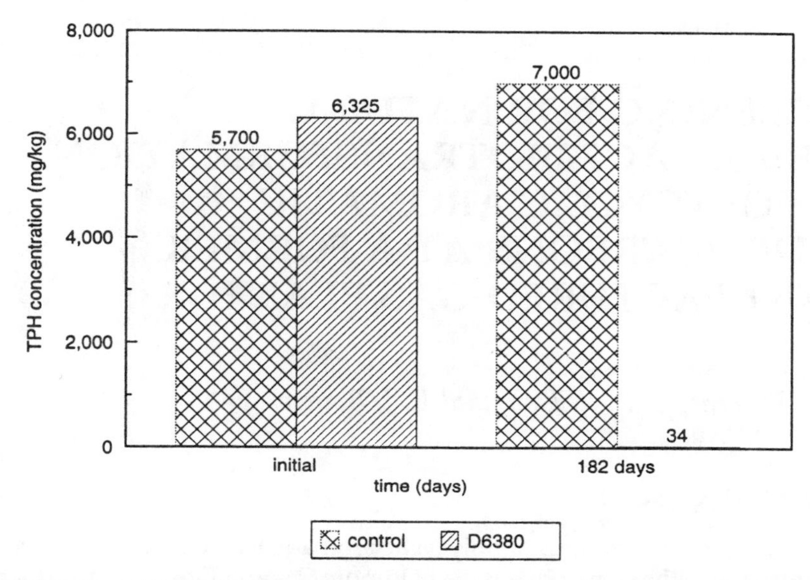

FIGURE 3. Residual TPH concentrations in industrial soil treated on-site with Dearborn organic amendment technology.

Edgehill, R. U., and R. K. Finn. 1983. "Microbial treatment of soil to remove pentachlorophenol." *Appl. Environ. Microbiol. 45*:1122-1125.

Kituenen, V. H., R. J. Valo, and M. S. Salkinoja-Salonen. 1987. "Contamination of soil around wood-preserving facilities by polychlorinated aromatic compounds." *Environ. Sci. Technol. 21*: 96-101.

Nelson, D. W., and L. E. Sommers. 1982. "Total carbon, organic carbon, and organic matter." In A. L. Page (Ed.), *Methods of Soil Analysis.* Agronomy No. 9, Part 2. Am. Soc. Agron., Madison, WI.

Peech, M. 1965. "Hydrogen-ion activity." In C. A. Black (Ed.), *Methods of Soil Analysis.* Agronomy No. 9, Part 2, pp. 914-926. Am. Soc. Agron., Madison, WI.

Rao, K. R. 1978. *Pentachlorophenol: Chemistry, Pharmacology, and Environmental Toxicology.* Plenum Publishing Corp., New York, NY.

SCREENING FOR NATURAL SUBSURFACE BIOTRANSFORMATION OF POLYCYCLIC AROMATIC HYDROCARBONS AT A FORMER MANUFACTURED GAS PLANT

N. D. Durant, L. P. Wilson, and E. J. Bouwer

INTRODUCTION

Elevated concentrations of polycyclic aromatic hydrocarbons (PAHs) have been detected in the subsurface at the Baltimore Gas and Electric Spring Gardens Facility (BG&E), the site of a former manufactured gas plant (MGP) in Baltimore, Maryland. An investigation has been initiated to evaluate the potential for natural and enhanced in situ biotransformation to remediate the subsurface contamination at the site. Specific objectives include (1) characterizing the distribution of subsurface microflora; (2) evaluating the capacity for indigenous populations to degrade certain PAHs; and (3) determining the optimal conditions for subsurface PAH biodegradation. Research on the first two objectives has been completed, and an overview of the results is presented herein.

MATERIAL AND METHODS

Sampling of Subsurface Sediments. A total of 57 subsurface samples of unconsolidated sediment were collected from five locations. A mud rotary drill rig equipped with a split-spoon sampler was used to collect sediment samples from depths ranging between the water table surface (1.6 m) and the top of bedrock (30.8 m). Sterile buterate liners were used to contain sediment cores within the split-spoon barrel. Rhodamine dye was added to the drilling mud as a tracer to confirm the absence of mud in samples.

Split-spoon sediment cores were subsampled in an on-site laboratory using an aseptic technique designed to exclude nonindigenous microflora (Madsen et al. 1991). Cores treated as anaerobic were subsampled in a portable anaerobic glove bag (Instruments for Science) filled with nitrogen gas and stored in sterile anaerobic gas packs (BBL).

Microbiological Enumeration. An acridine orange direct count (AODC) method, modified from Wilson et al. 1983, was used to estimate the total number of bacteria in subsurface sediments. Viable cells were estimated by colony-forming unit (CFU) growth on 5% PTYG agar (peptone, tryptone, yeast extract, and glucose) (Balkwill & Ghiorse 1985); and potato dextrose agar (Difco) in serial dilution pour plates. Anaerobic CFUs were developed on a 5% anaerobic nutrient agar and incubated in anaerobic gas packs. Samples were incubated at 22°C, and CFUs were counted after 2 weeks.

Mineralization Assay. Sediment-water microcosms prepared in 15-mL serum vials were used to assay the ability of indigenous microflora to mineralize ^{14}C-labeled acetate, naphthalene, and phenanthrene (Sigma Chemical Co.). Microcosms consisting of 5 g sediment and 3 mL filter-sterilized groundwater were prepared for 50 sediment samples. Groundwater used in the microcosms was obtained from monitoring wells located within 16.4 m of each borehole and screened at depths that corresponded with the sediment sample depth.

Each microcosm received approximately 6.76×10^{-2} µCi of ^{14}C-labeled and 1,000 µg/L unlabeled acetate, naphthalene, or phenanthrene. For each sediment sample, microcosms of the three substrates were prepared in quadruplicate; the fourth microcosm became the control by adding 1 mL of 0.5 N $HgCl_2$. A sterile, 2-mL glass vial containing 0.5 mL of 1 N KOH was placed in each microcosm to trap $^{14}CO_2$ generated from the mineralization of the ^{14}C-labeled substrate. Vials were capped with Teflon™ septa and aluminum crimps, and sealed with a layer of silicone. All microcosms were incubated at a constant temperature of 22°C. Anaerobic microcosms were prepared in an anaerobic glove box (COY) containing 95% N_2, 3% CO_2, and 2% H_2.

The trapping solution (KOH) was retrieved at 3 days, 7 days, and weekly thereafter, and replaced using a syringe inserted through the septa. Septa were resealed with silicone after each sampling. The $^{14}CO_2$ captured by the KOH was measured using liquid scintillation counting corrected for quench (Beckman Instruments, Model LS 3801).

RESULTS AND DISCUSSION

Enumeration. Total cell counts were typically low, ranging from 8.6×10^3 to 5×10^6 cells/g dry sediment (Table 1). Population densities did not appear to vary significantly as a function of depth, with the highest counts occurring at 11.8 to 12.5 m. Fine-grained samples tended to contain low or undetectable populations.

Aerobic plate counts also were low (between 0 and 10^4 CFUs g/dry sediment), with about half the samples yielding culturable populations. Possible causes of this are that the site microflora are not amenable to media used or surface conditions, or that these microflora are extremely slow-growing. For those samples with culturable populations, aerobic viable counts were generally 10- to 100-fold less than the total counts, which is typical for subsurface microflora (Madsen et al. 1991). Aerobic CFUs were detected in samples from both relatively clean and contaminated (2,300 µg/L naphthalene) zones (Table 1 gives the aqueous naphthalene concentration of the groundwater in sediment sampling locations). Viable anaerobic populations between 10^2 and 8×10^3 CFUs/g dry sediment were detected

TABLE 1. Summary of enumeration and mineralization results (excludes data from vadose zone and anaerobic samples).

Location	Sample Depth (m)	Sediment Type	Aqueous Naph. (ug/l)	Viable Count (CFUs/g-dw)	Total Count (Cells/g-dw)	Percent Biotic Mineralization Acetate	Naph.	Phen.	Biotic Mineralization Half-life (days)[a] Naph.	r^2 [b]	Phen	r^2
A	2.6-3.3	silty sand	0.6	NG[c]	$3x10^5$	27 (± 5)[d]	2 (±2)	0				
	6.6-7.2	clayey silt	BDL[e]	NG	BDL	NM[f]	NM	NM				
	11.8-12.5	gravel	BDL	NG	$5x10^6$	30 (±2)	8 (±2)	10 (±3)	480	0.78	140	0.97
	15.7-16.4	sand	BDL	$7x10^3$	$4x10^6$	20 (±2)	8 (±6)	11 (±11)				
	16.4-17.1	gravel	BDL	$1x10^4$	$3x10^5$	16 (±3)	2 (±8)	6 (±5)				
	22.3-22.9	gravel	4200	NG	$5x10^5$	24 (±3)	20 (±4)	NM				
B	3.3-3.9	sand	< 20	$2.4x10^3$	$8.6x10^3$	47 (±13)	55 (±5)	23 (±3)	18	0.91	75	0.86
	4.6-5.2	clayey silt	< 20	NG	$9.5x10^3$	54 (±49)	13 (±6)	12 (±11)	150	0.90	130	0.99
	30.5-30.8	gravel	2300	$1.4x10^4$	$1.4x10^5$	NM	4 (±3)	0				
C	15.7-16.4	sand	160	NG	$4x10^5$	5 (±2)	5 (±2)	7 (±3)			470	0.92
	22.3-22.9	gravel	100	$2x10^3$	$7x10^4$	4 (±2)	0	13 (±5)			130	0.87
D	2.6-3.3	silty clay	BDL	$3.2x10^3$	$1.4x10^5$	NM	29 (±5)	2 (±1)	67	0.89		
	13.8-14.4	gravel	BDL	$5.7x10^2$	$3.1x10^4$	63 (±5)	25 (±7)	5 (±5)	100	0.72		
	17.7-18.4	sand	BDL	$1.8x10^3$	$5.2x10^4$	45 (±25)	29 (±1)	9 (±10)	83	0.79		

(a) values computed only for samples with significant biodegradation and first order rates
(b) r^2 -- linear regression correlation coefficient
(c) NG -- no growth
(d) ± values represent one standard deviation from replicate mean
(e) BDL -- below detection limit
(f) NM -- not measured

at a variety of depths ranging from 3.3 to 22.3 m, occurring primarily in contaminated zones (2,300 to 30,000 μg/L naphthalene). Some samples contained both anaerobic and aerobic cells, suggesting the presence of facultative anaerobes.

Mineralization Assay. Table 1 reports the mean percent (± the standard deviation) of acetate, naphthalene, and phenanthrene biotically mineralized over 4 weeks of aerobic incubation. These data indicate that the majority of samples exhibited significant biodegradation of acetate under aerobic conditions. The most effective biodegradation of naphthalene (55% (± 5%)) and phenanthrene (23% (± 3%)) was observed in a relatively shallow (3.3 to 3.9 m) sample from Location B. Sediments from deeper zones at Location A also exhibited biodegradation of these compounds, but to a lesser extent. Samples from 22.3 to 22.9 m and 11.8 to 12.5 m biodegraded 20% (±4%) naphthalene and 10% (±3%) phenanthrene, respectively. Samples capable of biodegrading significant amounts of PAH did not exhibit any lag time, suggesting that many of the microflora at the site are acclimated to the compounds. Of the 10 sediments tested for anaerobic mineralization (not shown in Table 1), only two exhibited the capacity to biodegrade naphthalene, but most were able to biodegrade significant amounts of acetate.

A first-order kinetic model was used to estimate mineralization half-lives (Table 1) for those aerobic samples exhibiting significant biodegradation of naphthalene and phenanthrene. Naphthalene half-life ranged from 18 to 480 days, whereas phenanthrene ranged from 75 to 470 days. Mineralization rates for two sediments are depicted in Figure 1 in which mineralization of naphthalene and phenanthrene is plotted against time. The mineralization rates typically were highest during the first 15 days. This is likely the result of (1) typical first-order degradation kinetics in which the degradation rate decreases with substrate concentration, (2) limited oxygen and/or nutrient availability, and (3) limited substrate availability due to sorption (phenanthrene) and volatilization (naphthalene).

The enumeration did not detect CFUs in some samples exhibiting significant biodegradation of test compounds. The biodegradation in these samples suggests that viable cells are present, and that limitations in the plate count technique prevented detection of CFUs.

SUMMARY

Metabolically active microflora have been detected beneath the water table at a former MGP site contaminated with coal tars. Aerobic and anaerobic populations were observed in various unconsolidated sediments from 2.6 to 30.8 m below ground surface. Population densities did not appear to diminish with depth, nor did they vary significantly with the presence of aqueous PAHs. The lowest microbial numbers tended to occur in finer grained sediments.

A portion of the subsurface microflora appear to be acclimated to the presence of PAHs, exhibiting capacity to mineralize significant amounts of naphthalene (8% (±2%) to 55% (±5%)) and phenanthrene (10% (±3%) to 23% (±3%)) in sediment-water microcosms under aerobic conditions. Naphthalene biodegradation half-lives ranged from 18 to 480 days, whereas phenanthrene ranged from 75 to 470 days.

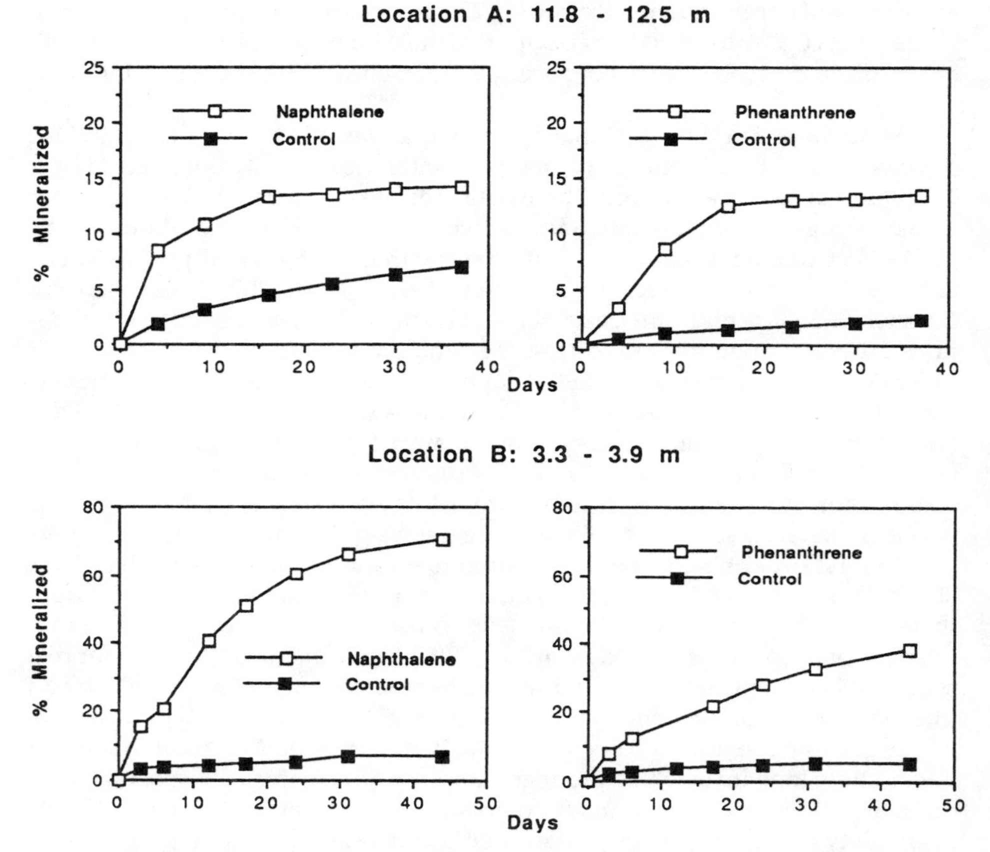

FIGURE 1. Mineralization of naphthalene and phenanthrene in two samples.

Data suggest that natural in situ biodegradation is occurring in aerobic zones of the BG&E subsurface, and that in situ oxygen addition may promote faster rates of PAH biotransformation.

ACKNOWLEDGMENT

The authors gratefully acknowledge the financial support of the Baltimore Gas & Electric Company, and the technical contributions of Herb Hoffman, Carla Logan, Ian MacFarlane, Frank Barranco, and Jean Kocornik.

REFERENCES

Balkwill, D. L., and W. C. Ghiorse. 1985. "Characterization of Subsurface Bacteria Associated with Two Shallow Aquifers in Oklahoma." *Applied and Environmental Microbiology* 50(3):580-588.

Madsen, E. L., S. N. Levine, and W. C. Ghiorse. 1991. *Microbiology of a Coal-Tar Disposal Site.* EPRI Final Report, EPRI EN-7319, EPRI, Palo Alto, CA.

Wilson, J. T., J. F. McNabb, D. L. Balkwill, and W. C. Ghiorse. 1983. "Enumeration of Bacteria Indigenous to a Shallow Water-Table Aquifer." *Ground Water* *21*(2): 134-142.

CONSIDERING MICROBIOLOGICAL CONDITIONS AT A FORMER MANUFACTURED GAS PLANT

I. D. MacFarlane, G. D. McCleary, H. L. Hoffman, and C. M. Logan

INTRODUCTION

Consideration of microbiological issues has become increasingly important in characterizing contaminated sites and evaluating remediation. This is the case for former manufactured gas plant (MGP) sites, where understanding of fate and effects and in situ bioremediation potential is in its infancy. Progress is being made through research focused on the microbiology of MGP and similar sites (Godsy et al. 1992, Madsen et al. 1991), as well as MGP bioremediation case studies (Werner 1991). Identifying and understanding the in situ redox conditions and likely terminal electron acceptors are important to understanding the ongoing natural fate processes, bioremediation potential, and potential reactions to altered conditions that may occur in response to a remedial action.

Researchers have shown the importance of redox conditions. Bouwer and McCarty (1984) assessed biotransformability by considering redox zones where one terminal electron acceptor (e.g., oxygen, nitrate, sulfate, or carbon dioxide) would prevail and control transformations. Lyngkilde and Christensen (1992a and 1992b) applied the redox zone concept to a case study of a landfill leachate plume and used the technique to assess the fate of organic constituents in various redox environments. The ferrogenic zone was found to contribute to the attenuation of organic contaminants. McFarland and Sims (1991) also suggest the importance of metals (iron and manganese) in providing reducing conditions for PAH degradation.

Our program tailors specific site investigations to these issues. For instance, groundwater samples are analyzed for redox-related parameters such as anions, oxidation reduction potential (ORP), and biogenic gases, in addition to standard EPA pollutant lists. We are conducting microbiological investigations to determine the most likely degradation processes at the site.

SITE SETTING

The study site at the Baltimore Gas and Electric Company's Spring Gardens facility is a 29-hectare (72-acre) former MGP, one of the largest former MGPs in the United States. Coal and fuel oil were stored on site. Soil and groundwater are contaminated to depths greater than 30 m with organic compounds, e.g., monocyclic and polycyclic aromatic hydrocarbons (MAHs and PAHs), and inorganics, such as cyanide, sulfate, and various metals including iron. Table 1 provides an overview of the groundwater chemistry. Also, subsurface accumulations of free-phase tar, fuel oil, ash, and box waste (spent manufactured gas filtering media) have been found. The tar, typically a dense, nonaqueous-phase liquid (DNAPL), is a complex mixture of 1 to 5% MAHs, such as benzene, toluene, ethylbenzene, and xylenes (BTEX), and 10 to 70% PAHs. Other hydrocarbons, such as aliphatics and heterocyclic aromatic hydrocarbons, typically comprise the remainder of the mixture. The box waste contains up to 600 mg/kg total cyanide, with no detectable free cyanide, and up to 9,000 mg/kg sulfate. Figure 1 is a schematic cross section of the site showing the complex distribution of tar

TABLE 1. Groundwater quality at Spring Gardens.

Analyte	Range	Typical	Background
pH	3 to 7	6	6.3
Oxidation Reduction Potential (mV)	-192 to 470	-20	0
Biological Oxygen Demand (mg/L)	<12 to 110	25	(c)
Chemical Oxygen Demand (mg/L)	<50 to 610	200	<50
BTEX (mg/L)	ND to 43	1	ND
Naphthalene (mg/L)	ND to 49	0.5	ND
PAH-Naphthalene (total)[a] (mg/L)	ND to 1	0.05	ND
Total Organic Carbon (mg/L)	<0.5 to 43	10	(c)
Cyanide (total) (mg/L)	<0.01 to 19.2	0.3	ND
Cyanide (amenable) (mg/L)	<0.01 to 6.1	0.1	ND
Cyanide (free) (mg/L)	<0.01 to 0.11	0.02	(c)
Dissolved Oxygen (mg/L)	<0.1 to 5	0.5	~2[b]
Nitrate (mg/L)	<0.5 to 4.6	<0.5	<0.5
Ammonia (mgN/L)	<0.05 to 83	20	(c)
Iron (mg/L)	0.1 to 2,350	50	16
Sulfate (mg/L)	<5 to 7,300	500	4
Carbon Dioxide (mg/L)	10 to 700	200	(c)

(a) EPA priority pollutant PAH minus naphthalene.
(b) Dissolved oxygen measured by probe — may read higher than Winkler titration method.
(c) Not measured.
BTEX = benzene, toluene, ethylbenzene, and xylenes; PAHs = polycyclic aromatic hydrocarbons.
ND = not detected.

DNAPL. The fill materials may contain ash, box waste, and other MGP-related residues.

MEASURING REDOX CONDITIONS

The first challenge in understanding redox conditions is to obtain accurate measurements. Besides atmospheric adulteration problems with sampling in low-redox environments (Barranco et al. 1994), analyses of ORP and dissolved oxygen (DO) are questionable. Potential inaccuracies with platinum electrodes are well documented by Peiffer et al. (1992). When analyzing site groundwater of relatively high ionic strength (sulfate and iron each >1,000 mg/L) and apparently low redox (undetectable DO) under acid conditions, a calibrated ORP probe can give a relatively high redox potential reading, e.g., >200 mV. DO probes also have shown variability compared to the Winkler titration method, particularly for DO conditions between 0 and 3 mg/L. Generally, the Winkler method has shown lower or undetectable DO, which is more plausible considering the confined hydrogeologic conditions and high oxygen demand of the groundwater.

DELINEATING REDOX ZONES

Assessing redox zones, as Lyngkilde and Christensen (1992a) have done, is complicated by the irregular distribution of aqueous-phase sources of potential

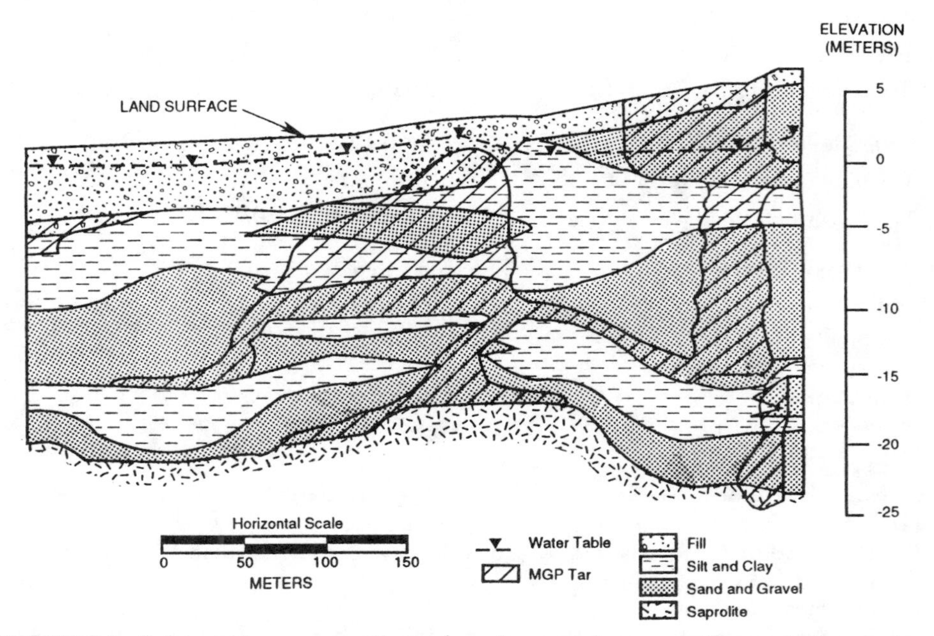

FIGURE 1. Schematic cross section of geology and sources through the center of the site.

electron donors and electron acceptors. As Figure 1 shows, tar DNAPL, a source of electron donors, is distributed in an intricate pattern relative to the soil stratigraphy. The situation is further complicated by the fact that three distinct aquifer units comprising the saturated portion of the cross section—the water table and two confined units—have some degree of hydraulic intercommunication. Sporadic occurrences of iron and sulfate (potential electron acceptors) in fill add another level of complexity. Although qualitative evidence exists for various oxidation and reduction processes, we have not yet delineated discrete redox zones emanating from source areas.

As Table 1 shows, DO and nitrate typically are depleted in the site groundwater, except in isolated circumstances. Flyvbjerg et al. (1991) have shown nitrite to be an electron acceptor for aromatic hydrocarbon degradation, but where measured, site groundwater also is depleted of this species, although ammonia is prevalent. However, iron, sulfate, and carbon dioxide typically are plentiful. Sulfate is associated with gas purification residues and ash. Elemental sulfur also has been shown to be a significant component of tar, e.g., 3 to 4% (Morgan & Stolzenbach 1934). Although some iron probably is leached from the natural aquifer media due to low redox and acid conditions, the anomalously high concentrations of iron found on site (typically associated with elevated sulfate) may be derived from gas purification and filter residue leaching. Iron oxide was a common filter medium used to sorb HCN and H_2S in the manufactured gas purification process (Harkins et al. 1987).

Field evidence (soil gas and well headspace analyses) points to sulfidogenesis and methanogenesis in areas in and around tar and oil accumulation. Furthermore, sulfate-reducing conditions apparently have been enhanced during an ongoing oil remediation project on site. An oil plume situated in MGP-related fill materials was being recovered by the pump-and-treat method. After pumping oil and water for several months, tar was mobilized to the recovery wells, and sulfidogenesis became evident. H_2S began to emanate from the carbon adsorption units, and what appeared to be FeS precipitant fouled several wells and the liquid-phase separation treatment.

It is reasonable to believe that aerobic biodegradation is occurring because of the surface infiltration of oxygenated water into aquifers. However, chemical oxygen demand (COD) cannot be discounted. If oxygen were used for aerobic respiration, then oxygen supply is a limiting factor based on low DO concentrations in groundwater over most of the site. Nitrate reduction also is possible due to the potential source of nitrogen from fossil fuel residues and the conspicuous absence of nitrate and nitrite in site groundwater.

DEGRADATION CHARACTERISTICS

Preferential degradation of various constituents may provide clues to microbial conditions. Degradation of aromatic hydrocarbons is related to redox conditions. Aerobic degradation of MAHs and PAHs generally has been possible under optimum conditions (Sims et al. 1989). However, numerous studies have shown the persistence of some compounds under nitrate- and sulfate-reducing conditions.

For instance, benzene typically is persistent in nitrate-reducing experiments, whereas other MAHs, particularly toluene, readily degrade (Acton & Barker 1992, Flyvbjerg et al. 1991).

Site groundwater quality patterns suggest preferential degradation, possibly implying biodegradation under reduced conditions. By calculating theoretical effective solubilities of the various tar constituents by methods described in Groher (1990) and Loehr et al. (1992), concentrations of MAHs and PAHs desorbing from tar can be estimated. In one case, toluene is calculated to be about 23% of the BTEX total. Groundwater samples from six wells within 100 to 200 ft of known tar source areas have an average toluene content of 26% of BTEX, which corresponds well with the theoretical effective solubility. The average toluene content of groundwater from 15 wells more distant from known sources is 3% of BTEX, which shows significant attenuation of toluene relative to benzene, ethylbenzene, and xylenes. Assuming sorption or other abiotic phenomena are insignificant, preferential biodegradation may be one cause of this phenomenon.

CYANIDE DECOMPOSITION

Low redox conditions caused by strong oxygen demand (e.g., microbial oxidation) may affect the fate of dissolved cyanide. Werner (1991) found that complexed iron cyanide in box wastes is prevalent at MGP sites, is not readily biodegradable, has a low solubility, and is not toxic. Hence, relative to other MGP problems, cyanide in the form of box waste typically is of much lower risk concern. Meeussen et al. (1992) found that dissolved, complexed iron cyanide tends to decompose to free cyanide, and the rate of decomposition is controlled by pH and redox. The half-life in low pH and Ep conditions is less than 1 year, which is 2 to 3 orders of magnitude less than the half-life under neutral pH and oxygenated conditions. Thus, under reduced and acidic conditions, as exist on parts of the subject site, free cyanide may be generated relatively quickly. Although free cyanide is more of an environmental concern than complexed cyanide, it is bioavailable.

Seven groundwater samples were analyzed for cyanide species, COD (a surrogate indicator parameter for ORP due to the inaccuracies described above), and pH. The highest amounts of cyanide (100%) amenable to chlorination (a standard method to quantify free to moderately complexed cyanide), as percentage of total cyanide, were associated with the highest COD and pH between 3 and 6. This observation supports the kinetic theory developed by Meeussen et al. (1992). The fact that free cyanide has been measured on site (Table 1), coupled with the possibility that some site conditions may exacerbate decomposition, makes considering redox effects on cyanide fate important.

CONCLUSIONS

Changes in redox conditions are important both for assessing existing natural in situ biodegradation conditions and for understanding how bioremediation may be applied and the effects of redox conditions on other environmentally related issues. Sims et al. (1989) state that "the goal of on-site bioremediation is degradation that results in detoxification of a parent compound to a product or product(s) that are no longer hazardous to human health and/or the environment." We add to this definition the concept that in performing bioremediation, we should not create more problems than are being solved. For instance, altering redox conditions for bioremediation could result in undesirable consequences. Adding oxygen to the subsurface environment that is rich in dissolved iron could cause iron oxide precipitation resulting in well or aquifer plugging. Enhancing reducing conditions could cause decomposition of complexed cyanide to the more toxic free cyanide form. Enhancing sulfidogenesis could cause complications to treatment operations and generate new environmental concerns if not considered in planning and design. "Souring" by sulfide is a well-known phenomenon in the oil and natural gas industry that can cause noxious or toxic accumulations of H_2S, equipment corrosion, treatment problems, and well formation fouling. Although the sulfide problem has been beneficial to building a case for natural, in situ sulfate-reducing biodegradation, significant expense has been incurred in the oil remediation project to compensate for the developed conditions.

In summary, all remedial actions, including bioremediation, should account for the consequences of affecting subsurface biological conditions.

ACKNOWLEDGMENTS

The authors gratefully acknowledge Edward Bouwer, Patricia Colberg, and Richard Luthy for contributing to the Spring Gardens microbial characterization program, and Philip Gschwend and Norbert Swoboda-Colberg for guidance on environmental chemistry issues.

REFERENCES

Acton, D. W., and J. F. Barker. 1992. "In situ biodegradation potential of aromatic hydrocarbons in anaerobic ground waters." *Journal of Contaminant Hydrology* 9: 325-352.

Barranco, F. T., J. L. Kocornik, I. D. MacFarlane, N. D. Durant, and L. P. Wilson. 1994. "Subsurface sampling techniques used for a microbiological investigation." In R. E. Hinchee, A. Leeson, L. Semprini, and S. K. Ong (Eds.), *Bioremediation of Chlorinated and Polycyclic Aromatic Hydrocarbon Compounds*. Lewis Publishers, Ann Arbor, MI.

Bouwer, E. J., and P. L. McCarty. 1984. "Modeling of trace organics biotransformation in the subsurface." *Ground Water* 22: 433-440.

Flyvbjerg, J., E. Arvin, B. K. Jensen, and S. K. Olsen. 1991. "Biodegradation of oil- and creosote-related aromatic compounds under nitrate-reducing conditions." In R. E. Hinchee and R. F. Olfenbuttel (Eds.), *In Situ Bioreclamation: Applications and Investigations for Hydrocarbon and Contaminated Site Remediation*, pp. 471-479. Butterworth-Heinemann, Stoneham, MA.

Godsy, E. M., D. F. Goerlitz, and D. Grbić-Galić. 1992. "Methanogenic biodegradation of creosote contaminants in natural and simulated ground-water ecosystems." *Ground Water* 30: 232-242.

Groher, D. M. 1990. "An Investigation of Factors Affecting the Concentration of Polycyclic Aromatic Hydrocarbons in Ground Water at Coal Tar Waste Sites." Masters Thesis, Massachusetts Institute of Technology, Cambridge, MA.

Harkins, S. M., R. S. Truesdale, R. Hill, P. Hoffman, and S. Winters. 1987. *U.S. Production of Manufactured Gases: Assessment of Past Disposal Practices.* EPA Contract No. 68-01-6826 D.O.#35, EPA Hazardous Waste Engineering Research Laboratory, Cincinnati, OH.

Loehr, R. C., P. S. C. Rao, L. S. Lee, and I. Okuda. 1992. *Estimating Release of Polycyclic Aromatic Hydrocarbons From Coal Tar at Manufactured-Gas Plant Sites.* Electric Power Research Institute Interim Report, EPRI TR-101060, EPRI, Palo Alto, CA.

Lyngkilde, J., and T. H. Christensen. 1992a. "Redox zones of a landfill leachate pollution plume (Vejen, Denmark)." *Journal of Contaminant Hydrology 10*: 273-289.

Lyngkilde, J., and T. H. Christensen. 1992b. "Fate of organic contaminants in the redox zones of a landfill leachate pollution plume (Vejen, Denmark)." *Journal of Contaminant Hydrology 10*: 291-307.

Madsen, E. L., S. N. Levine, and W. C. Ghiorse. 1991. *Microbiology of a Coal-Tar Disposal Site,* Final Report, EPRI EN-7319, Electric Power Research Institute, Palo Alto, CA.

McFarland, M. J., and R. C. Sims. 1991. "Thermodynamic framework for evaluating PAH degradation in the subsurface." *Ground Water 29*: 885-896.

Meeussen, J. C. L., M. G. Keizer, and F. A. M. de Haan. 1992. "Chemical stability and decomposition rate of iron cyanide complexes in soil solutions." *Environmental Science and Technology 26* (3): 511-516.

Morgan, J. J., and C. F. Stolzenbach. 1934. "Heavy oil tar emulsions in the water gas process." *American Gas Association Monthly 7*: 277-280.

Peiffer, S., O. Klemm, K. Pecher, and R. Hollerung. 1992. "Redox measurements in aqueous solutions — a theoretical approach to data interpretation, based on electrode kinetics." *Journal of Contaminant Hydrology 10*: 1-18.

Sims, J. L., R. C. Sims, and J. E. Matthews. 1989. *Bioremediation of Contaminated Surface Soils,* EPA-600/9-89/073, U.S. Environmental Protection Agency, Office of Research and Development, Ada, OK.

Werner, P. 1991. "German experiences in biodegradation of creosote and gaswork-specific substances." In R. E. Hinchee and R. F. Olfenbuttel (Eds.), *In Situ Bioreclamation: Applications and Investigations for Hydrocarbon, and Contaminated Site Remediation* pp. 496-517. Butterworth-Heinemann, Stoneham, MA.

PRELIMINARY EVIDENCE OF ANAEROBIC MICROBIAL ACTIVITY IN THE SUBSURFACE OF A FORMER MANUFACTURED GAS PLANT

P. J. S. Colberg, M. E. Bedessem, N. G. Swoboda-Colberg, H. L. Hoffman, and I. D. MacFarlane

INTRODUCTION

Even though microbially mediated reactions involving aromatic compounds are often thermodynamically favorable under highly reducing conditions, our understanding of the extent to which anaerobic microbial processes contribute to the turnover of environmental contaminants in situ is limited. This may be due, in part, either to the long-held view that anaerobic microorganisms are able to degrade only a limited number of substrates or to the tendency of microbiologists to work with pure cultures. We now know that many anaerobic transformations are carried out by microbial consortia (Ferry & Wolfe 1976; Kaiser & Hanselmann 1982; Shelton & Tiedje 1984; Zhang & Weigel 1990). Recent progress in elucidating anaerobic biotransformations of a variety of hazardous substances has enhanced our expectations for exploitation of indigenous microbial communities that exist in low redox environments.

Gas chromatographic (GC) analysis of the headspace of sampling wells installed at the Baltimore Gas and Electric (BG&E) Spring Gardens Facility, a former manufactured gas plant located in Baltimore, Maryland, suggests that anaerobic biotransformations may be significant to in situ processes. Chemical analyses of groundwater indicate high levels of sulfate (300 to 7,800 mg/L) and iron (0.5 to 3,000 mg/L), both of whose microbially mediated reductions are known to be coupled with organic transformations (see, e.g., Colberg 1990; Lovley & Lonergan 1990), whereas the evolution of methane (CH_4) from monitoring wells suggests that methanogenic conditions exist in some regions of the subsurface at the site.

Three major objectives of this newly initiated research are (1) to assess the potential for anaerobic microbial transformations of representative organic constituents in shallow and deep subsurface samples collected from the Spring Gardens site using laboratory microcosms established under iron-reducing, sulfidogenic, and methanogenic conditions; (2) to provide degradation rate data that may be used to estimate the contribution of low redox microbial processes to turnover of the contaminant pool in situ; and (3) to determine environmental conditions for enhancing in situ biotransformation activity in the subsurface.

This paper summarizes preliminary data on biogenic gases evolved from the field site and from sulfidogenic laboratory microcosms that contain shallow subsurface soils from four representative locations on the site.

MATERIAL AND METHODS

Sampling of Shallow Subsurface Soils. A total of 16 shallow subsurface soil samples were collected from the site. Samples were obtained from trenched test pits dug with a steam-cleaned backhoe and taken with a clean spoon from the soil pile. Sample depths ranged from 0.15 to 1.5 m. Contaminant concentrations in these soils range from 0 to 70,000 mg/kg for total polycyclic aromatic hydrocarbons (PAHs) and 0-5,000 mg/kg for monoaromatic hydrocarbons. Concentrations of iron vary from 1,600 to 52,300 mg/kg. The samples were shipped overnight on ice and stored at 4°C until use in the microcosm experiments.

Establishment of Sulfate-Reducing Microcosms. The soils were added as 5% (vol/vol) inoculum to a prereduced mineral salts medium for a total volume of 100 mL in 120-mL serum bottles that had been flushed with nitrogen. The medium was that of Häggblom and Young (1990) and contained 20 mM sulfate. The pH of the medium was approximately 7.0, and the initial composition of headspace gas was 100% N_2. All bottles were established in duplicate and sealed with butyl rubber stoppers and aluminum crimp seals. Two separate sets of abiotic control microcosms were established. One set contained 2 mL of glutaraldehyde; the other was repeatedly autoclaved. For comparison, microcosms were also established with an anoxic river sediment known to exhibit high levels of sulfidogenic activity to which supplementary carbon substrates were added (5 mg/L each of lactic and pyruvic acids).

Gas Analysis of Monitoring Wells and Sulfidogenic Microcosms. Gas samples from five locations on the site were extracted from airtight well caps through septa sealed with Swagelok™ fittings. These fittings were tapped into the well caps and sealed with Teflon™ tape. Teflon™ tubing was suspended from the well caps to several feet above the water table. Prior to sampling, at least one tubing volume was extracted to ensure collection of a representative sample. Following purging, gas samples from the wells were retrieved with a 500-µL gastight syringe.

Sulfidogenic microcosms were periodically sampled by removing gas samples from the headspace with a 1.0-mL gastight syringe and injecting the sample directly into a GC. The instrument used for all analyses with a Shimadzu 8A GC equipped was a thermal conductivity detector set at 100°C. The stainless steel column (1.83 m × 0.32 cm) was packed with Porapak Q (80/100 mesh). The carrier gas was helium (30 mL/min), and the column temperature was 50°C. Certified gas standards containing various mixtures of N_2, CO_2, CH_4, and H_2S were used for reference.

RESULTS AND DISCUSSION

Analyses of gas samples from the field suggest that both sulfate reduction and methanogenesis are occurring at some of the subsurface locations. Table 1 summarizes the results of gas measurements from five monitoring wells on the site.

The results of headspace gas analyses from several of the sulfate-reducing microcosms are presented in Figures 1 and 2. It was difficult to accurately determine overpressure in the serum bottles, so the CO_2 values are expressed as percentages of the total headspace gas, and H_2S is plotted as chromatographic peak areas. All values are conservative, in that they do not reflect solubility of the gases in the liquid phase. In addition, all microcosms (except controls) exhibited macroscopic evidence of sulfate-reducing conditions by the precipitations of metal sulfides in the medium.

Although it is not possible to directly compare the gas analyses from the field site (Table 1) with data obtained from the headspace of the microcosms, it is fair to conclude that sulfidogenic conditions were successfully established in the laboratory using shallow subsurface soils obtained from the site. Free-energy change calculations suggest that there are no thermodynamic barriers to the oxidation of many of the coal tar constituents under low redox conditions.

Shallow subsurface soils from the site produced levels of CO_2 comparable to those evolved from microcosms that contained the active sulfate-reducing river sediment (see Figure 1). H_2S levels detected in the microcosm headspace were generally lower in the coal tar microcosms (see Figure 2). This may have been due to the grater availability of metals in the soils which would have precipitated metal sulfides and reduced the amount of H_2S entering the gas phase. Control microcosms (data not shown) produced no H_2S and only trace levels of CO_2. Sulfidogenic activity in a subset of the active coal tar microcosms was inhibited after 103 days by the addition of sodium molybdate (20 mM). These microcosms exhibited an immediate cessation of both CO_2 and H_2S evolution suggesting that the observed microbial activity in the coal tar microcosms was due to sulfate reduction.

TABLE 1. On-site analysis of biogenic gases from five monitoring wells.

Sample Location	CH_4 (%)	CO_2 (%)	H_2S (mg/L)
1	4.2	1.6	0.5
2	0.03	3.3	<0.005
3	<0.005	3.1	<0.005
4	0.03	0.01	0.2
5	<0.005	0.02	<0.005

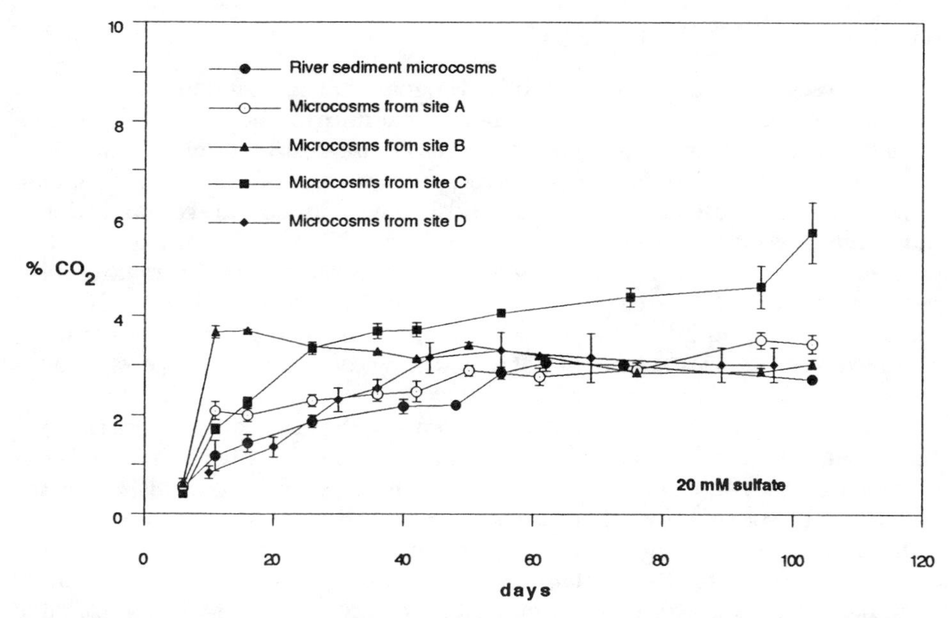

FIGURE 1. Percentage of CO_2 in the headspace of sulfidogenic laboratory microcosms. Gas data from microcosms containing an actively sulfate-reducing river sediment are shown for comparison.

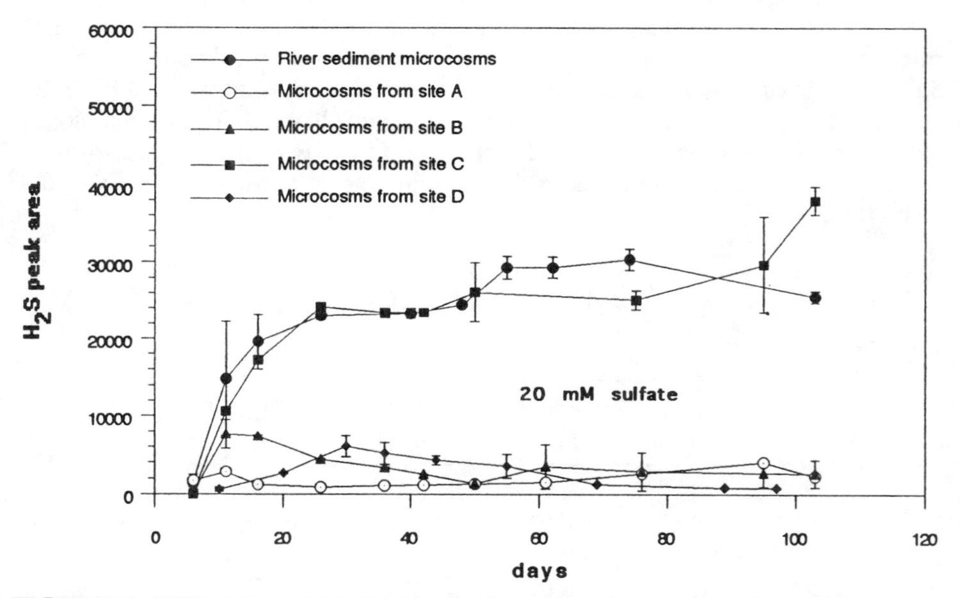

FIGURE 2. H_2S detected in the headspace of sulfidogenic laboratory microcosms. Gas data from microcosms containing an actively sulfate-reducing river sediment are shown for comparison.

SUMMARY

Preliminary characterization of shallow subsurface soils from the BG&E Spring Gardens Facility suggests that anaerobic biotransformations may be significant, naturally occurring in situ processes. Both H_2S and CH_4 are routinely detected in the soil vapor. High sulfate levels in the pore water from the site indicate the availability of SO_4^{2-} as an electron acceptor, whereas the evolution of H_2S confirms that sulfidogenesis is occurring in some locations. The detection of CH_4 gas on site implies that methanogenic conditions also exist. The high levels of dissolved iron could be supportive of iron reduction which is known to be coupled with some organic biotransformations.

Initial laboratory microcosm experiments were designed to determine whether or not anaerobic biotransformation activity could be established with materials from the site and to monitor evidence of any such activity by analyses of changes in the headspace gases over time. Studies involving the use of radiolabeled compounds and determination of the fate of specific organic contaminants in the microcosms are in progress.

Analyses of the initial laboratory microcosms established with shallow subsurface soils from the site (0.15 to 1.5 m) suggest that anaerobic conditions may be readily established in vitro. In the absence of any exogenous sources of carbon, levels of CO_2 and H_2S produced in the headspace of experimental microcosms are the products of sulfidogenic activity and are comparable to an active sulfate-reducing river sediment.

ACKNOWLEDGMENTS

This research is supported by a grant from BG&E, Inc. to P. J. S. Colberg. The technical assistance of Jean L. Kocornik and Frank Barranco of EA Engineering, Science, and Technology, Inc. is gratefully acknowledged.

REFERENCES

Colberg, P. S. J. 1990. "Role of Sulfate in Microbial Transformations of Environmental Contaminants: Chlorinated Aromatic Compounds." *Geomicrobial. J. 8*: 147-165.

Ferry, J. G., and R. S. Wolfe. 1976. "Anaerobic Degradation of Benzoate to Methane by a Microbial Consortium." *Arch. Microbiol. 56*: 33-40.

Häggblom, M. M., and L. Y. Young. 1990. "Chlorophenol Degradation Coupled to Sulfate Reduction." *Appl. Environ. Microbiol. 56*: 3255-3260.

Kaiser, J. P., and K. W. Hanselmann. 1982. "Fermentative Metabolism of Substituted Monoaromatic Compounds by a Bacterial Community from Anaerobic Sediments." *Arch. Microbiol. 133*: 185-194.

Lovley, D. R., and D. J. Lonergan. 1990. "Anaerobic Oxidation of Toluene, Phenol, and *p*-Cresol by the Dissimilatory Iron-Reducing Organism, GS-15." *Appl. Environ. Microbiol. 56*: 1858-1864.

Shelton, D. R., and J. M. Tiedje. 1984. "Isolation and Partial Characterization of Bacteria in an Anaerobic Consortium that Mineralizes 3-Chlorobenzoic Acid." *Appl. Environ. Microbiol. 48*: 840-848.

Zhang, X., and J. Weigel. 1990. "Sequential Anaerobic Degradation of 2,4-Dichlorophenol in Freshwater Sediments." *Appl. Environ. Microbiol. 56*: 1119-1127.

SUBSURFACE SAMPLING TECHNIQUES USED FOR A MICROBIOLOGICAL INVESTIGATION

F. T. Barranco, J. L. Kocornik, I. D. MacFarlane, N. D. Durant, and L. P. Wilson

INTRODUCTION

A microbiological investigation was conducted at a former manufactured gas plant (MGP) to evaluate the presence and extent of subsurface indigenous microorganisms. Proper collection and handling of sampled site media were paramount to assuring quality data. Soil and groundwater were collected for microbiological and chemical analyses. Chemical analyses of both media were beneficial to assessing microbiology-related conditions, e.g., aeration state, redox conditions, and potential electron acceptors/donors. During sampling of these media, significant opportunity for adulteration arose. For example, introduction of foreign material, i.e., by sampling devices and/or drilling residues, could have contaminated soil with nonindigenous organisms, rendering subsequent analyses inaccurate. Also, soil and groundwater sample exposure to the atmosphere was a concern due to oxygenation of anoxic media. Field procedures developed, tested, and used at this site to minimize the chance of adulteration are discussed herein.

SOIL SAMPLING PROCEDURES

Mud rotary drilling was necessary to sample unconsolidated sedimentary strata from near-surface depths to bedrock (30 m [approximately 100 ft]). The use of this standard drilling technique was a concern due to potential fluid penetration into sampled soil. Microbiological analyses of drilling fluid alone indicated high-microbial-density population by the enumeration technique and modest degradation potential of contaminant constituents by mineralization assay. Therefore, attempts were made to prevent drilling fluid permeation.

Several modifications to conventional split-spoon sampling were used to maximize unadulterated sample volume. First, 7.6-cm (3-in.)-diameter spoons were fitted with sterilized (ultraviolet light) buterate liners. Second, spoons were equipped with sludge barrel extensions to lengthen the spoon drive by an additional 15.2 cm (6 in.). Third, spoons were sealed to prevent entry of fluid prior to sample penetration. A sterile, watertight, 1-mm-thick plastic cap at the spoon tip and a ball valve above the sludge barrel prevented fluid entry. When penetrating in situ soil at the base of the fluid-filled borehole, the cap would rupture, allowing entry into the sample. To perform sampling aseptically, the drill rig and tools were steam-cleaned before commencing work at each borehole. Split spoons were cleaned after each use by standard decontamination procedures (Aller et al. 1991).

Following sample collection, the liner was capped and transported to an on-site field laboratory. By visual assessment, mud-affected sample was separated, using a flame-sterilized hacksaw, from unaffected sample. The interior of the unaffected portion was aseptically subsampled with a sterile syringe (Madsen et al. 1991). Aerobic samples were extruded into sterile Nasco whirlpack bags. Anaerobic samples were prepared similar to aerobic samples, except all work was performed in an oxygen-free environment, i.e., a nitrogen-filled glove bag. Anaerobic samples were stored in Gas Pak™ pouches with oxygen indication strips. Prior to bagging each sample, a split was obtained and evaluated on site for drilling fluid presence by tracer fluorescence.

FLUORESCENT TRACER

Rhodamine WT dye was used as a drilling fluid tracer for three borings. The use of this tracer provided a real-time method of testing for drilling fluid adulteration in samples. An initial bench-scale laboratory study investigated fluorescence characteristics of naturally occurring soil and drilling fluid (Quik-Gel™, NL Industries, Inc.), with various combinations of rhodamine WT and MGP contaminants of concern. Sample mixtures were prepared to reproduce the range of naturally occurring soil types and contaminant conditions from samples adulterated and unadulterated by drilling fluid. Polycyclic aromatic hydrocarbons (PAHs) fluoresce and are typical contaminants of MGP sites. PAH standard (Supelco 4-8905 stock containing 16 compounds) was added to the soil and drilling fluid in a concentration (2,000 µg/L) similar to field observations. Deionized water was added to simulate saturated conditions and drilling fluid consistency. Rhodamine WT was added to drilling fluid at concentrations of 20 and 100 µg/L. Samples were prepared for fluorescence evaluation by combining the various components, mixing until homogenized, and centrifuging until liquid/solid phase separation. The liquid fraction was decanted and assayed for relative fluorescence intensity, a unitless value, with a Model 10 Turner fluorometer.

Bench-scale laboratory testing indicated that mixtures of site soil, drilling fluid, and rhodamine WT fluoresce at significantly higher intensities than site soils alone (Figure 1). Furthermore, silt and clay in soils inhibit fluorescence. Other studies have confirmed these phenomena (Russell et al. 1992). Fluorescence was also inhibited by drilling fluid, which contained 85% montmorillonite clay. Detection of rhodamine fluorescence was not affected by PAH fluorescence. Rhodamine WT fluoresced at a relative intensity 10 times that of the PAH standard due to inherent differences in excitation and emission spectra (Acree et al. 1991, Pritchard & Zertuche-Gonzalez 1982).

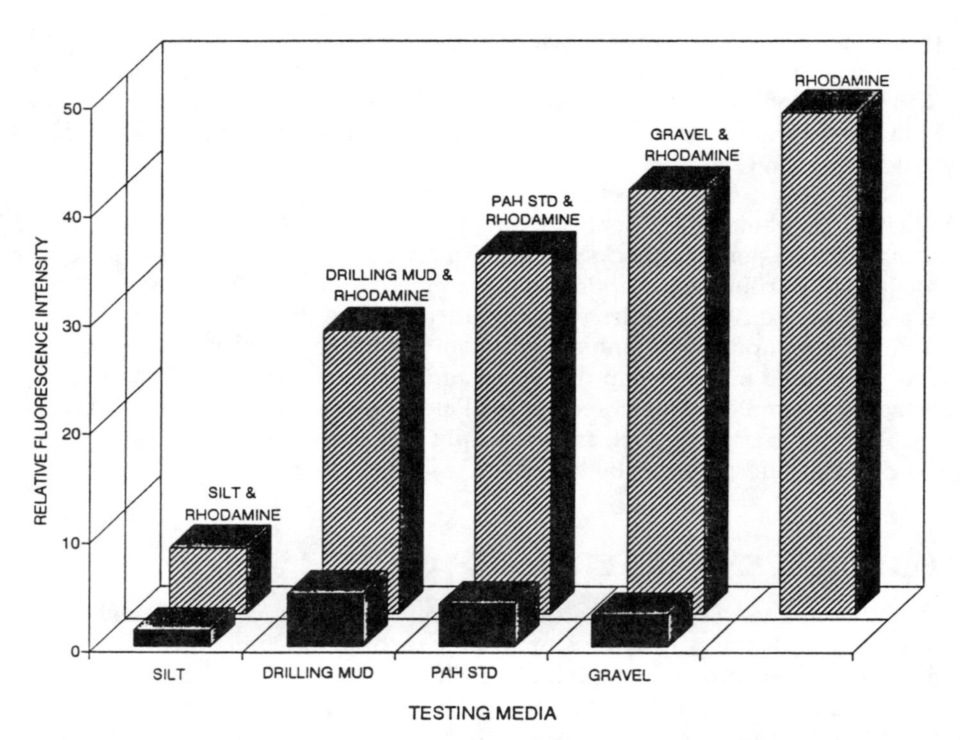

FIGURE 1. Histogram indicating relative fluorescence intensity of typically encountered media from bench-scale laboratory testing. (Rhodamine WT and PAH concentration were 20 and 2,000 µg/L, respectively.)

A higher tracer concentration was used in the field than in the laboratory as a conservative measure to ensure fluorescence detection. Field application involved homogenizing tracer and drilling fluid in a 1,600-L (425-gal) mud pan to produce a 5-mg/L rhodamine WT concentration prior to drilling advancement. The drilling fluid was assayed periodically for fluorescence. This assay was necessary to assess fluorescence inhibition by clayey silt strata because mud drilling is a downhole recirculation process. Additional rhodamine WT was added as necessary to maintain fluorescence. Soil sample splits were weighed and mixed with an equal weight of deionized water in the on-site field laboratory. Mixtures were assayed for fluorescence by bench-scale methods.

Soil samples for microbial analyses were obtained from a variety of depths from each boring. As an example, Figure 2 shows the results of fluorescence plotted against sample depth for Boring A. The fluorescence intensity of each sample was compared and contrasted with the fluorescence intensity of its laboratory-prepared counterpart. Under laboratory conditions, naturally occurring soils unaffected by drilling fluid displayed a relative fluorescence intensity background between 1.5 (silt and clay) and 4 (gravel and sand). Field samples

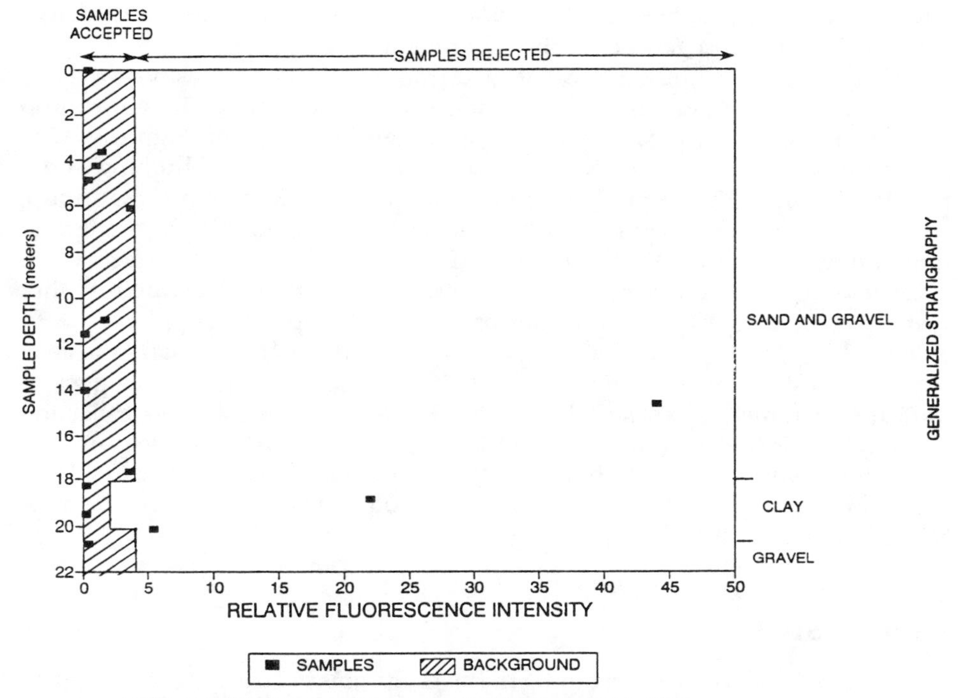

FIGURE 2. Plot of relative fluorescence intensity for actual and laboratory-prepared soil samples from Boring A.

that fluoresced considerably greater than their laboratory-prepared counterpart were disregarded from subsequent analyses based on the likelihood of drilling fluid presence. Two samples, 14.6 to 15.2 m and 18.9 to 19.5 m depth, fluoresced 5 to 10 times greater than their laboratory-prepared counterparts and were disregarded from subsequent microbial analyses.

GROUNDWATER SAMPLING PROCEDURES

A slow-pumping technique was used to obtain groundwater samples. Research performed by the Massachusetts Institute of Technology and the Electric Power Research Institute (Groher et al. 1990) was adapted for this site investigation (MacFarlane et al. 1992). Slow pumping is a low-yield well purging and sampling technique intended to facilitate influx of ambient water, rather than turbid, pump, or bail water containing naturally immobile particulate. Sample collection occurred in an enclosed, zero-headspace environment to minimize chemical alteration from atmospheric contact and provide a more accurate assessment of the true aqueous and geochemical environment.

Groundwater samples were collected from monitoring wells near the soil sampling boreholes and from corresponding soil sample depths. The slow-pump system consisted of two parts, pumping and sampling apparati (Figure 3). The pumping apparatus was comprised of downhole, stainless steel tubing and a peristaltic pump. The sampling apparatus, placed in line before the pump, comprised a self-contained box with a zero-headspace aliquot-filling sampling unit, a flowthrough Hydrolab Surveyor II water-quality meter, on/off valves, and vacuum gauges interconnected by stainless steel tubing. The water-quality meter was used to estimate indicator parameters (temperature, pH, dissolved oxygen [DO], specific conductance, and oxidation reduction potential) to achieve real-time evaluation of stable water quality during purging. Other general indicator parameters measured with Hach Kits following in-line collection included turbidity, alkalinity, carbon dioxide, and iron. DO was measured by two other in-line collection techniques, Winkler titration (Clesceri et al. 1989) and the Yellow Springs Instruments electrode method, the former yielding the most reliable results.

DISCUSSION

Further evidence collected during the investigations indicated that the adopted methodology was successful in minimizing sample adulteration. High microbial activity in drilling mud alone compared to low microbial activity in some field

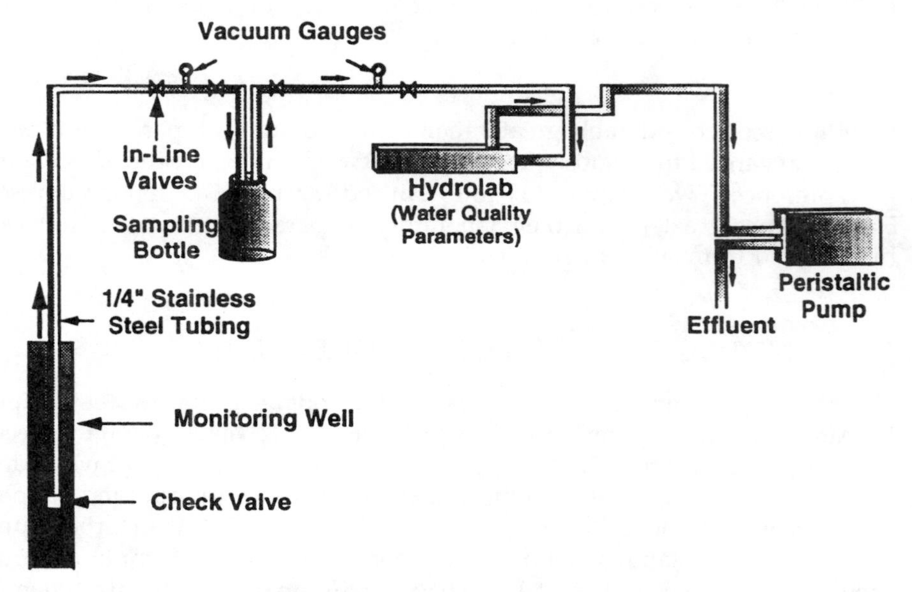

FIGURE 3. Schematic of the slow-pump sampling apparati.

samples indicated that soil sampling procedures minimized sample adulteration by the drilling fluid. Furthermore, sample screening by tracer testing ensured sample integrity. Lower DO values measured for given wells sampled by the slow-pumping versus traditional technique (e.g., bailer) indicated minimization of chemical alteration by atmospheric contact.

ACKNOWLEDGMENTS

Our thanks for sponsoring this work goes to the Baltimore Gas & Electric Company, in particular, Herb Hoffman and Carla Logan. The authors' appreciation also is extended to Ed Bouwer (Johns Hopkins University), Phil Gschwend (Massachusetts Institute of Technology), and Ishwar Murarka (Electric Power Research Institute).

REFERENCES

Acree, W. E., S. A. Tucker, and J. C. Fetzer. 1991. "Fluorescence Emission Properties of Polycyclic Aromatic Compounds in Review." *Polycyclic Aromatic Hydrocarbons* 2: 75-105.

Aller, L., T. W. Bennett, G. Hackett, R. J. Petty, J. H. Lehr, H. Sedoris, D. M. Nielson, and J. E. Denne (Eds.). 1991. *Handbook of Suggested Practices for the Design and Installation of Ground-Water Monitoring Wells.* EPA/600/4-89/034, rev. and enl. U.S. Environmental Protection Agency, Las Vegas, NV.

Clesceri, L. S., A. E. Greenberg, and R. R. Trussel (Eds.). 1989. *Standard Methods for the Examination of Water and Wastewater*, 17th ed., rev. and enl. American Public Health Association, Washington, DC.

Groher, D. M., P. M. Gschwend, D. Backhus, and J. MacFarlane. 1990. "Colloids and Sampling Ground Water to Determine Subsurface Mobile Loads." In I. P. Murarka and S. Cordle (Eds.), *Proceedings: Environmental Research Conference on Ground Water Quality and Waste Disposal*, pp. 19-1 to 19-10. EN-6749, Electric Power Research Institute, Palo Alto, CA.

MacFarlane, I. D., J. L. Kocornik, F. T. Barranco, and A. R. Bonas. 1992. "The Application of Slow Pumping at a Manufactured Gas Plant." In A. Stanley (Ed.), *The Proceedings of the Sixth National Outdoor Action Conference on Aquifer Restoration, Ground-Water Monitoring and Geophysical Methods*, pp. 413-427. Water Well Journal Publishing Company, Dublin, OH.

Madsen, E. L., S. N. Levine, and W. C. Ghiorse. 1991. *Microbiology of a Coal-Tar Disposal Site.* Final Report, EPRI EN-7319, Electric Power Research Institute, Palo Alto, CA.

Pritchard, D. W., and J. A. Zertuche-Gonzalez. 1982. *Mitigation of Natural Background Interference with Fluorometric Dye Measurements in Natural Waters.* Maryland Department of Natural Resources Power Plant Site Program Report, State University of New York Marine Sciences Research Center, Stony Brook, NY.

Russell, B. F., T. J. Phelps, W. T. Griffin, and K. A. Sargent. 1992. "Procedures for Sampling Deep Subsurface Microbial Communities in the Unconsolidated Sediments." *Ground-Water Monitoring Review* 12(4): 96-104.

FULL-SCALE SLURRY-PHASE BIOLOGICAL TREATMENT OF WOOD-PRESERVING WASTES

D. E. Jerger, D. J. Cady, and J. H. Exner

INTRODUCTION

OHM Remediation Services Corporation is presently conducting the full-scale remediation of creosote-contaminated wastes at the Southeastern Wood Preserving Superfund site in Canton, Mississippi. This site is an abandoned wood-preserving facility that operated from 1928 to 1979. The U.S. Environmental Protection Agency (EPA) initiated an emergency response action at the site in June 1986, and excavated approximately 8,000 yd^3 (6,100 m^3) of sludge and contaminated soil from lagoons, treatment facilities, and storage areas. The lagoon waste material was considered to be bottom sediment sludge from the treatment of wastewaters from wood-preserving processes using creosote. This material is Resource Conservation and Recovery Act (RCRA)-listed waste number K001. The excavated materials were stabilized with kiln dust and stockpiled for treatment.

Extensive laboratory treatability studies were performed to optimize the kinetics of the biological process for field pilot-scale testing and full-scale treatment operation. The final system configuration included a soil classification/washing process that concentrated the contaminants into a smaller volume of material for slurry reactor treatment and produced a washed soil product.

SYSTEM OPERATION

The stockpiled waste material at the site was power-screened through a 0.5-in (1.3-cm) grizzly screen. The screening process removed debris such as large stones, plastic sheeting, and railroad ties, and produced material appropriate for soil washing, slurry preparation, and biological treatment. Slurry preparation consisted of soil screening, soil washing, and classification to create a 20 to 25 wt% slurry containing soil particles less than 80 µ. The slurry was pumped into circular, steel, biological slurry reactors with an operating capacity of 180,000 gal (680,000 L). Each reactor was equipped with a direct-drive mixer and air diffuser system to maintain the slurry in suspension. Following treatment, the slurry was transferred to a high-density polyethylene (HDPE)-lined cell where excess water was collected via a leachate collection system. The water was transferred into a 350,000-gal (1,320,000 L) water management tank for reuse in soil washing and slurry operations.

The slurry reactors were operated in batch mode at a 20- to 30-day slurry retention time. Process monitoring consisted of dissolved oxygen, dissolved oxygen uptake rates, pH, temperature, total solids, ammonium-nitrogen, orthophosphate-phosphorus, bacterial biomass, and PAH constituent measurements by EPA SW-846 Method 8270. Foam production was minimized by amendment addition during slurry preparation.

SYSTEM PERFORMANCE

The power-screened material contained 5% gravel (>2.0 mm), 40% sand (>0.08 mm), and 55% silts and clays (<0.08 mm). Soil washing resulted in 85% of the particles treated in the slurry-phase reactors being less than 30 µ. The total organic carbon (Standard Methods Water and Wastewater 505B) and humic acid concentrations (Methods of Soil Analysis, SSSA, Section 30) of 32,000 mg/kg and 45,000 mg/kg, respectively, were in the upper range for these soil types. An oil and grease concentration (EPA SW-846 Method 9070) of 11,000 mg/kg consisted of the petroleum hydrocarbon-based carrier for the creosote material and the PAHs. Soil characteristics such as the high organic carbon fraction and fines content observed at the Southeastern Wood Preserving site have been reported to significantly influence and possibly limit the biological treatment of PAHs (Middleton et al. 1991). Aerobic heterotrophic and PAH-degrading bacterial populations enumerated from site material ranged from 8.2 to 9.4 and 6.0 to 7.8 log colony-forming units (CFU)/mL, respectively, indicating an indigenous population available for enhancement and degradation of the creosote constituents.

During reactor operation, ammonium-nitrogen and orthophosphate concentrations averaged 65 mg/L and 8 mg/L, respectively. The average temperature of the slurry was 33°C, and the pH was adjusted to 7.0 as necessary. The dissolved oxygen concentrations in the slurry ranged from 1.0 to 6.0 mg/L. Dissolved oxygen uptake rates ranged from 0.2 to 0.8 mg/L/min during the initial 5 days of the treatment cycle.

Treatment efficiencies approaching 95% were achieved for total PAHs in three representative full-scale batches (Figure 1). The majority of the PAH biodegradation occurred during the initial 5 to 10 days of treatment. Three-ring PAHs such as phenanthrene, anthracene, and fluorene were extensively biodegraded in the full-scale slurry reactors as reductions of 98 to 99% from initial concentrations were achieved. All of these PAHs have been reported to be utilized by microorganisms as sole carbon and energy sources (Heitkamp & Cerniglia 1988). The four-ring PAHs such as fluoranthene, pyrene, chrysene, and benzo(a)anthracene also were readily biodegradable. The overall treatment efficiency for this group of compounds ranged from 85 to 95%. In general, as the ring number increases, the aqueous solubility of the compound decreases, which adversely impacts bioavailability with a concomitant effect on biodegradability (Mueller et al. 1991). Treatment of the carcinogenic PAHs, including benzo(a)anthracene, chrysene, benzo(b)fluoranthene, benzo(k)fluoranthene, benzo(a)pyrene, dibenzo(a,h)anthracene, and indeno(1,2,3 cd) pyrene, ranged from 55 to 85% based on initial

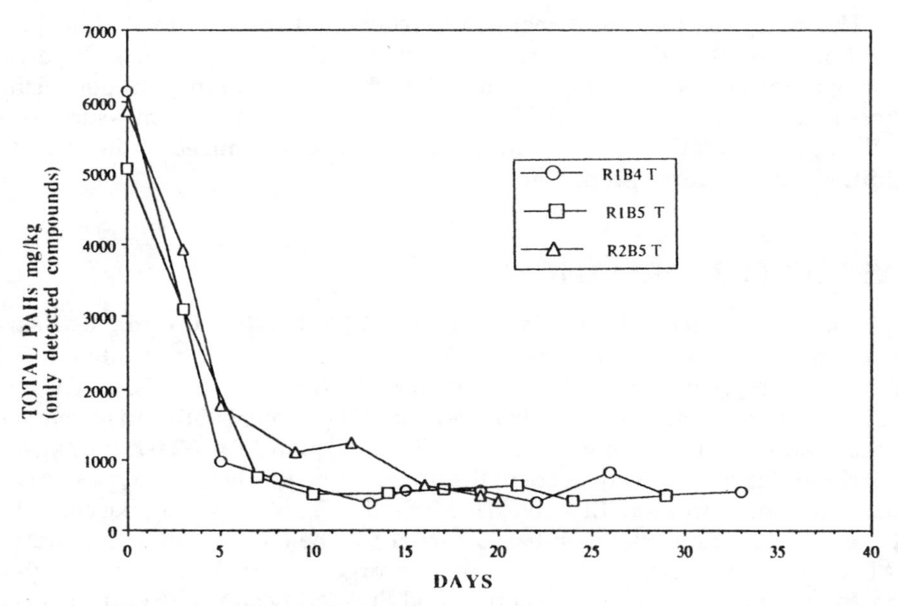

FIGURE 1. Full-scale slurry-phase biological treatment of polycyclic aromatic hydrocarbons in K001 wood-preserving waste.

concentrations (Figure 2). Carcinogenic PAH concentrations increased in two of the three batches following 4 days of incubation. This apparent increase in soil PAH concentration has been attributed to an increased PAH extraction efficiency (Lauch et al. 1992).

SUMMARY

A full-scale slurry-phase biological treatment process has been operated to treat K001 wastes at a former wood-preserving site. A treatment train approach coupling soil classification and soil washing with slurry-phase biological treatment achieved substantial reductions in both total and carcinogenic polycyclic aromatic hydrocarbons. The materials- handling and soil characterization process components were critical for successful operation of the slurry-phase reactor. Effective biological treatment was accomplished by enhancement of an indigenous population of PAH-degrading microorganisms.

The PAH degradation rates and efficiencies achieved at the Southeastern Wood site in a difficult soil matrix were consistent with reported values for treatment of creosote-contaminated soils and K001 wastes with lower organics and larger particle sizes in laboratory and pilot-scale bioslurry reactors (Lauch et al. 1992, Mueller et al. 1991).

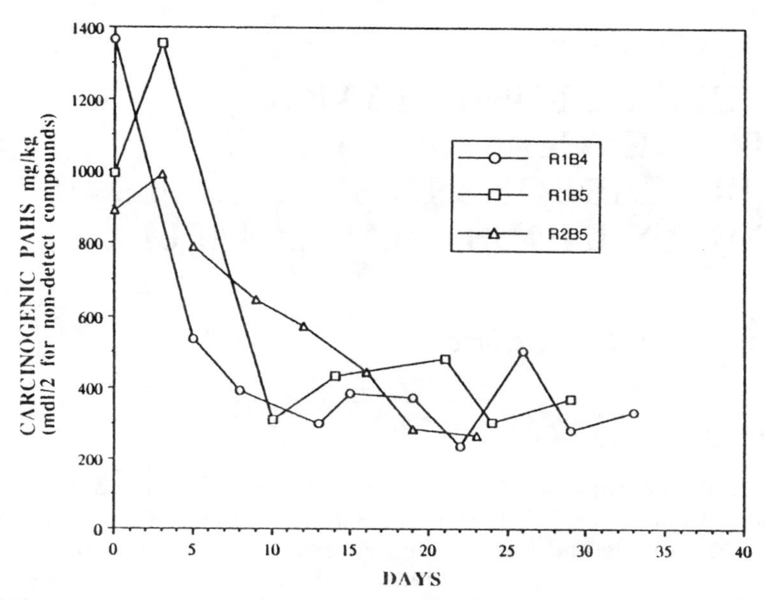

FIGURE 2. Full-scale slurry-phase biological treatment of carcinogenic polycyclic aromatic hydrocarbons in K001 wood-preserving waste.

REFERENCES

Heitkamp, M. A., and C. E. Cerniglia. 1988. "Mineralization of Polycyclic Aromatic Hydrocarbons by a Bacterium Isolated from Sediment Below an Oil Field." *Applied and Environ. Micro. 54*: 1612-1614.

Lauch, R. P., J. G. Herrmann, W. R. Mahaffey, A. B. Jones, M. Donsani, and J. Hessling. 1992. "Removal of Creosote from Soil by Bioslurry Reactors." *Environmental Progress 11*(4): 265-271.

Middleton, A. C., D. V. Nakles, and D. G. Linz. 1991. "The Influence of Soil Composition on Bioremediation of PAH-Contaminated Soils." *Remediation 1*: 391-407.

Mueller, J. G., S. E. Lantz, B. O. Blattmann, and P. J. Chapman. 1991. "Bench-Scale Evaluation of Alternative Biological Treatment Processes for the Remediation of Pentachlorophenol- and Creosote-Contaminated Materials: Slurry-Phase Bioremediation." *Environ. Sci. Technol. 25*: 1055-1061.

SLURRY BIOREMEDIATION OF POLYCYCLIC AROMATIC HYDROCARBONS IN SEDIMENTS FROM AN INDUSTRIAL COMPLEX

T. J. Simpkin and G. Giesbrecht

INTRODUCTION

The long-term operation of a water retention pond at an industrial complex in western Canada resulted in sediments in the pond being contaminated with a variety of volatile and nonvolatile hydrocarbons. A treatability study was undertaken to determine the effectiveness of a slurry bioreactor combined with land treatment in reducing contamination levels. This technical note focuses on certain aspects of this study.

The sediment in the retention pond was primarily silts and clays with a high oil and grease content (from 1 to 5%). The sediment contained a variety of hydrocarbon compounds including polycyclic aromatic hydrocarbons (PAHs) and odor-causing compounds.

Slurry bioreactor technology is a potential remediation alternative for the pond sediment, primarily due to its ability to capture and control releases of the volatile odorous compounds. The most economical application of slurry bioreactor technology is to couple it with land treatment. A slurry bioreactor with a relatively short retention time would be used to remove the volatile compounds and to initiate biodegradation. After the sediments are no longer odorous, they would be applied to land treatment beds where they would dewater and further biodegrade. Operation of the slurry bioreactors in a semicontinuous mode offers a number of potential benefits, as discussed below.

METHODS

The treatability test consisted of a laboratory batch "shake flask" test and a field pilot test. The shake flask test consisted of placing sediment with a variety of additives in sealed flasks, allowing the flasks to shake to provide aeration, and sacrificing flasks for analysis at regular intervals (seven times). PAHs were analyzed by U.S. Environmental Protection Agency (EPA) Method 8100. Solvent was added to the flasks to extract all possible organics. The additives included a nutrient solution only to one set of flasks, a biomass supplement to the second set, and an organic supplement to the third set.

The field pilot plant facility included two 60-L bioreactors and land treatment/drying beds. The reactors were operated in parallel, in a semicontinuous (fill and draw) mode. The semicontinuous operation involved removing an appropriate amount of contents from each reactor every day and refilling the reactor with the same volume of raw sediment. The sediment removed from the reactor was placed directly on the land treatment/drying beds. The volume of sediment removed and fed to the reactors was based on the target hydraulic retention time (HRT, 3 or 6 days). The sediment was diluted to approximatelv 25% solids before being fed to the reactors. Mixing in the reactors was provided by air diffusers and air lift pumps. Three HRTs (9 or 18 days) were allowed before the intense data collection. Foam was an operational problem that was dealt with by using antifoaming agents and foam traps.

RESULTS

The batch flask tests confirmed that significant reduction in the compounds of interest occurred in a reasonable time frame. The biomass supplement did not modify the rate of contaminant reduction. The organic supplement (acetate) appeared to decrease the rate of contaminant reduction, possibly due to the preferential degradation of the acetate. Table 1 summarizes the removals observed in the 74 days of the batch test for a selected group of compounds along with the first-order degradation rate constants determined from the batch data. Because the flasks were sealed, biodegradation was the primary process of the contaminant removal.

Removal of total oil and grease was low (about 4%), suggesting that the oil present was probably made up of high-molecular-weight, recalcitrant, extractable organics. It should also be noted that there appeared to be a lag in the degradation of pyrene of about 10 days. This lag corresponded relatively well with the time required to remove the low-molecular-weight PAHs.

To evaluate the pilot slurry reactors, their operations were divided into two periods. Period 1 covered the first 14 days, and Period 2 covered the last 27 days. Figures 1, 2, and 3 present the results for three representative compounds. The bars in these figures represent plus or minus the standard deviation. Removal

TABLE 1. Laboratory batch test results.

Compound	Concentration (mg/kg)		First-Order Rate Constant (day^{-1})
	Initial	Final	
Naphthalene	720	4	2.5
Pyrene	150	2	0.05
Total PAH	1,960	84	

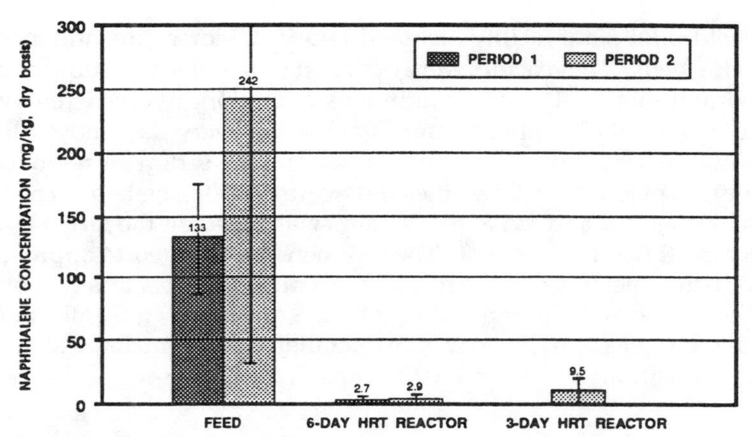

FIGURE 1. Naphthalene performance.

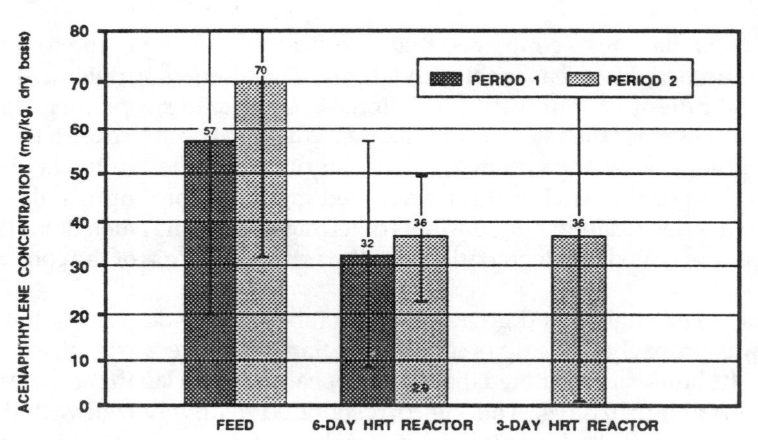

FIGURE 2. Acenaphthalene performance.

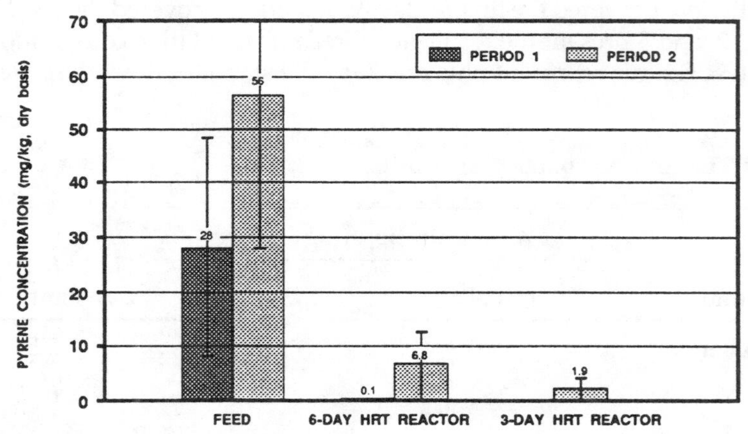

FIGURE 3. Pyrene performance.

of naphthalene, pyrene, and most other PAH compounds was very good (at least 95% for naphthalene and 88% for pyrene). Acenaphthylene was the exception to the good removal (44%). This is contrary to the lag time observed in the batch laboratory study, possibly due to the differences between batch and continuous operation. We have noted this in other studies. Pyrene removal was effective even in the 3-day HRT reactor. As in the batch study, oil and grease removal was poor. It should be noted that small tar balls formed in the reactors, which may have impeded the removal of oil and grease.

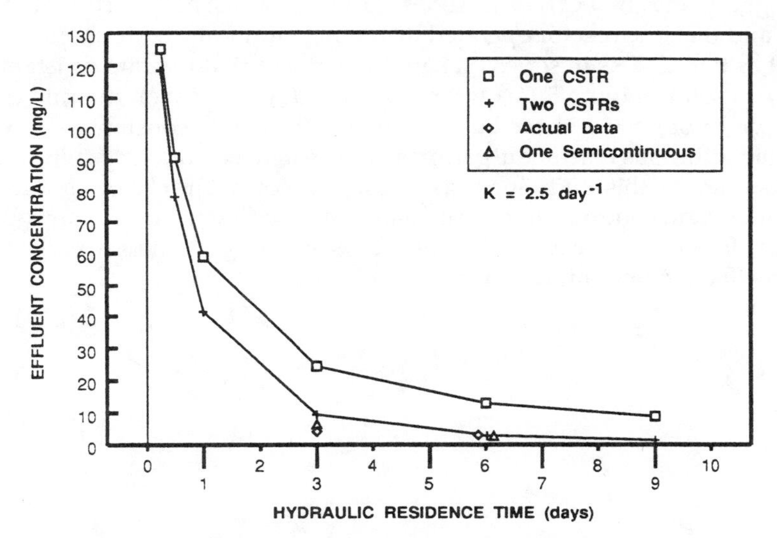

FIGURE 4. Slurry bioreactor model output — naphthalene.

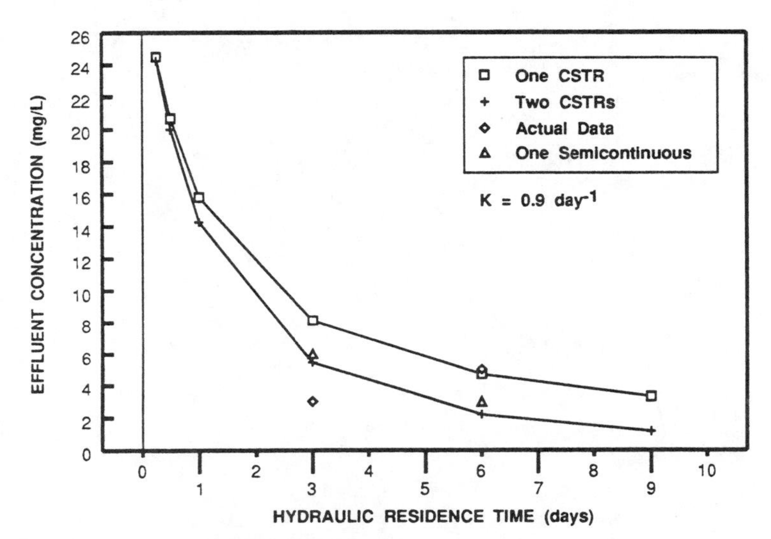

FIGURE 5. Slurry reactor model output — pyrene.

Although it is impossible to develop kinetic parameters from only two data points, it is possible to compare the performance achieved against theoretical models using the first-order rate constants developed from the batch flasks tests. Figures 4 and 5 present the output of the first-order models for naphthalene and pyrene, as well as the actual data. Three different models were used: one continuously stirred tank reactor (CSTR), two CSTRs in series, and one semicontinuously fed reactor (batch fed once per day). The semicontinuous reactor and the two CSTRs in series are in good agreement with the actual naphthalene data. Figure 4 also suggests that two CSTRs in series should produce better effluent than one CSTR, and that one reactor operated in a semicontinuous mode should perform as well as two CSTRs in series. It should be noted that the rate constant (K) of 0.9 day^{-1} used in Figure 5 for pyrene was much greater then the rate constant of 0.05 day^{-1} observed in the batch studies. These rate constants suggest that the semicontinuous reactor outperformed the batch reactors for pyrene removal.

In summary, this study suggests that slurry bioreactors can be used to reduce the contaminant concentrations in contaminated sediments while controlling the potential for odor releases. Semicontinuous operation of the slurry bioreactors was an effective operational mode.

WASHING AND SLURRY-PHASE BIOTREATMENT OF CREOSOTE-CONTAMINATED SOIL

J. D. Berg, B. Nesgård, R. Gundersen, A. Lorentsen, and T. E. Bennett

INTRODUCTION

Creosote-based wood impregnation has been practiced on an industrial scale for more than a century. Contamination of soil and groundwater at the sites has often resulted from existing impregnation practices as well as accidental spills. Polycyclic aromatic hydrocarbons (PAHs), some of which are persistent, toxic, and carcinogenic, comprise the majority of compounds found in coal tar-derived creosote (Mueller et al. 1989).

Aerobic biological treatment of creosote has been successful, in both solid and slurry phases, in a number of recent studies (Blais 1987, Matthews 1992, Mueller et al. 1991, Reineman 1987). Slurry-phase treatment offers several advantages over solid phase, including significantly shorter treatment times, more efficient removal of high-molecular-weight PAHs, and better process control.

Soil-washing technologies, originally developed for mining processes, are used for soil surface cleaning and volume reduction. When soil washing precedes slurry bioremediation, the integrated processes can be a very effective and economical means of soil reclamation. A number of vendors of soil washing technology report 80 to 98% reduction in hydrophobic organic contaminants such as PAHs (U.S. EPA 1989, 1991). However, particle-size distribution plays an important role in washing effectiveness in soils following soil washing. Removal of semivolatile compounds has been reported to be as low as 60% for the soil fraction < 250 µm, whereas the > 250-mm fraction showed > 98% removal for the same soil (Esposito et al. 1989).

STUDY DESCRIPTION

Soil washing and bioslurry treatment were evaluated at bench and pilot scale for the Norwegian State Railways (NSB), which owns the former wood impregnation study site. Some aspects of the bench-scale study were reported earlier (Berg et al. 1993). Detailed microbiological characterization of the soils has been conducted by SBP Technologies, Inc. and the U.S. Environmental Protection Agency (U.S. EPA) at Gulf Breeze, Florida. Results of the microbiological study are presented elsewhere in this volume (Mueller et al. 1994).

The pilot-scale washing was done using a 1,000-kg/h system designed in collaboration with Sala International of Sweden and Maskin Argo of Norway. The slurry reactor is a 454-L Biolift™ by EIMCO, Salt Lake City, Utah, USA.

The results presented herein summarize the bench-scale biodegradation tests and the ongoing pilot-scale washing and bioslurry treatment studies. A schematic of the pilot-scale process configuration is shown in Figure 1. The washing system features a rod mill for crushing and surface washing, as well as a spiral classifier and air flotation cell where particle sizing and separation occur. Fines from the classifier and the froth from the flotation cells are flocculated and thickened. The resulting sludge is pumped to the bioreactor for final treatment.

METHODS

The bench-scale biotreatability tests included two steps. The first was a screening test which used a Micro Oxymax™ respirometer (Columbus Instruments, Columbus, Ohio, USA). Headspace levels of carbon dioxide (CO_2) and oxygen (O_2) were monitored continuously; 5% solids in 100 mL of pH 7 buffered dilution water were used, and samples were mixed by moderate magnetic stirring. The purpose of these tests was to screen for inhibition of microbiological activity and the capability for degradation of contaminant by native microorganisms (10^6-10^7 CFU/mL). The nutrients were comprised of $(NH_4)_2SO_4$, NH_4NO_3, KH_2PO_4, and Na_2HPO_4 at concentrations of 1500, 750, 150, and 25 mg/L, respectively. The second step employed bench-scale, aerated slurry reactors (2 L) with pH control

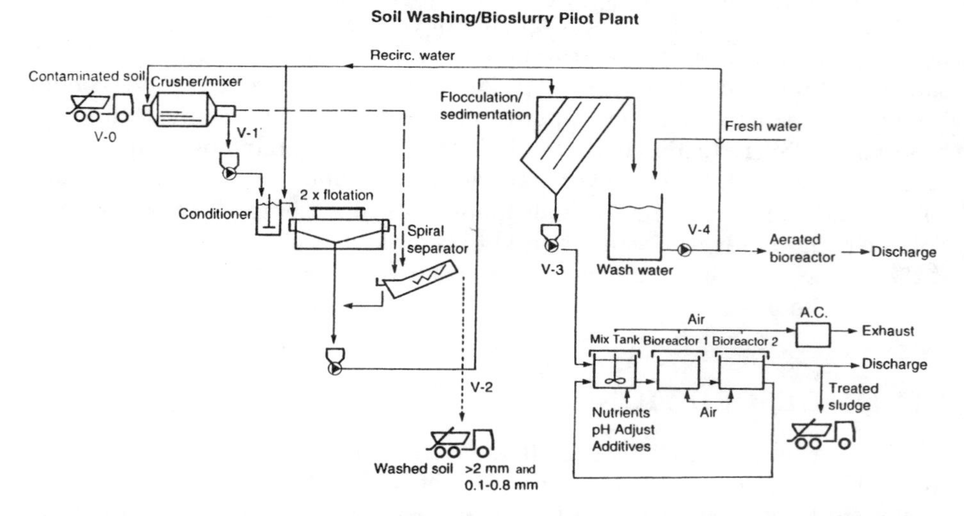

FIGURE 1. Schematic of the soil washing and bioslurry pilot plants. The washing plant was loaded at 500-800 kg/h. Sampling points are designated V-0 (soil in), V-1 (after attrition), V-2 (clean soil), V-3 (settled fines), and V-4 (clarified recycle water). Only bioreactor 1 was operated, in batch mode.

in which operational parameters, such as aeration rate and base requirements, could be evaluated using 20% soil slurries in 1 L of pH 7-buffered dilution water. All tests were conducted at 20°C. Untreated soil samples for the bench-scale tests were sieved to 2 mm and stored at 4°C. PAHs were extracted from the soil in four serial extractions using boiling cyclohexane. The total PAH values derived by this method and reported herein may be 2 to 5 times greater than those derived from the 16-component standard used by U.S. EPA (Method 8100). Work is under way to calibrate the two procedures so that the values can be compared. The compounds were categorized into four groups by the number of aromatic rings, which corresponds with their biodegradability, i.e., as the group number increases the biodegradability decreases (Mueller et al. 1991). In this study, Group 1 has 2 rings, Group 2 has 3 rings, Group 3 has 4 rings, and Group 4 has 5 rings or more.

RESULTS

Soil Characterization. The total PAH concentration in the soils ranged from 270 to 51,000 mg/kg (d.w.). The distribution of contaminant as a function of soil particle size for two surface soils is shown in Figure 2. A predominantly clayey soil from 5 m depth was excavated and subsequently used in the pilot-scale studies. This would represent the particle sizes < 212 μm as shown in Figure 2. The subsurface soil was chosen because it represented a worst-case scenario for the washing system, i.e., a predominance of fine material. PAH Groups 2 and 3,

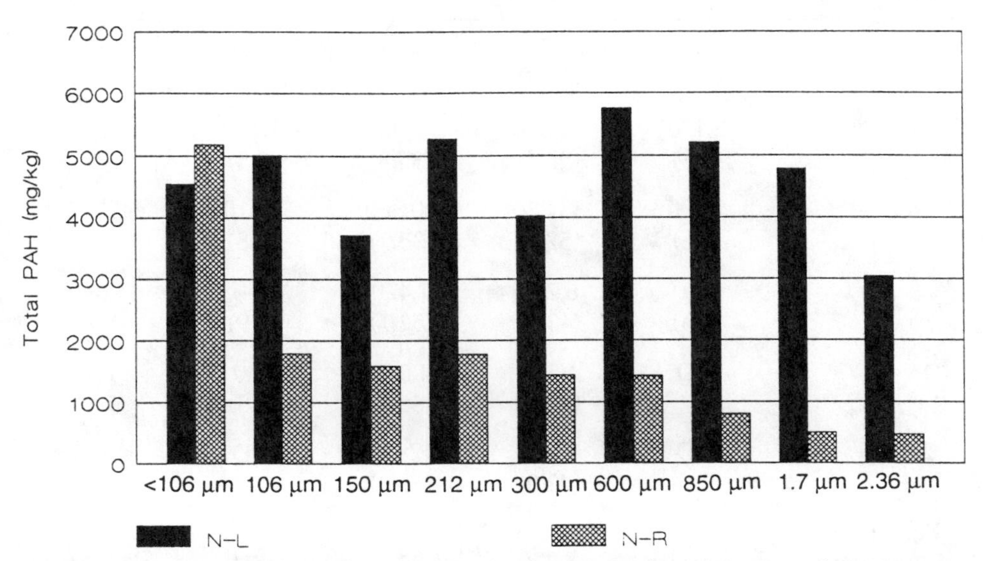

FIGURE 2. Distribution of total PAH as a function of particle sizes from 106 μm to 2.36 mm in surface soils from two wood impregnation sites. "N-L" is soil from the study site, "N-R" is soil from another creosote site.

i.e., 3- and 4-ring PAHs, were predominant, comprising up to 72% of total PAHs. Group 4 made up to 27% and Group 1 was up to 1%.

Bench-Scale Biotreatability Tests. The respirometric screening tests indicated that indigenous populations with proper nitrogen and phosphorus additions and pH control could successfully degrade the PAHs (Berg et al. 1993). An example of results obtained with a 1-L 20% slurry using a heavily contaminated sample (approximately 58,000 mg/kg PAH) is presented in Table 1. In this case, degradation of all four PAH groups was achieved, yielding an average of 97% reduction of total PAH in 21 days when the experiment was terminated. The heavier Group 4 compounds were the most refractory, as expected (Mueller et al. 1991). The apparent increase in PAH concentration between day 0 and 6 also has been observed at pilot scale on several occasions. There seems to be a release of extractable PAH as the slurry is mixed, perhaps due to scouring and physical destruction of larger particles.

TABLE 1. PAH degradation in a bench-scale slurry reactor showing the four PAH groups. Results are an average of duplicate runs using duplicate samples.

Day	PAH Concentration (mg/kg d.w.) by Group (Mass Percent in Parentheses)				Total Conc. (mg/kg)
	1	2	3	4	
0	2,650 (5.5)	25,740 (53.8)	17,740 (37.1)	1,710 (3.6)	47,840
2	2,170 (3.8)	33,600 (58.4)	20,320 (35.3)	1,430 (2.5)	57,520
6	2,100 (3.6)	33,820 (57.9)	20,860 (35.7)	1,610 (2.8)	58,390
9	70 (0.5)	6,250 (43.6)	7,470 (52.1)	550 (3.9)	14,340
14	ND	1,870 (28.1)	4,340 (65.1)	450 (6.8)	6,660
21	ND	580 (42.0)	780 (58.0)	ND	1,360
Percent Degradation	100.00	97.8	95.6	>74	97.2

ND = Not detectable

Pilot-Scale Soil Washing. Figure 3 shows the results from a run in which only recycled water (i.e., no additives) was used as the extractant. In that run, heavy clayey soil was loaded at a rate of 550 kg/h. The incoming soil had approximately 9,800 mg/kg total PAH. The "washed" fraction had about 1,000 mg/kg for a 90% reduction in PAH. The sludge for biotreatment had about 7,400 mg/kg PAH, whereas the soluble fraction of PAH in the washwater was about 5 mg/L. Unsettled fines in the washwater would account for the remaining mass of PAH. The addition of polymer to the washwater in subsequent tests has resulted in greater recovery of PAHs in the settled sludge such that spent washwater contained <1 mg/L PAH and could be discharged into the sewage system.

Pilot-Scale Bioslurry Treatment. The gravity settled sludge resulting from the water washing process was pumped to a 600-L mixing tank where pH was adjusted to between 7 and 7.5 using 5 M NaOH. Nutrients were added as NH_3-N, NO_3-N, and PO_4-P at approximately 300 mg/L, 400 mg/L, and 10 mg/L final concentrations, respectively. The slurry, approximately 14% solids, was then pumped to the Biolift reactor for batch operation. Samples were taken daily for analyses of solids, PAH, nitrogen, and phosphorus. The oxygen uptake rates (OURs), aeration rates, and mixing rates were monitored daily. Temperature was maintained between 28 and 32°C.

The results of one run are shown in Figure 4, where again the apparent increase in PAH concentration is observed on Day 2, followed by rapid degradation. In this test, the initial concentration of 5,700 mg/kg was reduced by 70% within 6 days and by 95% by Day 8. The greatest OUR measured at the peak of PAH removal (Day 4) was 0.40 mg/L-min.

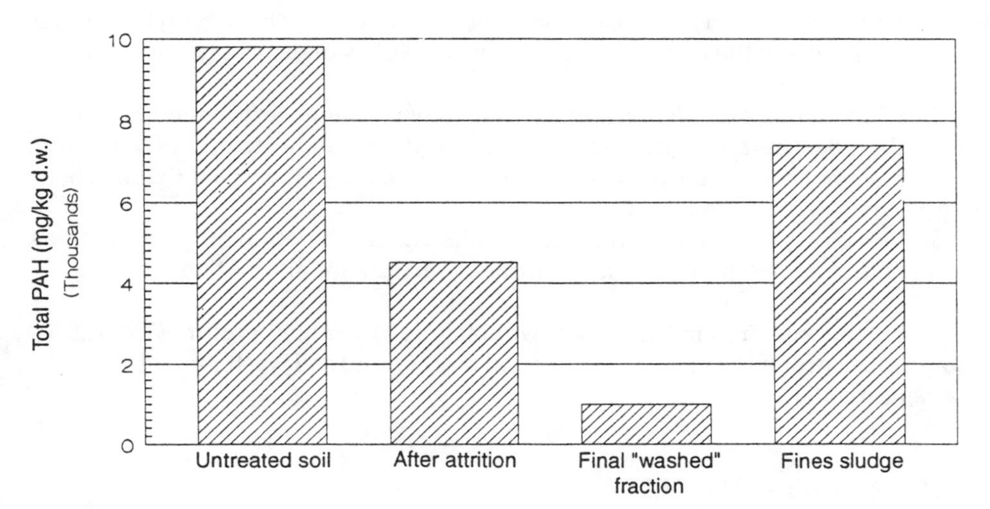

FIGURE 3. Effect of water washing a clayey subsurface soil. Soil was loaded at a rate of 550 kg/h and run for 6 hours using cold water only.

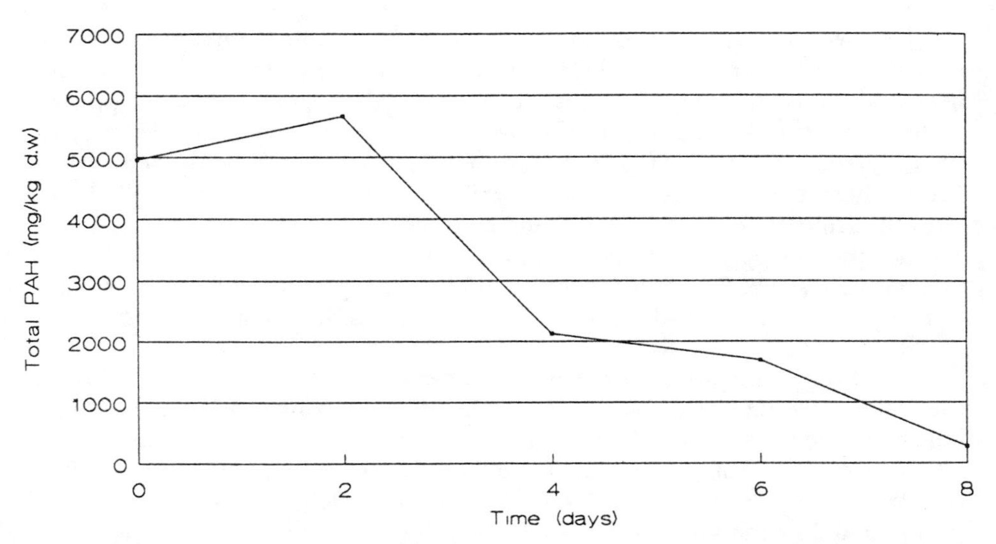

FIGURE 4. Effect of pilot-scale bioslurry treatment in a batch mode. The reactor was loaded with 14% solids.

SUMMARY

1. Bench- and pilot-scale studies showed that indigenous populations of microbes, with nutrient amendments and pH control, were capable of significant degradation of PAHs in creosote-containing soil slurries at approximately 30°C. Only slurries containing 14 to 20% solids have been tested thus far, however the manufacturer claims that 40% slurries can be treated in the equipment.
2. Soil washing provides an effective means of washing particles and increasing PAH bioavailability in slurries. The particle separation features of the system have not been used yet, due to the predominantly clayey soil that has been excavated and treated thus far.
3. Based on preliminary pilot-scale results, the combination of water washing and bioslurry treatment is an attractive treatment alternative.

Further research in the project will include lower temperature (15 to 20°C) biotreatment at higher solids loadings (30 to 40%), and soil washing with different amendments.

ACKNOWLEDGMENTS

The study is supported by the Norwegian State Railways (NSB), the Norwegian State Pollution Control Agency (SFT), and Norwegian Applied Technology A/S. The University of Washington, Valle Foundation provided support for T. Bennett. The collaboration with Dr. J. Mueller, SBP Technologies, Inc. and assistance of Drs. H. Pritchard and R. Devereaux of U.S. EPA Gulf Breeze Environmental Research Laboratory are gratefully acknowledged.

REFERENCES

Berg, J. D., T. E. Bennett, B. S. Nesgård, and J. G. Mueller. 1993. "Treatment of creosote-contaminated soil by soil washing and slurry-phase bioreactors." *Proceedings of International Symposium on Environmental Contamination in Central and Eastern Europe.* Budapest, Hungary, October 12-16, 1992. (In press).

Blais, L. 1987. "Pilot-scale biological treatment of contaminated groundwater at an abandoned wood treatment plant." In G. S. Omenn (Ed.), *Environmental Biotechnology: Reducing the Risk from Environmental Chemicals through Biotechnology*, pp. 445-446. Plenum Press, New York, NY.

Esposito, P., J. Hessling, B. B. Locke, M. Taylor, M. Szalo, R. Thurman, C. Rogers, R. Traver, and E. Barth. 1989. "Results of treatment evaluations of a contaminated synthetic soil." *JAPCA-Jour. Air and Waste Management Assoc. 39*: 294.

Matthews, J. E. 1992. "Evaluation of full-scale in situ and ex situ bioremediation of creosote wastes." *Symposium on Bioremediation of Hazardous Wastes*, EPA's Biosystems Technology Development Program, Chicago, IL, May 5-6.

Mueller, J. G., P. J. Chapman, and P. H. Pritchard. 1989. "Creosote-contaminated sites." *Environ. Sci. Technol.* 23: 1197-1201.

Mueller, J. G., P. J. Chapman, and P. H. Pritchard. 1991. "Biodegradation of creosote and pentachlorophenol in contaminated groundwater: Chemical and biological assessment." *Appl. Environ. Microbiol. 57*: 1277-1285.

Mueller, T. G., S. E. Lantz, R. Devereux, J. D. Berg, and P. H. Pritchard. 1994. "Studies on the microbial ecology of PAH biodegradation." In R. E. Hinchee, A. Leeson, L. Semprini, and S. K. Ong (Eds.), *Bioremediation of Chlorinated and Polycyclic Aromatic Hydrocarbons.* Lewis Publishers, Ann Arbor, MA.

Reineman, J. A. 1987. "Field demonstration of the effectiveness of land application for the treatment of creosote-contaminated wastes." In G. S. Omenn (Ed.), *Environmental Biotechnology: Reducing the Risk from Environmental Chemicals through Biotechnology*, pp. 459-460. Plenum Press, New York, NY.

U.S. EPA. 1991. "Engineering bulletin — Soil washing treatment." *Innovative Treatment Technologies: Overview and Guide to Information Sources*, pp. 4-15 to 4-25. EPA/540/9-91/002. U.S. Environmental Protection Agency, Washington, DC.

U.S. EPA. 1989. *Cleaning Excavated Soil Using Extraction Agents: A State-of-the-art Review.* EPA/600/2-89/034. U.S. Environmental Protection Agency, Cincinnati, OH.

DEMONSTRATION OF SOIL BIOREMEDIATION AND TOXICITY REDUCTION BY FUNGAL TREATMENT

F. Baud-Grasset, S. I. Safferman, S. Baud-Grasset, and R. T. Lamar

INTRODUCTION

A wide range of environmentally persistent pollutants, including polycyclic aromatic hydrocarbons (PAHs), have been shown to be biotransformed or completely degraded to carbon dioxide by a lignin-degrading fungus, *Phanerochaete chrysosporium*, under laboratory conditions (Aust 1990, Hammel 1989, Hammel et al. 1986, Lamar et al. 1990). A protocol to assess treatment of contaminated soils using lignin-degrading fungi has been developed (Baud-Grasset et al. 1992a, Safferman et al. 1992) and was tested on a PAH-contaminated soil from a hazardous waste site. The target compounds were acenaphthene, acenaphthylene, dibenzofuran, fluorene, phenanthrene, anthracene, fluoranthrene, pyrene, and benzo(a)anthracene. Standardized bioremediation protocols that mimic full-scale treatments allow for consistent and economical evaluations and comparisons of remediation options under controlled conditions. These protocols are essential for developing technologies such as lignin-degrading fungi processes, where the database of experience is limited. They not only help decrease the chance of failure of full-scale systems but also can be a research tool for further technical development. The objective of this study was to evaluate a pilot-scale procedure designed to assess the bioremediation and toxicity reduction of a PAH-contaminated soil by lignin-degrading fungi.

METHODOLOGY AND EXPERIMENTAL DESIGN

Fungi. Laboratory analyses were conducted to select fungal species with the lowest sensitivity to the PAH present in the soil and the best growth characteristics within the contaminated soil. *Phanerochaete chrysosporium* (BKM F-1767, ATCC 24725) and *Phanerochaete sordida* (HBB-8922-sp) showed the best potential to degrade the target pollutants and were selected for pilot-scale soil pan studies. Fungal strains were obtained from the Forest Products Laboratory, U.S. Department of Agriculture, Madison, Wisconsin. Inoculum medium consisted of a commercially available sawdust-grain mix from Lambert Spawn Co., Inc., Coatsville, Pennsylvania.

Pilot-Scale Soil Pan Study. The pilot-scale soil pan study was designed to mimic full-scale land treatment processes. This study was conducted in 30 × 30 × 45 cm stainless steel soil pans with a leachate collection system, all of which were housed in aerobic, ventilated glove boxes to protect researchers and to control air quality. The glove box temperature was maintained at 25°C ± 2°C. All air entering the glove boxes was filtered to remove microorganisms. A PAH-contaminated sandy soil from a hazardous waste site, where wood had been treated with preservative over a period of 40 years, was used for this study. The soil was well mixed and sieved to remove particles larger than 1.25 cm prior to utilization.

The soil was amended with sterile wood chips (2.5% dry weight basis) and inoculated with fungi-infested sawdust-grain mix at ratios of 6 and 12 g of mix per 100 g of soil for *Phanerochaete chrysosporium* (PC6% and PC12%) and at 12 g of mix per 100 g of soil for *Phanerochaete sordida* (PS12%) on a dry weight basis. These treatments were performed in duplicate. The study was conducted over a period of 2 months. The soil was tilled weekly to provide aeration, and its moisture content was maintained at field capacity by adding water daily. In addition, the removal of the pollutant from the soil by the indigenous microbial population or through abiotic mechanisms was assessed in nonsterile control pans. Two different controls were conducted: a control containing soil only (control) and a control containing soil and noncontaminated sterilized wood chips (control with wood chips).

Chemical Analyses. Samples were collected and extracted for PAH immediately after filling the pans with soil and after 2, 4, and 8 weeks. The soil was mixed before each sampling. For each sample, four subsamples were collected from different locations in accordance to a statistically designed plan and mixed together. Soil samples were stored at 4°C. The method used to extract the compounds from the soil was a modification of EPA SW-846 Method 3550. The modification consisted of using three sonication extractions with three different solvents. A mixture of 90% acetone and 10% hexane was mixed with the soil and sonicated for 15 min. The second extraction used an equal mix of both hexane and acetone, and the third contained 10% acetone and 90% hexane. All extracts were filtered through anhydrous sodium sulfate and combined in a separatory funnel. The hexane layer was then removed and concentrated using a Kuderna-Danish apparatus. EPA SW-846 Method 8100 was used to analyze the extract employing a Hewlett Packard Series II gas chromatograph with a flame ionization detector (GC/FID), and a DB5 capillary column at the initial temperature of 125°C for 3 min with a ramp of 5°C per min. The final temperature was 290°C for 25 min.

Biological Assessment. In addition to monitoring the disappearance of the target parent compounds, the actual detoxification of the soil was monitored by employing several toxicity assays using higher plants (Baud-Grasset et al. 1992b, Baud-Grasset et al., in press), and microorganisms (Microtox™ assay, described below). They were selected because toxicity of a complex chemical mixture often is not easily measured by chemical analyses and the disappearance of parent compounds may not indicate detoxification of the soil. The Microtox™ system was selected to evaluate the toxicity of the soil before and after fungal treatment using two different procedures: the aqueous-phase test (Microbics Corp. 1991a) and the solid-phase test (Microbics Corp. 1991b). The aqueous extracts were obtained by mixing 100 g of soil with 400 mL of water. The mixture was shaken in darkness at 20°C for 48 h. Resulting suspensions were centrifuged at 10,000 rpm for 10 min at 4°C and filtered through a 0.45-µm-pore-diameter filter. In the solid-phase test, luminescent microorganisms contained in Microtox™ reagent were exposed directly to the solid phase (0.3 g of soil) in a water suspension. After exposure to the sample, the organisms were separated by a filtration system provided by Microbics.

RESULTS AND DISCUSSION

PAH Depletion. Before treatment, total PAHs were present at concentrations ranging from 5,139 mg kg^{-1} to 6,582 mg kg^{-1}. These initial concentrations were not significantly different from one pan to another. The analysis of variance (ANOVA) gave an $F=0.42<2.85$. The three- and four-ring compounds ranged from 3,227 to 4,544 mg kg^{-1} and from 1,843 to 2,551 mg kg^{-1}, respectively. The PAH concentrations in the different fungal treated systems decreased with rates significantly higher than in the controls (those with indigenous microorganisms only) (Figure 1). After 8 weeks, the PAH concentrations in the soil were reduced by only 49% and 57% for the controls without and with the sterile wood chips, respectively. The concentration of PAHs after 8 weeks in the controls with sterile

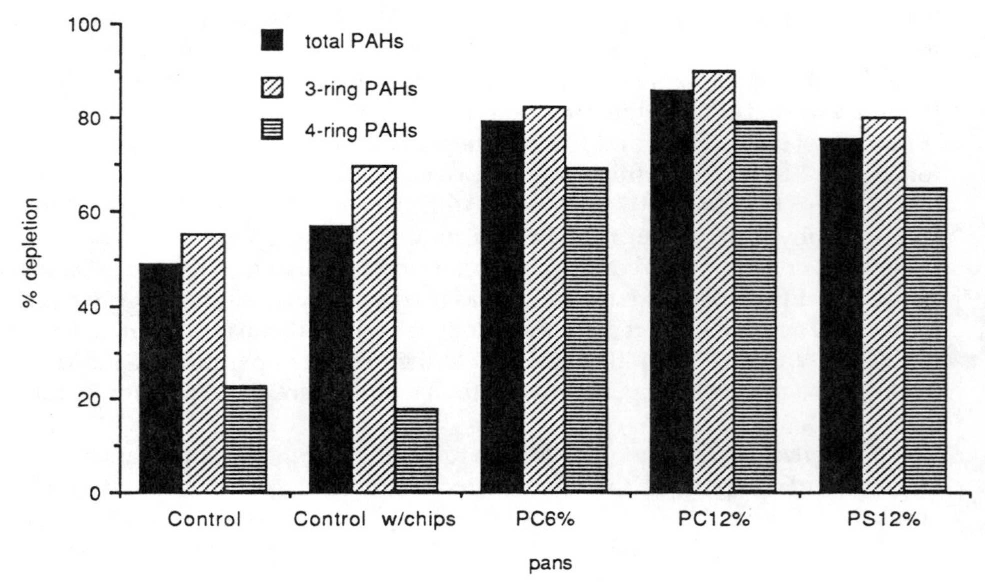

FIGURE 1. Percentages of PAH depletion after 8 weeks.

wood chips was significantly lower than in the contaminated soil alone, which could be explained by the adsorption of PAHs on the wood chips. PAH depletion in fungi-inoculated soils were 79%, 86%, and 76% for PC6%, PC12%, and PS12%, respectively. The PAH concentrations in the fungal treated systems were statistically different from those of the controls (the ANOVA gave an $F = 10.75 > 2.60$), and concentrations in PC12% also were statistically lower than in PS12% (mean comparison, Fisher and Yates test). The difference between the controls and the fungal treated systems was more significant for the four-ring compounds than for other PAHs, with 22% removal in the control treatment and 69% (PC6%), 79% (PC12%), and 65% (PS12%) with the fungal treatments. The depletion of the three-ring compounds, for example, also was high in the controls. The three-ring compounds were probably degraded by indigenous microorganisms.

Biological Assessment. The EC50 values measured for the Microtox™ aqueous-phase and solid-phase tests are summarized in Table 1. They represent the percentage of soil extract or contaminated soil that is estimated to give a 50% inhibition of the bioluminescent bacteria. Prior to treatment, the bioluminescent bacteria were very sensitive to the contaminated soil using both tests. A significant reduction in toxicity occurred after fungal treatment (PC12%) for both tests. The bioluminescent bacteria were more sensitive to toxic compounds in the solid-phase test than in the aqueous-phase test, probably because of the very low water solubility of most PAHs. Similar results were obtained with phytotoxicity tests conducted on soil samples (seed germination) or aqueous extracts (root elongation) (Baud-Grasset et al., in press). In addition, the tradescantia-micronucleus test showed a significant decrease of genotoxicity in the water extracts of the contaminated soil after fungal treatment (PC12%) (Baud-Grasset et al. 1992b). Although detoxification occurred, the soil still had a measurable toxic effect after treatment. The occurrence of nondegraded parent compounds or toxic metabolites could explain this remaining toxicity.

TABLE 1. EC50 values for the aqueous-phase test (percent of soil extract) and the solid-phase test (percent of contaminated soil).

	Aqueous-Phase Test		Solid-Phase Test	
	EC50	95% C.I.	EC50	95% C.I.
CBT (T=0)	7.06	5.98-8.39	0.19	0.17-0.22
CAT (T=8 weeks)	16.17	13.30-19.87	0.34	0.25-0.48
AFT (T=8 weeks)	28.80	23.66-35.27	2.81	2.44-3.24

CBT = control before treatment. CAT = control after treatment.
AFT = after fungal treatment. C.I. = confidence interval.

SUMMARY

This study was conducted to demonstrate the efficacy of a standardized protocol for testing bioremediation technologies. Evaluation of a fungal-based soil remediation technology using the protocol demonstrated distinct pollutant removal efficiencies between the controls and the different fungal inoculations, especially for the four-ring compounds. Toxicity tests revealed a significant detoxification of the soil during fungal treatment. This specific application of the standardized protocol predicts that a fungal treatment with the selected species would be capable of bioremediating and detoxifying PAH-contaminated soil in a full-scale land-farming treatment process.

REFERENCES

Aust, S. D. 1990. "Degradation of Environmental Pollutants by *Phanerochaete chrysosporium*." *Microbiol. Ecol. 20*: 197-209.

Baud-Grasset, F., S. I. Safferman, and R. T. Lamar. 1992a. "Development of Small-Scale Evaluation Techniques for Fungal Treatment of Soils." Abstract in *18th Annual RREL Research Symposium*, U.S. Environmental Protection Agency, Cincinnati, OH, April 14-16.

Baud-Grasset, F., S. Baud-Grasset, and S. I. Safferman. In Press. "Evaluation of the Bioremediation of a Contaminated Soil with Phytotoxicity Tests." *Chemosphere*.

Baud-Grasset, S., F. Baud-Grasset, J. M. Bifulco, J. R. Meier, and T. H. Ma. 1992b. "Bioremediation of a Contaminated Soil Evaluated with a Set of Toxicity Tests: Tradescantia-Micronucleus Test." In *Abstracts of the SETAC 13th Annual Meeting*, p. 197. Cincinnati, OH, November 8-12.

Hammel, K. E. 1989. "Organopollutant Degradation by Ligninolytic Fungi." *Enzyme Microb. Technol. 11*: 776-777.

Hammel, K. E., B. Kalyanaraman, and T. K. Kirk. 1986. "Oxidation of Polycyclic Aromatic Hydrocarbons and Dibenzo (p) Dioxins by *Phanerochaete chrysosporium* Ligninase." *J. Biol. Chem. 261*: 16948-16952.

Lamar, R. T., M. J. Larsen, and T. K. Kirk. 1990. "Sensitivity to and Degradation of Pentachlorophenol by *Phanerochaete* spp." *Appl. Environ. Microbiol. 56*: 3519-3526.

Microbics Corporation. 1991a. *Microtox Manual. Preliminary Release.* Carlsbad, CA.

Microbics Corporation. 1991b. *Solid Phase Test Protocols.* Carlsbad, CA.

Safferman, S. I., F. Baud-Grasset, S. Baud-Grasset, and R. T. Lamar. 1992. "Development of Small-Scale Evaluation Techniques for Fungal Treatment of Soils." In *Bioremediation of Hazardous Wastes Annual Symposium, EPA's Biosystems Technology Program*, pp. 135-138. Chicago, IL, May 5-6.

AUTHOR LIST

D. A. Abramowicz
GE Corporate Research and
Development Center
K-1, 3A25, P.O. Box 8
Schenectady, NY 12301 USA

D. W. Acton
Beak Consultants Limited
42 Arrow Road
Guelph, Ontario
CANADA N1K 1S6

S. D. Aust
Utah State University
Biotechnology Center
Logan, UT 84322-4705 USA

M. T. Balba
TreaTek-CRA Company
2801 Long Road
Grand Island, NY 14072 USA

D. P. Barr
Utah State University
Biotechnology Center
Logan, UT 84322-4705 USA

F. T. Barranco
EA Engineering, Science and
Technology, Inc.
15 Loveton Circle
Sparks, MD 21152 USA

F. Baud-Grasset
Rhône-Poulenc Industrialisation
Safety & Environment Department
24 Avenue Jean Jaurès
69153 Décines-Charpieu Cedex
FRANCE

S. Baud-Grasset
IT Corporation
U.S. EPA Test and Evaluation Facility
Cincinnati, OH 45208 USA

M. E. Bedessem
University of Wyoming
Department of Zoology & Physiology
P.O. Box 3166
Laramie, WY 82071 USA

R. E. Beeman
DuPont/Conoco
Conoco, Inc.
1000 South Pine, 131 RDE
Ponca City, OK 74603 USA

P. E. Bell
HYDROSYSTEMS, Inc.
2340 Commonwealth Drive, Suite 202
Charlottesville, VA 22901 USA

T. E. Bennett
W. D. Purnell and Associates
2138 Humboldt Street
Bellingham, WA 98227 USA

J. D. Berg
U.S. EPA Gulf Breeze Environmental
Research Laboratory
One Sabine Island Drive
Gulf Breeze, FL 32561-3999 USA

J. D. Berg
Aquateam A/S
Post Office Box 6326
Etterstad, 0604 Oslo NORWAY

V. H. Bess
BBC Laboratory
3220 South Fair Lane, Suite 18
Tempe, AZ 85008 USA

J. Blake
U.S. Environmental Protection Agency
Office of Pollution Prevention and
Toxics (TS-798)
401 M Street, S.W.
Washington, DC 20460 USA

R. Bos
Department of Biochemistry
University of Groningen
Nijenborgh 4
9747 AG Groningen
THE NETHERLANDS

E. J. Bouwer
The Johns Hopkins University
Department of Geography and Environmental Engineering
313 Ames Hall
34th and Charles Streets
Baltimore, MD 21218 USA

J. P. Bowman
Center for Environmental Biotechnology
University of Tennessee
10515 Research Drive, Suite 100
Knoxville, TN 37932 USA

A. A. Bracco
GE Corporate Research and Development Center
P.O. Box 8
Schenectady, NY 12301 USA

M. J. Brennan
GE Corporate Research and Development Center
P.O. Box 8
Schenectady, NY 12301 USA

F. J. Brockman
Battelle Pacific Northwest Laboratories
900 Battelle Boulevard
Mail Stop K4-06
Richland, WA 99352 USA

T. M. Brouns
Battelle Pacific Northwest Laboratories
P.O. Box 999
Richland, WA 99352 USA

R. G. Burns
Biological Laboratory
University of Kent at Canterbury
Kent CT2 7NJ, England UK

W. A. Butler
Du Pont Environmental Biotechnology Program
Glasgow Site, Building 300
P.O. 6101
Newark, DE 19714-6101 USA

J. R. Buttram
DuPont, Inc.
P.O. Box 2626
Victoria, TX 77902 USA

D. J. Cady
OHM Remediation Services Corporation
5335 Triangle Parkway, Suite 450
Norcross, GA 30092 USA

S. M. Caley
Battelle Pacific Northwest Laboratories
P.O. Box 999
Mail Stop P7-41
Richland, WA 99352 USA

R. Campbell
Center for Environmental Diagnostics and Bioremediation
Department of Biology
University of West Florida
Pensacola, FL 32514 USA

K. M. Carroll
GE Corporate Research and Development Center
P.O. Box 8
Schenectady, NY 12301 USA

G. O. Chieruzzi
Keystone Environmental Resources
Chester Environmental
3000 Tech Center Drive
Monroeville, PA 15146 USA

T. H. Christensen
Technical University of Denmark
Department of Environmental Engineering
Building 115
DK-2800 Lyngby DENMARK

N. Chung
Utah State University
Biotechnology Center
Logan, UT 84322-4705 USA

C. D. Chunn
Envirogen
4100 Quakerbridge Road
Lawrenceville, NJ 08648 USA

P.J.S. Colberg
University of Wyoming
Department of Zoology & Physiology
P.O. Box 3166
Laramie, WY 82071 USA

Y. Comeau
Department of Civil Engineering
École Polytechnique de Montréal
Post Office Box 6079
Station A
Montreal, Quebec
CANADA H3C 3A7

M. Y. Corapcioglu
Department of Civil Engineering
Texas A&M University
College Station, TX 77843-3136 USA

E. E. Cox
Beak Consultants Limited
42 Arrow Road
Guelph, Ontario
CANADA N1K 1S6

C. G. Coyle
U.S. Army Corps of Engineers
Kansas City District (ED-GE)
601 East 12th Street
Kansas City, MO 64106-2896 USA

S. L. De Wys
Malcolm Pirnie, Inc.
445 Hutchinson Avenue, Suite 350
Columbus, OH 43235-5622 USA

L. J. DeJong
Department of Civil Engineering, FX-10
University of Washington
Seattle, WA 98195 USA

R. E. Devereux
U.S. Environmental Protection Agency
Gulf Breeze Environmental Research Laboratory
One Sabine Island Drive
Gulf Breeze, FL 32561-3999 USA

D. K. Dietrich
GE Corporate Research and Development Center
P.O. Box 8
Schenectady, NY 12301 USA

M. E. Dolan
Stanford University
Department of Civil Engineering
Stanford, CA 94305-4020 USA

G. S. Douglas
Battelle Ocean Sciences
397 Washington Street
Duxbury, MA 02332 USA

J. Duffy
Occidental Chemical Corporation
Grand Island, NY 14072 USA

N. D. Durant
The Johns Hopkins University
Department of Geography and Environmental Engineering
313 Ames Hall
34th and Charles Street
Baltimore, MD 21218 USA

D. L. Elmendorf
Exxon Research and Engineering Company
Clinton Township, Route 22E
Annandale, NJ 08801 USA

J. H. Exner
JHE Technologies Systems
2 Waverly Court
Alamo, CA 94507 USA

G. A. Ferguson
ENSR Consulting and Engineering
740 Pasquinelli Drive
Westmont, IL 60559 USA

J. F. Ferguson
Department of Environmental Engineering & Science, FX-10
University of Washington
Seattle, WA 98195 USA

S. Fiorenza
Amoco Corporation
7201 East 38th Street, Space 7253
Tulsa, OK 74145 USA

K. M. Fish
GE Corporate Research and Development Center
P.O. Box 8
Schenectady, NY 12301 USA

G. A. Fisher
Motorola, Inc.
5005 East McDowell Road
Phoenix, AZ 85008 USA

W. P. Flanagan
GE Corporate Research and Development Center
P.O. Box 8
Schenectady, NY 12301 USA

D. R. Foster
Chester Environmental
3000 Tech Center Drive
Monroeville, PA 15146 USA

J. K. Fredrickson
Battelle Pacific Northwest Laboratories
P.O. Box 999
Richland, WA 99352 USA

F. H. Frimmel
DVGW - Forschungsstelle
University of Karlsruhe
Richard - Willstatter Allee 5
7500 Karlsruhe 1 GERMANY

C. W. Fulton
CH2M HILL Engineering Ltd.
1600, 555 - 4 Avenue S.W.
Calgary, Alberta
CANADA T2P 3E7

J. Gagnon
Department of Civil Engineering
École Polytechnique de Montréal
Post Office Box 6079
Station A
Montreal, Quebec
CANADA H3C 3A7

W. L. Gately
GE Corporate Research and Development Center
P.O. Box 8
Schenectady, NY 12301 USA

R. R. Gerger
U.S. Environmental Protection Agency
One Sabine Island Drive
Gulf Breeze, FL 32561-3999 USA

G. Giesbrecht
CH2M HILL Engineering Ltd.
1600, 555 4th Avenue
Calgary, Alberta
CANADA T2P 3E7

J. A. Glaser
U.S. Environmental Protection Agency
Risk Reduction Engineering Laboratory
26 W. Martin Luther King Drive
Cincinnati, OH 45268 USA

M. H. Gold
Chemical & Biological Sciences
Oregon Graduate Institute of Science & Technology
19600 N.W. von Neumann Drive
Beaverton, OR 97006-1999 USA

C. W. Greer
Biotechnology Research Institute
National Research Council Canada
6100 Royalmount Avenue
Montreal, Quebec
CANADA H4P 2R2

R. Gundersen
Aquateam A/S
Post Office Box 6326
Etterstad, 0604 Oslo NORWAY

C. E. Haith
Exxon Research and Engineering Company
Clinton Township, Route 22E
Annandale, NJ 08801 USA

E. J. Hansen
New Mexico Environment Department
1190 St. Francis Drive
Santa Fe, NM 87503 USA

M. R. Harkness
GE Corporate Research and Development Center
CEB 406, P.O. Box 8
Schenectady, NY 12301 USA

S. E. Herbes
Environmental Sciences Division
Oak Ridge National Laboratory
Oak Ridge, TN 37831 USA

E. L. Hockman, Jr.
Amoco Corporation
7201 East 38th Street, Space 7253
Tulsa, OK 74145 USA

H. L. Hoffman
Baltimore Gas & Electric Company
P.O. Box 1475
Baltimore, MD 21203 USA

R. E. Hoffmann
Alberta Research Council
Environmental Engineering and Engineering Department
P.O. Box 8330, Station F
Edmonton, Alberta
CANADA T6H 5X2

B. S. Hooker
Chemical Engineering Department
Tri-State University
Angola, IN 46703 USA

G. D. Hopkins
Department of Civil Engineering
Stanford University
Stanford, CA 94305-4020 USA

M. A. Hossain
Atlantic Environmental
188 Norwich Avenue
Colchester, CT 06415-0297 USA

H. Hötzl
Department of Applied Geology
University of Karlsruhe
Post Office Box 6980
D-7500 Karlsruhe
GERMANY

J. E. Howell
DuPont/Conoco
Conoco, Inc.
1000 South Pine, 131 RDE
Ponca City, OK 74603 USA

J. Hu-Primmer
Center for Environmental Diagnostics and Bioremediation
Department of Biology
University of West Florida
Pensacola, FL 32514 USA

A. Hüttermann
Forstbotanisches Institut der Universität Göttingen
Büsgenweg 2, D-3400 Göttingen
GERMANY

K. Ishizuka
Tsukuba University
1-1 Tennoudai Tsukuba Ibaraki, 305
JAPAN

K. Iwasaki
National Institute for Environmental Studies
16-2 Onogawa Tsukuba Ibaraki, 305
JAPAN

D. B. Janssen
University of Groningen
Laboratory of Biochemistry
Nijenborgh 4
9747 AG Groningen
THE NETHERLANDS

K. T. Järvinen
Tampere University of Technology
Institute of Water and Environmental Engineering
Post Office Box 699, SF-33101
Tampere FINLAND

D. E. Jerger
OHM Remediation Services Corporation
16406 U.S. State Route 224 East
Findlay, OH 45840 USA

L. Jimenez
University of Tennessee
Center for Environmental Biotechnology
Knoxville, TN 37920 USA

D. L. Johnstone
Washington State University
Sloan Hall
Civil and Environmental Engineering
Pullman, WA 99164-2910 USA

D. K. Joshi
Chemical & Biological Sciences
Oregon Graduate Institute of Science & Technology
19600 N.W. von Neumann Drive
Beaverton, OR 97006-1999 USA

D. H. Kampbell
U.S. Environmental Protection Agency
Robert S. Kerr Environmental Laboratory
P.O. Box 1198
Ada, OK 74820 USA

M. Kästner
Department of Biotechnology II
Technical University of Hamburg Harburg
Denicke Street 15
2100 Hamburg 90 GERMANY

J. E. Keenan
Department of Civil Engineering, FX-10
University of Washington
Seattle, WA 98195 USA

M. Kikuma
Tsukuba University
1-1 Tennoudai Tsukuba Ibaraki, 305
JAPAN

M. T. Kingsley
Battelle Pacific Northwest Laboratories
P.O. Box 999, P7-54
Richland, WA 99352 USA

J. L. Kocornik
EA Engineering, Science and Technology, Inc.
15 Loveton Circle
Sparks, MD 21152 USA

S.-C. Koh
Center for Environmental Biotechnology
University of Tennessee
10515 Research Drive, Suite 100
Knoxville, TN 37932 USA

S. J. Komisar
Department of Environmental Engineering & Science, FX-10
University of Washington
Seattle, WA 98195 USA

R. T. Lamar
U.S. Department of Agriculture
Forest Products Laboratory
1 Grifford Pinchot Drive
Madison, WI 53705-2398 USA

S. E. Lantz
U.S. Environmental Protection Agency
Gulf Breeze Environmental Research Laboratory
One Sabine Island Drive
Gulf Breeze, FL 32561-3999 USA

L. T. LaPat-Polasko
Woodword-Clyde Consultants
426 North 44th Street, Suite 300
Phoenix, AZ 85008 USA

M. D. Lee
Du Pont Environmental Biotechnology Program
Glasgow Site, Building 300
P.O. 6101
Newark, DE 19714-6101 USA

R. Legrand
Radian Corporation
P.O. Box 201088
Austin, TX 78720-1088 USA

D. T. Lin
GE Corporate Research and Development Center
P.O. Box 8
Schenectady, NY 12301 USA

C. D. Litchfield
Chester Environmental
3000 Tech Center Drive
Monroeville, PA 15146 USA

J. H. Lobos
GE Corporate Research and Development Center
CEB 406, P.O. Box 8
Schenectady, NY 12301 USA

C. M. Logan
Baltimore Gas & Electric Company
1000 Brandon Shores Road
Baltimore, MD 21226 USA

A. Lorentsen
Aquateam A/S
Post Office Box 6326
Etterstad, 0604 Oslo NORWAY

I. D. MacFarlane
EA Engineering, Science, and Technology, Inc.
15 Loveton Circle
Sparks, MD 21152 USA

R. Mackowski
Center for Environmental Biotechnology
University of Tennessee
Knoxville, TN 37932-2567 USA

B. Mahro
Department of Biotechnology II
Technical University of Hamburg Harburg
Denicke Street 15
2100 Hamburg 90
GERMANY

A. Majcherczyk
Forstbotanisches Institut der Universität Göttingen
Büsgenweg 2, D-3400 Göttingen
GERMANY

D. W. Major
Beak Consultants Limited
42 Arrow Road
Guelph, Ontario
CANADA N1K 1S6

L. Makowski
Florida State University
Institute of Molecular Biophysics
Tallahassee, FL 32306 USA

J. M. Marowitch
Alberta Research Council
Post Office Box 8330, Station F
Edmonton, Alberta
CANADA T6H 5X2

I. J. Marvan
Deaborn Chemical Company Ltd.
3451 Erindale Station Road
Post Office Box 3060, Station A
Mississauga, Ontario
CANADA L5A 3T5

N. B. Matolak
HYDROSYSTEMS, Inc.
2340 Commonwealth Drive, Suite 202
Charlottesville, VA 22901 USA

N. Matte
Biotechnology Research Institute — NRCC
6100 Royalmount Avenue
Montreal, Quebec
CANADA H4P 2R2

R. J. May
GE Corporate Research and
Development Center
P.O. Box 8
Schenectady, NY 12301 USA

P. L. McCarty
Department of Civil Engineering
Stanford University
Stanford, CA 94305-4020 USA

G. D. McCleary
EA Engineering, Science, and
Technology, Inc.
15 Loveton Circle
Sparks, MD 21152 USA

J. B. McDermott
GE Corporate Research and
Development Center
CEB 420, P.O. Box 8
Schenectady, NY 12301 USA

I. McGhee
Biological Laboratory
University of Kent at Canterbury
Kent CT2 7NJ, England UK

P. E. McIntire
HYDROSYSTEMS, Inc.
2340 Commonwealth Drive, Suite 202
Charlottesville, VA 22901 USA

F. B. Metting, Jr.
Battelle Pacific Northwest
Laboratories
P.O. Box 999
Richland, WA 99352 USA

D. L. Middleton
Chester Environmental
3000 Tech Center Drive
Monroeville, PA 15146 USA

K. Mishima
Ebara Research Company, Ltd.
2-1 Honfujisawa 4-chome
Fujisawa-shi 251 JAPAN

T. F. Mistretta
Du Pont Environmental Biotechnology
Program
Glasgow Site, Building 300
P.O. Box 6101
Newark, DE 19714-6101 USA

F. J. Mondello
GE Corporate Research and
Development Center
P.O. Box 8
Schenectady, NY 12301 USA

C. B. Morgan
GE Corporate Research and
Development Center
P.O. Box 8
Schenectady, NY 12301 USA

P. J. Morris
U.S. Environmental Protection Agency
Environmental Research Laboratory
One Sabine Island Drive
Gulf Breeze, FL 32561 USA

J. G. Mueller
U.S. Environmental Protection Agency
Gulf Breeze Environmental Research
Laboratory
One Sabine Island Drive
Gulf Breeze, FL 32561-3999 USA

M. Nahold
Department of Applied Geology
University of Karlsruhe GERMANY

B. Nesgård
Aquateam Norwegian Water
Technology Centre A/S
Post Office Box 6326
Etterstad, 0604 Oslo NORWAY

P. H. Nielsen
Technical University of Denmark
Department of Environmental
Engineering
Building 115
DK-2800 Lyngby DENMARK

N. A. Nimo
HYDROSYSTEMS, Inc.
100 Carpenter Drive, Suite 200
Sterling, VA 22170 USA

A. Ogram
Washington State University
Crops & Soils Services
Pullman, WA 99164-6420 USA

F. Okada
Ebara Research Company, Ltd.
2-1 Honfujisawa 4-chome
Fujisawa-shi 251 JAPAN

M. P. Otte
Environmental Sciences Institute
University of Quebec in Montreal
Post Office Box 8888, Station A
Montreal, Quebec
CANADA H3C 3P8

A. V. Palumbo
Environmental Sciences Division
Oak Ridge National Laboratory
Oak Ridge, TN 37831 USA

T. W. Payne
Southwest Missouri State University
901 South National
Springfield, MO 65804-0094 USA

R. E. Perkins
Du Pont Environmental Biotechnology Program
Glasgow Site, Building 300
P.O. Box 6160
Newark, DE 19714-6101 USA

J. N. Petersen
Chemical Engineering Department
Washington State University
Pullman, WA 99164-2710 USA

V. A. Petrenko
Institute of Biologically Active Substances "Vektor" NPO
Berdsk, Novosibirsk Region
RUSSIA

S. M. Pfiffner
Center for Environmental Biotechnology
University of Tennessee
Knoxville, TN 37932-2567 USA

T. J. Phelps
Environmental Sciences Division
Oak Ridge National Laboratory
Oak Ridge, TN 37831 USA

K. R. Piontek
CH2M HILL
10 South Broadway, Suite 450
St. Louis, MO 63102 USA

F. Pries
Department of Biochemistry
University of Groningen
Nijenborgh 4
9747 AG Groningen
THE NETHERLANDS

R. C. Prince
Exxon Research and Engineering Company
Clinton Township, Route 22E
Annandale, NJ 08801 USA

P. H. Pritchard
U.S. Environmental Protection Agency
Gulf Breeze Environmental Research Laboratory
One Sabine Island Drive
Gulf Breeze, FL 32561-3999 USA

J. A. Puhakka
Department of Civil Engineering, FX-10
University of Washington
Seattle, WA 98195 USA

K. Ramanand
TreaTek-CRA Company
Grand Island, NY 14072 USA

G. Rasmussen
Department of Environmental Engineering & Science, FX-10
University of Washington
Seattle, WA 98195 USA

M. J. Reagin
Technical Resources Inc.
c/o U.S. Environmental Protection Agency
Sabine Island
Gulf Breeze, FL 32561-3999 USA

D. Ringelberg
Center for Environmental Biotechnology
University of Tennessee
Knoxville, TN 37932-2567 USA

M. F. Romine
Battelle Pacific Northwest Laboratories
900 Battelle Boulevard
Mail Stop K4-06
Richland, WA 99352 USA

I. Rosario
University of Tennessee
Center for Environmental Biotechnology
Knoxville, TN 37920 USA

R. Rothmel
Envirogen
4100 Quakerbridge Road
Lawrenceville, NJ 08648 USA

S. I. Safferman
U.S. Environmental Protection Agency
Risk Reduction Engineering Laboratory
26 W. Martin Luther King Drive
Cincinnati, OH 45268 USA

E. A. Salazar
DuPont/Conoco
Conoco, Inc.
1000 South Pine, 131 RDE
Ponca City, OK 74603 USA

J. J. Salvo
GE Corporate Research and Development Center
K-1 3C21, P.O. Box 8
Schenectady, NY 12301 USA

R. Samson
Biotechnology Research Institute
National Research Council Canada
6100 Royalmount
Montréal, P.Q.
CANADA H4P 2R2

G. S. Sayler
Center for Environmental Biotechnology
University of Tennessee
10515 Research Drive, Suite 100
Knoxville, TN 37932 USA

G. Schaefer
Department of Biotechnology II
Technical University of Hamburg Harburg
Denicke Street 15
21 Hamburg 90
GERMANY

R. Schaubhut
Center for Environmental Diagnostics and Bioremediation
Department of Biology
University of West Florida
Pensacola, FL 32514 USA

A. G. Seech
Dearborn Chemical Company, Ltd.
3451 Erindale Station Road
Post Office Box 3060, Station A
Mississauga, Ontario
CANADA L5A 3T5

R. L. Segar, Jr.
University of Texas at Austin
Dept. of Civil Engineering, ECJ 8.6
Austin, TX 78712 USA

R. J. Seidler
U.S. Environmental Protection Agency
Environmental Research Laboratory
Microbial Ecology/Biotechnology Program
Corvallis, OR 97333 USA

L. Semprini
Department of Civil Engineering
Oregon State University
Corvallis, OR 97331-2302 USA

M. M. Shah
Utah State University
Biotechnology Center
Logan, UT 84322-4705 USA

M. J. R. Shannon
Envirogen
4100 Quakerbridge Road
Lawrenceville, NJ 08648 USA

M. S. Shields
Center for Environmental Diagnostics and Bioremediation
Department of Biology
University of West Florida
11000 University Parkway
Pensacola, FL 32514 USA

T. Shimomura
Ebara Research Company, Limited
2-1 Honfujisawa 4-chome
Fujisawa-shi 251
JAPAN

S. H. Shoemaker
DuPont/Conoco
Conoco, Inc.
1000 South Pine, 131 RDE
Ponca City, OK 74603 USA

A. K. Siler
HYDROSYSTEMS, Inc.
100 Carpenter Drive, Suite 200
Sterling, VA 22170 USA

T. J. Simpkin
CH2M HILL, Inc.
P.O. Box 22058
Denver, CO 80222 USA

R. S. Skeen
Battelle Pacific Northwest Laboratories
P.O. Box 999
Richland, WA 99352 USA

G. J. Smith
ENSR Consulting and Engineering
1220 Avenida Acaso
Camarillo, CA 93012 USA

C. Somerville
Technical Resources Inc.
c/o U.S. Environmental Protection Agency
Sabine Island
Gulf Breeze, FL 32561-3999 USA

G. E. Speitel, Jr.
University of Texas
Dept. of Civil Engineering, ECJ8.6
Austin, TX 78712 USA

H. D. Stensel
Department of Civil Engineering, FX-10
University of Washington
Seattle, WA 98195 USA

M. L. Stephens
GE Corporate Research and Development Center
P.O. Box 8
Schenectady, NY 12301 USA

M. Stieber
DVGW - Forschungsstelle
University of Karlsruhe
Richard-Willstatter-Allee 5
7500 Karlsruhe 1
GERMANY

S. E. Strand
College of Forest Resources, AR-10
University of Washington
Seattle, WA 98195 USA

W. Sun
Washington State University
Pullman, WA 99164 USA

N. G. Swoboda-Colberg
University of Wyoming
Department of Zoology & Physiology
P.O. Box 3166
Laramie, WY 82701 USA

S. Szojka
Komex International
4500 16th Avenue N.W., Suite 100
Calgary, Alberta
CANADA T3B 0M6

D. G. Thomson
Alberta Research Council
Post Office Box 8330, Station F
Edmonton, Alberta
CANADA T6H 5X2

S. C. Tremaine
HYDROSYSTEMS, Inc.
2340 Commonwealth Drive, Suite 202
Charlottesville, VA 22901 USA

J. T. Trevors
University of Guelph
Department of Environmental Biology
Room 3220, Edmund C. Bovey Building
Ontario Agricultural College
Guelph, Ontario
CANADA N1G 2W1

M. R. Trudell
Alberta Research Council
Post Office Box 8330, Station F
Edmonton, Alberta
CANADA T6H 5X2

M. J. Truex
Battelle Pacific Northwest
Laboratories
P.O. Box 999
Mail Stop P7-41
Richland, WA 99352 USA

H. Uchiyama
National Institute for Environmental
Studies
16-2 Onogawa Tsukuba Ibaraki, 305
JAPAN

R. Unterman
Envirogen
4100 Quakerbridge Road
Lawrenceville, NJ 08648 USA

K. Valli
Chemical & Biological Sciences
Oregon Graduate Institute of Science
& Technology
19600 N.W. von Neumann Drive
Beaverton, OR 97006-1999 USA

A. J. van den Wijngaard
Department of Biochemistry
University of Groningen
Nijenborgh 4
9747 AG Groningen
THE NETHERLANDS

J. R. van der Ploeg
Department of Biochemistry
University of Groningen
Nijenborgh 4
9747 AG Groningen
THE NETHERLANDS

H. Wariishi
Chemical & Biological Sciences
Oregon Graduate Institute of Science
& Technology
19600 N.W. von Neumann Drive
Beaverton, OR 97006-1999 USA

G. L. Warner
GE Corporate Research and
Development Center
CEB 406, P.O. Box 8
Schenectady, NY 12301 USA

P. Werner
DVGW - Forschungsstelle
University of Karlsruhe (TH)
Richard-Willstatter Allee 5
D-7500 Karlsruhe 1
GERMANY

D. C. White
Institute of Applied
Microbiology
University of Tennessee
10105 Research Drive
Knoxville, TN 37932-2567 USA

J. W. Wigger
Amoco Corporation
Groundwater Management Section
7201 E. 38th Street, Space 7253
Tulsa, OK 74145 USA

B. H. Wilson
U.S. Environmental Protection Agency
Robert S. Kerr Environmental
Laboratory
P.O. Box 1198
Ada, OK 74820 USA

L. P. Wilson
The Johns Hopkins University
Department of Geography and
Environmental Engineering
313 Ames Hall
34th and Charles Street
Baltimore, MD 21218 USA

P. R. Wilson
GE Corporate Research and
Development Center
P.O. Box 8
Schenectady, NY 12301 USA

R. M. Woeller
Water and Earth Science
Associates, Ltd.
Box 430
Carp, Ontario
CANADA K0A 1L0

D. J. Workman
Battelle Pacific Northwest
Laboratories
P.O. Box 999
Mail Stop K4-06
Richland, WA 99352 USA

O. Yagi
National Institute for Environmental
Studies
16-2 Onogawa Tsukuba Ibaraki, 305
JAPAN

I. J. Zanikos
Du Pont Environmental Biotechnology
Program
Glasgow Site, Building 300
P.O. Box 6101
Newark, DE 10714-6101 USA

A. Zeddel
Forstbotanisches Institut der Universität
Göttingen
Büsgenweg 2, D-3400 Göttingen
GERMANY

INDEX